Data Analysis and Regression

A SECOND COURSE IN STATISTICS

Data Analysis and Regression

A SECOND COURSE IN STATISTICS

FREDERICK MOSTELLER
Harvard University

JOHN W. TUKEY
Princeton University and Bell Telephone Laboratories

ADDISON-WESLEY PUBLISHING COMPANY

Reading, Massachusetts · Menlo Park, California
London · Amsterdam · Don Mills, Ontario · Sydney

This book is in the
ADDISON-WESLEY SERIES IN
BEHAVIORAL SCIENCE: QUANTITATIVE METHODS

Consulting editor:
Frederick Mosteller

The writing of this work, together with some of the research reported herein, was supported in part by the United States Government.

To Virginia and Elizabeth

Preface

Two mainstreams intermingle in this treatment of practical statistics: (a) a sequence of philosophical attitudes the student needs for effective data analysis, and (b) a flow of useful and adaptable techniques that make it possible to put these attitudes to work.

After a first course in statistics, the student needs further preparation for tackling the more difficult problems that commonly arise in data analysis, even some of those that appear simple. Many first courses concentrate on introducing the student to notions of confirmatory data analysis (primarily tests of significance), to some probability distributions, and perhaps also to regression and analysis of variance. In a first course, the instructor rarely has time to discuss the philosophy of data analysis or of exploratory approaches, let alone time to bring out the vital ifs, ands, and buts of multiple regression. Much can be said about applying statistical methods to get effective and reliable answers from real data, and we take these matters up in some detail.

We assume that the student has had a first course in statistics. (Having other courses can help, but nothing beyond a first course is assumed.) Depending on the kind and length of preparation, a student will move faster or slower through this book's chapters, but most of the material is likely to be new, whatever the preparation. Our emphasis is not mathematical, although some techniques are inevitably heavy with formulas. In most spots the mathematical details are handled by way of examples.

Calculus is not often used. In the two or three spots where it arises, the reader can skip details and accept results without losing much.

What a Reader can Get from this Book

More important than the techniques that this book can teach are attitudes and approaches. As they work their way through, our readers can learn to identify at least the following attitudes, understandings, and approaches, which they will want to return to and ponder after reading parts of the book:

⋄   An approach to the formulation of statistical and data-analytical problems such that, for example, (a) Student's shortcut to inference can be properly understood, and (b) the role of vague concepts becomes clear.

⬦   The role of indications (of pointers to behavior not necessarily on prechosen scales—in contrast to conclusions or decisions about prechosen quantities or alternatives).

⬦   The importance of displays and the value of graphs in forcing the unexpected upon the reader.

⬦   The importance of re-expression (as well as how to do it).

⬦   The need to seek out the real uncertainty as a nontrivial task.

⬦   The importance of iterated calculation, where we go round and round till the answer settles down (a once-through calculation is fine, when one can be found to meet our needs).

⬦   How the ideas of robustness (specifically, robustness of efficiency) and resistance can change both what we do and what we think.

⬦   What regression is all about.

⬦   What regression coefficients can and can't do for us.

⬦   That the behavior of our data can often be used to guide its analysis (through both guided regression and re-expression, to name but two routes).

⬦   The importance of looking at and drawing information from residuals.

⬦   The idea that data analysis, like calculations, can profit from repeated starts and fresh approaches; there is not just one analysis for a substantial problem.

The reader who has not yet begun this book is likely to be unclear just what some of these items mean; the reader who has worked through the sixteen chapters ought to know what they mean, and will have developed views about their relative importance.

Let us turn to the chapters themselves. In many cases, they need not be read in consecutive order, as the following description makes clear. The first two chapters offer some practical philosophy for data analysis and should lead off, almost certainly followed by Chapters 3 and 4 (unless the student has had preparation in exploratory data analysis, as most have not in 1977). (When we refer to the book *Exploratory Data Analysis* for fuller treatments, the references point to the 1977 first edition (Addison-Wesley Publishing Company, 1977), and not to the limited preliminary edition). Chapter 4 provides a proper background for simple linear regression.

Opinions seem to differ as to the nature and importance of re-expression. We put it where it is (Chapters 5 and 6) because to us it is both elementary and valuable. But it may be postponed, even placed well down among the regression chapters. We have postponed most details of Chapter 6 to an appendix following Chapter 16.

Chapters 7 and 8 are extremely general in their philosophy and are impressive in the variety and difficulty of the problems their techniques tackle. The ideas (a) that finding the true uncertainty may not be easy and (b) that it can still be done, even in messy situations where classical mathematical attacks seem almost impossible, can illuminate all we may go on to do in the analysis of data—and ought to have a chance to do this. The student who has mastered the techniques of Chapter 8 has made profound progress in confirmatory data analysis as well as in cross-validation. The techniques are as all-purpose for confirmatory data analysis as regression and analysis of variance are for their jobs. Chapter 8 illustrates their uses in multiple regression. If these chapters, which are not explicitly used in later chapters, are omitted, the reader will suffer a grave loss.

Chapters 9 through 11 form a special bundle of three distinct techniques that any statistician or data analyst needs, though they are not widely available: the direct and flexible approach to two-way tables (Chapter 9); an up-to-date look at resistant/robust techniques in the simpler applications (Chapter 10); and a reasonably extensive account of standardization (Chapter 11). Many students pass through several statistics courses without learning about standardization, and then find themselves at a loss when they meet practical problems. We feel that the material of Chapter 11 (and Chapters 9 and 10) is an essential part of a statistician's or data analyst's education. Again, these chapters need not be read before the regression chapters, except that at least part of Chapter 10 is needed before—or in conjunction with—the latter part of Chapter 14.

Some subjects are naturally so interrelated and interlinked as to make a simple straight-through account at least difficult. We fear that today the bundle of issues and topics labeled "regression" provides an instance. We have tried to make things as orderly as we can. Chapter 12 covers the ABC's of regression. Much of this material is often passed by without notice. We think it is important, deserving to come first as regression is studied. Next comes Chapter 13 on the woes of regression coefficients. Chapter 14 explains how the machinery underlying the usual fitting of regression works and how it can be used to make robust and resistant fits as well as the weak fits of usual least squares.

Chapter 14 offers a mathematical approach to understanding regression that many students have found instructive and helpful. Not all will want to study it in full detail. Detailed study requires an ability to follow the summation notation closely and an appreciation that certain operations can be carried out, but not necessarily by the reader. The reader who is able to take statements of fact on faith can get much from this chapter without following all the formal details. The two preceding chapters give essential background, especially Chapter 12.

The idea that there is only one regression that we could consider fitting to a given set of data is often false. Chapter 15, which makes strong use of Chapters 12 and 13, suggests ways to deal with a recognition that it is false for our specific data. Chapter 16 tells us something of how to hunt out additional fits to try next.

For those who want a fast dash to the regression chapters, the route is Chapter 4, Chapter 10, and then on to the Chapter 12 through 16 sequence, backtracking sometime to Chapter 5, and to Chapters 1 and 2.

Special Features

Although some special features have already been alluded to, a more complete listing is useful. Among others, we present (in the order of their appearance):

- ⋄   An introduction to stem-and-leaf displays.

- ⋄   Use of running medians for smoothing.

- ⋄   The ladder of re-expressions for straightening curves.

- ⋄   Methods of re-expression for analysis, and a way to assess their worthwhileness in specific instances.

- ⋄   Special tables to make re-expression easy in hand calculations.

- ⋄   The method of direct assessment.

- ⋄   The jackknife, extensively illustrated.

- ⋄   Robust analyses of two-way tables.

- ⋄   Robust and resistant measures of location and scale.

- ⋄   Direct and indirect methods of standardization, as indication or as confirmatory data analysis.

- ⋄   Techniques of standardization that include allowance for broad categories.

- ⋄   A residual-emphasizing approach to regression.

- ⋄   Problems of collinearity in regression.

- ⋄   Regression with errors of measurement.

- ⋄   The interpretation of regression coefficients.

- ⋄   Effects of proxy variables.

- ⋄   Ordinary least squares.

- ⋄   Weighted least squares of several kinds.

- ⋄   Least absolute deviations.

- ⋄   Choosing among regression models, especially in stepwise fitting.

- ⋄   Choosing new variables and appraising old ones in multiple regression.

Problems to be Done

This book contains many homework problems and projects, all grouped at the end of the book and organized by section.

Many of these problems, especially for the chapters on regression (and for Chapter 10), will take more time than one might like. Teachers should be sparing in their assignments; students should be brave in their efforts. Everyone should recognize that, as is true in practical data analysis, problems may often not have neat answers or a single correct solution.

If you can be sure of what is happening, having the heavy calculations done on a computer can save you much effort. But be sure that the computer is doing exactly what you suppose. Not all of the many "packages" of statistical computer programs can be trusted to do exactly what one might hope they will do. More treacherous by far are isolated programs, often written by those who understand the computer and its programming far better than they understand the traps, classical or newly discovered, of statistical algorithms. Take extra care when using them.

How this Book Came to be

About 1963, Gardner Lindzey and Eliot Aronson, who were preparing to edit a new edition of the *Handbook of Social Psychology* (1968), invited the present authors to write a section to replace the Mosteller and Bush article in the first edition. Expansive ideas and editorial planning produced (a) "Data Analysis—Including Statistics" as Chapter 10 (pages 80 to 203 of volume 2) of that revised handbook and (b) an understanding that we were free to use material from that article as part of a free-standing book. A variety of topics were then worked on, many reaching fairly polished form, but only when the goals reflected in the present volume were recognized and seized upon did the present book begin to coalesce. Chapters 1, 2, 7, and 8 herein cling closely to the *Handbook* article, while the others reflect some or all of the following insights.

1. Students need a book that combines exploratory and confirmatory approaches, and that is aimed, more or less directly, at analyzing data with a consciousness of the sorts of things that are being done in practice.

2. Students need a book that tells the truth about regression, as to aims, as to techniques, as to what can be done and what can't.

3. Students need materials that emphasize many of the important issues or techniques that so frequently have fallen between the cracks in statistical instruction.

While we recognize that the continuing evolution of knowledge and insight makes it impossible to completely meet these three goals, we hope the reader will be pleased to discover how far it has been possible to go.

Acknowledgments

We take pleasure in expressing our appreciation to the many people and organizations who have contributed to this venture, including those acknowledged in the *Handbook of Social Psychology* article (Mosteller and Tukey, 1968). Among colleagues and students who have read drafts of chapters and helped with comments, special thanks go to David Hoaglin and Colin Mallows. The whole book and its examples and tables have been checked over by Anita Parunak. Earlier, Cleo Youtz and Gale Mosteller reviewed many examples and tables. We had considerable aid in identifying and drafting the problems from Miriam Gasko-Green, Keith Soper, Michael Stoto, and George Wong. Steven Fosburg programmed most of the extensive calculations of Chapter 8. We have had typing support from Holly Grano, Marjorie Olson, and Mary E. Bittrich (Chapter 10).

Preparation of this work has been facilitated by National Science Foundation grants SOC72-05257 and SOC75-15702 (Harvard), Army Research Office (Durham) contracts and grants (Princeton) DA-31-124-ARO(D)-215, No. PO11 DAHC-04-73-C-0031, DAHC-04-74-G-0178, DAHC-04-75-G-0188, DAAG-29-76-G-0298, Energy Research and Development Administration grant E(11-1)-2310 (formerly AEC AT(11-1)-2310 (Princeton), and by Bell Telephone Laboratories, Inc. A substantial amount of the work was done during 1974–75 at the University of California at Berkeley while Mosteller served as Miller Research Professor, a post supported by the Miller Institute for Basic Research in Science.

All the exhibits from *Exploratory Data Analysis* are reprinted by permission of the Addison-Wesley Publishing Company.

We much appreciate the generosity of the many scholars and publishers who have given us permission to use their data as examples. Data from genuine investigations bring freshness and uncontrived challenge to the student of data analysis, and so the realism of the problem material is itself a contribution to our profession.

Cambridge, Massachusetts   F. M.
Princeton, New Jersey   J. W. T.
February, 1977

Contents

Chapter 1/Approaching Data Analysis

Introduction

Every student of the art of data analysis repeatedly needs to build upon his previous statistical knowledge and to reform that foundation through fresh insights and emphases. This account of data analysis assumes that all of its readers are familiar with elementary statistical concepts and techniques; a few starred sections and paragraphs assume that some readers already have moderately advanced knowledge of both statistics and data analysis.

Applications of mathematics are always complicated by the obligation to be true to the subject matter treated as well as to the mathematics. The structure of the art of data analysis makes its exposition especially difficult because it does not branch neatly like a tree, nor does it seem susceptible to an orderly progressive treatment. Consequently, we repeatedly deal with new insights and old fundamentals. Our purpose is to develop ideas profitable to both practitioners and critics of the analysis of data.

Many mathematical and philosophical discussions begin with a general theory, from which are derived general principles; from these, in turn, specific procedures are produced and finally exemplified. In discussing data analysis, we find the following somewhat opposite order more practical.

a) First, what to do. (What treatment to apply to the data of a given sort of problem, arithmetically or graphically.)

b) Then, why choose that treatment. (What reasons are there for choosing this treatment, from among so many, to meet this sort of problem? Usually the explanation must be made in terms of specific mathematical models.)

c) Next, what to include in the why. (What can be said about types of models, and classes of reasons, that have proved useful in carrying out this choice, and which types and classes, by contrast, have proved misleading.)

d) Last, how to structure our thinking about the overall process of data analysis. (What general theories produce the reasons of (c) as deductions rather than as inductions from experience.)

Great overturns at the (d) level have had little effect on what is done in practice, in comparison with the effects of less imposing changes at the (b) level.

In this chapter, we strike a number of the themes: the staircase of primary, secondary, tertiary, ... statistics and Student's shortcut to inference; the distributional properties of observations and measurements; the need for vague

1

concepts in the evaluation of more definite concepts and criteria; the distinction between indication, determination, and inference. Each theme recurs.

1A. The Staircase and the Shortcut to Inference

Before Student's time, every analysis of data that considered "what might have been" resembled a long staircase from the near foreground to the misty heights. One began by calculating a primary statistic, a number that indicated quite directly what the data seemed to say about the point at issue. The primary statistic might, for instance, have been a sample mean. Then one faced the question of "How much different might its value have been?" and calculated a secondary statistic, a number that indicated quite directly how variable (or perhaps how stable and invariable) the primary statistic seemed to be. The secondary statistic might have been an estimate of the standard deviation of such a sample mean. After this step, one again needed to face the question of "How much different?", this time for the secondary statistic, which again and again turned out to be less stable (of itself) than the primary statistic whose stability it indicated. In principle, one should have gone on to a tertiary statistic, which indicated the variability or stability of the secondary statistic, then to a quaternary statistic, . . . , and so on up and up a staircase which, since the tertiary was a poorer indicator than the secondary, and the quaternary was even worse, could only be pictured as becoming mistier and mistier. In practice, workers usually stopped with primary and secondary statistics.

Student (1908) broke new ground by asking essentially "What if I have n observations randomly drawn from a normal distribution about whose average and variance I wish to assume nothing?" Let $\bar{y}$ be the sample mean, μ the population mean, y_i a measurement, n the sample size. We shall think of $\bar{y}$ and y_i as random variables for the moment, rather than as the numerical values realized in an investigation. Student's name identifies the *ratio:*

$$t = \frac{\text{sample mean} - \text{distribution mean}}{\sqrt{(\text{sample estimate of distribution variance})/n}}$$

$$= \frac{\bar{y} - \mu}{\sqrt{\dfrac{s^2}{n}}} = \frac{\bar{y} - \mu}{\sqrt{\left(\dfrac{\sum(y_i - \bar{y})^2}{n-1}\right)\bigg/ n}}$$

$$= \frac{\text{sample mean} - \text{distribution mean}}{\sqrt{\text{sample estimate of variance of numerator}}}$$

$$= \frac{\bar{y} - \mu}{s_{\bar{y}}} = \frac{\bar{y} - \mu}{\sqrt{\dfrac{\sum(y_i - \bar{y})^2}{(n-1)n}}}$$

whose distribution, based on normal theory, was found to depend only upon n.

Student calculated some numerical aspects of the distribution of t. Then, after applying empirical sampling to 3000 measurements of finger lengths and 3000 heights, which had been collected as an aid in identifying criminals, he succeeded in correctly guessing the mathematical form of the distribution of t. R. A. Fisher (1925) verified Student's guess 17 years later.

This approach cut off the misty staircase after the third step—indeed, almost after the second step. For, in order to tell us about the population mean, the data were asked to provide only:

1. the sample mean—a primary statistic,

2. the sample estimate of variance—a secondary statistic,

3. the sample size—a tertiary statistic, one that was easy to obtain and remarkably stable, at least so long as one compared this sample with other possible samples of the same size.

All else was provided by the assumption of exact normality.

Note that, while the sample mean, sample variance, and sample size are primary, secondary, and tertiary statistics for this purpose, they may play other roles in other circumstances.

Given the normality assumption, these three numbers, and any contemplated value μ_C for the distribution mean, we can calculate

$$t = \frac{\text{sample mean} - \text{contemplated value}}{\sqrt{\text{sample estimate of variance of numerator}}} = \frac{\bar{y} - \mu_C}{s_{\bar{y}}}.$$

The contemplated value could be any number, and could be changed many times for a single set of data. It might be zero if we were studying group differences and thought, either seriously or as a straw man, that we might find none. It might be 500, if we were comparing a group of students, perhaps the freshman class of some college, with the standard offered by a national average on a test standardized as the Educational Testing Service often does (mean 500, standard deviation 100). It might take on, in turn, all possible values, as when we seek a confidence interval.

Each time we insert a contemplated value into the formula for t, we take the first step in making a significance test. When the contemplated value exactly equals the population mean from which the y_i are drawn, the distribution of t is the one given by the usual tables. When the contemplated value is far from the true population mean, t is likely to yield a number large in absolute value. These ideas, applied to this and other key statistics, made a bit more precise with tables of critical values, and extended by the concept of operating characteristics or power of tests (see, for example, Mosteller and Bush, 1954), provide the whole machinery of significance testing and almost all the machinery used in practice to set confidence intervals.

By the 1930's and through the 1940's, people were learning to short-cut the staircase without making such strong assumptions. They introduced "non-parametric" or "distribution-free" procedures, thus eliminating dependence on the normal distribution for making "5%" really 5%, thus providing yet another stage in a continuing revolution.

1B. Student's True Contribution

In the $\frac{3}{4}$-century following its introduction, Student's t has been used extremely often in practice, as have many and diverse techniques that have evolved out of the same chain of development. Its impact on the usage of words and on the development of statistical theory have been equally striking.

The value of Student's work was not that it led to great changes in the numbers obtained in the analysis of data, because in the main it did not. To see this, look at Exhibit **1**, where we show some % points that would be used to set

Exhibit **1** of Chapter 1

Standard Confidence Points for Student's t

f = degrees of freedom	1/40 = 2.5%	1/6 = $16\frac{2}{3}$%	1/2 = 50%	5/6 = $83\frac{1}{3}$%	39/40 = 97.5%	24/f (for interpolating)
1	−12.71	−1.73	0.00	1.73	12.71	
2	−4.30	−1.26	0.00	1.26	4.30	
3	−3.18	−1.15	0.00	1.15	3.18	
4	−2.78	−1.10	0.00	1.10	2.78	
5	−2.57	−1.07	0.00	1.07	2.57	
6	−2.45	−1.05	0.00	1.05	2.45	4
8	−2.31	−1.03	0.00	1.03	2.31	3
12	−2.18	−1.01	0.00	1.01	2.18	2
24	−2.06	−.99	0.00	.99	2.06	1
∞	−1.96	−.97	0.00	.97	1.96	0

Notes:

1. 2.5% and 97.5% points combine to give two-sided 95% confidence limits, or a two-sided 5% significance test.
2. Interpolation in the reciprocal of the degrees of freedom gives good accuracy. For example, for 48 degrees of freedom 24/f = 0.5. Consequently the corresponding 97.5% point is halfway between 2.06 and 1.96 at 2.01.
3. When we expect to use a t in a symmetrical two-sided way, it is often convenient to think of the distribution of $|t|$, the absolute value of t. We write $|t|_{.95}$ for the two-sided 95% point of $|t|$, which is given above in the column headed 39/40. We write $|t|_{2/3}$ for the two-sided 2/3 point, which is given above in the column headed 5/6. We use similar notations for any two-sided % points of t.

two-sided 95% = 19/20 = 38/40 confidence limits (and also those needed for two-sided 2/3 = $66\frac{2}{3}$% confidence limits, a less usual level selected for reasons we shall explain in Section 1C. One measure of the effect of Student's t is the ratio of the length of confidence interval obtained by using it to the length obtained when the standard normal distribution, which is equivalent to t for infinite degrees of freedom, is used. Long before Student came along, many people calculated the ratio now known as Student's t (except for a slight change in the definition of the sample standard deviation as discussed below), namely,

$$\frac{\text{sample mean} - \text{contemplated value}}{\text{sample standard deviation}/\sqrt{n}}.$$

Not having Student's table, all they then knew how to do was to refer this ratio to a table of the standard normal distribution and to use varying amounts of verbal caution in interpreting the result.

How great was the change? We find from Exhibit 1 that, for example, if 95% confidence limits are set using the standard normal distribution, the multiplier of the standard error $s_{\bar{y}}$ is 1.96, while, using a t distribution with 12 degrees of freedom, the multiplier is 2.18. Since $2.18/1.96 \approx 1.11$, the use of the t distribution adds only 11% to the length of the 95% confidence interval when we have 12 degrees of freedom. **(We use the symbol "≈" to stand for "is approximately equal to" or "approximates" or similar phrases implying approximation rather than strict equality.)** If a slightly different definition of $s_{\bar{y}}$ were used, again with the standard normal distribution, namely $\sqrt{\sum(y_i - \bar{y})^2/n^2}$, the ratio would rise to 1.15.) For levels of confidence less extreme, the effect is less marked. For example, for two-sided 2/3 confidence, we can go as low as 5 degrees of freedom before the ratio of lengths of confidence intervals based on t to that based on the normal is as much as 1.1 ($\approx 1.07/0.97$).

We are not trying to sweep under the rug the gigantic ratio of 6.5 $\approx$ 12.7/1.96 associated with 95% confidence and 1 degree of freedom. But most investigators use more degrees of freedom; indeed, an important impact of the t table has been to encourage investigators to get more degrees of freedom so as to avoid such terrible ratios as the 6.5, 2.2, and 1.6 offered at 95% by 1, 2, and 3 degrees of freedom, respectively. Note that, for 2/3 confidence, the corresponding ratios are only 1.8, 1.3, and 1.2.

The value of Student's work lay not in a great numerical change, but in:

⬦ recognition that one could, if appropriate assumptions held, make allowance for the "uncertainties" of small samples, not only in Student's original problem but in others as well;

⬦ provision of a numerical assessment of how small the necessary numerical adjustments of confidence points were in Student's problem and, as we have just seen, how they depended on the extremeness of the probabilities involved;

◇presentation of tables that could be used—in setting confidence limits, in making significance tests—to assess the uncertainty associated with even very small samples.

Besides its values, Student's contribution had its drawbacks, notably:

◇it made it too easy to neglect the proviso "if appropriate assumptions held";

◇it overemphasized the "exactness" of Student's solution for his idealized problem;

◇it helped to divert the attention of theoretical statisticians to the development of "exact" ways of treating other problems; and

◇it failed to attack "problems of multiplicity": the difficulties and temptations associated with the application of large numbers of tests to the same data.

The great importance given to exactness of treatment is even more surprising when we consider how much of the small differences between the critical values of the normal approximation and Student's t disappears (see Exhibit **2**), especially at and near the much-used two-sided 5% point, when, as

Exhibit **2** of Chapter 1

Standard Confidence Points for Burrau's Modification of Student's t

$$\sqrt{\frac{f-2}{f}} \cdot t = \frac{\bar{y} - \mu}{\sqrt{\sum(y_i - \bar{y})^2/n(n-3)}} \qquad \text{where} \qquad f = n - 1$$

f = degrees of freedom	1/40 = 2.5%	1/6 = $16\frac{2}{3}$%	1/2 = 50%	5/6 = $83\frac{1}{3}$%	39/40 = 97.5%	24/f (for interpolating)
3	−1.84	−0.66	0.00	0.66	1.84	
4	−1.96	−0.78	0.00	0.78	1.96	
5	−1.99	−0.83	0.00	0.83	1.99	
6	−2.00	−0.86	0.00	0.86	2.00	4
8	−2.00	−0.89	0.00	0.89	2.00	3
12	−1.99	−0.92	0.00	0.92	1.99	2
24	−1.98	−0.95	0.00	0.95	1.98	1
∞	−1.96	−0.97	0.00	0.97	1.96	0

Notes:
1. The variance of t is infinite for 2 or fewer degrees of freedom, so that Burrau's formula is inapplicable for f = 1 and 2.
2. 2.5% and 97.5% points combine to give two-sided 95% confidence limits, or a two-sided 5% significance test.

suggested by Burrau (1943), we multiply t by the constant required to bring its variance to 1. (Regardless of its practical pros and cons, Burrau's modification frees our insight from the possibly misleading effects of the quite nonconstant variance of t.)

The time has long since come to pay attention to the advantages of Student's work and to recognize its drawbacks for what they are, skimping neither in comparison to the other.

1C. Distributions and Their Troubles

Student himself always remembered that observations and measurements were never distributed in magic bell-shaped curves, even when they were chemical determinations of commercial importance made under his own supervision (Student, 1927). The history of statistics and data analysis is a messy mixture of healthy skepticism and naive optimism about the exact shapes of the distributions of observations. Such optimism has often been inflated by the wonderful properties of a single family of distributions, the "normal" distributions, whose probability densities are given by

$$f(x) = \frac{1}{\sqrt{2\pi}\sigma} e^{-\frac{1}{2}(x-\mu)^2/\sigma^2} \qquad \text{for} \qquad -\infty < x < \infty,$$

where μ and σ are the population mean and standard deviation, respectively, e is the base of the natural logarithms $2.7182818\ldots$, and π is our old friend $3.1415926\ldots$. Exhibit **3** portrays three instances of normal distributions with differing combinations of μ and σ.

We use the adjectives "normal" and "Gaussian" interchangeably for distributions that fit this formula *exactly*. Neither term is wholly satisfactory. Some misinterpret the word "normal" to mean "the ordinarily occurring"— but, so far as we know, distributions that exactly fit this formula never occur in practice—not for individual observations, not for sample means, not for other derived quantities—though we have both theory and evidence that many empirical distributions do approximate this shape, sometimes quite usefully, but sometimes only apparently rather than meaningfully. (The characteristics that matter most are often those that are concealed, not revealed, by conventional histograms.)

The connection of the distributions of derived statistics, such as sample means and Student's t, to the underlying distribution of individual values is often subtle, as we shall illustrate shortly. Since distributions of the derived statistics usually determine the adequacy of our statements of uncertainty, significance, and confidence, the particular aspects of approximation that matter in practice differ from situation to situation and are often not easily checked by examining a single sample, or even the whole body of data before us, which may consist of hundreds of observations.

We say of the three distributions in Exhibit 3 that they all have the same "shape". Let us discuss the concept of shape.

Suppose that we have many observations, so that the histogram representing actual counts closely approximates the underlying distribution. Suppose that this histogram has been plotted on graph paper, but that someone has forgotten to put numbers along the axes. What can we learn from what we have? What have we lost? Without numbers on the vertical axis, we cannot tell how large the sample was. Since our interest centers on the distribution, not the sample, and the sample was large, we may be able to forget this. Without numbers on the horizontal axis, we cannot tell near what value the observations fell, or how widespread or concentrated they were; we can tell nothing about location or about scale, to use technical terms. These are important things to have lost. What remains?

By giving up location and scale, by classing together all distributions that differ only by a linear transformation, we have given up only 2 numbers; accordingly much remains: all, indeed, that is usually meant by the word **shape**.

Exhibit **3** of Chapter 1

A collection of three normal or Gaussian probability density functions, with differing means and differing standard deviations: $\mu_1 = -2$, $\sigma_1 = 1$; $\mu_2 = 0$, $\sigma_2 = 0.5$; $\mu_3 = 4$, $\sigma_3 = 2$.

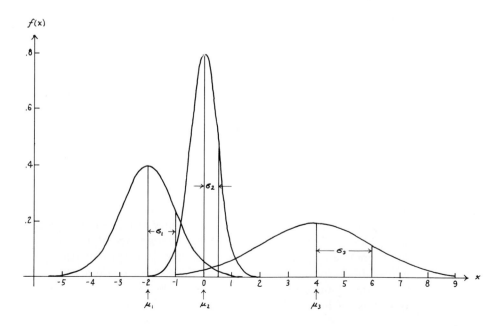

Exhibit **4** of Chapter 1

A collection of beta densities

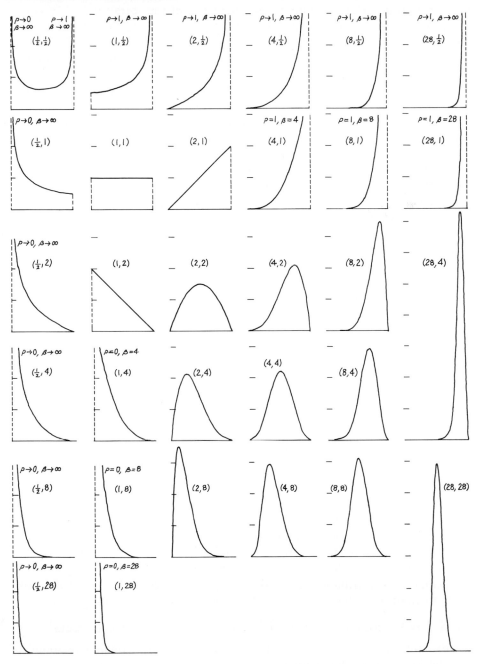

Distributions, even distributions belonging to the same mathematical family, can differ widely in shape. The family of beta functions, for instance, though still a very special case, includes distributions of many shapes, as Exhibit **4** illustrates. It gives a collection of beta densities of the form

$$\beta(p) = cp^{a-1}(1 - p)^{b-1} \qquad (0 \leqq p \leqq 1),$$

where c is the constant making the total area 1, and where the parameters (a, b) are indicated in the figure.

Given the histogram, we can locate the central 2/3 of all cases and find the number of spaces along the horizontal axis required to cover them. We can do the same for the central 19/20, and compare the two lengths. If this ratio is almost exactly 2 to 1, we have confirmed one characteristic of a normal distribution. If it is appreciably larger than 2 to 1, we have evidence that the two-sided 5% points of our distribution are more widespread than we would have expected on the basis of how the central part of the distribution is spread out and how normal distributions behave. If the sample is large enough, this is already clear evidence that the population is not normal. (Other indications can be more revealing in small or moderate-sized samples, as we illustrate in Chapter 2.)

We can look at such ratios farther and farther into the tails. If, when we do this, either for the whole population or for sufficiently large samples, these ratios eventually tend to be too large for normal distributions, we know that we are dealing with tails that are **more straggling** or more stretched out than those of a normal distribution. This statement has nothing to do with the overall spread of a distribution. Rather it concerns the spread of the extreme tails in comparison with the spread of the central body of the distribution.

A glance at Exhibit 1 shows that the distributions of Student's t appear more straggling than the Gaussian, except, of course, for the case with ∞ degrees of freedom, which has exactly the Gaussian shape. The ratio just mentioned (length of middle 19/20 to length of middle 2/3), which is about 2.02 for the Gaussian, is 2.16 for 12 degrees of freedom, 2.4 for 5, 3.4 for 2, and a whopping 7.3 for 1 degree of freedom. If you know that 2/3 of a Gaussian distribution lies in a certain interval, and you consider an interval 7 times as long but with the same center, you know that all but an extremely tiny part (1 in a hundred billion) of that Gaussian distribution lies in the larger interval. If you know that 2/3 of a "Student's t with one degree of freedom" distribution lies in a certain interval, and you consider an interval 7 times as long but with the same center, we have just seen that more than 5% = 1/20 of the distribution can fall outside the longer interval. When tails straggle a lot, the center of a distribution can fool the eye about the behavior of the tails. Real distributions often straggle a lot compared to a normal distribution.

1D. A Classical Example: Wilson and Hilferty's Analysis of Peirce's Data

An example investigating unusually extensive data for normality of shape of distribution may be illuminating.

In an empirical study intended to test the appropriateness of the normal distribution, C. S. Peirce (1873) analyzed the times elapsed between a sharp tone stimulus and the response by an observer, who made about 500 responses each day for 24 days. Though Peirce seems to have concluded that the normal shape of this distribution was on the whole verified, when Wilson and Hilferty (1929) reanalyzed Peirce's published material sixty years later, they came to very different conclusions.

They calculated many different statistics from Peirce's data, treating each day's measurements separately. We select a few of the 23 statistics Wilson and

Exhibit **5** of Chapter 1

Daily Statistics from Wilson and Hilferty's Analysis of C. S. Peirce's Data

	(1) $\bar{x} \pm s_{\bar{x}}$	(2) $\dfrac{Q_3 - Q_1}{2(0.6745s)}$	(3) No. within 0.25s of $\bar{x}$		(4)	(5) Kurtosis	(6) Errors $> 3.1s$		
Day	(milliseconds)		Obs.	Exp.	Skewness	$\hat{\beta}_2 - 3$	Neg.	Pos.	Total
1	475.6 ± 4.2	0.932	110	98	1.18	3.1	1	3	4
2	241.5 ± 2.1	0.842	113	97	0.43	0.9	1	0	1
3	203.1 ± 2.0	0.905	113	97	1.09	3.6	0	7	7
4	205.6 ± 1.8	0.730	134	99	1.82	9.7	1	7	8
5	148.5 ± 1.6	0.912	110	97	0.39	1.3	0	4	4
6	175.6 ± 1.8	0.744	119	97	1.48	6.4	0	6	6
7	186.9 ± 2.2	0.753	132	98	2.96	24.9	0	6	6
8	194.1 ± 1.4	0.840	120	97	0.48	4.1	2	6	8
9	195.8 ± 1.6	0.756	132	98	1.71	13.8	2	4	6
10	215.5 ± 1.3	0.850	120	99	0.84	8.8	2	1	3
11	216.6 ± 1.7	0.782	135	99	1.69	9.8	1	5	6
12	235.6 ± 1.7	0.759	103	78	0.63	4.7	3	5	8
13	244.5 ± 1.2	0.922	101	97	−0.22	2.6	6	1	7
14	236.7 ± 1.8	0.529	192	99	5.74	63.6	2	3	5
15	236.0 ± 1.4	0.662	162	98	1.68	27.9	4	4	8
16	233.2 ± 1.7	0.612	162	98	6.39	90.6	4	2	6
17	265.5 ± 1.7	0.792	123	100	0.21	4.3	3	5	8
18	253.0 ± 1.1	0.959	114	98	0.27	1.8	0	4	4
19	258.7 ± 1.8	0.502	187	99	10.94	143.9	1	3	4
20	255.4 ± 2.0	0.521	179	98	7.71	91.4	0	3	3
21	245.0 ± 1.2	0.790	120	99	0.23	8.2	3	4	7
22	255.6 ± 1.4	0.688	142	99	5.27	68.1	2	4	6
23	251.4 ± 1.6	0.610	158	98	2.73	31.1	0	3	3
24	243.4 ± 1.1	0.730	113	98	−0.02	5.4	3	3	6
Averages							1.7	3.9	5.6

Hilferty reported and display them in Exhibit **5**. Column (2) gives an estimate of the distance from the 25% (Q_1) to the 75% point (Q_3) of that day's distribution (based on the corresponding % points of the distribution of that day's sample), divided by a suitable multiple of the s computed from the very same observations. For a normal distribution, these numbers should vary around 1.00, roughly symmetrically. In Peirce's data, however, all 24 numbers are less than 1.00, many substantially. Column (6) shows that too many measurements deviate from their mean by more than 3.1 sample standard deviations ($3.1s$) in each direction. For a normal distribution, 1 observation in 500 on the average would be more than $3.1s$ from $\bar{x}$, half in either direction, whereas Peirce's data average 5.6 per 500. Similarly, measures of skewness and kurtosis are consistently positive instead of varying around zero, as they should for a normal distribution. Furthermore, according to column (3), too many measurements are near the mean for normally distributed data, again, by comparison with s. Thus Peirce's data clearly do not follow the normal or Gaussian "law".

Some would try to ascribe this to the choice of reaction *time*, properly asserting that today an informed worker would not suppose that such times were likely to be normally distributed, and arguing that perhaps the logarithm of the reaction times might well be nearly normally distributed. A glance at the table in Exhibit 5 shows that this approach cannot succeed. Indeed, the number of negative errors beyond $3.1s$ is already 3 times that for a normal distribution, and a logarithmic transformation would *increase* this number.

The important deviations from normality in Peirce's data are not matters of discreteness, nor are they matters of gross differences in shape. Columns (2) and (3) agree generally with each other both as to typical value and in day-to-day changes. As Winsor's "principle" predicts for the distributions that arise in practice, the distributions of these very large samples are, except for discreteness, reasonably "normal in the middle". However, the normal distribution that would fit the center of the observed distribution would have a spread only about 3/4 of that of the distribution fitted to the observed value of the standard deviation.

The median value in column (2), omitting day 1, is 0.756, which corresponds to a variance due to the "normal body" of the distribution of about $(.756)^2 \approx 0.57$ of that observed. Thus more than 40% ($\approx 100\% - 57\%$) of the observed variance comes from the fact that the tails straggle more than for a normal distribution. Again we see that Peirce's observations are glaringly far from being normally distributed.

1E. Kinds of Nonnormality and of Robustness

When a distribution does not have the Gaussian shape, its failure to be Gaussian may arise from:

1. *Discreteness and irregularity*. (i) Ordinarily the value of an actual observation or measurement can never be just any number from minus infinity to plus infinity—its possible values will be limited, often, in particular, to multiples of some least count. (For example, children and marriages come in whole numbers, prices on many stock markets come in 1/8's, and many machine-shop measurements are in thousandths of an inch.) (ii) Beyond this, both real and theoretical distributions are subject to further irregularities, such as those caused by observers' digit preferences and those that appear even in theoretical distributions, such as the sampling distribution of the rank-correlation coefficient in the null situation (when the variables being correlated are actually independent) (see Exhibit **6**). To help guide the eye, the successive ordinates of this discrete distribution have been connected by straight lines.

Exhibit **6** of Chapter 1

Distribution of rank correlation coefficient for $n = 7$.

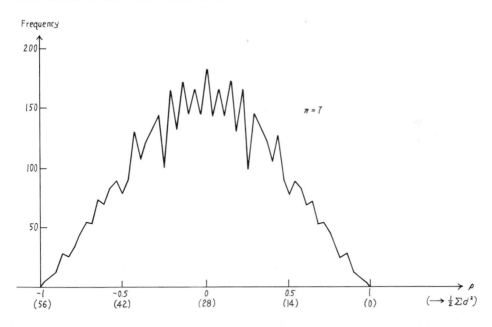

To obtain probabilities, ordinates should be divided by 7!. Probabilities are zero unless half the sum of squares of differences, $\frac{1}{2}\sum d^2$, is an integer. *Example:* Given the two matched rankings $\frac{1234567}{3214567}$, the sum of squares $\sum d^2 = 2^2 + 0^2 + 2^2 = 8$ and $\frac{1}{2}\sum d^2 = 4$. (After Kendall, Kendall, and Smith, 1938.)

2. Gross differences in shape. Among these we include those differences that can be reliably detected in samples of, say, 50 to 100 such as the gross skewness of χ^2 (or F) with few degrees of freedom. The abrupt ends of a rectangular distribution, which spreads its probability evenly over a fixed range, are barely this gross. Exhibit **7** represents the densities of χ^2 distributions with 1 and 3 degrees of freedom; Exhibit **8** represents the density of a rectangular distribution; Exhibit **9** represents the density of a symmetrical triangular distribution, which, incidentally, represents the distribution of the sum of two independent and rectangularly distributed (uniform) random variables, each with range $L/2$, whose means sum to μ.

3. Minor differences in central shape. Difficult to separate from either of the first two and rarely of importance on their own.

4. Behavior in the tails. Hard to detect, yet often important because a few straggling values scattered far from the bulk of the measurements can, for example, alter a sample mean drastically, and a sample s^2 catastrophically.

Exhibit **7** of Chapter 1

Probability density functions for the χ^2 distribution with 1 degree of freedom and 3 degrees of freedom, respectively.

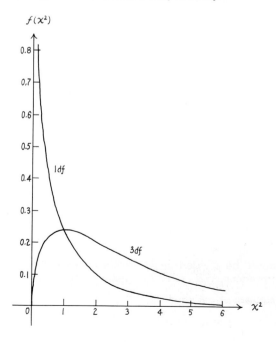

In the usual measurement situations, behavior in the tails is both the most vital and the least securely tied down of all. It is not well tied down because changes in tail size by large and important factors need involve only a few percent of all observations, and are therefore both difficult to detect by sampling and easy for transient causes to produce. It is vital because the extreme tails of the distribution of individual values are likely to influence large portions of the distributions of many derived quantities.

Consider the situation where, perhaps because of human actions, there is one chance in a thousand of a huge deviation (an observation strikingly far from some central value like a mean or median). In samples of size 100, there will be one such huge deviation per 10 samples, or about 100 huge deviations

Exhibit **8** of Chapter 1

Probability density function for the rectangular distribution with range L and mean μ.

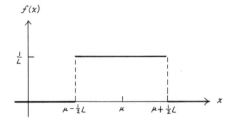

Exhibit **9** of Chapter 1

Probability density function of a triangular distribution.

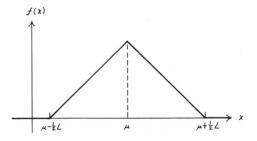

per 1000 samples. Thus 10% of all sample means will be affected. This can be enough to distort 5% points seriously. If these "wild shots" are extreme enough, they will clearly perturb sample means seriously. Student's t, on the other hand, is a function of the observations that behaves in quite a different way than a sample mean does. The effect of one observation differing very greatly, both from the mean of the other observations and from the contemplated value, is to make the value of t fall close to $+1$ or -1. When working at 95%, this is going to cause reduction of % risk, thus "failing safe", but can make many instances that would otherwise be "significant" come out "not significant". Our real need is accurate answers, not just safe statements. Often the use of other significance tests will avoid such ill effects from outliers and bring us closer to describing the real state of affairs.

Both of these desirable properties are kinds of **robustness**, kinds of lack of susceptibility to the effects of nonnormality. The first property, tolerance of nonnormal tails, has been called **robustness of validity**, and is exemplified by confidence intervals for μ that have a 95% chance of covering μ, whatever population may have been sampled. We get this sort of robustness exactly, for example, when a confidence interval for the population median is based on the sign test. The second property, high effectiveness in the face of nonnormal tails, is called **robustness of efficiency**, and is exemplified by confidence intervals for μ that tend to be almost as narrow, for any roughly normal distribution (that is, any distribution "approximately normal in the middle") as the best that could be done if the true shape of the distribution were known to a close approximation. Such procedures are now available (e.g., Gross, 1976).

You will find that most analytical studies of the effects of "nonnormality" consider only what happens when behavior in the tails of the distribution agrees with the results of naive extrapolation from the behavior of the body of the distribution. In the real world things are usually different, since many causes that could have contributed to each of the values that we had observed, like *gross* human error, act only infrequently. As a result, real tails rarely match the body of the distribution, and are likely, when very large samples begin to reveal their true behavior, to appear poorly pasted on.

Both sorts of tail misbehavior can be of considerable importance. The difference between grossly nonnormal distributions and distributions "normal in the middle" but with straggling tails is not mainly in the consequences, which are rather similar, often equally unfortunate. Rather it is that gross nonnormality is likely to be detected, or even brought forcibly to one's notice, but straggling tails tend to escape both notice and careful examination.

One way to give specific examples of distributions with straggling tails is to consider mixtures of normal distributions. For example, we might consider a distribution formed from observations drawn from two normal distributions having identical means, but unequal standard deviations. Almost all of the measurements would come from the normal distribution with the smaller standard deviation, but a small fraction, say 1%, comes from a distribution with 3 times the standard deviation of the first. Such mixed distributions are sometimes called **contaminated distributions**. This one would be called 1% contaminated at scale 3.

We shall also need the notion of **relative efficiency**. Roughly speaking, when two estimates of the same quantity have unequal variances, the ratio of the smaller variance to the larger is called the relative efficiency of the estimate with the larger variance. To be more careful in our definition, the distributions of the two statistics should have roughly similar shapes, and their variances should approximate constant multiples of $1/n$. Fortunately, very good estimates often do this. Under these conditions, the relative efficiency satisfactorily gives the ratio of the sample sizes required for two statistics to do the same job. As an example, for large samples from normal distributions, the sample median has variance approximately $(\pi/2)\sigma^2/n$, and the sample mean has variance σ^2/n. We say that the median has efficiency $2/\pi \approx 0.64$. In other words, a sample of size 100 using the sample median estimates the population center about as well as a sample of 64 does using the sample mean.

Consider an extreme case, where we compare a normal distribution with a distribution 1% contaminated at scale 3. This represents a lengthening of tails of the basic normal by so little as to require thousands of observations for reliable detection. Yet the effect on the distributions of derived quantities can be very serious. In large samples, we can compare two estimates of spread: (i) that based on the standard deviation, s, where squares of deviations are summed, and (ii) that based on the mean deviation, where absolute values of deviations are summed $(\sum |y_i - \bar{y}|/n)$. For the relative efficiencies we find (Tukey, 1960):

For normality:	Mean deviation	88% as good as s
For 1% contamination:	Mean deviation	144% as good as s

Clearly, quite small differences in shape of distribution can greatly affect the relative efficiency and therefore the comparative palatability of different procedures.

1F. The Role of Vague Concepts

Effective data analysis requires us to consider *vague concepts*, concepts that can be made definite in many ways. To help understand many definite concepts, we need to go back to more primitive and less definite concepts and then work our way forward. Let us take a simple example.

Most beginning students of statistics know that the standard deviation of a distribution is one of the useful things to estimate. The concept of standard deviation is hard enough to grasp and use; students have not, as a rule, had much time to consider why and when to use it. Let us discuss these questions.

One approach from the vague toward the particular begins with the crude and qualitative idea that some distributions are more widely spread out, others are more tightly packed, and with the insight that it might be well to measure this by some numerical measure of **spread**. This leads us to the sequence of ideas:

◇ Spreads differ.

◇ A numerical measure may be useful.

◇ One numerical measure is the standard deviation.

◇ How do we judge whether this is a good choice?

Note that we have so far been discussing the problem only for a complete distribution or population, and that no questions of sampling or estimation have yet been considered. We are still interested in what questions we want answered, not yet with how to get their approximate answers.

The usual answers to the last question that are both acceptable and favorable to the standard deviation include the following:

1. The definition of a standard deviation (of a complete distribution) is relatively clear and widely applicable.

2. So long as one is concerned with simple or weighted sample means, or with many of the other conventional linear combinations of observations (such as $5 + 2x + 3y$, where x and y are observations), the standard deviation (or its square, the variance) is a peculiarly useful measure because there are convenient relationships between the variance of the defined statistic and the variance of the underlying distribution that do not depend upon distributional shape. For example, with x and y as observations, a and b as constants, σ_x, σ_y, and σ_{ax+by} the standard deviations of the indicated quantities, and ρ the correlation coefficient between x and y, we have the relation:

$$\sigma_{ax+by}^2 = a^2\sigma_x^2 + 2ab\rho\sigma_x\sigma_y + b^2\sigma_y^2.$$

3. Certain algebraically convenient relations hold "on the average" between sample and population for the variance (and hence for its square root, the standard deviation); for example, the expected or average value of s^2 is σ^2, where s^2 is the sample variance, $s^2 = \sum(y_i - \bar{y})^2/(n - 1)$, and σ^2 is the population variance.

4. If one can confine one's attention to distributions of given shape (all normal, all rectangular, all symmetric triangular, etc.), then any two measures of spread are connected by multiplication by fixed constants, and so it doesn't really matter which population measures of spread one chooses. (Recall from Section 1C that many larger families of closely related distributions, for example, all beta distributions, do not have a single shape.) Consequently, why not choose the standard deviation?

On the other hand,

◇Some distributions have infinite standard deviations; indeed, we have seen the examples of the *t* distribution with 1 and 2 degrees of freedom in the first footnote to Exhibit 2. When such a distribution is just twice as widely spread as another, their standard deviations do not tell us this, although a variety of other measures of spread, including the interquartile range, do.

◇Means are a poor way to summarize samples from distributions with infinite standard deviations, so that (2) is irrelevant when working with such distributions.

◇Even though the population variance is infinite, all the sample variances are finite, so that (3) buys us little or nothing in such cases.

◇If, for a distribution with finite variance, we change shape enough to make the variance infinite, the relation of the standard deviation to other measures of spread that remain finite changes drastically.

◇An imperceptible probability of sufficiently deviant values can make the variance infinite.

◇Consequently, the difference, in most practice, between distributions with infinite variances and distributions with finite variances and long straggling tails cannot be great, so that (4) buys us little or nothing in practice.

We do not wish to conclude from these arguments that the standard deviation is a poor choice. In many situations, it is still an ideal choice as a description of a population—and in some it is far from ideal.

We do wish to conclude:

◇that the standard deviation is *a choice*, and

◇that, in order to understand that it is a choice and to decide whether it is a good choice, we need the vaguer idea of *measures of spread.*

We shall repeatedly need vague concepts like spread to help us understand and appraise the usefulness of definite concepts like standard deviation, range, and interquartile range. Sometimes the specific concepts come first, and the vague concept is sought out to help in dealing with them. More often, however, the vague concept comes first and guides us in identifying corresponding specific concepts.

1G. Other Vague Concepts

The student of elementary statistics may have learned to classify some experimental results as "significant" and others as "not significant". And he or she may have learned that observed sample means, as well as many other expressions calculated from the sample, can generate "confidence intervals" for population means. These are specific ways of realizing another vague concept: "expressing amounts of uncertainty".

Here, the sequence of ideas is:

⋄ Observed numerical summaries, such as sample means, from small or medium-sized samples do not coincide with what would have been found "by the same methods" from either extremely large samples or complete populations.

⋄ We do not know the difference between sample summary value and population summary value in any one instance (if we did, we would make the adjustment), but we may be able to assess about how large it might reasonably be.

⋄ The plausible size thus assessed will be different in different circumstances; obviously, some sample summary values are subject to greater uncertainties than others.

⋄ Thus there is a place for ways of assessing and expressing amounts of uncertainty.

Different specific ways of "expressing amounts of uncertainty" are in daily use. One whole class of choices, often closely related to one another, is provided by confidence intervals. (To give a 90% confidence interval is not the same as to give a 99% confidence interval, but in simple problems, the information conveyed by the one is often roughly translatable into that given by the other.) Besides confidence intervals, there are not only significance tests but a number of other ways of expressing extent of uncertainty.

We cannot be effective in thinking comparatively about these ways, to say nothing of being effective in using them, without some attention to a more primitive vague notion: assessing the uncertainty of a numerical summary.

We have introduced above, perhaps without your notice, an approximation to another important but vague concept. The term "numerical summary" was illustrated by the example of a sample mean. It was hoped that the usual meanings of "numerical" and "summary" would be enough to seem to give it meaning for the moment. A still more general notion covers any of those things that have been drawn from a body of data to exhibit at least one aspect of what that body can suggest. This may be a numerical summary—or a graphical summary.

We are going to call any such summary an *indication*, and we shall insist on its meeting only two of the statistician's obligations:

⋄ It must differ from an anecdote by allowing each of the observations to contribute to it. (Anecdotes usually involve one or a few observations.)

⋄ It must be expressed in such a way that at least some of those who are interested in the subject can think about its interpretation.

These conditions are important, but should not be interpreted too stringently. The mean of a sample of 15 is an indication, but so is the median of that sample, since each observation contributes to the median in the sense that, in the absence of ties at the median, if that one observation had been sufficiently different, the median would have been different. Each observation had its effect, even though small changes in a single observation (each of which would have an effect on the mean, smaller than on the individual value, but still present) need not change the median in the slightest.

Both mean and median belong to a special class of indications one might call **typical values**, or *centers*, or in more technical language, *measures of location*. In particular, for nicely behaved data where the observations follow a crudely bell-shaped distribution, both mean and median usually indicate, roughly, where the center or body of that distribution falls. (If the tails are straggling and unsymmetrical enough, the mean may fail to do this.)

Measures of spread, measures of association, all the results of most of the standard procedures of statistical analysis are indications, as are interesting bumps on curves, greater wiggliness of a graph on its lefthand side than on its right, or apparent segregation of a scatter diagram into clumps.

Statistics courses tend not to emphasize indication; instead, they concentrate upon how to express the uncertainties of particular kinds of indications. In impressing the student with the importance of assessing uncertainties when one can (and can afford the effort), it is perhaps inevitable that they may give the impression that indications whose uncertainties have not been assessed are worthless. This is, of course, just not so. We need to assess uncertainties vigorously and often, but we also need to look at indications of unassessed uncertainty, especially where assessment is impossible or uneconomical.

1H. Indication, Determination, or Inference

Since Student's revolution, many kinds of formal inference have been developed; today we have our choice of three levels of statistical analysis. We can be concerned with any or all of the following:

◇ pure indication in which, for example, we attend only to primary statistics, such as means and percentages, with no attempt to assess uncertainties;

◇ determination, or augmented measurement, where both primary and secondary statistics are to be calculated and considered, for example, a mean and an estimate of its standard deviation (and where, though nothing is said about inference, the secondary statistics are most often interesting only as tools for appraising uncertainty);

◇ formal inference, where some relatively precise specification or description of a manageable mathematical model allows us, at least apparently, to tie up our uncertainties in a neat package—for example, through confidence limits, significance tests, fiducial inference, posterior distributions, or likelihood functions.

In some specific situations, the third course is not open to us, perhaps for some of the following reasons:

◇ the data do not allow some of the important causes of variation to show their effects;

◇ the causes show themselves in a distinctly nonrandom way; or

◇ no formal inference procedure is available because none has been developed, or those that have been worked out all employ such stringent assumptions as to make their use unwise or misleading (or even perhaps because the computational effort required would be impractically great).

Sometimes, indeed, we may appear to lack, as in the second of these, for instance, even any reasonable way to compute secondary statistics. Then we are likely to appear forced to resort to pure indication.

When our concern is with indication alone, there is an important distinction between indications that point toward something definite and those that point usefully but not toward anything, as when a graph causes its viewers to say "Look, there's a kink in the curve!" Indications of the former kind are also called estimates. Their indicators are also estimators. We discuss estimates and estimators further in Section 2D.

This distinction between estimators and nonestimators is closely similar to, but definitely not the same as, the distinction between quantitative indicators and qualitative indicators. A large value of chi-square, for example, often indicates that some hypothesis fits poorly, yet the value of chi-square itself points toward no number that has a real meaning for the underlying situation.

Summary: Data Analysis

Student, in 1908, introduced an alternative to the infinite staircase in which assessing the variability of each statistic introduced a further statistic, whose variability was harder to assess.

Three decades later, statisticians learned to shortcut the staircase without making such strong assumptions as Student's, thereby introducing the heyday of "nonparametric" or "distribution-free" procedures.

The main value of Student's works lay in (i) showing that one could deal explicitly with small samples, (ii) giving numbers for how much this mattered in an important case, and (iii) providing tables for that case. Against this, his work, not by his own choice, made it too easy (iv) to neglect "if . . . assumptions hold," (v) to overstress the "exactness" in other problems, and (vi) to fail to attack the problems of multiplicity.

We may have to be concerned with shapes of distribution whose tails are longer than might be anticipated from their middles or shoulders (we say "straggling tails" or "stretched-out tails"). **Real data**, as illustrated by C. S. Peirce's data on response time, **often has stretched-out tails**.

Among the usual sorts of deviation from normality, we need to include: (i) discreteness and irregularity, (ii) gross differences in shape, (iii) minor differences in central shape, (iv) behavior in the tails; while these are listed in order of increasing difficulty of detection, the most important is (iv), and the least (iii).

We need to be concerned not only with Gaussian efficiency but even more with robustness of efficiency.

Vague concepts often control the value and relevance of precise concepts, a point illustrated in some detail by discussing the standard deviation.

Tracing back "significance" or "confidence" leads us to "expressing amounts of uncertainty" and to "a numerical summary" as key vague concepts, the latter leading to the concept of "indication."

Three important levels of data analysis can reasonably be called "indication", "determination", and "formal inference."

References

Burrau, Ø. (1943). "Middelfejlen som Usikkerhedsmaal." *Mat. Tidskr. B.*, 1943, 9–16. (See *Mathematical tables and other aids to computation*, **2**, 1946, 74–75.)

Cohen, J. (1969). *Statistical power analysis for the behavioral sciences.* New York: Academic Press.

Fisher, R. A. (1925). "Applications of "Student's" distribution." *Metron*, **5**(3), 90–104. (References to this paper often use the date 1926, as Fisher did in his bibliography in *Statistical methods for research workers*; the journal gives 1 Dec., 1925.)

Gross, A. M. (1976). "Confidence interval robustness with long-tailed symmetric distributions." *J. Amer. Statist. Assoc.*, **71**, 409–416.

Kendall, M. G., S. F. H. Kendall, and B. B. Smith (1938). "The distribution of Spearman's coefficient of rank correlation in a universe in which all rankings occur an equal number of times." *Biometrika*, **30**, 251–273.

Mosteller, F., and R. R. Bush (1954). "Selected quantitative techniques." In G. Lindzey (Ed.), *Handbook of social psychology.* Cambridge: Addison-Wesley; pp. 289–334.

Odeh, R. E., and M. Fox (1975). *Sample size choice: charts for experiments with linear models.* New York: Marcel Dekker, Inc.

Peirce, C. S. (1873). "Theory of errors of observations." *Report of Superintendent of U.S. Coast Survey* (for the year ending Nov. 1, 1870). Washington, D.C.:

Government Printing Office. Appendix No. 21, pp. 200–224 and Plate No. 27.

"Student" (W. S. Gosset) (1908). "The probable error of a mean." *Biometrika*, **6**, 1–25. Also in *"Student's" collected papers* (edited by E. S. Pearson and J. Wishart), issued by the Biometrika Office, University College, London, 1942. Paper 2; pp. 11–34.

————, (1927). "Errors of routine analysis." *Biometrika*, **19**, 151–164. Also in *"Student's" collected papers* (edited by E. S. Pearson and J. Wishart), issued by the Biometrika Office, University College, London, 1942. Paper 14; pp. 135–149.

Tukey, J. W. (1960). "A survey of sampling from contaminated distributions." In I. Olkin, S. G. Ghurye, W. Hoeffding, W. G. Madow, and H. B. Mann (Eds.), *Contributions to probability and statistics.* Essays in honor of Harold Hotelling. Stanford: Stanford Univ. Press; pp. 448–485.

Wilson, E. B., and M. M. Hilferty (1929). "Note on C. S. Peirce's experimental discussion of the law of errors." *Proc. Nat. Acad. Sci.*, **15**(2), 120–125.

Chapter 2/Indication and Indicators

Chapter index on next page

Indication is elementary, important, and neglected.

In beginning to remedy this neglect, we have to show indication as valuable and often the best that can be done. We have to illustrate how indicators may be reasonably chosen; how indication can require care, as illustrated by cross-validation; and how graphs are of great value, almost entirely because of their function as indicators.

2A. The Value of Indication

One hallmark of the statistically conscious investigator is a firm belief that, however the survey, experiment, or observational program actually turned out, it could have turned out somewhat differently. Holding such a belief and taking appropriate actions make effective use of data possible. We need not always ask explicitly "About how much differently?", but we should be aware of such questions. Most of us find uncertainty uncomfortable; the history of data analysis can be read as a succession of searches for certainty about uncertainty. All of us who deal with the analysis and interpretation of data must learn to cope with uncertainty in a way that meets our own needs.

A caricature of one recipe might read: Apply a significance test to each result, believe the result implicitly if the conventional level of significance is reached, believe the null hypothesis otherwise. Such a complete flight from reality and its uncertainties is fortunately rare, but periodically considering its extremism may help us keep our balance.

To the researcher, the primary value of data lies in what they **indicate,** what they appear to show. Such an appearance may be quite correct and exact, may be a matter of chance, and probably is a mixture. In some fields, when it is only a matter of counting noses or counting punched cards, assessing the indications tends to be overlooked or regarded as an unimportant detail, though demographic and population studies offer good exceptions.

Nose-counting and reporting counts form only one extreme of indication. At the other, the most essential parts of our most complicated schemes of analyzing data—complex analyses of variance, multiple regression analyses, factor analyses, latent structure analyses—have as their main function the assessment of indication.

The word indication is a vague concept intended to include, at one extreme, all of the classical descriptive statistics (for example: mean, median,

mode, quantile, standard deviation, correlation), but also, at another extreme, to include any hints and suggestions obtained from data by an understandable process, suggestions that might prove informative to a reasonable man. Examples are appearances of similarity (all the curves look S-shaped or ogival) or general behavior (the frequencies seem to be falling off roughly exponentially, or, again, although the means of the groups vary widely, the standard deviations are all close to those usually found in these experiments). Indication includes not only isolated numbers (differences, slopes, and other indices), but also inequalities and trends (the female appears to be more deadly than the male, blood pressure appears to rise with age), and appearances of graphs and diagrams (these scatter diagrams appear doughnut-shaped).

What indication is *not* is inference or treatment of uncertainty; indication does not include confidence limits, significance tests, posterior and fiducial distributions, or even standard errors.

The treatment of variation or uncertainty looms large and that of indication small in most discussions of data analysis. This imbalance arises naturally: important considerations in assessing indications are often specific to a particular problem and therefore difficult to generalize about; since the study of variation is often neglected because of the beginner's eagerness to find regularities and simple appearances, we need to focus our attention on problems of uncertainty; and, conversely, for many, it is psychologically satisfactory to seek some kind of certainty. These are some of the many reasons that have combined to generate a bulky literature about measuring and allowing for variation, and to make most introductory statistics texts give their main attention to what is done once the indication is found. Naturally, we all desire an adequate assessment of both the indications and their uncertainties, but we shouldn't refuse good cake only because we can't have frosting too.

2B. Examples of Stopping with Indication

Often, the analyst of data stops calculating after reaching pure indication. (Some further thought may still give judgmental impressions of the uncertainties involved.) Let us illustrate a few such circumstances. Suppose, for example, that, in a haphazardly assembled remedial-reading class of 23 taught by a single teacher using one special method, the reading of 5 students substantially improved and that of 18 was substantially unchanged. The indication is that the treatment of the students improves perhaps 1/4 of the students. This figure is fraught with uncertainties about which the data give us little aid. All the same, if other teachers and methods are getting improvement for 85% of the students, we ignore at our peril the indication that this method is not as good as general practice.

If we sought to assess the uncertainties of our 5/23, what could we do with such problems as:

⋄ From what population was this class something like a random sample?

⋄ Is that population something like those from which the samples taught by other methods were drawn?

⋄ How much do the teacher's personality and tendencies to deviate from the assigned method matter?

In this example, a good assessment of uncertainty seems most unlikely. Nevertheless, we have an indication that can be the basis for action.

Matters need not be so simple as this example of indication suggests. To get data to yield a sensible indication may require subtle and complex analysis chosen with sophistication. As an example, let us simplify a problem that was much discussed in scientific quarters in the early 1960's. Consider choosing the adjustments required to take equitable and defendable account of background characteristics in a comparison of blacks and whites on a standard intelligence test. (To illustrate difficulties of indication, we need not raise either the hard issue of proper measures of intelligence or the biological problem of purity of strains.) Age, family size, socioeconomic status, population density, years of schooling, kind of schooling, strength of home life and its orientation toward things of the mind might provide a basis for handling this question. We still have the tough question of how these data shall be used. Obviously multiple-control methods are required, and we may be delighted to stop with indication, once these methods have been effectively applied. The techniques of indication often go far beyond the descriptive statistics found in first courses in quantitative methods.

Problems of Multiplicity

Another kind of problem arises when we want to distinguish two groups of individuals by a difference in variability for at least one of a large number, say 100, of measured characteristics, having one measurement on each characteristic for each individual. We can, if we are careful, compare the variances on any one aspect of the two populations in a way that is reasonably reliable at, say, 95% confidence. Even if the null hypothesis of equal variability were true, the average number of tests individually significant at 5% is just the product of 5% and the number of tests, namely $(.05)(100) = 5$.

By the same token, the average number of tests individually significant at 0.05% (1/20 of 1%) is $(.0005)(100) = 0.05$, which can be interpreted to mean that one or more such will occur in about 5% of such situations (a situation means a repetition of the whole: here 100 tests each at a 0.05% level). This means that if the apparently most-extreme variance ratio is to be significant at the 5% level **in its role as the largest of 100,** it must be **individually** significant at about 0.05%!

Such an extreme value must fall in a region where the exact but unknown shape of the underlying distribution may have large effects on the probability of our test ratio, whether this be $(s_1^2/\sigma_1^2)/(s_2^2/\sigma_2^2)$ or something more robust, falling beyond a given value. No sort of statistical procedure is yet sure to deal well with this difficulty. The only way to obtain adequate control of such uncertainty may be to make a repeat study confined to those few characteristics that appeared in the first study to differ a lot from one group to another. In the interim, even though the data may have given the relevant sources of variation full opportunity to reveal themselves, it may be quite unwise to do anything beyond indication.

An example farther along, but in a somewhat more familiar direction, arises when we are searching for interesting indications that may serve as hints in approaching further data. The intent here is not one of attaining a conclusion, nor of making measurements for record, but only of hunting out interesting indications. Suppose that we have looked at many aspects, say 1000 to 10,000, instead of only 100, and have selected from this exploration a list of aspects whose values appear interesting. Now the dangers of implicit belief in very extreme % points are even greater. Here we must stop our calculations with indications and be careful to think of our results only as hints as to what to study next, rather than as established results.

A seemingly slight modification arises when exploration and hint-searching has been carried out on a portion, say one-third, of the total data, with the intention of coming up with approximately a given number of hints, say 10 or 20, which are to be tried out, either for measurement unaffected by selection or as subjects for conclusions, on the remaining two-thirds of the total data. Here, at the transition from exploration of the first third to confirmation by the other two thirds, we need nothing but indication; our problem is to pick the best-looking hints for immediate trial. (Compare this process with cross-validation discussed in Section 2F below.)

(Indeed, on the closely analogous problem of selecting the "best" of many new strains of a crop, it is often wise to make the first selections in circumstances where one knows that observed differences will almost certainly not be provably real (Yates, 1950). Some find this paradoxical. Breeders can measure accurately if they raise many specimens of a very few strains with, of course, little chance that an especially good strain has been included. Or they can raise many strains and measure them relatively inaccurately because they have few individuals of each strain. Following the latter procedure increases the chance of raising a good strain, but, because of poorer measurement, decreases the chance of detecting it. Attaining a desirable balance produces the paradox.)

Some of the important reasons for stopping with indications are:

a) The form of the data (its structure) masks the important sources of variability.

b) We have no good way to deal with what may be substantial differences in variability of this sort of indication, differences that experience suggests may occur but to an extent not detectable in a single set of data. Usually this difficulty arises from the multiplicity of variables or aspects being examined.

c) Preliminary exploration of data of high multiplicity has led to selection of a few indications for later confirmation.

Whatever the reason or reasons for stopping with indication, they should be reported.

Consider, as a final example, comparing short and long forms of a projective test in order to study the effect of form length on reliability of scoring. Each form has 32 different ways of scoring a single protocol. Suppose that, as would be the case for the Rorschach test, the scores are available on a split-half basis, so that the calculation of reliability coefficients is quite feasible. To reduce the effects of sampling upon the final comparison, the investigator gives the short and long forms to the same people at the same time, perhaps by having the short form part of the long form. He finds it easy: to calculate 32 reliability coefficients for the short form, and their average; to do the same with the reliability coefficients for the long form; and to look at the difference (or perhaps the ratio) of these average reliability coefficients. This indication is, he hopes, reasonably responsive to the original question of the comparative reliability of long and short forms.

What if he seeks to pass from indication to inference? His statistical advisor is apt to tell him that each of his 64 basic scores is correlated, to unknown and probably differing extents, with each of the others. Here are $\frac{1}{2}(63)(64) = 2016$ correlation coefficients to consider. And each correlation coefficient between basic scores requires conversion, not a trivial task, to a correlation coefficient between the corresponding split-half reliabilities. When this is done, the variances and covariances of the reliabilities can be found, then those of the two average reliabilities, and finally the variance of their difference. If large-sample theory is adequate, as it may be if he happens to have 10,000 protocols, the uncertainty of the difference can now be assessed. After thinking about the effort involved, what would you do?

Under the conditions described, indication would be the end of the road for most investigators. (We shall introduce in Chapter 8 a technique that makes inference feasible even in problems this gory.)

Sometimes results that individually might reasonably have arisen from purely chance variation are worthy of report because they strengthen one another. Sometimes we have parallel results based on independent bodies of data, all contained in a single study where the assessment of mutual support is easy. Or there may be parallel results on overlapping or interrelated data, as where seven questions have all been answered by the same respondents, where the results agree in direction but the amount of mutual support is hazy. Or it may be that one result in today's study is likely to be paralleled in due course by other results in other studies.

In all these situations, it can be quite wasteful to suppress results either because they are not individually significant or because their joint significance cannot be satisfactorily appraised.

2C. Concealed Inference

When we study data for the answers to a specific question, we sometimes find the evidence so strong as to obviously resolve the question. When matters are so clear-cut, quite informal inference is usually adequate, both in analysis and in communication.

If the data are only obviously strong, the reader or listener is often asked to look at the indications, and "see" the situation. The investigator typically says "No statistics is necessary to see that . . .", meaning "simple ultraconservative statistical methods are overwhelming". Back of such statements lies some notion of statistical analysis that is not being displayed. As examples: "85 successes out of 100 cases could scarcely be compatible with a probability of success of 0.1" or "the standard deviation of the difference is obviously over 100, but even if it were as small as 50, a difference in sample means of 10 could scarcely be strong evidence in favor of either method". Although such discussions are common, considerable sophistication or experience may be required to be able to dispense with formal inference, and many of us have been brought up short by a neophyte's saying that he or she is not convinced, sometimes because the obvious has required a rather lengthy demonstration.

If the data are much stronger than this, only the indications are given, and nothing at all is said to the reader or listener.

Such instances are not instances of stopping with indication, as we use this term. They are merely cases of inference where there has been no need for any of the formalities of inference. Informal inference, perhaps expressed by "obviously", or by "looking at . . ., we see that", or by absence of any remark at all, is the only sort of inference felt necessary. Yet, even if no word be written or spoken, it is inference.

When we speak of stopping with indication, we do not intend to include these pleasant cases of informal inference. Rather we refer, in the main, to cases of indication, without even the barest sort of appraisal of the indicator's stability, variability, or reliability to provide a basis for informal or formal inference.

2D. Choice of Indicators

The analyst often chooses which indicator to use. He or she compares the responsiveness of the different indicators to the specific question reached. The analyst is likely to compare their ease of calculation and frequently asks whether the results of one will be more, or less, stable than those of another.

Making such judgments takes more than looking things up in the proper book. Wholly inferior indicators are indeed subject to the struggle for existence and tend to die out. But, as we saw at the close of Section 1C, a difference in distributional shape so small as to be difficult to detect in samples of substantial size (like a few thousand) can alter the balance between two indicators drastically. In the example cited, the indicators happened to be estimators. One, under ideal conditions, did only nine-tenths as well as the other; under more realistic conditions it did almost half-again better.

Such difficulties and delicacies must be faced even, perhaps especially, if the indicator selected is to be used for pure indication alone.

Before discussing choices among estimating indicators—among estimators—we need to clarify the meaning of the verb "estimate". When does an indicator estimate something? What does it estimate?

One naive answer is: An estimate estimates a parameter. Historically, the word "parameter" has meant two quite different things:

◇ The numerical value of a particular symbol in a particular way of specifying a family of distributions, as when the family of normal distributions is parametrized either by μ and σ^2 or by μ and σ.

◇ A numerical characteristic of a distribution, as when any numerical population is partially characterized by its median.

To restrict "estimation" to estimating a particular coefficient in a family of distributions would lose much of the usefulness of this vague concept. The idea of estimation should be as general as possible. Thus, while it is usually helpful to know about the comparative performance of two estimators when the individual fluctuations in the data follow distributions of Gaussian shape, it can be misleading not to also know how they behave for other shapes.

Accordingly, even if the analysts know just what to estimate, a contingency not as frequent as is commonly thought, they need to be guided in choosing an estimate by all that is known—or felt—to be true under a variety of alternative circumstances.

An estimator is a function of the observations, a specific way of putting them together. It may be specified by an arithmetic formula, like $\bar{y} = \sum x_i/n$, or by words alone, as in directions for finding a sample median by ordering and counting. We distinguish between the estimator and its value, an estimate, obtained from a specific set of data. The variance estimator, $s^2 =$

$\sum (x_i - \bar{x})^2/(n - 1)$, yields the estimate 7 from the three observations 2, 3, 7. We say s^2 is an estimator for σ^2, and we call σ^2 the estimand. In the numerical example, 7 estimates σ^2.

Sometimes the estimator comes first, and we then ask what it points to. We speak of the estimator's target as an *estimand* (as something to be estimated) rather than as just a parameter. Our problem is to match estimands to estimators. When can we do this? How precisely? Are there general circumstances when we can expect a match?

These questions have not had the research attention they deserve. The question of sample size plays a central role, although we might wish that it did not. In actual investigations, we use our estimator on a sample of a particular size. Yet today's approaches to the choice of matching estimands to estimators depend on what we would do with larger samples, a somewhat unsatisfactory state of affairs.

As a device to aid discussion of this question at this time, we divide the estimators we use for a particular sample size into three classes:

1. those we would use for all larger samples,

2. those we would use for much larger, but not arbitrarily large, samples,

3. those we would not care to use for much larger samples.

Class 3 is hard to deal with in the present framework. It may never provide us with satisfying matches between estimator and estimand.

Class 2 occurs frequently in actual practice. The sample range is often an example. Within our framework, we can often get practical answers by forgetting that we would not use this estimator for arbitrarily large samples, by pretending that we have class 1.

Class 1, then, unrealistic though it may be, is the class we now act as if we have. Even in this Utopia we cannot lay down simple rules that will pick an estimand to go with our estimator. All that we can do is to list some alternative circumstances where we are likely to be satisfied with the match between estimator and estimand. These include:

◇ If the average value of the estimate (the result of averaging over all samples of a given size) has the same value for all sample sizes, this value is a good candidate for the matching estimand, as is the case for s^2 and σ^2, so long as σ^2 is finite. If, in addition, the distributions of the estimate condense around this estimand as the sample size grows, most would find the match more gratifying. Exceptions arise, for example, when the average value of the estimator is not at all typical of its distribution, as when considering s^2 for a distribution with infinite σ^2.

◇ If the average values depend on sample size, but converge to a limit as the sample size increases, as is the case for s when σ is finite, much the same can be said. Most statisticians are more pleased when the dependence on sample size is slight, as in the s, σ example for normal distributions for $n \geq 10$, say.

◇ In a similar vein, if the median of the distribution of sample estimates is the same for all sample sizes, or if its value converges to a limit as the sample size grows, this common or limiting value is likely to be a satisfactorily matching estimand. For, especially in large samples, the median of an estimate distribution rarely fails to be reasonably typical of that distribution.

◇ The limit, as sample size grows, of any reasonable typical value (see Section 1G), other than the mean or median, of the distribution of estimates for a fixed sample size is also likely to be a satisfactorily matching estimand.

These rules are not neat and detailed, but they do identify some matching estimands that many would regard as satisfactory. Our own diffident and tentative attitude toward this problem stems partly from the limited studies on which these remarks have been based, and partly from dissatisfaction with having the acceptability of estimands depend on properties of collections of sample sizes not present in the investigation.

2E. An Example of Choice of Indicator

Suppose that one can make repeated observations of some quantity, observations which behave rather like a sample from a population with long, straggling tails. Having collected these observations, the observer wishes to summarize them in a way that indicates their location with as much precision as can be simply obtained. He or she considers the arithmetic mean of all the observations but finds that the long straggling tails impart so much variance that this indicator is unduly imprecise. The analyst next considers using the sample median, because it recovers about 2/3 (actually, in large samples, as we noted in Section 1E, $2/\pi$) of the information about location in a sample from a normal distribution and is likely to do even better than this in samples from distributions with more straggling tails.

The stability of the sample median depends on the height of the density near the population median. Exhibit **1** represents a density for which the median is a poor choice for measuring location. As long as the straggling-tailed distribution has a reasonable density in the middle, our observer prefers the median to the mean (might there be a still better choice?). Some may think the median unduly variable; it is not likely to be a very good candidate for use of the jackknife (Chapter 8); it may miss enough information to matter.

An alternative to the median is the trimmed mean. The sample is trimmed of its possibly straggling tails by setting aside some fraction of the measurements from each tail of the sample. Specifically, let us suppose that we decide to set aside the lowest 10% and the highest 10% of the observations, and to take the arithmetic mean of the remaining 80%.

How well would this do for a normal distribution? Keeping only the median gives an efficiency of about 2/3 (more precisely, $2/\pi \approx 63.7\%$). Using the trimmed mean regains a fraction of the remaining 1/3 of the information about μ contained in the sample. For symmetrical trimming, this fractional gain obviously runs from 0% (for 50% trimmed from each tail, leaving only the sample median) to 100% (for 0% trimmed). Investigations we do not give here show that the gain is somewhat faster than linear, and so we will be conservative if we assign an efficiency for the proportion α trimmed from each tail of

$$\frac{2}{\pi} + (1 - 2\alpha)\left(1 - \frac{2}{\pi}\right) \approx \frac{2}{3} + (1 - 2\alpha)\frac{1}{3},$$

which gives, in our example,

$$\frac{2}{3} + 0.8\left(\frac{1}{3}\right) \approx 0.93 = 93\%$$

of the information about location. In samples from distributions whose tails straggle moderately more than those of Gaussian distributions, trimmed means do even better than this. Trimmed means can be safely jacknifed, a technique described in Chapter 8.

Our observer has chosen an indication. It indicates "location", for if we add a constant to all the observations, the trimmed mean changes by the same amount and in the same direction. Has there so far been a definite distributional model? No, except for the existence of a parent distribution. Or has a

Exhibit **1** of Chapter 2

Example of a density for which the sample median would be an extremely poor choice for measuring location.

If the area under each part is close to 1/2, then from sample to sample, the median will flop back and forth between the left and right half, producing substantial variability.

In the figure, P is the probability to the left of a given value x.

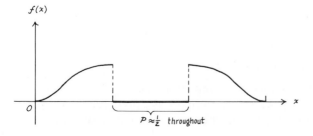

definite parameter to be estimated been explicitly identified? Again, no, though a class is implied by the specific choice, namely, the typical values of means from samples of this size trimmed 10% on each tail.

In our trimmed-mean example, as the sample grows larger and larger, the central 80% of the sample will more and more nearly match the central 80% of the population. This makes the natural choice of estimand the mean of the central 80% of the population.

(Optional)

We have reached our answer and expressed it understandably. The more mathematically inclined may like to see the result expressed more formally. Let θ be the parameter being estimated, the estimand; let F be the cumulative. Then

$$\theta = \int_{-\infty}^{\infty} y\phi(F(y)) \cdot dF(y),$$

where

$$\phi(u) = \begin{cases} 1.25 = 1/0.8 & \text{for } 0.1 < u < 0.9, \\ 0 & \text{elsewhere.} \end{cases}$$

We gain even more from thinking about the finite-sample formulation:

$$\text{ave }\{10\% \text{ trimmed mean}\} = \int_{-\infty}^{\infty} y\phi_n(F(y)) \cdot dF(y),$$

where, writing $[n/10]$ for the number of observations (approximately 10% of n) set aside from each tail:

$$\phi_n(u) \begin{cases} \text{vanishes like } u^{[n/10]} \text{ or } (1-u)^{[n/10]} \text{ near 0 and 1, and} \\ \text{is mainly concentrated over and near the interval } (0.1, 0.9). \end{cases}$$

This formulation can be used to tell how large a sample is needed to suitably reduce the contributions to this average value from the extreme tails of the distribution.

We now understand what the estimand is, and a bit about when trouble from quite straggling tails is likely—*provided we are dealing with a sample of independent observations.*

2F. Indications of Quality: Cross-Validation

Even though they are not always used for prediction, the results provided by some statistical techniques can be loosely described as "predictive" or "forecasting". A multiple regression of one variable upon a number of others, for example, may set forth what is hoped to be a standard property that holds for other data.

Users have often been disappointed by procedures, such as multiple regression equations, that "forecast" quite well for the data on which they were built. When tried on fresh data, the predictive power of these procedures fell dismally.

Let us discuss this situation in terms of a single example, multiple regression, while understanding that our insights and conclusions apply to many other techniques as well. In dealing with multiple regression, when we speak of a "procedure", we mean, for example, a specific regression equation:

$$z = 3.4x + 2.5y - 5.4.$$

When we speak of the "form", we mean, for example, choosing which variables from among many shall enter the regression and deciding individually for each variable whether we use the original measure, its logarithm, square root, or other re-expression. We also have to choose how the variables shall be grouped and combined—sums, products, ratios, or how? In this discussion, then, a "procedure" consists of a "form" and numerical values for its coefficients.

When we use data to determine some procedure, we want to be able to answer the question: How well may I expect the chosen procedure to behave in use? Even when the specific variables to be used in a multiple regression have been picked in advance, so that the form is determined, the coefficients are chosen from infinitely many combinations of possibilities to make the results of substituting in the formula fit the data as closely as possible. Testing the procedure on the data that gave it birth is almost certain to overestimate performance, for the optimizing process that chose it from among many possible procedures will have made the greatest use possible of any and all idiosyncrasies of those particular data. Sometimes we say that "Optimization capitalizes on chance!" As a result, the procedure will likely work better for these data than for almost any other data that will arise in practice. The apparent degree of fit will be closer than the true fit, on the average.

No one knows how to appraise a procedure safely except by using different bodies of data from those that determined it. In other words, appraisal requires some form of cross-validation. We recognize two levels of cross-validation, simple and double, the simple being more widely recognized. They are:

⬦ **Simple cross-validation.** Test the procedure on data different from those used to choose its numerical coefficients.

⬦ **Double cross-validation.** Test the procedure on data different both from those used to guide the choice of its form and from those used to choose its numerical coefficients.

The second level of cross-validation, which, by analogy with the physician's "double-blind" study, we have called "double cross-validation", is to be had only by going to fresh data. These fresh data are best gathered after choosing form and coefficients. When fresh gathering is not feasible, good results can come from going to a body of data that has been kept in a locked safe where it has rested untouched and unscanned during all the choices and optimizations (as in the studies of Macdonald and Ward, 1963, Miller, 1962, and Mosteller and Wallace, 1964). For the full validating effect, the data placed in the safe must differ from those used to choose the procedure in ways that adequately represent the sources of variation anticipated in practice. For example, they may need to involve distinct school systems, distinct investigators, or distinct years of observation. Despite the high merit of double cross-validation, we cannot always afford it.

Whether we are content with, or stuck with, simple cross-validation, today the computer offers us new freedom and power. In the classical approach to simple cross-validation, the available body of data was divided into two (sometimes more) pieces of similar size. One was used for optimization, the other for testing. Some energetic workers would then interchange the two pieces and repeat, thus gaining more information from the same data. Although they learned more by this process, a subtle miasma of suspicion arose from the unknown correlation between the two estimates of quality.

Such suspicions show again how insistence on inference as *the* goal tends to distort attitudes toward indication. Unknown correlations among the component estimates do, indeed, destroy the possibility of using degree of mutual agreement to assess **stability** of the combined result precisely. On the other hand, whenever each estimate by itself is sound, a weighted combination of two or more, however much or little correlated, is both equally sound and more precise than the individual values.

The man who has halved his data and cross-validated in both directions has used *all* of his data to assess the quality of what is to be had by optimizing upon a body of data **half** as big as his total collection. If he has so much data that halving or doubling has little effect on the quality of the optimized procedure, well and good. He can do little more than he has done. Few are so fortunate.

When routine computation was expensive, even *doing* the cross-validation in both directions seemed an effort. Today, we can face much more.

Frank Yates (1957) proposed and P. J. McCarthy (1976) has recently expounded the use of more than one way to halve the data to obtain additional information. Or suppose that we divide the data into ten parts of similar size. Then we can combine any nine parts, optimize the procedure for this nine-tenths, and validate on the remaining tenth. Once we have done this ten times, separating a different tenth each time, we have used **all** the data to assess the quality of what is to be had by optimizing upon a body of data **nine-tenths** the size of the total collection. This often comes appreciably closer to answering the most usual question: Approximately what performance may I expect from the result of optimizing upon **all** of my data?

With a computer, the ten calculations are often little more effort than the conventional one or two, because we have only to repeat the same pattern of computation, with little more programming and debugging time.

Often we can profitably go much further. Suppose that we set aside one individual case, optimize for what is left, then test on the set-aside case. Repeating this for every case squeezes the data almost dry. If we have to go through the full optimization calculation every time, the extra computation may be hard to face. Occasionally, one can easily calculate, either exactly or to an adequate approximation, what the effect of dropping a specific and very small part of the data will be on the optimized result. This adjusted optimized result can then be compared with the values for the omitted individual. That is, we make one optimization for all the data, followed by one repetition per case of a much simpler calculation, a calculation of the effect of dropping each individual, followed by one test of that individual. When practical, this approach is attractive.

M. Stone (1974) gives a generalized form of a cross-validation criterion applied to the choice and assessment of statistical prediction.

One drawback of all kinds of single cross-validation is that the test sample is all too often much more like the optimization sample than is typical of the population of individuals or situations to which we wish our indication to refer. Accordingly, single cross-validation is all too often weaker, by an unknown amount, than it appears to be.

The possibility of cross-validation is one of the main advantages of most automatic schemes of optimization. Approximate, cut-and-try, judgment-guided selection of procedures can often come very close in quality of result, or even be superior, to those that formally optimize, but because one cannot fully specify what the procedure is and how it chooses the results, one cannot be sure that it is being tested on "independent" data. On the other hand, at the simple cross-validation level, every procedure generated by automatic optimization is easily tagged with the body of data used to determine the numerical values of its coefficients.

The difficulty with determination by subjective judgment recurs at the level of double cross-validation. When we choose the form of a procedure, the full body of data that was used may not be clear to anyone. Often the choice leans, sometimes very usefully, on a lifetime of experience, fact, and folklore, as well as on the convenience of the specific analysis. The experienced investigator will be especially hard-pressed to behave "double blind" because the form chosen may have been picked with the idiosyncrasies of the sets of data yet to be gathered already in mind. Satisfactory cross-validation on these next few sets will verify the investigator's foresight and satisfy us about its use on that sort of data, but can over-assure us about the cross-validation for uses further into the future that may not have been considered. Obviously we are not discussing an issue of "bad" versus "good". The investigator wants to be sure to consider as far as possible the whole range of uses, and thus be able to expect different performances under different circumstances.

Summary: Indication and Indicators

Indicate means **appear** to show; sometimes indication is as far as we need carry an analysis; indication may be numerical or qualitative.

To go beyond indication requires an assessment of uncertainty of the indication, preferably a trustworthy assessment.

We must be concerned about problems of multiplicity, because having 1 result out of 1000 results tested turn out nominally just significant at 5% means something quite different from having 1 of 1 do this.

In selection problems, a well-designed experiment may be one that almost always produces only indications, and where statistical significance of the results as a whole is only rarely to be anticipated.

We often deal with concealed inference, where no formalism and no arithmetic need be used by a skilled analyst of data to see that either (i) there is no doubt of significance, or (ii) there is no hope of significance.

One way to judge the quality of indicators is to consider how well they estimate what they are supposed to estimate.

While sometimes we have something in mind for an indicator to "estimate" when we pick the indicator, it is often better to ask the indicator, in terms of the answers it gives, what it is trying to estimate—what is its *estimand.*

One way to do this is to ask what happens when the indicator is applied to larger and larger samples.

Cross-validation is a natural route to an indication of the quality of any data-derived quantity, be it estimate or pure indication or something else.

Cross-validation can be single (on data not used to fit numerical coefficients) or double (on data not used either to choose coefficients **or** to choose the form of the analysis or fit). Double is always safer.

A major value of automatic schemes of optimization, fitting, and so forth, is that they can be *reliably* cross-validated.

We take seriously indications or indicators not taken further down the inference trail; we take them seriously but with a pinch of salt labelled "indication only".

We plan to cross-validate carefully wherever we can.

References

Macdonald, N. J., and F. Ward (1963). "The prediction of geomagnetic disturbance indices: 1. The elimination of internally predictable variations." *J. Geophys. Res.,* **68,** 3351–3373.

McCarthy, P. J. (1976). "The use of balanced half-sample replication in cross-validation studies." *J. Amer. Stat. Assoc.*, **44,** 596–604.

Miller, R. G. (1962). "Statistical prediction by discriminant analysis." *Meteorol. Monogr.*, **4**(25), 1–54.

Mosteller, F., and D. L. Wallace (1964). *Inference and disputed authorship: The Federalist.* Reading, Mass: Addison-Wesley.

Stone, M. (1974). "Cross-validatory choice and assessment of statistical predictions." *J. Roy. Stat. Soc.*, Series B, **36,** 111–147.

Yates, F. (1950). "Recent applications of biometrical methods in genetics: 1. Experimental techniques in plant improvement." *Biometrics*, **6,** 200–207 (especially pp. 204–205).

Yates, F. (1957). Personal communication (to J. W. Tukey).

Chapter 3/Displays and Summaries for Batches

3A. Stem-and-Leaf

Quick ways to write down batches of numbers are useful, especially if they reveal a fair amount about patterns of distribution. So are ways that let us "count in" easily, as when we are asked to find the 15th value from each end. All of these things can be done with **stem-and-leaf displays**, though we may need to use two, or more, styles of display to do two things well enough. These displays go further than the classical tallying devices for producing histograms.

Consider the 1971 preliminary figures for dollar value of mineral production in the United States, by states. Alabama produced 291.4 millions of dollars worth. This we shall break into three parts as follows:

	Stem	Leaf	Forget	(millions)
Ala.	2	9	14	(291.4)
Alask.	3	3	28	(332.8)
Ariz.	9	8	10	(981.0)
Ark.	2	5	32	(253.2)
Calif.	19	2	06	(1920.6)

where we have included, as well, the data for the next four states in alphabetic order. The result, for these five states only, is the skeleton stem-and-leaf display shown in Panel A of Exhibit **1**.

A number of ways to ring the changes on stem-and-leaf displays are discussed in *EDA*,* Chapter 1. The only other versions we illustrate here are shown in Exhibit **2**.

3B. Medians, Hinges, etc.

When values are arranged in increasing order, the middle one is called the median, as has previously been mentioned. If we have already written the values out in order, the median is easy to find. Given, for example, 25 values (counting ties, if any), the 13th from either end is the median, which we will call M. Given 24 values in all, the median is the $12\frac{1}{2}$th from either end. (Here we define $12\frac{1}{2}$th as halfway between the 12th and 13th.) The general rule is

$$\text{depth of median} = \tfrac{1}{2}(1 + \text{batch size}).$$

---

* *Exploratory Data Analysis*, J. W. Tukey (Reading, Mass.: Addison-Wesley, 1977).

We can also readily find values at prescribed depths once we have a stem-and-leaf display. Its cumulative column shows us that Panel B of Exhibit 2 contains $76 + 5 + 74 = 155$ values. The median value, which is the 78th from each end, is to be found by counting within $18 \mid 17001$ as if we had it written down as

(76 lower)
180
180 M
181
181
187
(74 higher)

Exhibit **1** of Chapter 3

A stem-and-leaf for value of mineral production in U.S. by state, 1971 preliminary (1 in a leaf is 10 million)

A) SKELETON STEM-AND-LEAF

stem	leaf
1	
2	95
3	3
4	
5	
6	
7	
8	
9	8
10	
11	
12	
13	
14	
15	
16	
17	
18	
19	2

Interpretation. 9 at 2 is 29, 5 at 2 is 25; thus 2|95 means 29 and 25 each once. 8 at 9 means 98, and 2 at 19 means 192.

B) ALL FIFTY STATES

17	0	20229571997066398
21	1	1261
8	2	95286893
21	3	3948
17	4	0
16	5	82
14	6	405
11	7	01
	8	
9	9	82
7	(H)	(192, 555, 104, 118, 114, 680, 127)

Notes

1. The 0-stem-0-leaf entries are for 2.241 millions in Delaware and 4.299 millions in Rhode Island.
2. The lefthand column contains cumulative counts of the number of states, counting in from the top and from the bottom, except that the 8 at the stem of 2 is the number of states with that stem.
3. The line labeled (H) gives the numbers larger than 99.

Interpretation. 3948 at 3 is one each 33, 39, 34, 38. 95286893 at 2 is 22, 23, 25, 26, 28, 28, 29.

S) Source

The World Almanac, 1973, page 423.

where the 78th value from either end is the value marked M. We call the median M a *letter value*. M tells us about the center of this batch of numbers in a way not much influenced by either one or a few "outliers", values much smaller or larger than most of the other values in the distribution. (We make a more specific study of this in Section 14G, especially Exhibit 1.)

Other letter values. We often gain by defining not only a value halfway from each end, but values that we count in a quarter of the way, an eighth of the way, and so on. These values tell us about the variability of our numbers and the shape of the distribution. An easy way to remember just what we are to do is to note that our rule for taking a step from one letter value to another looks

Exhibit **2** of Chapter 3

Splitting stems and covering two decades to display telephones/city for North American cities with at least 100,000 phones in 1972 (1 in a leaf is 10,000 phones)

A) UNSPLIT STEMS, but CHANGE IN STEM WIDTH (using roughly half of the cases that appear in the first two columns of the source)

```
0  |
1* | 3971692860137462118107480310661313614478105321 4
2  | 5312173869
3  | 0142
4  | 83200
5* | 4
6  | 823
7  | 4
8  | 71
9* | 6

1* * | 06, 37,
2    | 35,
3    |
4    | 85,
5* * |
```

Note
1. Leaves in top portion are 1-digit but are 2-digit in lower portion. The single asterisks following the numbers 1, 5, and 9 actually apply to all the stems in the top portion and indicate that we are dealing with a 2-digit number, a one-digit stem and a one-digit leaf. But the asterisks are given for only 1, 5, and 9 to avoid clutter. Similarly, the two asterisks in the lower portion remind us that we are dealing with 3-digit numbers, a one-digit stem and a two-digit leaf, and again only the 1 and 5 have been marked.
2. The first entry 1* | 3 stands for 130,000 phones.
3. The first entry in the bottom portion 1** | 06 stands for 1,060,000 phones.

→

Exhibit **2** of Chapter 3 (continued)

B) SPLIT STEMS, *et al.* (one digit per leaf, everywhere)

0*

14	**10**	13787253637324
28	**11**	72416415612651
36	**12**	23953110
46	**13**	3706140514
54	**14**	53289349
60	**15**	357939
68	**16**	11737448
76	**17**	23717516
5	**18**	17001
74	**19**	131

one to two
hundred
thousand

1 in leaf
is 1,000

71	**2***	31213332003434
57	**2**	57869888759
46	**3***	014204246148
34	**4**	832000
28	**5***	4305625
21	**6**	8239
17	**7***	41110
12	**8**	717
9	**9***	6

two to nine
hundred
thousands

1 in leaf
is 10,000

8	**1*x**	03205
3	**2*x**	3
	3*x	
2	**4*x**	8
1	**5*x**	9

one to five
millions

1 in leaf
is 100,000

Notes

1. Three digits carried in top section, one less in bottom section. Thus 19 | 131 means two 191's (191,000 to 191,999 phones) and one 193, while 7* | 41110 means one 74 (740,000 to 749,999 phones) three 71's, and one 70. Moreover 4*x | 8 means one 48x (4,800,000 phones).

2. Two lines of 2 |, first for 0 to 4, second for 5 to 9.

3. Counting in from each end gives rise to the leftmost column. Thus, for example, there is one 5*x and nothing more extreme, hence a count of 1. There is one 4*x, which makes a total of 2 at 400 or higher. And so on, from each end. Note further that we want to count in from either end no more than half the total.

The asterisks have an interpretation as given in Part A, Note 1 of this exhibit. The **0*** at the top of the stem column makes a place for 5-digit numbers starting with a zero and continuing with a one-digit leaf and three unspecified digits.

S) Source

The World Almanac, 1973, page 419.

much like the rule that relates the median to the total count. What we shall do is to use

$$\text{next letter depth} = \tfrac{1}{2}(1 + \text{previous letter depth*}),$$

where the * reminds us to throw away any excess over the next smaller integer. If the previous letter depth is $12\frac{1}{2}$, we use 12, and the next letter depth is

$$\tfrac{1}{2}(1 + 12) = 6\tfrac{1}{2}.$$

We label the successive letter depths (after M for median) as H (for hinge, or use Q for quarter if you prefer), E (for eighth), and then D, C, B, A, Z, Y, X, W ... (for inverse alphabetic labeling).

For our telephone example, we have

Batch	Count 155	Calculation
M	depth 78	$78 = \tfrac{1}{2}(1 + 155)$
H	depth $39\tfrac{1}{2}$	$39\tfrac{1}{2} = \tfrac{1}{2}(1 + 78)$
E	depth 20	$20 = \tfrac{1}{2}(1 + 39)$
D	depth $10\tfrac{1}{2}$	$10\tfrac{1}{2} = \tfrac{1}{2}(1 + 20)$
	etc.	

Note the use of $1 + 39$ and not $1 + 39\tfrac{1}{2}$ at depth E.

Next we look up the 78th (already done), $39\tfrac{1}{2}$th, 20th, and $10\tfrac{1}{2}$th from each end. Where is the $39\tfrac{1}{2}$th from the bottom? The 36th is 129, the highest value for stem 12, and the next 10 go 130, 130, 131, 131, 133, 134, 134, 135, 136, 137. The $39\tfrac{1}{2}$th will be halfway between the 39th and the 40th, or the average of the 3rd and 4th above the 36th, hence 131.

We also want the $39\tfrac{1}{2}$th highest. There are 34 occurrences of 40 or more, and moving down shows us 38, 36, 34, 34, 34, 34, 32, 32, 31, 31, 30, 30, where we need to go halfway between the third and fourth 34 to reach 34 at depth $39\tfrac{1}{2}$ (really 340,000 to 349,999).

Depth 20 is fairly easy to find. We have

$$
\begin{aligned}
&\text{(21st)} \quad 62 \\
&\text{(20th)} \quad 63 \\
&\text{(19th)} \quad 68 \\
&\text{(18th)} \quad 69 \\
&\text{(17 larger)}
\end{aligned}
$$

so that the 20th from the top is 63.

Exhibit **3** shows a simple standard form that includes the depth asked for so that these can be checked if desired.

Even if we had done no more than write down the letter values, as in Exhibit 3, our results would show something of the shape of distribution of these 155 numbers. At the upper end, the letter values keep on rising rapidly as we go from E to D to C and on. At the lower end, the corresponding letter values hardly move at all. Clearly, the distribution of this batch of 155 is quite skew.

3C. Mids and Spreads

If we want to work further with our letter values, convenient statistics, based on pairs of letter values at the same depth, are their means (here also medians) and the differences of such pairs. Thus, at depth D, where we had

$$87x \quad \text{and} \quad 106\tfrac{1}{2},$$

we now go on to find

$$48x = \tfrac{1}{2}(87x + 106\tfrac{1}{2}) \quad \text{(the "mid")}$$

and

$$76x = 87x - 106\tfrac{1}{2} \quad \text{(the "spread")}.$$

The spread calculations tell us about the "spread" of numbers about the median. We can make such calculations for any letter. Ways of using these further will be found in *EDA*, Chapter 19.

When, as in Exhibit 21, Panel J, of Chapter 9, we wish to give the spreads as part of the letter-value display, we write our values down like this:

#	155			
M	78	180		spread
H	$39\tfrac{1}{2}$	34x	131	209
E	20	63x	112	527
D	$10\tfrac{1}{2}$	87x	$106\tfrac{1}{2}$	767

Exhibit **3** of Chapter 3

Skeleton letter-value display for the data of Exhibit 2

	# 155		Letter values (in thousands)	
	Depths		From top	From bottom
M	78		180	
H	$39\tfrac{1}{2}$		34x	131
E	20		63x	112
D	$10\tfrac{1}{2}$		87x	$106\tfrac{1}{2}$
C	$5\tfrac{1}{2}$		125x	103
B	3		23xx	102
A	2		48xx	102
Z	$1\tfrac{1}{2}$		535x	$101\tfrac{1}{2}$
Y	1		59xx	101

Here x stands for an unspecified digit. The actual numbers are 340 thousand for 34x, 639 thousand for 63x, 873.5 thousand for 87x, and so on, as can be checked by going to the source.

The spreads shown were obtained from the original numbers in the source, rather than the abbreviated ones using the x in the third digit.

3D. Subsampling

Sometimes we have such large sets of data that sheer bulk stands in the way of careful analysis. In such straits, which have their pleasant side, subsampling, or even sequences of subsamples together with successive analyses, may be both economical and instructive. Section 12H discusses this for regression, but the idea is equally applicable for other techniques, beginning with stem-and-leaf analysis.

3E. Exploratory Plotting

If we want to diagnose relations further, we may be aided by graphical work. In this section we describe a flexible approach that may be profitable when we have a substantial amount of data. The general idea is to obtain a fairly smooth regression relating the response variable y to its matched x-values without making strong assumptions about the form of the relation.

When we are lucky enough to have a very large number of data points, we may not wish to plot them all, and subsampling may be useful. Sometimes other omissions may be more helpful.

For 20 to, say, 400 data sets, a good shortcut is to plot the k points with the highest y's and the k points with the lowest y's, where

$$k = \text{the larger of 10 and } \sqrt{n}$$

where n is the number of data sets.

For $n > 400$, k will be over 20, and we would like to cut it down a little. Starting with the largest k and smallest k values and then taking about 10 out of each of the two sets of k, roughly uniformly spread by y value, may be effective.

The 1962 *County and City Data Book* provides information for 88 unincorporated urban places (with populations of at least 25,000), including the 1959 median family income for each. Exhibit **4** gives a variety of information for (a) the 10 places (of 88) with the highest family income and (b) the 10 places with the lowest family income. Exhibit **5** shows plots, for these 20 places, of family income against 4 of the 8 quantities given in Exhibit 4.

The plot against median age shows strong dependence, and a few strays, which have been identified. The plot against percent using public transportation may surprise us, if it means, as it seems to indicate, that the more affluent use more public transportation. The plot against percent in the same house after 5 years shows a modest relation, with one New Jersey extreme stray. The plot against percent of housing units in single-unit structures shows at most a weak relationship.

Exhibit **4** of Chapter 3

The 10 urban unincorporated places, 1960, with highest and 10 with lowest incomes, 1959, for families living there in 1960.

A) Headings are column numbers in *County and City Data Book*, 1962, Table 5, pages 468–475.

Place	Median family income	(212)	(229)	(244)	(246)	(248)	(249)	(256)	(264)
New Hanover, N.J.	4572	21.8	1.1	42.2	0.4	4.7	78.8	82.2	24.1
Florence–Graham, Calif.	4904	25.7	40.2	17.1	16.8	3.9	86.6	41.2	1.3
Kannapolis, N.C.	5182	29.0	54.5	20.4	6.1	4.6	96.7	27.1	6.5
Brownsville, Fla.	5306	22.6	35.3	39.1	3.2	4.6	93.5	46.4	26.0
East Los Angeles, Calif.	5439	25.1	44.8	26.7	19.7	4.2	79.8	37.1	6.0
Bell Gardens, Calif.	5567	24.4	26.1	67.6	1.5	3.8	89.8	59.6	30.3
Hempfield, Pa.	5909	29.3	58.4	37.0	5.3	5.2	95.0	24.9	3.0
South San Gabriel, Calif.	6076	29.3	40.9	36.9	6.3	4.3	90.1	41.5	10.4
Essex, Md.	6160	24.8	46.7	34.5	6.7	5.4	78.5	37.5	12.0
Methuen, Mass.	6278	34.6	63.1	38.2	6.7	5.3	72.6	19.9	3.6
Needham, Mass.	9282	32.5	56.0	69.3	14.0	6.2	92.3	22.5	11.2
Teaneck, N. J.	9518	33.0	63.0	62.9	29.4	6.1	80.6	17.9	26.9
Silver Springs, Md.	9540	31.7	48.6	76.2	11.0	5.7	68.9	33.7	40.6
Greenwich, Conn.	9588	35.6	55.8	54.5	9.7	5.8	69.9	20.7	11.6
West Hartford, Conn.	9712	37.4	50.4	72.1	13.5	6.2	79.8	22.1	16.0
Cheltenham, Pa.	9985	36.6	57.9	75.0	24.0	6.5	73.5	23.8	41.8
Mount Lebanon, Pa.	11108	36.9	49.1	62.8	25.5	6.2	81.4	25.4	14.0
Wellesley, Mass.	11478	31.3	52.4	70.8	15.4	6.7	94.7	23.2	9.8
Lower Merion, Pa.	12204	32.6	57.4	69.0	18.8	7.1	78.2	21.2	37.9
Bethesda, Md.	12357	31.4	36.3	82.6	8.5	6.5	82.3	37.4	41.7

I) Column identification

(212) = median age,
(229) = % in same houses, 1955 and 1960,
(244) = % in white collar occupations,
(246) = % using public transportation,
(248) = median rooms/unit,
(249) = % units in one-unit structures,
(256) = % moved in, 1958 to 1960,
(264) = % with air conditioning.

Exhibit **5** of Chapter 3

Plots, for the 10-PLUS-10 places with high or low family income, of family income against each of four other quantities

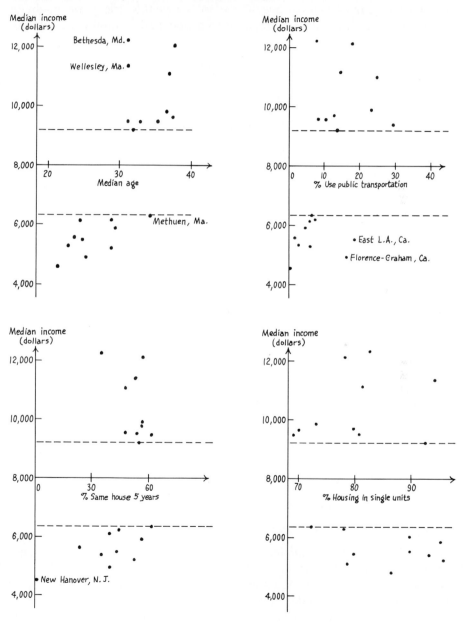

We have explored only a few of the 67 columns of data given in our source. Even so, we can see that, if we wished to describe median family income in terms of other variables, we would have a variety of places to begin.

Let us begin with median age, where the median family income might rise $300 for each year of median age. Once we know this, the y we ought to consider is changed, for we have prepared to divide median family income into two parts

fit PLUS residual.

Here

$$\text{fit} = (\$300)(\text{median age in years}),$$

and so fit has a clear description. The second part is

$$\text{residual} = (\text{median family income}) - (\$300)(\text{median age in years})$$

which remains undescribed and ought to serve us as our new y for further exploration.

It would be attractive to calculate all 88 values of the new y and pick out a new 10 highest and a new 10 lowest. When the data are in a computer where this is easy, we gladly do this. Otherwise, we might use the same 10 high and 10 low for two or three steps of exploration and then recompute.

3F. Trends and Running Medians

In relating a response variable y to its matched x value, we may want to know the shape of the relationship primarily for its own sake, or for a further analysis of departures from the observed regression line, of residuals. Sometimes the values of x form a grid of equally spaced points. But we may have other circumstances as well. When we have no very strong views about the form of the relation and do not expect it to be linear, many procedures suggest themselves. Among these are:

1. drawing a freehand curve through the points to give an eye-satisfying fit;

2. chopping up the x-axis into chunks and taking medians or means of y-values and of x-values within chunks, and then somehow passing a curve near the resulting points, and

3. using running means or medians.

In this section we discuss running medians, and in later sections we discuss other possibilities.

Let us orient our discussion toward time, but only to fix ideas, not to restrict the use of the method. Let us suppose that an observation is composed of two parts, the regression curve we wish we knew and an independent error.

Thus the response at time t $(= 1, 2, 3, \ldots, T)$ can be represented as

$$\text{response} = \text{regression} + \text{error},$$

or, if $f(t)$ is the value of the regression at time t, we can write more formally

$$y(t) = f(t) + \text{error}_t.$$

If the error is substantial compared to the variation in f, then it is tempting to estimate $f(t)$, not just by $y(t)$, but by the average of the y's in the neighborhood of t. This average gives a better estimate of $f(t)$ than does $y(t)$ alone, provided the errors are not much correlated with one another. Indeed, to the extent that the values of $f(t)$ fall along a straight line, not necessarily a horizontal one, the arithmetic average of values centered on t is an unbiased estimate of $f(t)$. We have then a strong urge to average results for several response values of time near and centered at t. Against this, insofar as $f(t)$ is not linear, we are, through this averaging or smoothing as we shall call it, muddling up the regression. If we average three adjacent values, $y(t-1)$, $y(t)$, $y(t+1)$, then we are estimating

$$\frac{f(t-1) + f(t) + f(t+1)}{3} \qquad \text{instead of } f(t).$$

We must not extend the number of points being used too far because, if we do, we will gradually lose the character of the function we are trying to estimate.

Such averaging has been widely used in time-series work. Naturally, many sorts of averages are used, often with unequal weights. In this section, we emphasize running medians where three adjacent values of t are employed. Our primary purpose here is to give some idea of what happens when such smoothing is used. In a later section, we use such smoothing on a real problem. We are not especially recommending the particular brand of smoothing described here, and other better methods are available in *EDA*, Chapters 7 and 16, and in Velleman (1975), to which we refer the reader.

The two main things we plan to illustrate in this section are what happens when running medians are applied to the simplest ideal situation, and what happens when they are applied in more complicated situations. We hope that not a great deal is lost when they are applied, and we hope that they will respond sensitively in circumstances where the regression function, $f(t)$, is curved—or jumps. In the course of exploring these issues through examples, we also see special features that we need to be aware of.

Running medians. We plan to compute running medians and then to continue computing them on the resulting sequence of medians until they stabilize (repeat themselves completely). We illustrate this with adjacent sets of three, but for any odd number we compute in a similar manner. (For even numbers, recall that we take the average of the two most middling values of y in a set: thus, the median of 7, 2, 10, 3 is $(7+3)/2 = 5$.)

Example. Computing running medians. Exhibit **6** shows 20 observations, one associated with each of the times 1, 2, . . . , 20. These happen to be random normal deviates drawn from a random-number table. Indeed, we secretly know that their population mean is 0 and their variance 1. Accordingly, for the process generating these y's, $f(t) = 0$ for all 20 values of t. Thus, we are dealing with a situation where the true regression is a straight horizontal line. In a practical situation we would rarely know all this.

We plan to smooth by running medians and then to fit a least-squares line to the resulting set of smoothed values. Then we will compare the result with what would have been obtained if we had directly fitted a least-squares straight line to the data, as we would if we believed in linearity and normality but did not know the slope and level of the line. Finally, we will be using these same data twice more in illustrating what happens when we fit in situations where the underlying regression is far from linear.

In Exhibit **6** the first column gives the time period and the second the associated values of y. Since we have to do something about the first position, we merely recopy it in taking the running median of three. Then we take the values for time 1, 2, and 3 and take their median (the median of −0.423, −0.602, and 1.703 is −0.423). This is written on the second line of the third column, and then running medians of each successive set of three are given on successive lines, until the line for $t = 20$, where again we copy the original number. This process is repeated until no more changes occur.

We can shorten the writing; it is enough to write down only the changing values rather than every value. Then the 0.401 of line $t = 9$ would be written only in the $y(t)$ column. The final smoothed results would be given by the rightmost number on each line.

Exhibit **7** shows plots of both the original set of data and the smoothed result. As one would expect, the smoothed data vary less than the raw. One way to think about smoothness is to count *turning points*. Given three successive numbers, if the middle one is largest or smallest, it is called a turning point. In a random sequence of length n, the expected number of turning points is $\frac{2}{3}(n - 2)$, assuming that each point in the triplet has a distinct y-value. (Out of the 6 possible orders of three different numbers, 4 have their largest or smallest number in the middle; therefore the probability of a turning point in a set of three is $\frac{4}{6} = \frac{2}{3}$. The −2 in the formula comes from end effects.)

In sample No. 1's raw data we find 10 instead of the expected 12. In the smoothed numbers we have to deal with ties, adjacent ties, but it seems reasonable to say that there are only 3 turning points, if we interpret them as local maxima or minima, corresponding to positions 4, 5, to 15, 16, and to 17, 18, 19. (In these flat spots, the issue is not the exact location of the turning point with respect to t. What we care about is that a turn is made.)

Exhibit **8** shows the results of smoothing for the first ten random samples of size 20 that we drew. They give some notion of the variability that actually occurs in the smoothed running medians. We secretly know that what we would like to see is a flat curve running along the horizontal axis, as the estimate of *f*. We see a warning in Sample No. 4, where the rise at the beginning and the drop at the end are not real in the original function *f*, though we might mistakenly tend to believe them because the rest of the points do hover so close to the horizontal axis.

Exhibit **6** of Chapter 3

Smoothing a set of 20 response values by repeatedly computing running medians of three adjacent values. These *y*(*t*) are Sample No. 1 of 10 samples of size 20.

t	*y*(*t*) random normal deviates	Running medians of 3 random normal deviates	Running medians of 3 medians 1st iteration	Running medians of 3 medians 2nd iteration *y'*
1	−.423	−.423	−.423	−.423
2	−.602	−.423	−.423	−.423
3	1.703	1.703	1.703	1.703
4	1.887	1.887	1.887	1.887
5	2.049	1.887	1.887	1.887
6	1.127	1.127	1.127	1.127
7	.651	.651	.651	.651
8	−.836	.401	.401	.401
9	.401	.401	.401	.401
10	.906	.410	.410	.410
11	.410	.410	.410	.410
12	.221	.410	.410	.410
13	.968	.426	.426	.426
14	.426	.426	.426	.426
15	−.844	−.844	−.844	−.844
16	−1.290	−.844	−.844	−.844
17	.657	.063	.063	.063
18	.063	.657	.063	.063
19	1.283	.063	.063	.063
20	−1.563	−1.563	−1.563	−1.563

S) Source

Rand Corporation (1955). *A Million Random Digits with 100,000 Normal Deviates*. New York: The Free Press. Page 154 in Table of Gaussian Deviates; first 20 values from first 4 columns.

In several of the examples we see flat tops or bottoms of length 2. For example, in Sample No. 1 the opening two points have a flat bottom of length one unit and there is another flat bottom between 15 and 16. This is a common feature of running medians of length three that often bothers us. Other methods in the reference already given (*EDA*, Chapter 7) take steps toward removing this feature.

Exhibit **7** of Chapter 3

Comparison of the raw data of Sample No. 1 and the data smoothed by repeated running medians of 3.

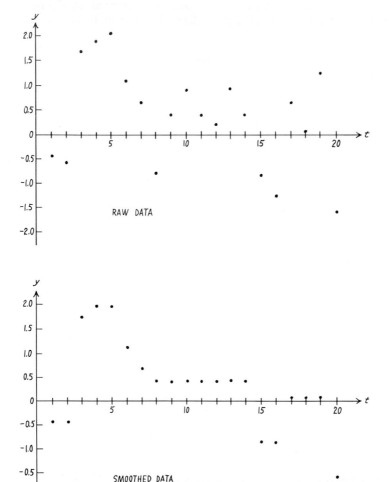

Exhibit **8** of Chapter 3

Smoothed plots of 10 samples of size 20 random normal deviates equally spaced horizontally. (The small x's indicate where a least-squares line fitted to the data passes.)

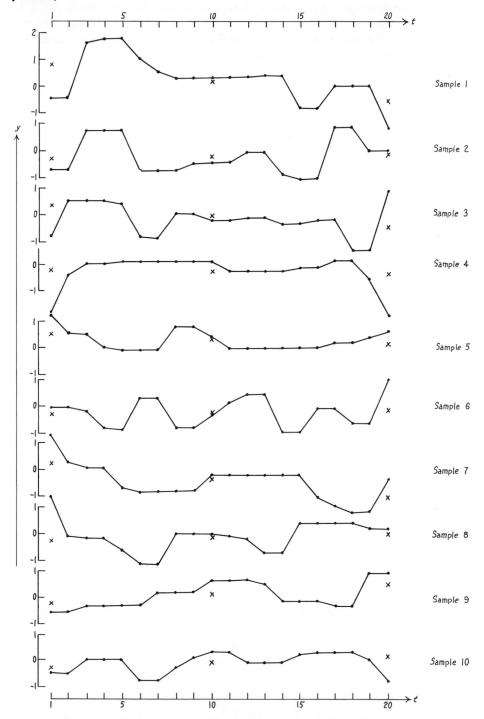

In Sample No. 4, the rise at the beginning and fall at the end are impressive to the eye, the more so because of the flatness from $t = 2$ to $t = 18$.

Similarly, the hump in the middle of Sample No. 5 stands out because it disturbs an otherwise eye-soothing smoothness.

Sample No. 7 has an impressive drop at the beginning and a sag at the end, almost compelling one to believe in the reality of these effects in the process generating the data.

Samples No. 2, 6, and 10 have a rather sinusoidal character.

Let us remember that all these results are supposed to be estimating points on the horizontal axis. And so the message is that indications of shape in smoothed data must be viewed with reserve. When we look back at the raw data in Exhibit 7, we are not much impressed with any feature of it because the variability of the points is substantial. It gives the appearance of trending downhill from left to right, but the three high points in the upper left and the two low ones in positions 15 and 16 don't impress us much because of the considerable overall variability. When we look at the smoothed data in Exhibit 7, we are more likely to be impressed with the bulge up on the left and the dip at 15 and 16, because the basic variation in the points is much less. Consequently one must be restrained in interpreting shapes from smoothed curves, because the restraints imposed by observing natural variability are no longer visually available, as we now explain.

Running medians give rather lumpy looking results—often a little more lumpy than running means. Since a particular piece of input data affects only a few values of a running median (or running mean), if the input data bobbles up and down around some value, we expect its running medians to do the same. But they do not do this as irregularly as the data. When a running median of 3, for instance, is quite high, this is because two of the three values are at least correspondingly high. Often these two will be adjacent to each other. When this happens, at least two successive running medians of 3 will have to be high. Thus, high values tend to come in pairs (or triples, or more).

This is particularly noticeable when we use repeated medians of 3, for if an early medianing leaves an isolated high value, the next medianing will eliminate it. Thus the peaks and valleys left after repeated smoothing by medians tend to be rather distinct lumps. Essentially, the high-frequency noise has been smoothed out, leaving the low-frequency noise behind, and looking more impressive.

Fitting straight lines. Next, let us see what happens when we fit straight lines by equally weighted least squares (ordinary least squares) to both the raw samples and the smoothed data. As an example, for Sample No. 1,

	Uncentered form	Centered form
raw data:	$y = -0.0588x + 0.9776$	$= -.0588(x - 10.5) + 0.360,$
smoothed data:	$y' = -0.0834x + 1.187$	$= -.0834(x - 10.5) + 0.311.$

Exhibit **9** shows the (slope, centered intercept) pairs for all ten samples, and Exhibit **10** shows the same results graphically. (We use the centered intercept since, in this ideal situation, it is distributed independently of the slope when we use the raw values.) Theory says the distribution of the smoothed results

Exhibit **9** of Chapter 3

Comparisons of slopes and intercepts of lines fitted to raw data and smoothed data from Exhibit 8.

	Slope		Centered intercept (at $t = 10.5$)	
Sample	Raw	Smoothed	Raw	Smoothed
1	−.0588	−.0834	.360	.311
2	.0054	.0138	−.148	−.148
3	−.0498	−.0306	−.188	−.184
4	−.0352	−.0082	−.125	−.255
5	−.0054	−.0133	.195	.373
6	−.0011	.0052	−.136	−.226
7	−.0720	−.0753	−.590	−.486
8	.0373	.0256	.120	−.030
9	.0505	.0448	.140	.095
10	.0154	.0209	−.059	−.144
Ideal	0.0	0.0	0.0	0.0

Exhibit **10** of Chapter 3

Graphical comparison of slopes and intercepts for raw and smoothed data.

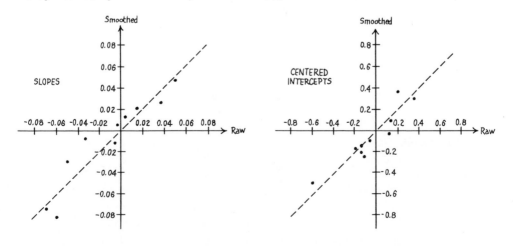

Exhibit **11** of Chapter 3

Stem-and-leaf comparing raw and smoothed slopes and centered intercepts in terms of both values and sizes.

A) VALUES

Slopes				Centered intercepts		
Raw		Smoothed		Raw		Smoothed
	.06				.5	
0	.05				.4	
	.04	4		6	.3	17
7	.03				.2	
	.02	50		924	.1	
5	.01	3			.0	9
5	.00	5				
---	---	---		---	---	---
15	−.00	8		5	−.0	3
	−.01	3		3284	−.1	448
	−.02				−.2	25
5	−.03	0			−.3	
9	−.04				−.4	8
8	−.05			9	−.5	
	−.06				−.6	
2	−.07	5				
	−.08	3				

B) SIZES

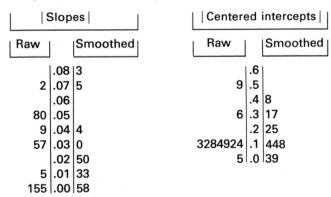

|Slopes|				|Centered intercepts|		
Raw		Smoothed		Raw		Smoothed
	.08	3			.6	
2	.07	5		9	.5	
	.06				.4	8
80	.05			6	.3	17
9	.04	4			.2	25
57	.03	0		3284924	.1	448
	.02	50		5	.0	39
5	.01	33				
155	.00	58				

should be slightly more variable than those for the raw values. Incidentally, in Exhibit 8 the small crosses at each end of each figure and in the middle show where the least-squares line fitted to the smoothed points would pass.

Finally, Exhibit **11** shows the stem-and-leaf diagrams for slope and for centered intercept, giving a direct comparison of the observed marginal distributions, and we see again that the variability of the results is much the same, even though theory tells us that the raw fit is preferable in this ideal situation (though not when the data are far from Gaussian).

This completes our illustration of what smoothing does in an ideal null situation, that is, one with no trend and with identically normally distributed error at each point. We turn now to illustrations where the regression curve f is nonlinear.

3G. Smoothing Nonlinear Regressions

Next we illustrate the smoothing of nonlinear regressions where the variation in the random component is nonnegligible, but not overwhelming. Exhibit **12** shows a regression function f_1 with its 20 points connected by a dotted line. When the curve oscillates rapidly, we cannot expect to pick up these oscillations if our grid of t values is coarse. Furthermore, if we smooth our data, we will to some extent smooth the basic function f_1. Exhibit 12 shows, as the heavy line, the smoothed function f_1^* using running medians of length 3 (smoothed until they repeat). It is this running median function f_1^* that we will actually be looking toward rather than the original function, when we smooth data in which f is perturbed by random errors. To repeat, smoothing the data smooths the function being estimated as well. Essentially it has smoothed out two turning points and lost what might be the scientifically important jag down at $t = 6$ and the spike at $t = 7$. Smoothing, then, can smooth away important effects.

Exhibit **13** shows what happens when the unsmoothed normally distributed errors in the samples of Section 3F are added to f_1, and then running medians are computed for the ten samples.

Except for Sample Nos. 5, 6, 7, and 8, the general shapes of the smoothed data agree with that of the smoothed regression function f_1^* shown in Exhibit 12. These four graphs smoothed away the smaller real hump at the left. All the smoothed results have trouble displaying the roundness of the righthand end.

By comparing the graphs of Exhibit 8 with the f_1^* function in Exhibit 12, one can, to some extent, forecast what happens in Exhibit 13. For example, in Exhibit 8 the initial drops in Sample Nos. 5, 6, 7, and 8 tend to destroy the hump of Exhibit 12.

Exhibit **12** of Chapter 3

The regression function f_1, shown dotted, and its smoothed version, f_1^*, smoothed by running medians of length 3. Note that $f_1 = f_1^*$ except at $t = 5$, 6, 7.

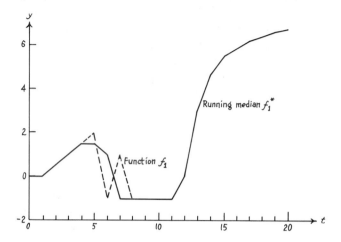

Exhibit **13** of Chapter 3

The smoothed data when the unsmoothed normally-distributed errors of the samples of Section 3F are added to the irregular function f_1 shown in Exhibit 12 and then smoothed by repeated running medians of length 3.

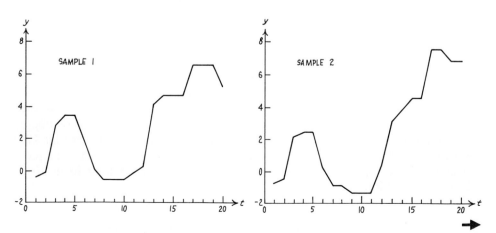

Exhibit **13** of Chapter 3 (continued)

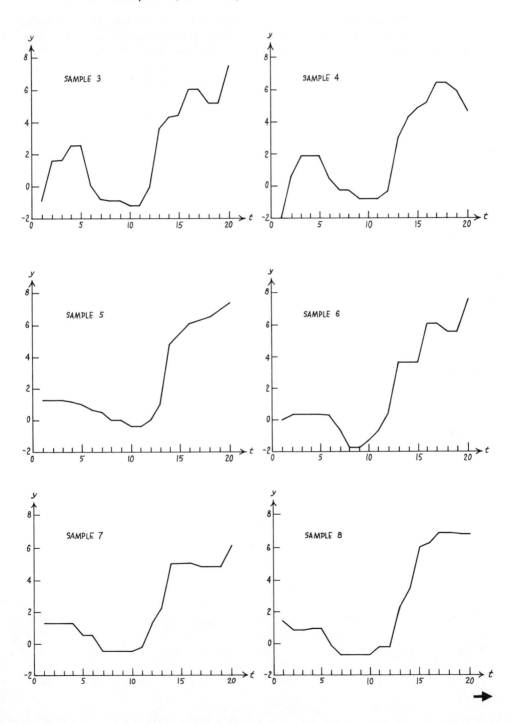

3H. Looking for Patterns

When we examine departures from the observed regression line, the residuals, we have many ways to detect patterns. Trends, bends, oscillations, and wedge shapes (pictured shortly) may be directly interpretable. We may want to re-express to eliminate the wedge. Perhaps patterns of parallel diagonal bars may reflect concentration on "permitted" numbers, or perhaps they reflect only a curiosity.

Sections 3F and 3G discussed the use of running medians to see relationships and deviations for linear and nonlinear regressions. This section explains another means of detecting patterns, the procedure suggested at the beginning of Section 3F, namely, passing a curve through the means or medians of the chunks of x-values and of y-values resulting from chopping up the x-axis.

To illustrate a number of these points, we study empirically some counts related to a famous mathematical problem that many mathematicians have worked on, among them L. Euler and G. H. Hardy. Hardy's work (1906) investigates the counts we describe.

The Goldbach counts. A deep problem in mathematics, still unsolved after more than 200 years, is the Goldbach Conjecture. It states that every even number, starting with 6, can be represented as the sum of two odd primes

Exhibit **13** of Chapter 3 (concluded)

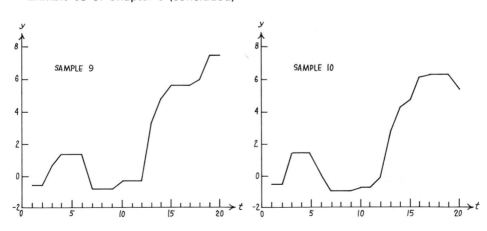

(6 = 3 + 3). (A prime is an integer that can be divided exactly only by itself and 1. The number 1 itself is not counted as a prime.) Much progress has been made on this conjecture. However, what we will think about is not its proof, but the *number of ways* it is possible to represent an *even number, E,* as the sum of two odd primes (including 0 as a possible count). We call this number the Goldbach count. Examples are:

$$20 = 3 + 17 = \; 7 + 13 \qquad \text{(Goldbach count = 2)}$$
$$30 = 7 + 23 = 11 + 19 = 13 + 17 \quad \text{(Goldbach count = 3)}$$
$$68 = 7 + 61 = 31 + 37 \qquad \text{(Goldbach count = 2)}$$

Exhibit **14** displays selected patches of Goldbach counts for the even numbers E from 2 to 508 and from 9500 to 10,000. The even numbers are listed by tens under the heading T; and under $T + 0$, $T + 2, \ldots, T + 8$, the Goldbach counts for the corresponding even numbers are given.

In Exhibit **15** we plot the Goldbach counts against half the even number to be represented, to see what sort of pattern emerges. Thus the number we plot horizontally is $n = \frac{1}{2}E$.

First we see a general trend up and to the right.

Second we see the counts spreading vertically as we move to the right, forming a wedge shape. We might ask whether we could transform the numbers so as to stabilize the variability, but we set this aside for the moment in favor of searching for other patterns.

We also see that some values of N have particularly low counts and others particularly high ones as we move to the right. Can we detect a pattern in this? We would do well to get a regression line or curve to act as a base line. Then we can look hard at deviations from the line. We can choose many ways to make clumps and get average horizontal values and average vertical values. We could take successive sets of 10 values of N: 0 to 9, 10 to 19, and so on, and center the horizontal value at the midpoint. As our vertical coordinate, we could take the median or the mean of the y-values, here Goldbach counts. We will call such points the centroids, though in physics that term would be used only for the point whose coordinates are the mean of the x's and the mean of the y's. The observations seem fairly well behaved in this region of N, and so let us take the arithmetic mean within each set. The points corresponding to the midpoint for N and the mean for y, the Goldbach count, are marked with a cross on Exhibit 15. The crosses are connected by line segments to give a crude regression curve.

The advantage of this crude regression is that it gives us a way of separating large counts from small ones. Until we get as far right as about $N = 36$, the large values do not stand out, but starting about here they do. And they tend to form a recognizable pattern in the graph. *Every third point* has a *high* count.

Exhibit **14** of Chapter 3

Goldbach counts for selected even numbers E, where E = T + 0, T + 2, ... , T + 8.

T\T+	0	2	4	6	8	T\T+	0	2	4	6	8
0	—	0	0	1	1	250	9	16	9	8	14
10	2	1	2	2	2	260	10	9	16	8	9
20	2	3	3	3	2	270	19	7	11	16	7
30	3	2	4	4	2	280	14	16	8	12	17
40	3	4	3	4	5	290	10	8	19	8	11
50	4	3	5	3	4	300	21	9	10	15	8
60	6	3	5	6	2	310	12	17	9	10	15
70	5	6	5	5	7	320	11	11	20	7	10
80	4	5	8	5	4	330	24	6	11	19	9
90	9	4	5	7	3	340	13	17	10	9	16
100	6	8	5	6	8	350	13	10	20	9	10
110	6	7	10	6	6	360	22	8	14	18	8
120	12	4	5	10	3	370	14	18	10	11	22
130	7	9	6	5	8	380	13	10	19	12	9
140	7	8	11	6	5	390	27	11	11	21	7
150	12	4	8	11	5	400	14	17	11	13	20
160	8	10	5	6	13	410	13	11	21	10	11
170	9	6	11	7	7	420	30	11	12	21	9
180	14	6	8	13	5	430	14	19	13	11	21
190	8	11	7	9	13	440	14	13	21	12	13
200	8	9	14	7	7	450	27	12	12	24	9
210	19	6	8	13	7	460	16	28	12	13	24
220	9	11	7	7	12	470	15	13	23	14	11
230	9	7	15	9	9	480	29	11	14	23	9
240	18	8	9	16	6	490	19	22	13	13	23
						500	13	15	27	15	14

T\T+	0	2	4	6	8	T\T+	0	2	4	6	8
9500	135	100	209	117	97	9750	286	108	99	194	124
9510	253	97	105	211	86	9760	135	185	98	90	219
9520	170	199	101	109	188	9770	132	117	191	108	99
9530	123	94	232	97	104	9780	260	103	99	230	98
9540	257	104	103	202	120	9790	151	205	102	110	200
9550	130	199	104	102	193	9800	147	118	205	95	101
9560	123	121	191	94	110	9810	258	101	118	206	94
9570	295	97	98	245	90	9820	135	206	98	100	259
9580	132	194	93	94	199	9830	124	98	219	96	91
9590	159	109	221	96	97	9840	264	129	104	196	100

Looking more carefully, we see that

$$39, \quad 42, \quad 45, \quad 48, \quad 51, \quad 54, \quad 57, \quad \text{and so on}$$

all have values above the line, and so we have detected a pattern. N's divisible by 3 seem to have larger Goldbach counts than those that are not divisible by 3. At any rate, they fall above the fitted regression.

Now that we have this hypothesis, we could look back and see whether we can detect it for earlier N. Note that 30, 27, 24, and 21 are all above the line segments we have fitted.

Admittedly, sometimes N's *not* divisible by 3 also have high Goldbach counts. They deserve exploration, but now that we have had success with large counts, let us look for especially small counts. One way to search might be to set aside the N's divisible by 3, fit a new curve, and then look carefully at the large and the small counts. The next exhibit does this.

Exhibit **16** shows a much more homogeneous set of vertical values than does Exhibit 15, still rising as we go to the right, but the vertical variation is much reduced. Can we detect any further pattern in the remaining points?

We might look again to see which ones are much above and which below a new regression curve and consider these. We take successive sets of 9 N's for this. Each set of 9 N's has 6 counts left. We use these to form the crude regression as before.

Exhibit **14** of Chapter 3 (continued)

9600	261	77	127	197	91		9850	130	195	103	130	202
9610	137	190	117	93	234		9860	144	104	204	86	99
9620	135	100	194	93	106		9870	316	102	104	208	110
9630	264	112	97	212	91		9880	156	200	117	101	196
9640	126	191	93	124	202		9890	146	103	214	96	118
9650	123	105	192	99	114		9900	301	98	102	211	94
9660	324	101	110	194	97		9910	134	233	100	109	223
9670	140	220	119	98	191		9920	141	112	200	122	108
9680	140	101	197	96	118		9930	266	105	103	202	103
9690	284	93	101	193	106		9940	162	200	113	113	196
9700	121	254	98	104	184		9950	126	95	248	98	105
9710	134	99	192	117	102		9960	269	113	99	217	120
9720	254	96	128	195	103		9970	139	194	93	104	195
9730	161	193	103	101	196		9980	136	135	211	103	110
9740	133	101	235	113	93		9990	269	102	98	255	99
							10000	127				

S) Source

Frederick Mosteller (1972). "A data-analytic look at Goldbach counts." *Statistica Neerlandica*, 26, No. 3, 227–242. Reprinted by permission of the editors of the journal.

Let us first look at those with high counts—high by more than one unit, just to pick a number. We find the *N*-axis values:

$$17, \quad 32, \quad 50, \quad 56, \quad 65, \quad 71, \quad 77, \quad 80$$

We know this problem deals with number theory, and so a pattern might depend upon which numbers divide the *N*'s, as we have already observed with 3's. We found that divisibility by 3 seems to raise the size of the count. Are these new *N*'s divisible by anything other than by 2 or 3, which we have already considered? Beneath each number *N*, let us write the divisors other than 2 or 3, or *N* itself.

```
17,   32,   50,   56,   65,   71,   77,   80
 —     —     5     7     5     —     7     5
                        13           11
```

Exhibit **15** of Chapter 3

Goldbach counts for even numbers—2*N*, *N* = 1, 2, 3, ... The line segments connect the centroids of successive bursts of 10 points, *N* = 0 − 9, 10 − 19, etc.

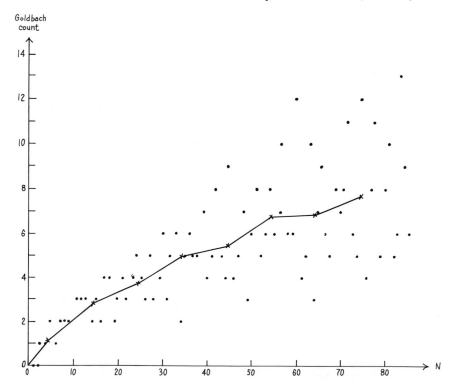

We note that 5, 7, 11, and 13 are all primes, the next small primes. And they all appear as divisors. Indeed, they are the successive primes after 3. Thus the suspicion is that divisibility of N by small odd primes raises the size of the Goldbach count on the average. In the spirit of mathematical experimentation, let us suspend judgment on this and look at the small counts, those low by more than one unit, to see what their N's reveal. They are, with their divisors other than 2 and 3,

$$34, \quad 49, \quad 61, \quad 64, \quad 74, \quad 76, \quad 79$$
$$17 \quad \ \ 7 \qquad\qquad\qquad 37 \quad 19$$

We find these N's for lower counts are divisible by fewer small primes (no 5's, one 7, no 11's, no 13's) than the higher numbers. We have the general idea now that N's divisible at least once by small odd primes produce larger counts. Perhaps the first divisibility by each small prime is more important than repeated divisibility.

If this were true, then we could, by multiplying successive distinct small odd primes together, construct a number that *should* have a high Goldbach count relative to its neighbors. Then we would see whether its Goldbach count were high compared with those of its neighbors. Let us do this.

Exhibit **16** of Chapter 3

Goldbach counts after removing the N's divisible by 3. The trend continues, as does the increase of spread with N.

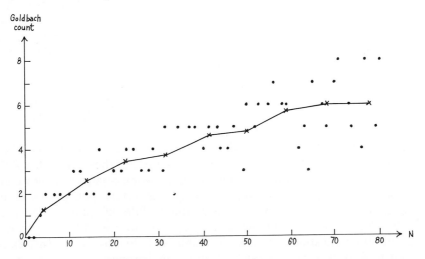

Let $N = 3 \times 5 \times 7 = 105$, so that the even number is $2N = 210$. Then the counts near and at 210 are:

200,	202,	204,	206,	208,	210,	213,	214,	216,	218,	220
8	9	14	7	7	19	6	8	13	7	9

and we see that we have constructed a number that produces a high count relative to its neighbors.

Although we could go on enjoying this example, what we have found is that, by looking carefully at the departures from an observed regression function, we could begin to detect patterns. Of course, we have not proved anything mathematically here. That was not our job. What we wanted to do was to attack a set of data empirically by looking at residuals and see whether we could detect any patterns. So far we have detected the idea that when the even number is divisible by several different small primes, we tend to get higher Goldbach counts. We also found that the regression curve rises as N increases and that the variability of the counts increases as N increases.

It is not our intention to press this issue further, though it could be studied through residuals and through extensive exploratory data analysis as in Mosteller (1972). But we have made an important point with this example. Rather simple data analysis can illuminate deep problems, as mathematical giants like Euler and Gauss well knew.

3I. Residuals More Generally

Now that we have wrestled informally with residuals in an example and seen a bit of what can be learned from studying residuals, we consider here, and again in Chapter 16, other approaches to analyzing residuals. And let us begin generally.

We can examine residuals arithmetically or graphically. Careful arithmetic examination is tedious by hand and could be easy by computer, but still we usually should have a graphical examination. In the Goldbach example, we did not actually graph the residuals, although we graphed the original data. But the residuals were clear in the graph. (For careful arithmetic examination, we refer the reader to Anscombe and Tukey, 1963; and to Anscombe *et al.*, 1974.) Calculation of letter values and identification of "outside" and "far out" values (see *EDA*, Chapter 5) may help, especially if the residuals come from a resistant fit.

A graphical look at residuals almost always reduces to making (x, y) plots, where the y's are the residuals. The main questions are thus: "Which x's?" and "Which plotting techniques?" The technique question relates to how we are to apply our efforts efficiently, since halving the effort per plot will often encourage us to do more than twice as many different plots, often a good tradeoff. That a single technique would serve us well is not to be expected. Sometimes the indications we seek are coarse and obvious. At another extreme we want to find even delicate and indistinct appearances.

We discussed exploring with economy in Section 3E, plotting for the high and low y's only, a dozen or so points each. This plot will help find gold nuggets at the grass roots. What should be done about the finer screenings?

Example. Temperatures and geography. We explore the residuals of maximum January temperatures against latitude. Exhibit **17** shows, for cities of the United States, their maximum January temperature plotted against latitude. The relation is fairly close, although a few cities stand out, especially Jacksonville, Fla., Seattle, Wash., and Juneau, Alaska. Essentially, they all look warm for their latitude.

Exhibit **17** of Chapter 3

Maximum January temperature against latitude (x indicates two coincident observations)

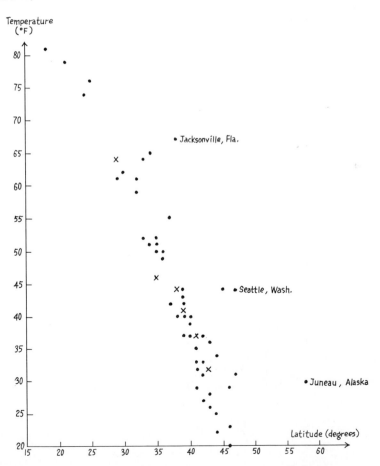

A least-squares line was fitted to these points; its slope is $b_1 = -1.94$. Then, letting temperature be y and latitude x_1, and letting $\bar{y}$ and $\bar{x}_1$ be their means, the residuals

$$y - \bar{y} - b_1(x_1 - \bar{x}_1)$$

were calculated. These residuals in turn were plotted in Exhibit **18** against the longitude of the city. The diurnal rotation of the earth would, of course, have an averaging effect on temperature for various longitudes, but the land masses and water bodies (which affect temperature) bear a relation, albeit complicated, to longitude. So longitude is useful as abscissa. The cities that were outliers

Exhibit **18** of Chapter 3

Residuals from (max [January temperature vs. latitude]) against longitude.

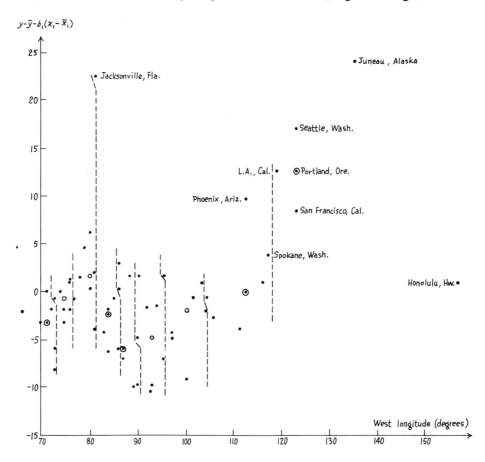

before are outliers again in the residual plot. But, in addition, we note, as we go from 100° to 130°, that the points take a general drift upward. What are these high-residual, far west cities? Examples are Los Angeles and San Francisco, Calif., Portland, Ore., Seattle and Spokane, Wash. Four of the five are on the Pacific. This suggests a warming effect from the ocean.

For a further indication, we can take medians for longitude of successive groups of seven places and the medians of their corresponding residuals. This yields the summary points shown by the circles in Exhibit 18. The slight rise of the median temperature around longitude 75° may be caused by the accident of having a number of southern cities clumped near that longitude. Even though the linear effect of the latitude has been removed, clumps of points can display seeming effects. And, of course, the warming effects of being near the ocean may in part account for the rise. Exhibit **19** shows the original data for the cities as well as the altitude and the residual from fitting temperature by latitude.

Exhibit **19** of Chapter 3

Maximum January temperatures in degrees Fahrenheit, from 1931–1960, for cities in the U.S., with latitude, longitude, and altitude.

Cities of the U.S.	Max. Jan. temperature	x_1 Lat.°	x_2 Long.°	x_3 Alt. (ft.)	Residuals $y - \bar{y} - b_1(x_1 - \bar{x}_1)$
Mobile, Ala.	61	30	88	5	2.0
Montgomery, Ala.	59	32	86	160	2.9
Juneau, Alaska	30	58	134	50	24.3
Phoenix, Ariz.	64	33	112	1090	9.8
Little Rock, Ark.	51	34	92	286	−1.2
Los Angeles, Calif.	65	34	118	340	12.8
San Francisco, Calif.	55	37	122	65	8.6
Denver, Col.	42	39	104	5280	−.5
New Haven, Conn.	37	41	72	40	−1.7
Wilmington, Del.	41	39	75	135	−1.5
Washington, D.C.	44	38	77	25	−.5
Jacksonville, Fla.	67	38	81	20	22.5
Key West, Fla.	74	24	81	5	2.4
Miami, Fla.	76	25	80	10	6.3
Atlanta, Ga.	52	33	84	1050	−2.2
Honolulu, Hawaii	79	21	157	21	1.5
Boise, Idaho	36	43	116	2704	1.2
Chicago, Ill.	33	41	87	595	−5.7
Indianapolis, Ind.	37	39	86	710	−5.5

➡

Exhibit **19** of Chapter 3 (continued)

Des Moines, Iowa	29	41	93	805	−9.7
Dubuque, Iowa	27	42	90	620	−9.7
Wichita, Kansas	42	37	97	1290	−4.4
Louisville, Ky.	44	38	85	450	−.5
New Orleans, La.	64	29	90	5	2.1
Portland, Maine	32	43	70	25	−2.8
Baltimore, Md.	44	39	76	20	1.5
Boston, Mass.	37	42	71	21	.3
Detroit, Mich.	33	42	83	585	−3.7
Sault Ste. Marie, Mich.	23	46	84	650	−6.0
Minn.–St. Paul, Minn.	22	44	93	815	−10.8
St. Louis, Missouri	40	38	90	455	−4.5
Helena, Montana	29	46	112	4155	0.0
Omaha, Nebraska	32	41	95	1040	−6.7
Concord, N.H.	32	43	71	290	−2.8
Atlantic City, N.J.	43	39	74	10	.5
Albuquerque, N.M.	46	35	106	4945	−4.3
Albany, N.Y.	31	42	73	20	−5.7
New York, N.Y.	40	40	73	55	−.6
Charlotte, N.C.	51	35	80	720	.7
Raleigh, N.C.	52	35	78	365	1.7
Bismarck, N.D.	20	46	100	1674	−9.0
Cincinnati, Ohio	41	39	84	550	−1.5
Cleveland, Ohio	35	41	81	660	−3.7
Oklahoma City, Okla.	46	35	97	1195	−4.3
Portland, Ore.	44	45	122	77	13.1
Harrisburg, Pa.	39	40	76	365	−1.6
Philadelphia, Pa.	40	39	75	100	−2.5
Charlestown, S.C.	61	32	79	9	4.9
Rapid City, S.D.	34	44	103	3230	1.2
Nashville, Tenn.	49	36	86	450	.6
Amarillo, Tx.	50	35	101	3685	−.3
Galveston, Tx.	61	29	94	5	−.9
Houston, Tx.	64	29	95	40	2.1
Salt Lake City, Utah	37	40	111	4390	−3.6
Burlington, Vt.	25	44	73	110	−7.8
Norfolk, Va.	50	36	76	10	1.6
Seattle–Tacoma, Wash.	44	47	122	10	17.0
Spokane, Wash.	31	47	117	1890	4.0
Madison, Wisc.	26	43	89	860	−8.8
Milwaukee, Wisc.	28	43	87	635	−6.8
Cheyenne, Wyoming	37	41	104	6100	−1.7
San Juan, Puerto Rico	81	18	66	35	−2.3

S) Source

Temperatures are from page 263, and the geographical positions from pages 704–705 of *The World Almanac and Book of Facts, 1974 edition*; Copyright © Newspaper Enterprise Association, New York. 1973. Reprinted by permission of the publisher.

3J. Plotting and Smoothing

How can we accommodate the extremes of economical and delicate exploration along with the intermediate cases? Exhibit **20** shows the data for 88
unincorporated urban places from the 1962 *County and City Data Book*. The
percent of housing units in one-unit structures is associated with the median
family income for 1959. We plan to regress the income on the housing. We
have chopped the "% one unit" data into 20 groups of 4 or 5, but sometimes
as few as 1 or as many as 7. In chopping up the variable, some attempt was
made to keep the groups nearly the same size without making the intervals
terribly different in size. The chopping is similar to that in the Goldbach
example, but we have not retained equal intervals.

For each group we have computed the median. And now, instead of
merely connecting these medians, we smooth them by the smoothing methods
described in Section 3F, repeated running medians of 3.

The general impression in plotting the smooth points and medians in
Exhibit **21** is of remarkably little relation between median family income and

Exhibit **20** of Chapter 3

**One choice of about 20 groups, and the corresponding median family incomes,
raw, medianed, and smoothed**

x % one unit	y Median family income	$y' =$ Median y	Smoothed* y'	Median x
23 to 30	7151, 8380, 6910	7151		25
40	7003	7003	7151	40
56 to 59	7538, 8372	7955		58
68 to 70	9588, 7451, 9540	9540	7955	69
72 to 75	7113, 6908, 6693, 7495, 6278, 9985	7010		74
78	6160, 4572, 6806, 12264	6483	7010	78
79 to 80	5434, 6539, 9712, 9518, 8088	8088	7662	79
81 to 82	6522, 12357, 7662, 7550, 11108	7662		82
83 to 85	7741, 7371, 7003, 8863, 8895, 7413	7577	7577	84
86	8596, 4904, 7475, 6489	7276	7577	86
87	8123, 6338, 7973, 7753	7863	7494	87
88	7474, 8561, 7494, 7276	7494		88
89	5567, 8265, 7260, 6908, 8446	7260	7420	89
90 to 91	6613, 6076, 8685, 7597, 7243, 8728	7420		90
92 to 93	7869, 7782, 7978, 5306, 9282, 8368	7880		92
94	7189, 11478, 8888, 6922	8039	7880	94
95	8191, 7908, 5909, 7169	7538		95
96 to 97	9043, 8336, 5182, 6602, 6615	6615	7538	96
98	8602, 7186, 7365, 8363, 8671, 9236	8482	7656	98
99	7936, 7467, 8998, 6987, 7656	7656		99

* By repeated running medians of 3; only changes shown.

% of housing units in one-unit structures. The present picture is more sensitive than a plot of the 10 high and 10 low points alone, and more sensitive than a plot of the 88 individual points would have been. If we examine it carefully, we are almost tempted to believe in a rise in median family income from perhaps 7200 at the left to 7600 at the right. But our experience with smoothing warns us not to take this seriously.

Effort. Putting Exhibit 21 together is not free of effort. But it is not much, if any, more work than plotting all the points would have been.

If we had had a deck of 88 index cards—one for each unincorporated urban place—bearing the data, doing the equivalent work directly from cards would have been easy. All that we would have done would be to order the cards by % one-unit, then to pick up in groups of reasonable size, writing down group medians as we go. Only the smoothing and plotting would remain. With

Exhibit **21** of Chapter 3

The 10 highest and 10 lowest income communities and the smoothed median from the 20 groups of Exhibit 20 of % housing units in single-unit structures and median income for the 88 unincorporated urban places.

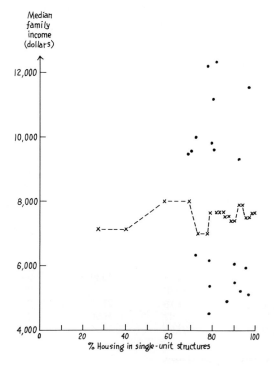

card decks, to which successive residuals are added as found, this type of careful plotting is workable even for fairly substantial bodies of data. (For more than a few hundred data sets, subsampling is likely to be desirable.)

Summary: Batch Summaries and Displays

We write down batches of numbers in two or three of the simplest stem-and-leaf forms.

Medians can be naturally supplemented by other letter values, whose depths are defined, one after another, in almost the same way that the depth of the median is related to the batch size.

With letter values we made letter-value displays, both skeleton displays, and those including one or both of spreads or mids.

In handling larger bodies of data, subsampling offers convenience and flexibility. Its use can be crucial. Without it, analysis might be defeated by sheer bulk of work.

We saw some of the advantages and disadvantages of one resistant smoothing procedure, chosen for simplicity.

To explore (x, y) data, we can do different things, including (i) plotting all points (often inadequate alone); (ii) plotting the 10 to 20 points with highest y-values and a similar number with the lowest y's; (iii) fitting a straight line and examining the residuals (perhaps as in (ii)); (iv) taking running medians of the y-values (after re-ordering by the x-values); (v) dividing the data into groups according to their x-values, finding medians of these groups, smoothing the medians (as in (iii), say) and plotting the results; and (vi) combinations of the above.

References

Anscombe, F. J., and J. W. Tukey (1963). "The examination and analysis of residuals." *Technometrics*, **5,** 141–160.

Anscombe, F. J., D. R. E. Bancroft, and J. G. Glynn (1974). "Tests of residuals in the additive analysis of a two-way table—a suggested computer program." Technical Report No. 32, November, 1974. Department of Statistics, Yale University.

EDA = Tukey, J. W. (1977). *Exploratory Data Analysis.* Reading, Mass.: Addison-Wesley.

Hardy, G. H. (1906). *Messenger Math.*, **35,** p. 145.

Mosteller, F. (1972). "A data-analytic look at Goldbach counts." *Statistic Neerlandica*, **26,** 227–242.

Velleman, P. F. (1975). "Robust nonlinear data smoothing." *Technical Report No. 89*, (*Series* 2), Department of Statistics, Princeton University (AEC).

Chapter 4/Straightening Curves and Plots

The Idea of Straightening out Curves

When we deal with closely related variables, some advantages occur if we can express their relationship linearly. Interpolation and interpretation are relatively easier, and departures from the fit are more clearly detected. In this chapter, we offer ways of straightening out curves.

When we have empirically determined relations between variables, we cannot hope that straightening the relation within the range of the observed data will also straighten it far beyond the observed range. Although luck might help us, ordinarily we need some sort of theory or previous experience to do that.

This chapter helps us re-express one or both of a pair of variables so that relations originally curved are straighter. If y and x are the variables, we consider re-expressing y or x or both. The primary tools are:

1. a ladder of re-expressions, and

2. rules for determining which direction to move on the ladder.

Learning the techniques will be simplified if at first we concentrate on straightening out a functional relation, and then later extend the idea to scatter plots and other empirical data.

4A. The Ladder of Re-expressions

Since we need a systematic set of re-expressions, the powers of a variable naturally suggest themselves. As a start, let the power p take the values

$$-3, \quad -2, \quad -1, \quad -\tfrac{1}{2}, \quad \#, \quad \tfrac{1}{2}, \quad 1, \quad 2, \quad 3.$$

(More about $\#$ later.) Let us think about only positive values of the variable, which for convenience we call t.

First, we want a set of re-expressions each of which is monotonic in the same direction. When $p > 0$, as t increases, t^p increases. When $p < 0$, as t increases, t^p decreases. To make them all increase as t increases, we can use $-t^p$ when p is negative.

Second, what shapes do these curves have? When $p > 1$, they are hollow upward $\smile$. When $p = 1$, the curve is straight. When $p < 1$, the curves are

79

hollow downward $\frown$. Exhibit **1** shows the shapes of these curves, although they have been rescaled as in Exhibit 8 of Chapter 5 and pulled apart by additive constants, so that they can be seen more clearly.

What do we choose for #? The value $p = 0$ leads to a constant, and so we cannot usefully choose it. Instead we choose log t. (We might think of these powers of t as coming from $\int t^{p-1} \, dt$. When $p = 0$, we get log t.) Some may want to do something else, and they might get a different answer. The log t curve fits in well and we wouldn't want to leave out the logarithm because it is the

Exhibit **1** of Chapter 4

Shapes of curves $z = t^p$ for $p = -3, -2, \#, 1, 2, 3$.

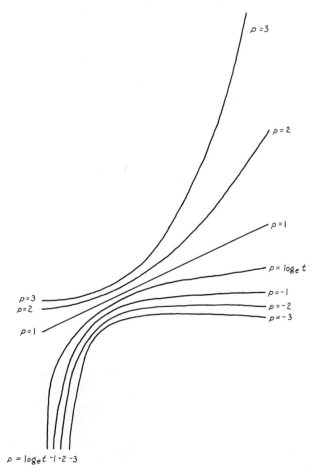

transformation most commonly used. Consequently, at # we do not use t^0, but log t.

When we are trying re-expressions, we will move up and down the ladder of re-expressions given here, searching for one that straightens well. To aid our intuition about which direction to move on the ladder, we will do one example where we have complete command of the information. Then we will give a set of rules.

4B. Re-expressing $y = x^2$

As our instructive example, we choose the functional relation

$$y = x^2, \qquad x \geq 0.$$

Its graph is shown in Exhibit **2,** and we note that it is hollow upward and increasing as x increases.

1. *What re-expression of y would straighten out the curve?* Let us consider what would happen if we replaced y by either y^2 or $y^{1/2}$. We would be helped by considering points in two intervals 0 to 1 and 1 to 4, because all t^p are equal at $t = 1$.

Using y^2. If y is replaced by y^2, then all the points for $0 < x < 1$ will be lower than they were before, because squaring a number between 0 and 1 makes it smaller. Squaring the y's where $x > 1$ makes the y's larger. We have, all told, then, bent the curve more than before.

Exhibit **2** of Chapter 4

Graph of $y = x^2$ hollow upwards, y increasing as x increases.

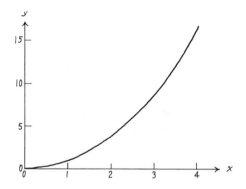

Three points, two slopes. Let's make this more quantitative by a rough device. We could compute the slope of the chord from the origin to the value at $x = 1$ and the slope from the point at $x = 1$ to $x = 4$ for both the original curve and the transformed one. Ideally the slopes are equal for the two intervals.

Original slopes: $0 < x < 1$, $1/1 = 1$; $1 < x < 4$, $15/3 = 5$;

New slopes: $0 < x < 1$, $1/1 = 1$; $1 < x < 4$, $255/3 = 85$.

The original slopes are 1 and 5, a ratio of 5, and the new slopes are 1 and 85, a ratio of 85. Using y^2 has not moved the slopes closer together but farther apart. We are moving in the wrong direction.

Next let us try $y^{1/2} = \sqrt{y}$. Now, points in the interval $0 < x < 1$ are **increased,** because the square root—indeed, *any* root with power between 0 and 1—*raises* such values; for example,

$$\sqrt{0.01} = 0.1, \qquad \sqrt[3]{0.001} = 0.1.$$

In the interval $1 < x < 4$, the numbers become **smaller.** Thus, we are raising the lefthand set of numbers and decreasing the righthand ones, a possible step in the correct direction. Thus, if we replace y by $y^* = \sqrt{y}$ we get the relation

$$y^* = \sqrt{x^2}$$

or
$$y^* = x, \qquad x > 0,$$

and this is an equation of a **straight line** through the origin.

The suggestion we want to draw from this example is that for hollow-upward, monotonically increasing curves, if we want to replace y, we should move down the ladder to a p smaller than 1. We do not draw the lesson that this will work well or that we know how far to go. In the present example, the idea of either squaring or square-rooting sticks out because we knew the formula.

Let us now try for a second lesson by going back to the original curve $y = x^2$.

2. *What re-expression of x would straighten out the curve?* Again, our knowledge of the functional form suggests that we replace x by either $x^* = x^2$ or $x^* = \sqrt{x}$. If we replace x by $x^* = x^2$, then we will have points (x^2, y), and since $y = x^2$, we again get a straight line because the points are (x^2, x^2). We have moved up the ladder. We are tacitly assuming that all powers p are available, but relatively few are used. Let us suppose that only the powers -3, -2, -1, #, 1, 2, 3 had been available. What would we try for y?

3. *Trying* log y. Since we want to go down the ladder for y, let us replace y by $\log y$ and see what happens. We use $\log_e$. We have $\log 0 = -\infty$, $\log 1 = 0$, $\log 4 = 1.39$.

Original slopes: $0 < x < 1$, 1; $1 < x < 4$, 5;

New slopes: $0 < x < 1$, ∞; $1 < x < 4$, 0.46.

The logarithm increased the slope for the lefthand interval and decreased it for the righthand one, both moves in the correct direction, but it overcorrected.

Starting. We could have avoided the infinities here if we had added a constant to y before we started. Let's ask what constant we could have added to make the slopes of the two chords equal. We want c so that

$$\frac{\log(1 + c) - \log c}{1} = \frac{\log(16 + c) - \log(1 + c)}{3}$$

$$\log\frac{1 + c}{c} = \tfrac{1}{3}\log\left(\frac{16 + c}{1 + c}\right).$$

Trying a few values suggests that $c = 0.95$ gives a close approximation. Ordinarily, we would round this c off to 1, but let us go ahead with 0.95. We are ready to replace

$$y \qquad \text{by} \qquad y^* = \log(y + 0.95).$$

Tabular values to two decimals for $x = 0, 1, 2, 3, 4$ are

x	0	1	2	3	4
y^*	−0.05	0.67	1.60	2.30	2.83

The graph is shown in Exhibit **3.** On the one hand, the curve is not straight, but on the other hand it is much straighter than it was to begin with, and it might

Exhibit **3** of chapter 4

Graph of $y^* = \log(x^2 + 0.95)$.

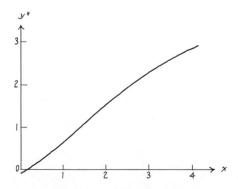

serve us very well. A straight line could be fitted to this curve that would not be off more than 0.1 in the vertical direction over the range $0 < x < 4$.

The suggestion here is that perfection may not be necessary. A coarse grid on p might be quite satisfactory. Usually we include $p = \frac{1}{2}$ and $p = -\frac{1}{2}$; sometimes $p = \frac{1}{3}$, and even other fractional powers. How much detail is worthwhile depends upon the example being treated.

Although we have not proved it, the direction of the hollowness—upward or downward—and the direction of the monotonicity lead us to a set of rules for making the fit. We give these rules without further discussion in the next section.

4C. The Bulging Rule

The fundamental rule is to move on the ladder in the direction in which the bulge points. The bulge points separately for x and for y. Exhibit **4** is a reminder of how to use the ladder of powers to aid re-expression. The arcs illustrate four combinations of slope and curvature, four kinds of bulging.

Let us try another example. Data for it are given in Exhibit **5,** and these data are plotted in Exhibit **6.**

Exhibit **4** of Chapter 4

Directions to move. The arrows point in the direction of the bulge for each type of curve and for each variable separately.

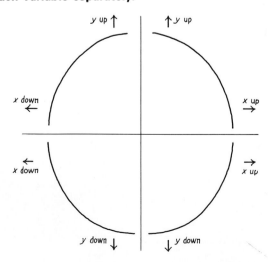

Example 1. Use the ladder to straighten out the curve in Exhibit 6 by re-expressing y.

Solution. From the point of view of y, the curve bulges upward, and we want to go up the ladder from $p = 1$.

Let us pick out three points and compute the pairs of slopes. Let us choose $x = 0.1, 3, 9$. The original slopes of the chords are

$$0.1 < x < 3, \quad \frac{2.88 - 0.93}{3 - 0.1} = 0.67; \quad 3 < x < 9, \quad \frac{4.16 - 2.88}{9 - 3} = 0.21;$$

$$\text{Ratio of slopes:} \quad 0.67/0.21 = 3.2$$

Exhibit **5** of Chapter 4

Data relating y and x for Example 1, Section 4C

x	y		x	y
.1	.93		1	2.00
.3	1.34		3	2.88
.5	1.59		5	3.42
.7	1.78		7	3.83
.9	1.93		9	4.16

Exhibit **6** of Chapter 4

Plot of data for Example 1.

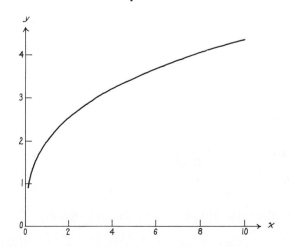

Let us now move up the y ladder to y^2.

x	0.1	3	9
$y^* = y^2$	0.86	8.29	17.31

The new slopes are

$$0.1 < x < 3, \quad \frac{8.29 - 0.86}{3 - 0.1} = 2.56; \quad 3 < x < 9, \quad \frac{17.31 - 8.29}{9 - 3} = 1.50;$$

Ratio of slopes: $2.56/1.50 = 1.7.$

The move has reduced the ratio, but we have not gone far enough. Let us try y^3.

x	0.1	3	9
$y^{**} = y^3$	0.80	23.89	72.0

$$0.1 < x < 3, \quad \frac{23.89 - 0.80}{3 - 0.1} = 7.97; \quad 3 < x < 9, \quad \frac{72.0 - 23.9}{9 - 3} = 8.01;$$

Ratio of slopes: $7.97/8.01 = 0.995.$

This is within rounding error of 1.00, and we conclude that replacing y by y^3 will straighten out the curve.

Since we secretly know that the original data were rounded values of $y = 2\sqrt[3]{x}$, we can see that the movement on the ladder has actually found exactly the right re-expression. Nevertheless, it will be instructive to see what would have happened if we had tried instead to re-express x.

Example 2. Straighten out the curve of Exhibit 6 by re-expressing x.

Solution. The curve bulges left, and so we want to go down the ladder to $\log x$ or $-1/x$. Let us try these values, using the same 3 choices of x:

y	0.93	2.88	4.16
x	0.1	3	9
$x^* = \log x$	-1	0.48	0.95
$x^{**} = -1/x$	-10	-0.33	-0.11

As the calculations below show, the ratio of the slopes increases steadily.

	Left interval		Right interval		Ratio	log ratio
x	$\dfrac{2.88 - 0.93}{3 - 0.1}$	$= 0.67$	$\dfrac{4.16 - 2.88}{9 - 3}$	$= 0.21$	0.31	-0.51
x^*	$\dfrac{2.88 - 0.93}{0.48 - (-1)}$	$= 1.32$	$\dfrac{4.16 - 2.88}{0.95 - 0.48}$	$= 2.72$	2.06	0.31
x^{**}	$\dfrac{2.88 - 0.93}{-0.33 - (-10)}$	$= 0.20$	$\dfrac{4.16 - 2.88}{-0.11 - (-0.33)}$	$= 5.82$	29.1	1.46

Moving to the logarithm was already too far.

If we plot the logs of the ratios against p for the three points, we can interpolate to get an estimate of the re-expression that straightens most. The estimate is 0.38, or about $\frac{1}{3}$ (the exact result).

4D. More Complicated Curves

If a curve had a lazy-S shape, we would not be likely to remove the curvature by the process we have described. We might conceivably break it up into two pieces at the inflection point ∮ and fit the two pieces separately. More complicated curves could be treated similarly. Sometimes we can do better than this.

Sometimes we may have theory to guide us, and then we would expect to use it. For example, if we knew that the number of primes among the integers less than x was $x/\log_e x$, approximately, then plotting the observed number y against $x/\log_e x$ would be likely to produce the desired linearity.

4E. Scatter Plots

When the data are not as smooth as those we have dealt with, we replace values in a narrow array by some average—the median or the mean of both y and of x, and then we work with three of these points at a time, as before. The median has some advantages here. The median of the re-expressed values is the re-expressed value of the original median, whereas the mean of the re-expressed values is not the re-expressed mean. Ordinarily it is the mean that possesses such attractive commutative properties, but nonlinearity is not one of those situations.

Summary: Straightening Curves

Straightening out the relation of y to x over the range where we have data need not ensure straightening out the relation outside that range.

"Straightening" is to be attacked initially in terms of three well-selected points.

We use the ratio of the slopes of the two segments connecting the middle (according to x) point to the upper and lower points, planning to bring this ratio close to 1.0 by trying various rungs on the ladder of re-expression, and, where helpful, interpolating.

If a point-cloud (scatter plot) appears otherwise too fuzzy for effective straightening, we divide the (x, y) points into groups according to their x-values, then find (x-median, y-median) for each group, and work with these latter points.

The simplest re-expressions for amounts form a ladder, mainly consisting of t^p for various p, but with $\log t$ taking the place that would otherwise be reserved for t^0 (which is everywhere = 1 and thus is unhelpful).

When our curve or point-cloud is still curved, we move on the ladder of re-expression in the direction that the bulge points, a rule that can be applied, with different consequences, to re-expressing either x or y.

Chapter 5/The Practice of Re-expression

Chapter index on next page

Numbers are primarily recorded or reported in a form that reflects habit or convenience rather than suitability for analysis. As a result, we often need to re-express data before analyzing it. This chapter suggests re-expressions and time-saving ways to make them, and then offers some general guidelines as to what to try first.

The aims of re-expression will not appear in this chapter. Some of them have already appeared (particularly in Chapter 4); others will appear later (for example, in Chapters 9 and 12). For all these particular aims, there are general principles about what sorts of re-expressions are likely to be effective with what kind of numbers. These are the subject of the present chapter.

Some readers may prefer to read only the first few sections (5A and 5B are a minimum) on first reading and refer to the others as need arises.

5A. Kinds of Numbers

Numbers come to us expressed in several different ways, but most kinds fall into a few broad classes. Assisted by rules of thumb about choosing familiar analogs when we have an unfamiliar expression, these broad classes give us a start on most data we want to analyze.

The broad classes are:

amounts and counts, neither of which can be negative.

◇balances, which can have either sign and can be almost arbitrarily large, either positively or negatively.

◇counted fractions, as in "I saw 37 magpies and 4 had yellow bills, so 4/37 belonged to the yellow-billed species."

◇ranks, where either 1 = largest, 2 = next to largest, . . . , or 1 = smallest, 2 = next to smallest, . . .

◇grades—ordered labels, as from A, B, C, D, E, F or from *, **, ***, ****.

(Note that all these expressions are ordered. Some data come as names (perhaps robin, blackbird, sparrow, and so forth), but these names are not put in numbers—though COUNTS of that occurrence may be—so we need not re-express the names. Numbers on professional football players' jerseys are really "names" also. Just writing in digits does not make a number.)

We try the analog rules that follow only when dealing with numbers that do not fit in any of the broad classes:

◇ if the number is free to move in one direction, but has a definite bound, say A, in the other, a natural analog is an amount or count. The rule is: Take $y - A$, if A is a lower bound, or $A - y$, if it is an upper bound, and treat it like an amount.

◇ if the number is definitely constrained in both directions, say $B \le y \le C$, a natural analog is the counted fraction. The rule is: Treat $(y - B)/(C - B)$ as if it were a counted fraction.

Re-expressing amounts or counts. If y is an amount or count, the more frequently used re-expressions are

$$\log (y + c)$$

or

$$(y + c)^p$$

where c and p are constants and c is often zero. When not zero, c avoids special difficulty in the logarithm when $y = 0$ and helps similarly with powers when $p < 0$—for example, when $p = -1$. We call such a c the "start"; and, when $c \ne 0$, we speak of "started logs" or of "started powers".

The most common powers are $p = \frac{1}{2}$, $p = -1$, $p = \frac{1}{3}$ in descending frequency of use. See Section 4A and the ladder of re-expression in Exhibit 1 of Chapter 4. ($p = 1$ is just the original expression.)

If the ratio of largest to smallest value of y is substantial, we usually begin by looking at $\log y$.

Re-expressing balances. We re-express balances themselves rather infrequently. Much more often, we plunge deeper into the data and find that the balance y arises as a difference

$$y = z - u,$$

where z and u are amounts or counts. Then we look at a new number, y^*, often a balance of the form

$$y^* = \text{(a re-expression of } z) - \text{(a re-expression of } u).$$

Since knowing the numerical value of y alone does not fix the numerical value of y^*, these steps do more than just re-express y.

If V is an amount, then $\log V$ is a balance, because logs of numbers between 0 and 1 are negative, and logs are unbounded in both directions. Thus one possible kind of re-expression for balances, y, is

$$e^{cy}.$$

Since

$$e^{cy} = (e^c)^y,$$

we could take logs to the base e^c, and write

$$\log_{e^c} e^{cy} = y,$$

and thus e^{cy} is an amount whose log (to the base e^c) is the given balance. Some of the rather infrequent re-expressions of balances follow this pattern.

Re-expressing counted fractions. To analyze counted fractions, we often treat the fraction in one class, say class A, and those in class not-A comparably—symmetrically EXCEPT for direction. To illustrate one method, we can use for our example of 4 in 37 magpies:

$$(\text{re-expression of } \tfrac{4}{37}) - (\text{same re-expression of } \tfrac{33}{37}).$$

We call such expressions "foldings". Some foldings use the sorts of re-expressions described above for amounts and counts. More complex foldings, which usually have the folding built into the expression, are sometimes useful, though we will say almost nothing about them here. Those already familiar with such devices may recognize such names as probits, normits, standard normal (or Gaussian) deviates, anglits, angular transforms, and so on. Logits, which might seem to belong to the complex case, are just folded logarithms, because the logit is

$$\log\left(\frac{p}{1-p}\right) = \log p - \log(1-p)$$

$$= (\text{re-expression of } p) - (\text{same re-expression of } (1-p))$$

Re-expressing ranks. As we will see later, folding a started log is one plausible and effective re-expression of ranks.

Re-expressing grades. The most useful idea here seems to be to fix on some standard distribution, and ask: "If we divide up the standard distribution according to the observed fractions, where does the center of gravity of each piece fall?" More in due course.

Note. We return to each of these points later in this chapter.

5B. Quick Logs

Although we live in the high-speed computer age, many will be astonished at how much can be done by hand in less time than it takes to debug a complicated program. Especially in the exploration of data, multiple analyses and reanalyses are the rule, with earlier tries frequently discarded. Thus, unless one has strong skill as a programmer and considerable resources in subroutines and computer access, a fair amount of pilot work by hand can be profitable before programming the computer with the selected plan. Consequently, even liking high-speed computation as well as we do, we still also appreciate the value of handwork. Therefore we provide materials to speed it.

If we have a hand-held electronic calculator that goes beyond $+$, $-$, $\times$, $\div$, we have only to press one button to turn the number we have entered into a log. Moreover, we have a choice as to how many decimals we are to keep. Usually we keep more than we need, slowing down our calculations. One good practice is to start by keeping two decimals, which is often one more than we need, planning to go back and redo the work if the analysis suggests that two are not enough. If our calculator takes two button presses for $\log_{10}$ and only one for $\log_e = \ln$, let us use $\log_e$ to 2 decimals.

For those without a hand-held calculator, a table can help greatly. Exhibit 1 gives a table of convenient form, taken from *EDA*, Chapter 3 (to which we refer you for further details, although the table is self-contained).

Using Exhibit 1 for quick logs to the base 10. After determining the characteristic, obtain the mantissa by locating the number between the two numbers in the first column, and choosing the number in the second column.

We call this a "break" table because it breaks the x-variable into intervals such that all numbers in the interval are assigned the same value $f(x)$ for any x in the interval. It differs, therefore, substantially from the usual table where, for a set of x's, values of $f(x)$ are given, and then we interpolate in the table to get $f(x)$.

Example. Thus 82.1 has characteristic 1 from the remainder triangle of Panel B (the left right triangle) of Exhibit 1. We now regard the number as a four-digit one, 8210, and read from the last column 0.91, and so the logarithm is 1.91. Had we had 82.23, we would have chosen 0.92, because our rule for ties is to choose the *even* answer.

Started logs. For

$$\log(y + c), \qquad c > 0,$$

we add c to y, for y an amount, and use Exhibit 1. For y a count, we may take advantage of having only integers for y's. Three cases arise, c small, middling, and large:

c small:	can be neglected for $y > 1$.
c middling:	use a standard fraction, say 1/6.
c large:	if large enough, c can be taken to be an integer.

Exhibit **1** of Chapter 5

Break table for two-decimal logs to the base 10

A) MAIN BREAK TABLE for mantissas

Break	log	Break	log	Break	log	Break	log	Break	log
9886	.00	1567	.20	2484	.40	3936	.60	6238	.80
1012	.01	1603	.21	2540	.41	4027	.61	6382	.81
1036	.02	1641	.22	2601	.42	4121	.62	6532	.82
1059	.03	1678	.23	2660	.43	4216	.63	6683	.83
1084	.04	1718	.24	2723	.44	4316	.64	6840	.84
1109	.05	1757	.25	2786	.45	4415	.65	6998	.85
1136	.06	1799	.26	2852	.46	4519	.66	7162	.86
1161	.07	1840	.27	2917	.47	4623	.67	7328	.87
1189	.08	1884	.28	2986	.48	4732	.68	7499	.88
1216	.09	1927	.29	3054	.49	4841	.69	7673	.89
1245	.10	1973	.30	3127	.50	4955	.70	7853	.90
1273	.11	2018	.31	3198	.51	5069	.71	8035	.91
1302	.12	2066	.32	3274	.52	5187	.72	8223	.92
1333	.13	2113	.33	3349	.53	5308	.73	8413	.93
1365	.14	2163	.34	3428	.54	5433	.74	8610	.94
1396	.15	2213	.35	3507	.55	5559	.75	8810	.95
1429	.16	2265	.36	3590	.56	5689	.76	9016	.96
1462	.17	2317	.37	3672	.57	5821	.77	9225	.97
1495	.18	2372	.38	3759	.58	5957	.78	9441	.98
1531	.19	2426	.39	3845	.59	6095	.79	9660	.99
1567		2484		3936		6238		9886	

When in doubt, use an even answer; thus, 1462 gives .16 and 1.495 gives .18.

B) SETTING DECIMAL POINTS

1	+0	−1	1
10	+1	−2	0.1
100	+2	−3	0.01
1000	+3	−4	0.001
10,000	+4	−5	0.0001
100,000	+5	−6	0.00001
1,000,000			0.000001

C) EXAMPLES

Number	**B**	**A**		log number
log 137.2	**2**	**+ .14**	=	2.14
log 0.03694	**−2**	**+ .57**	=	−1.43
log 0.896	**−1**	**+ .95**	=	−0.05
log 174,321	**+5**	**+ .24**	=	5.24

For small and large c, we have only (except for the smallest y's in the first case) to refer to a conventional log table, which is always effective in supplying logs of integers. The following supplementary table of log $(y + c)$ shows why:

Short table of log $(y + c)$, $c \geq 0$

	$y = 0$	$y = 1$	$y = 2$	$y = 3$	$y = 4$
$(c = 0)$	$(-\infty)$	(0.00)	(0.30)	(0.48)	(0.60)
$c = 0.01$	-2.00	0.00	0.30	0.48	0.60
$c = 0.03$	-1.52	0.01	0.31	0.48	0.61
$c = 0.1$	-1.00	0.04	0.32	0.49	0.61
$(c = 0.25)$	(-0.60)	(0.10)	(0.35)	(0.51)	(0.63)

Clearly we hardly need this table for $y \geq 1$ for $c = 0.01$ and 0.03, and for $y \geq 2$ for $c = 0.1$ and even for $c = 0.25$. The entries scarcely vary down a column for such y's.

For a moderate c, it is convenient to standardize on a single c. The choice $c = 1/6$ has rather recondite reasons in its favor. More importantly, it seems to work just about as well as the more classical $1/10$ or $1/4$. So Exhibit **2** gives some 2-decimal values of log $(y + \frac{1}{6})$. Once $y \geq 10$, log $(y + \frac{1}{6})$ matches log y so closely we might as well move to Exhibit 1.

Some readers may now wish to skip to Chapter 7, returning to the skipped material when necessary.

Exhibit **2** of Chapter 5

Some two-decimal values of $\log_{10} (y + \frac{1}{6})$

Tens	Units									
	0	1	2	3	4	5	6	7	8	9
00	$-.78$	.07	.34	.50	.62	.71	.79	.86	.91	.96
10	1.01	1.05	1.09	1.12	1.15	1.18	1.21	1.23	1.26	1.28
20	1.30	1.33	1.35	1.36	1.38	1.40	1.42	1.43	1.45	1.46

Example

The cell in the row labeled 10 and column labeled 2 represents $y = 10 + 2 = 12$. The entry in the cell is log $12\frac{1}{6} = 1.09$.

Exhibit **3** of Chapter 5

Break table for (square) roots

A) EXAMPLES

Starting from the decimal point, divide the number into periods of two digits. Thus, 124.2 is 1 24 2, but 1242 is 12 42. Likewise, 0.00654 is 00 65 4 or 65 4.

|Number|	|Periods|	|From B|	|From C|	|Number|
124.2	1 24 2	ab.	112	11.2
1242	12 42	ab.	35	35.
.00654	00 65 4	.0x	80	.080

B) BREAK TABLES to SET DECIMAL POINT—enter between bold figures, leave with light figures.

1			**1**
1 00	a.	.x	.01
1 00 00	ab.	.0x	.00 01
1 00 00 00	abc.	.00x	.00 00 01
1 00 00 00 00	abcd.	.000x	.00 00 00 01

C) MAIN BREAK TABLE—in and out as in panel B

|Break|	|Root|	|Break|	|Root|	|Break|	|Root|	|Break|	|Root|	|Break|	|Root|
98 01		**2 49 64**		**5 66**		**15 60**		**35 40**	
1 02 01	100	**2 62 44**	160	**5 90**	240	**16 40**	40	**37 21**	60
1 06 09	102	**2 75 56**	164	**6 20**	246	**17 22**	41	**39 69**	62
1 10 25	104	**2 89 00**	168	**6 50**	252	**18 06**	42	**42 25**	64
1 14 49	106	**3 02 76**	172	**6 81**	258	**18 92**	43	**44 89**	66
1 18 81	108	**3 16 84**	176	**7 13**	264	**19 80**	44	**47 61**	68
1 23 21	110	**3 31 24**	180	**7 45**	270	**20 70**	45	**50 41**	70
1 27 69	112	**3 45 96**	184	**7 78**	276	**21 62**	46	**53 29**	72
1 32 25	114	**3 61 00**	188	**8 12**	282	**22 56**	47	**56 25**	74
1 36 89	116	**3 76 36**	192	**8 47**	288	**23 52**	48	**59 29**	76
1 41 61	118	**3 92 04**	196	**8 82**	294	**24 50**	49	**62 41**	78
1 48 84	120	**4 08 04**	200	**9 30**	30	**25 50**	50	**65 61**	80
1 58 76	124	**4 24 36**	204	**9 92**	31	**26 52**	51	**68 89**	82
1 69 00	128	**4 41 00**	208	**10 56**	32	**27 56**	52	**72 25**	84
1 79 56	132	**4 57 96**	212	**11 22**	33	**28 62**	53	**75 69**	86
1 90 44	136	**4 75 24**	216	**11 90**	34	**29 70**	54	**79 21**	88
2 01 64	140	**4 92 84**	220	**12 60**	35	**30 80**	55	**82 81**	90
2 13 16	144	**5 10 76**	224	**13 32**	36	**31 92**	56	**86 49**	92
2 25 00	148	**5 29 00**	228	**14 06**	37	**33 06**	57	**90 25**	94
2 37 16	152	**5 47 56**	232	**14 82**	38	**34 22**	58	**94 09**	96
2 49 64	156	**5 66 44**	236	**15 60**	39	**35 40**	59	**98 01**	98

5C. Quick (Square) Roots and Reciprocals

After logs, we want most frequently to use (square) roots or reciprocals. Again the hand-held calculator can do the trick. On the other hand, we might not have one, and would then want a table.

For roots we have to be a little more careful since $\sqrt{20} = 4.47$ does not look like $\sqrt{2} = 1.41$ or $\sqrt{200} = 14.1$. The details are given and illustrated by examples in Panel A of Exhibit **3.**

Examples. Both 124.2 and 1242 point off into two two-digit periods to the left of the decimal. Each period "produces" one digit in the root, indicated as a and b here.

The pattern of 124.2 is 1 24.2; therefore, it goes with an isolated digit in its leftmost period, and so we locate it in the first break column of Panel C. Since 1242 has two digits in its leftmost period (12 42), we locate it in the third break column.

The number 0.00654 points off as .00 65 4 and so has two digits in its leftmost nonzero period and is located in column 5. Each 00 period contributes a 0 to the answer. And so .00 00 00 65 4 would give .00080.

The compact table for (square) roots shown in Exhibit 3 is taken from *EDA*, Chapter 3, which discusses its use in rather more detail.

Reciprocals. When we work with reciprocals, it is often convenient to use

$$-\frac{D}{y}, \quad \text{or} \quad C - \frac{D}{y},$$

where C and D are positive constants, so that our working values increase as the raw values increase. Exhibit **4,** also from *EDA*, Chapter 3, gives a compact table of $-1000/y$.

Started roots and reciprocals. Again, when dealing with counts, we are likely to "start" our roots or reciprocals. Again, if the start is either negligible or an integer, we can profit from conventional tables of (square) roots or reciprocals. For small values of c, we can see—from this short table of started (square) roots of $y + c$;

Short table of started (square) roots of $y + c$

	$y = 0$	$y = 1$	$y = 2$	$y = 3$	$y = 4$	$y = 5$
$(c = 0)$	(0.00)	(1.00)	(1.41)	(1.73)	(2.00)	(2.24)
$c = 0.01$	0.10	1.00	1.42	1.73	2.00	2.24
$c = 0.03$	0.17	1.01	1.42	1.74	2.01	2.24
$c = 0.1$	0.32	1.05	1.45	1.76	2.02	2.26
$(c = 0.3)$	(0.55)	(1.14)	(1.52)	(1.82)	(2.07)	(2.30)

Exhibit **4** of Chapter 5

Break table for (negative) reciprocal (using −1000/number)

A) BREAK TABLES for SETTING DECIMAL POINT

Break	Setting	Setting	Break
1000			1000
10,000	.x	a.	100
100,000	.0x	ab.	10
1,000,000	.00x	abc.	1.
10,000,000	.000x	abcd.	0.1
100,000,000	.0000x	abcde.	0.01

B) MAIN BREAK TABLE—digits of negative reciprocal

Break	Value	Break	Value	Break	Value	Break	Value	Break	Value
990	−100	1639	−60	2469	−40	4115	−240	617	−160
1010	−98	1681	−59	2532	−39	4202	−236	633	−156
1031	−96	1709	−58	2597	−38	4274	−232	649	−152
1053	−94	1739	−57	2667	−37	4348	−228	667	−148
1075	−92	1770	−56	2740	−36	4425	−224	685	−144
1099	−90	1802	−55	2817	−35	4505	−220	704	−140
1124	−88	1835	−54	2899	−34	4587	−216	725	−136
1149	−86	1869	−53	2985	−33	4673	−212	746	−132
1176	−84	1905	−52	3077	−32	4762	−208	769	−128
1205	−82	1942	−51	3175	−31	4854	−204	794	−124
1235	−80	1980	−50	3279	−30	4950	−200	820	−120
1266	−78	2020	−49	3367	−294	505	−196	840	−118
1299	−76	2062	−48	3436	−288	515	−192	855	−116
1333	−74	2105	−47	3509	−282	526	−188	870	−114
1370	−72	2151	−46	3584	−276	538	−184	885	−112
1408	−70	2198	−45	3663	−270	549	−180	901	−110
1449	−68	2247	−44	3745	−264	562	−176	917	−108
1493	−66	2299	−43	3831	−258	575	−172	935	−106
1538	−64	2353	−42	3922	−252	588	−168	952	−104
1587	−62	2410	−41	4016	−246	602	−164	971	−102
1639		2469		4115		617		990	

Examples

Number	A	B	−1000/number
124.2	a.	−80	−8.0
.04739	abcde.	−212	−212**.
1242.	.x	−80	−.80

or this, of started (negative) reciprocals:

Short table of started (negative) reciprocals of $y + c$

	$y = 0$	$y = 1$	$y = 2$	$y = 3$	$y = 4$	$y = 5$
$(c = 0)$	$(-\infty)$	(-1000)	(-500)	(-333)	(-250)	(-200)
$c = 0.01$	-100000	-990	-498	-332	-249	-200
$c = 0.03$	-33333	-971	-493	-330	-248	-199
$c = 0.1$	-10000	-909	-476	-323	-244	-196
$(c = 0.3)$	(-3333)	-769	(-435)	(-303)	(-233)	(-189)

—that we only have to bother with very small c's for quite small y's.

The situation for $c = \frac{1}{6}$ is set out in Exhibits **5** and **6**. Again we can use rather small tables (sometimes with a unit change in the last place) for larger y.

Exhibit **5** of Chapter 5

Some two-decimal values of $\sqrt{y + \frac{1}{6}}$.

	Units									
Tens	0	1	2	3	4	5	6	7	8	9
00	.41	1.08	1.47	1.78	2.04	2.27	2.48	2.68	2.86	3.03
10	3.19	3.34	3.49	3.63	3.76	3.89	4.02	4.14	4.26	4.38
20	4.49	4.60	4.71	4.81	4.92	5.02	5.12	5.21	5.31	5.40

For $30 \leq y \leq 288$, add 0.01 to $\sqrt{y}$; for $y > 288$, use $\sqrt{y}$ as in Exhibit 3.

Exhibit **6** of Chapter 5

Some two-decimal values of $-1000/(y + \frac{1}{6})$

	Units									
Tens	0	1	2	3	4	5	6	7	8	9
00	−6000	−857	−462	−376	−240	−194	−162	−140	−122	−109
10	−98.4	−89.6	−82.2	−75.9	−70.6	−65.9	−61.9	−58.3	−55.0	−52.2
20	−49.6	−47.2	−45.1	−43.2	−41.4	−39.7	−38.2	−36.8	−35.5	−34.3
30	−33.1	−32.1	−31.1	−30.2						

For $34 \leq y \leq 56$, decrease $-1000/y$ by 0.1; for $57 \leq y$, use $-1000/y$, as in Exhibit 4.

5D. Quick Re-expressions of Counted Fractions, Percentages, Etc.

We have proposed to fold our re-expressions of fractions, so that 50% will always be expressed as 0.00. How much further can we carry this agreement? The size of the plurality is $p - (1 - p) = 2p - 1$. Can we have 51% as 0.02 and 48% as -0.04 for all the re-expressions we want to consider? As Exhibit **7** shows, we can do such matching, to the two decimals we are routinely keeping, from 38% to 62%, after which the re-expressions diverge, at first slowly and then more rapidly.

Exhibit **7** of Chapter 5

Re-expressions of fractions, matched at 50%

Pluralities, folded roots, folded logarithms—Alternative expressions for counted fractions (take sign of answer from head of column giving %).

A) MAIN TABLE

+	Plur.	froot	flog	–	+	Plur.	froot	flog	–
50%	use	.00	use	50%	85%	.70	.76	.87	15%
51	→	.02	←	49	86	.72	.78	.91	14
52		.04		48	87	.74	.81	.95	13
53		.06		47	88	.76	.84	1.00	12
54		.08		46	89	.78	.87	1.05	11
55%	use	.10	use	45%	90.0%	.80	.89	1.10	10.0%
56	→	.12	←	44	90.5	.81	.91	1.13	9.5
57		.14		43	91	.82	.92	1.16	9
58		.16		42	91.5	.83	.94	1.19	8.5
59		.18		41	92	.84	.96	1.22	8
60%	use	.20	use	40%	92.5%	.85	.97	1.26	7.5%
61	→	.22	←	39	93	.86	.99	1.29	7
62		.24		38	93.5	.87	1.01	1.33	6.5
63	.26	.26	.27	37	94	.88	1.02	1.38	6
64	.28	.28	.29	36	94.5	.89	1.04	1.42	5.5
65%	.30	.30	.31	35%	95.0%	.90	1.06	1.47	5.0%
66	.32	.32	.33	34	95.5	.91	1.08	1.53	4.5
67	.34	.35	.35	33	96	.92	1.10	1.59	4
68	.36	.37	.38	32	96.5	.93	1.12	1.66	3.5
69	.38	.39	.40	31	97	.94	1.15	1.74	3

Exhibit 7 of Chapter 5(continued)

70%	.40	.41	.42	30%	97.2%	.94	1.16	1.77	2.8%
71	.42	.43	.45	29	97.4	.95	1.17	1.81	2.6
72	.44	.45	.47	28	97.6	.95	1.18	1.85	2.4
73	.46	.47	.50	27	97.8	.96	1.19	1.90	2.2
74	.48	.50	.52	26	98.0	.96	1.20	1.95	2.0
75%	.50	.52	.55	25%	98.2%	.96	1.21	2.00	1.8%
76	.52	.54	.58	24	98.4	.97	1.22	2.06	1.6
77	.54	.56	.60	23	98.6	.97	1.24	2.13	1.4
78	.56	.59	.63	22	98.8	.98	1.25	2.21	1.2
79	.58	.61	.66	21	99.0	.98	1.27	2.30	1.0
80%	.60	.63	.69	20%	99.2%	.98	1.28	2.41	0.8%
81	.62	.66	.73	19	99.4	.99	1.30	2.55	0.6
82	.64	.68	.76	18	99.6	.99	1.32	2.76	0.4
83	.66	.71	.79	17	99.8	1.00	1.35	3.11	0.2
84	.68	.73	.83	16	100.0%	1.00	1.41	∞	0.0

B) SUPPLEMENTARY TABLE—for flogs of fractions beyond 1% or 99%

+	flog	−		+	flog	−
99.0%	2.30	1.0%		99.80%	3.11	.20%
.1	2.35	.9		.82	3.16	.18
.2	2.41	.8		.84	3.22	.16
.3	2.48	.7		.86	3.28	.14
.4	2.55	.6		.88	3.36	.12
99.50	2.65	.50		99.90	3.45	.10
.52	2.67	.48		.91	3.51	.09
.54	2.69	.46		.92	3.57	.08
.56	2.71	.44		.93	3.63	.07
.58	2.73	.42		.94	3.71	.06
99.60	2.76	.40		99.95	3.80	.05
.62	2.78	.38		.96	3.91	.04
.64	2.81	.36		.97	4.06	.03
.66	2.84	.34		.98	4.26	.02
.68	2.87	.32		.99	4.61	.01
99.70	2.90	.30				
.72	2.94	.28				
.74	2.97	.26				
.76	3.01	.24				
.78	3.06	.22				

Examples
_____

99.29% gives 2.47
0.37% gives −2.80

All we have to do is to take as the definition of froots (folded roots)

$$\sqrt{2}(\sqrt{f} - \sqrt{1 - f})$$

instead of

$$\sqrt{f} - \sqrt{1 - f}$$

and to take as the definition of flogs (folded logs)

$$\tfrac{1}{2}(\log_e f - \log_e (1 - f)) = 1.1513 (\log_{10} f - \log_{10} (1 - f))$$

instead of either

$$\log_e f - \log_e (1 - f) = \log_e \frac{f}{1 - f}$$

or

$$\log_{10} f - \log_{10} (1 - f) = \log_{10} \frac{f}{1 - f}.$$

For total counts up to a thousand or so, two-decimal values are satisfactory. For larger values, it may be worthwhile going to more decimals.

Starting counted fractions. To start re-expressions of counted fractions, we begin with (let one count be x, the other count $(n - x)$, and c the start)

$$\frac{\text{(one count)} + \text{(start)}}{\text{(total count)} + \text{(start)} + \text{(start)}} = \frac{x + c}{n + 2c},$$

so that this fraction is identical with

$$1 - \frac{\text{(the other count)} + \text{(start)}}{\text{(total count)} + \text{(start)} + \text{(start)}} = 1 - \frac{n - x + c}{n + 2c} = \frac{x + c}{n + 2c}.$$

Then in the original folded-log expression, if $c = \tfrac{1}{6}$, we have

$$\log \text{(one count)} - \log (\text{(the other count)})$$

re-expressed as

$$\log (\text{one count} + \tfrac{1}{6}) - \log (\text{the other count} + \tfrac{1}{6}).$$

Note that we started with $p = x/n$ and $(1 - p) = (n - x)/n$, and that the two terms in $\log n$ added to zero. In the same way, in the started logs, the terms in $\log (n + 2c)$ added to zero. Consequently, we can use Exhibit 2 for our values of logs of $y + \tfrac{1}{6}$, where y is an integer. If we gain by matching at 50%, we put in the factor 1.1513, to make the match, even when we start the counts in our counted fractions. We forget about this factor when it makes little difference.

5E. Matching for Powers and Logs

Having matched at 50%–50% when re-expressing counted fractions, one asks about matching powers and logs when re-expressing amounts or counts. All we have to do is to decide at just what value A of y, we want the match to come.

Once we have done this, we can use any of the following six expressions (see Exhibit **8**):

$$(p = 2) \qquad \frac{A}{2}\left(\frac{y}{A}\right)^2 + \frac{A}{2}$$

$$(p = 1) \qquad y = A\left(\frac{y}{A}\right)$$

$$(p = \tfrac{1}{2}) \qquad 2A\left(\frac{\sqrt{y}}{\sqrt{A}}\right) - A$$

$$(p = \text{pseudo } 0) \qquad A \log_e \frac{y}{A} + A$$

$$(p = -1) \qquad -A\left(\frac{y}{A}\right)^{-1} + 2A$$

$$(p = -2) \qquad -\frac{A}{2}\left(\frac{y}{A}\right)^{-2} + \tfrac{3}{2}A$$

or, more generally,

$$\frac{A}{p}\left(\frac{y}{A}\right)^p + \left(1 - \frac{1}{p}\right)A,$$

for any $p \neq 0$, as a re-expression of y which is matched to y at and near $y = A$.

For a convenient example, we can take $A = 300$, obtaining the results shown in Exhibit 8. The values of p go smoothly from left to right across the columns. The fourth column of Exhibit 8 shows clearly how, for powers matched to y at $y = 300$, the matched logarithm takes the place left open by the failure of y^p with $p = 0$ to be a useful re-expression.

Exhibit **9** tabulates these same expressions more finely, and to two more decimals, in the neighborhood of $y = 300$, so as to bring out the closeness of the matching near that value.

5F. Re-expressions for Grades

Suppose that we have individuals classified as A, B, C, D, or E, and that the frequencies of occurrence are as in Exhibit **10**. For each grade we can calculate cumulative proportions, as in that exhibit:

⋄a fraction p of individuals beyond the grade,
⋄a fraction P of individuals beyond and including the grade,

and then use either the table or formula of Exhibit **11** to calculate the corresponding values of $\phi(\ )$. The tentative re-expression of each grade is then the value of

$$\frac{\phi(P) - \phi(p)}{P - p},$$

where P and p are the two fractions.

Panels B and C of Exhibit 10 show the same calculation for hypothetical groups of "good" and "poor" students, which we suppose to have been sorted out without any use of the grade we are re-expressing. It would be well if the re-expressions suggested by these groups agreed with each other (and with the

Exhibit **8** of Chapter 5

Some values of re-expressions of logs and powers matched at and near $y = A = 300$.

Values of p

2	1	$\frac{1}{2}$	[pseudo 0]	-1	-2
$\dfrac{y^2}{2A} + \dfrac{A}{2}$	y	$\sqrt{4Ay} - A$	$A \log_e\left(\dfrac{y}{A}\right) + A$	$-\dfrac{A^2}{y} + 2A$	$-\dfrac{A^3}{2y^2} + \dfrac{3A}{2}$
150	0	−300	−∞	−∞	−∞
151	25	−127	−445	−3000	−21150
154	50	−55	−238	−1200	−4950
167	100	46	−30	−300	−900
217	200	190	178	150	112
254	250	248	245	240	234
281	280	280	279	279	278
290	290	290	290	290	289
299	299	299	299	299	299
300	300	300	300	300	300
301	301	301	301	301	301
321	320	320	319	319	318
354	350	348	346	343	340
417	400	393	386	375	366
567	500	475	453	420	396
1817	1000	795	661	510	436
6817	2000	1249	869	555	447
∞	∞	∞	∞	600	450
$p = 2$	1	$\frac{1}{2}$	[pseudo-zero]	−1	−2

figures for the whole group). Agreement here would be excellent if the re-expressions suggested by the two groups were to agree up to an additive constant. Nor would we be bothered if it took a multiplicative constant, instead, before the re-expressions agreed. So our question is: Is the one suggested re-expression nearly a linear function of the other?

Exhibit **12** shows the plot of one suggested re-expression against the other. The result is not a straight line, but neither is it badly bent or jagged. Accordingly, we think either re-expression is reasonable—and, from an excess of caution, plan to use the mean of the two, namely:

A	B	C	D	E
−4.48	−2.54	0.08	2.80	5.20

Exhibit **9** of Chapter 5

Finer tabulation of Exhibit 8 near $y = A = 300$, where the matching is closest.

$\dfrac{y^2}{2A} + \dfrac{A}{2}$	y	$\sqrt{4Ay} - A$	$A \log_e\left(\dfrac{y}{A}\right) + A$	$-\dfrac{A^2}{y} + 2A$	$-\dfrac{A^3}{2y^2} + \dfrac{3A}{2}$
280.67	**280**	279.66	279.30	278.57	277.81
285.38	**285**	284.81	284.61	284.21	283.80
290.17	**290**	289.92	289.83	289.66	289.48
292.11	**292**	291.95	291.89	291.78	291.67
294.06	**294**	293.97	293.94	293.88	293.82
296.03	**296**	295.99	295.97	295.95	295.92
298.01	**298**	298.00	297.99	297.99	297.98
299.00	**299**	299.00	299.00	299.00	299.00
300.00	**300**	300.00	300.00	300.00	300.00
301.00	**301**	301.00	301.00	301.00	301.00
302.01	**302**	302.00	301.99	301.99	301.98
304.03	**304**	303.99	303.97	303.95	303.92
306.06	**306**	305.97	305.94	305.88	305.82
308.11	**308**	307.95	307.90	307.79	307.69
310.17	**310**	309.92	309.84	309.68	309.52
315.38	**315**	314.82	314.64	314.29	313.95
320.67	**320**	319.68	319.36	318.75	318.16
$p = 2$	1	$\frac{1}{2}$	[pseudo-zero]	−1	−2

Actually, these are nearly enough equally spaced for us to expect to do well by using -2, -1, 0, 1, 2. We can see this by first adding -0.09 and then dividing by 2.6, to get:

A	B	C	D	E
-1.75	-1.01	0.00	1.05	1.97

Exhibit **10** of Chapter 5

Example of scoring grades

A) OVERALL EXAMPLE

Grade	# = count	p = fraction >	P = fraction ≥	$\phi(p)$	$\phi(P)$	$\dfrac{\lvert\phi(P) - \phi(p)\rvert}{P - p}$
A	127	.0000	.0304	.0000	−.1361	−4.48
B	497	.0304	.1496	−.1361	−.4220	−2.40
C	3243	.1496	.9269	−.4220	−.2616	.21
D	231	.9269	.9823	−.2616	−.0889	3.12
E	74	.9823	1.0000	−.0889	.0000	5.02
(*Total*)	(4172)					

Notes: .0304 = 127/4172,   .1496 = (127 + 497)/4172,   etc.   Values   of $\phi(p)$ and $\phi(P)$ from Exhibit 11.

B) HYPOTHETICAL GOOD STUDENTS

A	64	.0000	.0792	.0000	−.2768	−3.49
B	127	.0792	.2364	−.2768	−.5469	−1.72
C	560	.2364	.9295	−.5469	−.2549	.42
D	54	.9295	.9963	−.2549	−.0244	3.45
E	3	.9963	1.0000	−.0244	.0000	6.59
	(808)					

C) HYPOTHETICAL POOR STUDENTS

A	12	.0000	.0114	.0000	−.0623	−5.46
B	53	.0114	.0617	−.0623	−.2316	−3.37
C	821	.0617	.8406	−.2316	−.4387	−.27
D	107	.8406	.9421	−.4387	−.2212	2.14
E	61	.9421	1.0000	−.2212	.0000	3.82
	(1054)					

Exhibit **11** of Chapter 5

Values—and formula—for $\phi(p)$ or $\phi(P)$

A) Values for selected p's or P's

					q (last digit) $=$					
p or P	0	1	2	3	4	5	6	7	8	9
.000q	.0000	−.0010	−.0019	−.0027	−.0035	−.0043	−.0051	−.0058	−.0065	−.0072
.00q	.0000	−.0079	−.0144	−.0204	−.0261	−.0315	−.0367	−.0417	−.0466	−.0514
.0q	.0000	−.0560	−.0980	−.1347	−.1679	−.1985	−.2270	−.2536	−.2788	−.3025
.1q	−.3251	−.3465	−.3669	−.3864	−.4050	−.4227	−.4397	−.4559	−.4714	−.4862
.2q	−.5004	−.5140	−.5269	−.5393	−.5511	−.5623	−.5731	−.5833	−.5930	−.6022
.3q	−.6109	−.6191	−.6269	−.6342	−.6410	−.6474	−.6534	−.6590	−.6641	−.6687
.4q	−.6730	−.6769	−.6803	−.6833	−.6859	−.6881	−.6899	−.6913	−.6923	−.6929
.5q	−.6931	−.6929	−.6923	−.6913	−.6899	−.6881	−.6859	−.6833	−.6803	−.6769
.6q	−.6730	−.6687	−.6641	−.6590	−.6534	−.6474	−.6410	−.6342	−.6269	−.6191
.7q	−.6109	−.6022	−.5930	−.5833	−.5731	−.5623	−.5511	−.5393	−.5269	−.5140
.8q	−.5004	−.4862	−.4714	−.4559	−.4397	−.4227	−.4050	−.3864	−.3669	−.3465
.9q	−.3251	−.3025	−.2788	−.2536	−.2270	−.1985	−.1679	−.1347	−.0980	−.0560
.99q	−.0560	−.0514	−.0466	−.0417	−.0367	−.0315	−.0261	−.0204	−.0144	−.0079
.999q	−0079	−.0072	−.0065	−.0058	−.0051	−.0043	−.0035	−.0027	−.0019	−.0010

B) Formula for $\phi(p)$

$$\phi(p) = p \log_e p + (1 - p) \log_e (1 - p)$$

Exhibit **12** of Chapter 5

The two suggested re-expressions interrelated

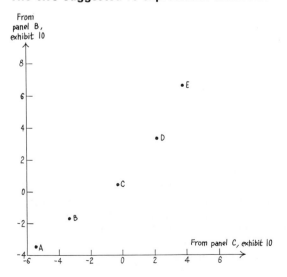

From panel B, exhibit 10

From panel C, exhibit 10

5G. Re-expressing Ranks

When we want to re-express ranks, we may gain by treating them like counted fractions. If an observation is 5th from one end of 37, let us consider that dividing all 37 observations in two parts by comparing them to a nearby division (cutting) value will give:

⋄4 in one class, 33 in the other, if the division (cutting) value is close on one side of the fifth observation,

⋄5 in one class, 32 in the other, if the division (cutting) value is close on the other side.

These two counted-fraction situations lead to started-and-folded logs

$$\log (4 + \tfrac{1}{6}) - \log (33 + \tfrac{1}{6})$$

and

$$\log (5 + \tfrac{1}{6}) - \log (32 + \tfrac{1}{6})$$

so that a natural related expression for rank 5 uses the average of the corresponding two arguments, and gives

$$\log (4\tfrac{1}{2} + \tfrac{1}{6}) - \log (32\tfrac{1}{2} + \tfrac{1}{6})$$

or, equivalently,

$$\log (5 - \tfrac{1}{3}) - \log (33 - \tfrac{1}{3})$$

where rank 5 from one end is rank 33 from the other.

Exhibit **13** of Chapter 5

Some two-decimal values of log $(i - \tfrac{1}{3})$ for integers $i < 100$.

q(last digit) =

i	0	1	2	3	4	5	6	7	8	9
0q	—	−.18	.22	.43	.56	.67	.75	.82	.88	.94
1q	.99	1.03	1.07	1.10	1.14	1.17	1.19	1.22	1.25	1.27
2q	1.29	1.32	1.34	1.36	1.37	1.39	1.41	1.43	1.44	1.46
3q	1.47	1.49	1.50	1.51	1.53	1.54	1.55	1.56	1.58	1.59
4q	1.60	1.61	1.62	1.63	1.64	1.65	1.66	1.67	1.68	1.69
5q	1.70	1.70	1.71	1.72	1.73	1.74	1.75	1.75	1.76	1.77
6q	1.78	1.78	1.79	1.80	1.80	1.81	1.82	1.82	1.83	1.84
7q	1.84	1.85	1.86	1.86	1.87	1.87	1.88	1.88	1.89	1.90
8q	1.90	1.91	1.91	1.92	1.92	1.93	1.93	1.94	1.94	1.95
9q	1.95	1.96	1.96	1.97	1.97	1.98	1.98	1.99	1.99	1.99

For $i \geq 30$, these can also be used as values of log i if desired.

More generally, we can use

$$\log\left(i - \tfrac{1}{3}\right) - \log\left(n + 1 - i - \tfrac{1}{3}\right) = \log\frac{i - \tfrac{1}{3}}{n + 1 - i - \tfrac{1}{3}}$$

for a rank of i (from the chosen end). Calculations are facilitated by using Exhibit **13** to provide the values of logs of integers less 1/3. Clearly, we can forget about the $-\tfrac{1}{3}$ as soon as i is greater than 30, so long as two decimals suffice.

5H. First Aid in Re-expression

Choosing exactly the right re-expression for a particular quantity may not be easy. To try to do a good job, we may have to (1) sense rather weak indications from the data in hand, (2) draw on experience with other bodies of data, or (3) lean on subject-matter knowledge. Even all three may not suffice. Both because we may not be prepared to try hard to choose our re-expression, or because we have too little information for anyone to choose reliably, we need rules of thumb that can provide "first aid", that can lead us to re-expressions that are almost always not bad—and usually pretty good.

Four rules will deal quite effectively with most of our needs, namely:

1. Take logs of an amount or count (if there are zeros or infinities, we may need to deal with them; see the next section).

2. Take logits or folded logs of fractions or percents; use some multiple of

$$\log\left(\frac{p}{1 - p}\right)$$

(zeros or ones call for special treatment). (If the data are restricted so that $A \le x \le B$ for some $A \ne 0$ and $B \ne 1$, use

$$\log\left(\frac{x - A}{B - x}\right).$$

3. Transform a rank i out of n, by

$$\log\left(\frac{i - \tfrac{1}{3}}{n - i + \tfrac{2}{3}}\right) = \log\left(\frac{3i - 1}{(3n + 1) - (3i - 1)}\right)$$

4. Let a balance stand.

These rules are not supposed to be a final answer—just as first aid for the injured is no substitute for a physician—but they offer a safe beginning.

Second aid. What if first aid is not good enough, and no careful guide can be found? In such cases we may revert to "second aid," along the following lines:

If we started with amounts or counts, first aid would have us take logs. We might need to undo this, going back from

$$x^* = \log x$$

to

$$x = \text{antilog } x^*.$$

If we are prepared to do this to x^*, we may also be prepared to do it, instead, to cx^*, reaching

$$x^{**} = \text{antilog } cx^*$$

which, in view of

$$\log (x^c) = c \cdot \log x = cx^*$$

will give us

$$x^{**} = x^c$$

for some exponent c.

If x is a fraction or a percent, so that first aid gives

$$x^* = \log \frac{x}{1 - x}$$

backtracking from cx^* gives

$$x^{**} = \frac{\left(\dfrac{x}{(1 - x)}\right)^c}{1 + \left(\dfrac{x}{(1 - x)}\right)^c} = \frac{x^c}{x^c + (1 - x)^c},$$

with whose use we have little experience. A related possibility is to start from

$$x^* = \log \frac{x}{1 - x} = \log x - \log (1 - x)$$

and undo the two logs separately, reaching

$$x^{**} = x^c - (1 - x)^c,$$

which has proved useful, for example, for $c = \frac{1}{2}$.

For ranks we can do similar things—again there is little experience. If we think of

$$\log \frac{i - \frac{1}{3}}{n - i + \frac{2}{3}} = \text{Gol} \left(\frac{i - \frac{1}{3}}{n + \frac{1}{3}}\right),$$

where

$$\text{Gol} (p) = \log p - \log (1 - p)$$

is the inverse of the cumulative logistic distribution, we might also consider

$$\text{Aug}\left(\frac{i - \frac{1}{3}}{n + \frac{1}{3}}\right),$$

where Aug is the inverse of the cumulative Gaussian (normal) distribution. (The result is quite close to what have been known as "rankits" or "normal scores".)

For balances, we can go via ranks, taking the n values of our balance, ordering them, assigning ranks, and forming the same re-expression of the ranks

$$\log\frac{3i - 1}{3(n + 1) - (3i + 1)} = \log\frac{i - \frac{1}{3}}{n - i + 1 - \frac{1}{3}}$$
$$= \log\left(i - \tfrac{1}{3}\right) - \log\left(n - i + 1 - \tfrac{1}{3}\right)$$

that we discussed in Section 5G.

None of these are guaranteed to organize the data well, but one or another often does. If none meets our needs, we may have to fit a sum of two or more terms, each involving a different re-expression of our x.

Ordered values. On occasion, we have a variable whose given values are not numbered but come in an order, as when grades A, B, C,... are assigned to performances. To bring such variables into a regression, we need to give numerical values to the levels. For each grade, we can regard the percentage assigned to that grade as a slice out of a logistic distribution. Then we assign each grade the numerical value of the corresponding center of gravity (CG) for the logistic.

On a computer, we take the CG of the slice from $p = A$ to $p = B$ to be

$$\frac{B \log B + (1 - B) \log(1 - B) - A \log A - (1 - A) \log(1 - A)}{B - A},$$

or, equivalently, we can use the analysis of Section 5F and the table of Exhibit 11.

For hand work we can often do well enough with

$$\frac{1}{6}\left(\log\frac{A}{1 - A} + 4\log\frac{(A + B)/2}{1 - (A + B)/2} + \log\frac{B}{1 - B}\right),$$

which is easier when we have only a table of

$$\log\left(\frac{A}{1 - A}\right).$$

We can do this either for the data as a whole or on each of several quite differently behaved parts of the data. In the latter case, we first check to see the results for the parts that behave similarly (see Exhibit 10, and its discussion in text, for an example), and then combine the results for the various parts.

5I. What to do with Zeros—and Infinities

If rules (1) or (2) in Section 5H tell us to take the log of zero, what should we do? To some extent, the answer depends on just what we are going to do with the variable—and on how many zeros we face.

If we face only a few zeros in y (or a small percent of zeros) AND we are to make a resistant fit so that any sufficiently discrepant value will be given zero weight, we can use

$$\log 0 = L,$$

where L is taken to be *less than any number* (but larger in absolute value). This approach gives the few zeros zero weight, without regard to how they relate to the previous fit. Although sometimes acceptable for a few zeros, this approach may not be acceptable for a substantial percentage of zeros.

Another simple solution is to "start" our logs, using

$$\log(x_j + c) \quad \text{or} \quad \log(y + c)$$

instead of $\log x_j$ or $\log y$. When dealing with counts, popular values for c are 1.00 and 0.25, although both smaller and larger starts are sometimes used. The analyst who wants to use only one adjustment, ranks and all included, can use 1/6 for a start, not 1/3, as we have explained. When dealing with amounts that include zeros, we have no clear practice, and it is hard to be sure what to do, although setting $\log 0 = L$ may still work.

Although starting is neat, we do not know whether it is better to deal with counts by starting or by assigning a suitable negative value in place of log 0 (since $\log 1 = 0$, any negative value goes in the right direction). If we wish to pick a value "out of a hat" to replace $\log 0$, there is something to be said for such choices as

$$-\log 4 \quad \text{or} \quad -\log 6 \quad \text{or} \quad -\log 8.$$

(Note: $-\log 6$ takes the logarithm of the average of the arguments of $-\log 4$ and $-\log 8$.) If we want to be more careful about matters, we can let the value we assign depend upon the frequency of zeros. Bohidar, Gruber, and Tukey have studied some aspects of this choice, and find that it is reasonable to use, $(\log p)/(1 - p)$, where p is estimated by the proportion of zero cells (if all cells are 0, calculate p as if one cell were not 0). Different portions of a table might use different estimates of p.

Infinities. Sometimes we have amounts that can be infinite, either actually or for practical purposes. The amount of time a rat takes to run a maze may be infinite, when the rat refuses to run. The amount of time taken by an apprentice to be promoted may be infinite, when he quits the job first. The amount of time taken by a fish to die in the presence of a certain amount of pollution may be practically infinite, when the fish is still alive after a substantial period and the results of the study must be pulled together, especially if most fish have died rather quickly.

An easy cure for such infinities is to re-express the result, not as logs but rather as reciprocals. The reciprocal time for the rat who does not run is merely zero. Indeed, in such an example we are now analyzing speed of running and not time to run.

Our speeds are now amounts safely away from infinity, but they do include some zeros. If we want to use the first-aid rules, we can now do any of the things described earlier in this section, including using log (speed PLUS start), that are appropriate.

Both zeros and infinities. What if we have *both* zeros and infinities? If we start first and then take reciprocals, we use

$$\frac{1}{\text{amount PLUS start}},$$

and if we then want to go to logs, we can start again, possibly using a new value, start*, reaching

$$\log\left(\frac{1}{\text{amount PLUS start}}\text{PLUS start*}\right).$$

Signs. To avoid negative signs, some use positive reciprocals, and, if the times or amounts are mostly—or always—greater than one and our start is small, they use

$$-\log\left(\frac{1}{\text{amount PLUS start}}\right),$$

since otherwise most logs would be negative. If the data measure "time to", they analyze speeds for raw reciprocals and slownesses for started log reciprocals.

Others like to keep re-expressions going in the same direction. If we began by thinking about slowness, their choice would be to use negative reciprocals

$$-\frac{1}{\text{amount}}$$

and to always use the minus sign in

$$-\log\left(\frac{1}{\text{amount PLUS start}}\right) \quad \text{or} \quad -\log\left(\frac{1}{\text{amount PLUS start}}+\text{start*}\right),$$

whether this leads to plus signs or to minus signs.

Choice of sign does not affect the results of most specified computations. It can affect one's sensitivity to the results of intermediate steps or to hints arising in the computation.

Fractions, ranks, and related quantities. If we begin with a fraction, and then go to

$$x^* = \log\left(\frac{p}{1-p}\right)$$

as first aid, we will have a zero–infinity problem when either $p = 0$ or $p = 1$. Because we can write

$$x^* = \log\frac{p}{1-p} = \log\frac{k}{n-k},$$

where k were observed out of n, it is natural to go to

$$x^* = \log\frac{k + \text{start}}{n - k + \text{start}}$$

with the sort of values for start that we use for counts.

Summary: Re-expression

We distinguish the following kinds of values: (i) amounts and counts, (ii) balances, (iii) counted fractions, (iv) ranks, (v) grades.

The most usual re-expressions, of an amount or a count, y, are $\log(y + c)$ and $(y + c)^p$ for various values of p and c.

Ordinarily we do not re-express balances, finding it wiser to re-express two (or more) quantities of which the balance is a difference (in Section 9F we will meet a contrary instance, where re-expression by e^{cy} seems natural).

Counted fractions seem well treated in terms of folded quantities, often folded logs (including logits) and folded roots.

Break tables can make re-expression quick and easy.

Our break tables give 2-decimal logs, roots, and reciprocals.

Matching at and near 50% can make pluralities (folded %), folded roots, and folded logarithms nearly the same from 38% to 62% and reasonably the same from 25% to 75%.

A natural approach to re-expression for grades is to focus on the interval covered by a given grade (the % below, the % at, and the % above) and then to assign a center of gravity to that grade either on all the data or on a well-defined part of it.

Exhibit 11 of this chapter helps assign centers of gravity, applicable to a part (or all) of the data, for any grade.

We combine information from doing this for several parts, in order to get a single, broadly-useful re-expression.

We may re-express ranks in terms of a "folded logarithm" with both counts started by $-1/3$.

We exercise "first aid to unexamined numbers" by (i) taking logs of amounts or counts, (ii) taking folded logs of fractions, percents, or other counted fractions, (iii) treating ranks as just explained, (iv) letting any balance stand.

We pick natural starts when necessary or desirable as part of first aid.

First-aid rules for re-expressing data modulate into second-aid re-expressions that are appreciably more flexible.

We extended the idea of starting to deal simultaneously with both "zeros" and "infinities."

We "start" our roots or logs by taking $c > 0$, then in particular, by taking $c = 1/6$.

We may "start" counted fractions by adding 1/6 to both counts.

We see how to match any two of the simple re-expressions both at and near any given A. (The same simple linear re-expression will match re-expressions with a common c at and near $(A - c)$. The reader should be able to work out how to match re-expressions with different values of c.)

References

Bohidar, N. R., D. G. Gruber, and J. W. Tukey, "Efficacy estimates for parasite-count data, including zero counts." Submitted to *Experimental Parasitology*.

EDA = J. W. Tukey (1977). *Exploratory Data Analysis*. Reading, Mass.: Addison-Wesley.

(For further details see Appendix, following Chapter 16.)

Chapter 6/Need We Re-express?

We may know, from experience with the type of data being analyzed, that a specific re-expression of the raw x's is helpful. Or we may have other reasons for believing that the re-expression is appropriate. We still might prefer not to bother, either because of the extra trouble, because of the difficulty of explaining and defending the analysis of the re-expressed data, or because little would be gained. We care, though, about the possible losses incurred by *not* re-expressing.

Although we have given considerable attention to tools for re-expression (in Chapter 5), sometimes the effort of re-expression is not likely to be rewarded—for instance, when choosing either between fitting y by

$$bx \qquad \text{or} \qquad b* \log x$$

or between fitting y by

$$a + bx \qquad \text{or} \qquad a* + b* \log x$$

when $1 \leq x \leq 1.001$.

These are special cases of choosing between the raw carrier, x in the example above, and the re-expressed carrier, $\log x$. Throughout, we will be considering only x's that are everywhere positive. The assumption in this chapter is that the re-expressed carrier will do a better job; the question is: "Will it be ENOUGH better to be worth the trouble of re-expression?" Here quality of performance is to be measured in terms of residuals. A poorer fit is exposed as such by the fact that it leaves unnecessarily large residuals.

To study this question, we would like to use as little information about the data as we can, thus simplifying our task. We provide a method of guidance that requires at the start three pieces of information:

◊the number of data points,

◊the closeness of relation between the raw and re-expressed carriers, and

◊the closeness of relation between the raw carrier and the response.

To describe the two closenesses, we use correlation coefficients. It is useful to recall that

$$\frac{\text{variance \{residuals\}}}{\text{variance \{response\}}} = 1 - r^2,$$

where r^2 is the correlation coefficient between the response and the carrier used in a simple regression.

While it is reasonable to suppose that we have at least plotted y against x_{Raw}, the raw carrier, so that we can assess, by eye if need be, the closeness of relation of the response to the raw carrier, we may well have related neither y nor x_{Raw} to $x_{Straightened}$, the re-expressed carrier. We can perhaps avoid the effort of re-expression if we can judge the closeness of x_{Raw} and $x_{Straightened}$, and thus the value of straightening, in some simple way. A most useful aspect of the data for judging the value of straightening is given by the ratio of the largest x_{Raw} to the smallest x_{Raw}. In brief, if the ratio is large compared with 1, we will be more likely to want to re-express, and if it is near 1, we may well not bother.

We give here only some general hints about how the answer comes out when the properly re-expressed carrier is the logarithm of the raw carrier, leaving to the Appendix (following Chapter 16) most details for this case and all details for the cases where the re-expressed carrier is the square root or the (negative) reciprocal of the raw carrier.

General Hints when Re-expressed Carrier is log x

◇If the largest value of x is twice the smallest value of x, we will usually need to re-express x if the correlation between the response and x is high (say >0.90). (If we are being very careful, we will re-express for smaller correlations; if we dare be offhand, we may not bother for correlations less than 0.95.)

◇If the ratio of the largest value of x to the smallest value of x is less than 2, we will not re-express unless the correlation between response and raw x is even higher than 0.95.

◇If the largest value of x is twenty or more times the smallest value, we are likely to want to re-express almost every time that the relation between response and raw x seems at all helpful.

◇If we are in doubt, either we should read the Appendix, or we should try re-expression and see how much better it works, or both.

Chapter 7/Hunting Out the Real Uncertainty

Chapter index on next page

To go beyond indication, we need to assess the uncertainty of our indications. Although precision of assessment has value, reality of assessment is more basic, because we can be easily misled by variables not represented or recognized in a study.

We assign contributions to uncertainty to two sources: those that might be judged from the data at hand—internal uncertainty; and those that come from causes whose effects are not revealed by the data—supplementary uncertainty. Thus internal and supplementary uncertainty are two vague concepts intended to aid our understanding of uncertainty, variation, and stability. Failure to attend to both sources can lead to serious underestimates of uncertainty and consequent overoptimism about the stability of the indication. To avoid these traps, we need to choose a satisfactory error term from the data, and we need to allow for sources of variation that are present but not made visible by the data-gathering process.

Good design in observational programs and experiments can reduce the impact of all kinds of variation upon the uncertainty of our results. Design can be especially valuable in helping to make sure that major sources of variation are introduced into the investigation. It is often wise to "broaden the base" of a narrowly focused investigation so that the internal uncertainty can properly represent the real variation and the supplementary uncertainty can be reduced.

Once a good design has been executed, we want to get a sound estimate of internal uncertainty from the data at hand—*direct assessment* of uncertainty. Beyond this, we still need to assess the likely magnitude of effects that cannot be examined in the data. Some sources might be: systematic errors of measurement (example: tendency to omit young children from a census); mismatch between sampled and target population (example: in public-opinion polling, the sampled population is not the voting population, and even if it were, the population of opinions prior to the official balloting need not be the population of opinions held in the voting booth); halo effect, where in repeated measurement of the same object the observer tends to agree too closely with his first measurement.

We begin our discussion of internal uncertainty by illustrating how the classical formula

$$\sigma_{\bar{y}} = \frac{\sigma}{\sqrt{n}}$$

may serve us poorly. Next, we note that real or conceptual randomness of some accessible subdivision is, in practice, the basis for the assessment of internal uncertainty. Different levels of grouping offer a basis for the direct assessment of variability, and we offer some guidelines for choosing the appropriate level. After discussing an example, we suggest some major difficulties with direct assessment and point to some solutions.

Turning to supplementary uncertainty, we stress its importance, consider its appraisal, and discuss its combination with internal uncertainty.

7A. How $\sigma/\sqrt{n}$ Can Mislead

Both nonmathematical and mathematical introductory treatments of statistics take pains to emphasize that the standard deviation of the sample mean is $\sigma/\sqrt{n}$, where σ is the population standard deviation and n is the sample size. This idea is most important, and it is part of the basis for the theory of sampling, but it leans, as an introduction must, on an oversimplified view of what is going on. Later on, the analysis of variance may introduce the idea of diverse sources of variation, but we should emphasize early and often the need to allow the data themselves to speak quite directly for their own variation. Peirce's study of reaction time (see Section 1D) again provides an example with both substantive and methodological interest.

From Exhibit 5 of Chapter 1 we draw the following values:

Day:	(1)	2	3	4	5	6 to 24
Observed mean:	(475.6)	241.5	203.1	205.6	148.5	175.6 to 265.5
$s/\sqrt{n}$:	(4.2)	2.1	2.0	1.8	1.6	1.1 to 2.2

Setting aside the first day's results, which are obviously discrepant, we observe that $s/\sqrt{n}$ varies from 1.1 to 2.2. If the values of $s/\sqrt{n}$ measured the standard deviations of the observed means, the variance of a difference from one day to the next would be between

$$(1.1)^2 + (1.1)^2 = 2.42 \qquad \text{for the smallest variability}$$

and

$$(2.2)^2 + (2.2)^2 = 9.68 \qquad \text{for the largest.}$$

These limits correspond to standard deviations of the difference between means for pairs of days of $\sqrt{2.42} \approx 1.6$ and $\sqrt{9.68} \approx 3.1$. If these standard deviations based upon $\sigma/\sqrt{n}$ were appropriate, then the magnitude of most day-to-day differences would have to be less than two standard deviations, or less than 3.2 to 6.2, and that of almost all differences would have to be less than three standard deviations, or less than 4.8 to 9.3. The actual differences for adjacent days, not including the first day (see Exhibit 5 of Chapter 1):

$-38, +2, -57, +27, +11, +7, +2, +20, +1, +19, +9,$

$\qquad\qquad -8, -1, -3, +32, -12, +6, -3, -10, +11, -4, -8$

impolitely pay little attention to such limitations.

In the language of analysis of variance, Peirce's data show considerable day-to-day variation. In the language of Walter Shewhart, such data are "out of control"—the within-day variation does not properly predict the between-days variation. Nor is it just a matter of the observer "settling down" in the beginning. Even after the 20th day he still wobbles.

The wavering in these data exemplifies the history of the "personal equation" problem of astronomy. The hope had been that each observer's systematic errors could be first stabilized and then adjusted for, thus improving accuracy. Unfortunately, attempts in this direction have failed repeatedly, as these data suggest they might. Thus the observer's daily idiosyncrasies need to have account taken of them, at least by assigning additional day-to-day variation. (What about hour-to-hour? Or week-to-week? We can only guess here, since these particular data are tabulated by whole days and do not stretch over many weeks.) The big change from first to all later days is also a common feature of many kinds of data, whose reduction in the main experiment by pilot work and training is often most important.

Wilson and Hilferty (see Section 1D) made it clear that Peirce's data illustrate "the principle that we must have a plurality of samples if we wish to estimate the variability of some statistical quantity, and that reliance on such formula as $\sigma/\sqrt{n}$ is not scientifically satisfactory in practice, even for estimating unreliability of means."

Even in dealing with so simple a statistic as the arithmetic mean, it is often vital to use as direct an assessment of its internal uncertainty as possible. Obtaining a valid measure of uncertainty is not just a matter of looking up a formula.

7B. A Further Example of the Need for Direct Assessment of Variability

Let us turn from a measurement problem to one of counting, where the binomial distribution suggests itself. We tend to think automatically of the binomial variation of counts, with standard deviation for the observed fraction $\sigma = \sqrt{pq/n}$, where n is the sample count and $p = 1 - q$ is the proportion of successes in the population. Again, the idea of this microcosmic standard deviation is important for many purposes, yet it may underestimate actual variation considerably. Let us turn to mass production for an example involving mixing individual differences with the behavior of machines.

If among thousands of manufactured piece parts the observed fraction of defective piece parts is p, and 1000 pieces are produced by one operator on one machine in one shift, it is risky business to suppose that the long-run average proportion of defectives will be between $p - 3\sqrt{pq/1000}$ and $p + 3\sqrt{pq/1000}$. The fraction defective is likely to depend on many things: the day of the week (Mondays being notorious), the operator, the machine, the shift, the supervisor, the inspector, and other plant matters we should not detail here. Appreciating this bramble of sources of variability led Walter Shewhart (1931) to devise methods of quality control with limits $p \pm 3\sqrt{pq/n}$ as an ideal to be nearly achieved only after the most strenuous and sophisticated engineering efforts. What mass production with all its control and measuring ability cannot attain, other fields cannot expect formulas to give. Belief in such formulas may produce fancied security and sad surprises.

Nothing can substitute for relatively direct assessment of variability. In direct assessment, we base differences on large observational groups, differences more nearly representative of the many sources of variation that we must face. In the manufacturing example, we might look at numbers of defectives for several combinations of operators, shifts, machines, and days to get a notion of the variability actual manufacturing produces. (As we discuss in Section 7G, we cannot ordinarily expect the data to tell us about all sources of variation.) In complicated investigations, the many sources of variability oblige us to assess variability directly.

Even if such a multiplicity of sources did not exist, the lack of an appropriate mathematical formula connecting micro-differences to macro-variability would often drive us to direct assessment. Even when one makes drastic oversimplifications (perhaps assuming independence, absence of many known sources of variability, and Gaussian distributions), the corresponding mathematical formula may never have been derived, may require an impractical effort to derive or locate, and may be misleading if found. The need for diverse and complex analyses also forces us to direct assessment.

7C. Choosing an Error Term

A large body of data offers considerable freedom for measuring its internal uncertainty. In a study of a group of educational tests, for instance, we may have had all ninth-grade students in a city school system as subjects. There is a natural hierarchy:

a) *student*   b) *class*   c) *school*   d) *city.*

We could, if we chose, regard those students who were, in the year in question, in the ninth grade in a particular school as a random sample of those who "might" have been there, considering, for instance, the socioeconomic, ethnic, and criminological background of the area from which this school draws its pupils. If we did this for each school in a city and regarded both the set of schools and the city as fixed, we would have an adequately specific model to support an assessment of internal uncertainty. Here we would turn to pupil-to-pupil differences within class as the basis of our measures of stability.

To the extent that our concern is with exactly these schools in this one city, such an assessment may be quite satisfactory. Alternatively, to the extent that broader, more encompassing, assessments are impossible (as when only one school in the city has ninth-grade pupils), such an assessment may be the best that we can do.

If our concern is with the general nature of a broad urban milieu, one in which no particular distribution of, say, socioeconomic and ethnic backgrounds has a distinguished role, then we will do better—in the sense of giving ourselves more useful information—by focusing our attention on school-to-school differences as a basis for assessing internal uncertainties. To do this means, in practice, to act as if the *schools* studied were a random sample from a larger population of schools.

By so doing, we bring into the assessment at least part of the neighborhood-to-neighborhood differences of our broad urban milieu. If there are regional differences, and "our" city belongs to a distinctive region, we will not have adequately represented neighborhood-to-neighborhood differences across regions in our assessment of instability. If, on the other hand, socioeconomic class is the dominant influence, and the fractions of socioeconomic classes for entire cities are nearly constant from city to city, then we may have overrepresented neighborhood-to-neighborhood variation. Our assessment may have imperfections of several kinds, but with data from only one city, it may be the best that can be done.

While it is easy to write down formulas based on other kinds of assumptions, assessment of variability in practice seems to be universally based upon treating suitable units—students, classes, schools, cities, or even the collection of tests being studied—as if they were a random sample. Recognizing this is important, both for making certain general techniques of assessing variability broadly applicable (see Chapter 8) and in making it clear that the practical choice is usually between "treat it as random" and "forget it, sweep it under the rug."

We have been discussing, in a very pragmatic vein, what the analysis-of-variance oriented worker usually calls "the choice of error term."

One might argue that it would be well to restrict the calculation of an indication of instability "as if so-and-so were random" to those cases where so-and-so was indeed random. The writers believe this position often leads to artificially lowered estimates of instability because of the exclusion of sources of variability that were sampled, though perhaps not very randomly or completely. Consequently, we encourage treating effects as randomly sampled in many circumstances where the randomness is at best dubious.

It would be a disservice to leave the impression that the most all-encompassing assessment of internal uncertainty is always the best. Our purposes may make a less encompassing assessment more relevant. If we have studied the *only* three large cities in some state, for instance, then so far as decisions about policies that apply uniformly to all that state's large cities are concerned, an appropriately weighted average of the results for these three cities is the natural guide. Only the uncertainties of the individual studies contribute uncertainty to this guide; city-to-city differences do not contribute further uncertainty because all the cities to be affected have been considered. (For a nationally oriented survey, conducted in Boston, New Orleans, and Seattle, the opposite might be likely.)

Then, too, in many instances, large-scale differences involve so few comparisons as to make any assessment that includes these differences quite unstable. If a study is made in exactly two cities, for instance, it may be best to give assessments of internal uncertainty for the results of each city separately, and for the results for the two cities combined, allowing only for "sampling" *within* the cities, but stating pointedly that further allowance for city-to-city variation must be made.

By following this plan, different readers can combine both informed judgment and the observed city-to-city differences differently so as to assess an appropriate city-to-city contribution to uncertainty. There need not be any one "right" answer. Different purposes may well demand different assessments of instability.

7D. More Detailed Choices of Error Terms

Given a number of groups (cities, years, etc.) the differences between which are to provide an error term, and given a result assessed separately for each, we are likely to proceed by using Student's t to get limits on the effects of internally expressed variability.

Often, we have little choice as to the number of groups to use, but sometimes we are freer to choose. If our principles for the selection of an error term leave us with 100 groups, all equivalent, we could consider assigning these 100 randomly into 20 supergroups of five each or randomly into five supergroups of 20 each. Making fewer supergroups saves computational effort; what will it cost elsewhere?

How many groups should one take? Generally speaking, the more the better, but usually economic and data pressures force one to take few. A look at Exhibit **1** showing the two-sided 5% levels for the t table is enlightening. Note that, when we get to three degrees of freedom, we are already about 89% on the way to an infinite number of degrees of freedom for the 5% point as measured in the t scale. And by 10 df, we are only about 10% worse off on this scale than at ∞ df. (Actually the loss in variance terms, which may be more relevant, nearly corresponds to the square of the ratio of t-table entries.) Consequently, numbers of groups from 4 to 10 are quite practical. Subject to calculational difficulties, larger numbers are preferable. We often use 10 as a rule of thumb.

If groups of the right sort were always evident in the data, and if it were always easy to calculate results based on each group alone, and always efficient to use such results, the discussion of internal uncertainty could end here.

7E. Making Direct Assessment Possible

When the choice of independent groups of data is not automatic, it is often helpful to deliberately construct equivalent subsamples. For example, in making a sample survey from a list, we might draw several systematic subsamples (each consisting of every mth individual), having starting points randomly chosen from the first m on the list. To estimate the population mean μ, we first compute the mean $\bar{y}_i$ for each separate subsample. The means of each of these k subsamples are then treated as single independent measurements. Their mean $\bar{\bar{y}} = \sum \bar{y}_i/k$ estimates the population mean, μ. The estimated variance of $\bar{\bar{y}}$, $s_{\bar{y}}^2 = \sum(\bar{y}_i - \bar{\bar{y}})^2/k(k-1)$, makes it easy to compute confidence limits for μ:

$$\bar{\bar{y}} \pm |t_{k-1}| \, s_{\bar{y}}, \tag{a}$$

where $|t_{k-1}|$ comes from the Student t table with $k - 1$ degrees of freedom for whatever two-sided confidence level has been chosen. In the limits given by expression (a), we have used a direct assessment of the variability of the $\bar{y}_i$.

Do not suppose that $s_{\bar{y}}$ is equivalent to the $s/\sqrt{n}$ deprecated in the Peirce example (Section 7A). True, we are still using the notion of variance to measure variability, but the appraisal of that variability in $s_{\bar{y}}$ comes from differences between good-sized chunks, perhaps "days" in the Peirce data, while that in the Peirce $s/\sqrt{n}$ comes from differences between single measurements.

In the same vein, several small stratified samples could be drawn, each a replicate sample drawn from the whole population by a method that specified a certain sort of stratification. Here, to estimate the population mean, a different $\bar{y}_i$, a weighted estimate for the population mean, might be constructed

Exhibit **1** of Chapter 7

Two-sided 5% Levels for the t Table, $|t_k|_{.95}$

df	$|t|_{.95}$	df	$|t|_{.95}$
1	12.71	6	2.45
2	4.30	7	2.36
3	3.18	8	2.31
4	2.78	9	2.26
5	2.57	10	2.23
		∞	1.96

for each of the stratified samples. Then one would compute confidence limits on the population mean μ based on these $\bar{y}_i$, just as for the limits given by expression (a) above. One advantage of the use of such equivalent subsamples (they are often called "interpenetrating") is the ease with which they can be used to allow for the variation represented, for example, by different interviewers and supervisors, whose services can be assigned to different samples.

Exactly similar techniques apply to many statistics. For an example dealing with estimated slopes (regression coefficients, here linear combinations of observed y's with coefficients depending on the x's), let us consider part of an experiment by Johnson and Tsao (1944; Johnson, 1949).

Johnson–Tsao experiment. The subject holds a ring pulled upwards by one of seven initial weights (100, 150, 200, 250, 300, 350, or 400 grams). By means of liquid and valves, the pull is increased at one of four rates (100, 200, 300, or 400 grams per minute). The subject announces "now" when he notices the increased heaviness. The observation (that is, the just noticeable change) is taken as the change in pull up to the instant the subject reports. Preliminary practice runs accustom the subject to the apparatus and procedure. Johnson and Tsao used 8 subjects, 4 sighted and 4 blind, 4 male and 4 female. Each subject executed the experiment twice. The experimenters randomized the order of procedure for 5 replications of each of the 28 ($= 4 \times 7$) measurements (for each of 8 observers at each session).

A graphical examination of variability as a function of level will show that the logarithms of the observations seem more suitable for the analysis than do the raw observations. The data suggest that, for a fixed rate of change, the observations may not depend much upon the initial weights. Let us look further at this.

For the four sighted subjects and the rate 300 g/min, let us compute the slopes of the regression lines of the common logarithms of the observations upon the coded initial weights (1 corresponds to 100 grams, 7 to 400 grams). Exhibit **2** shows the raw and logarithmic observations. The slopes on x for the four subjects, measured in $\log_{10}$ units per 50g change in initial weight, are: 0.0029, 0.0154, −0.0064, 0.0021. The average slope is $\bar{b} = 0.0035$, and $s_{\bar{b}} = 0.0045$. Consequently, entering a t table with three degrees of freedom, we find the 95% confidence limits for the slope to be 0.0178 and −0.0108.

For the whole change from 100 grams to 400 grams, we would multiply these numbers by 6 ($= 7 - 1$) to get, for example, $6\bar{b} = 0.0210$, 95% interval from −0.065 to +0.107. This corresponds to an estimated change in the raw observation by only about 5% (a factor of antilog$_{10}$ 0.0210 $= 1.05$) with a 95% confidence interval from −14% to +28% (a factor between 0.86 and 1.28). In view of the variation in the data from other sources, there are many purposes for which the variation owing to initial weights is not only not significantly different from zero, but clearly not very important.

Note that the results here obtained by direct assessment of internal uncertainty are exactly those that would have followed from a straightforward complete analysis of variance, in which "slope × subjects" was used as the error term for "slope". Aside from the systematic attitudes it produces, perhaps the greatest single virtue of the analysis of variance is its provision of direct assessments of internal uncertainty.

We can carry out simple direct assessment of internal uncertainty for results that depend on the data in more complex ways than as linear combinations with fixed weights. In some problems such an approach is quite satisfactory; in others the difficulties about to be discussed are serious.

Exhibit **2** of Chapter 7

Johnson–Tsao Experiment

Coded weight	(1) Male	(2) Male	(3) Female	(4) Female

Original data: sighted subjects, 300 g/min

	Grams added before subject says "now"			
1	15.8	35.0	27.2	12.2
2	18.6	39.3	41.1	9.6
3	12.2	47.8	32.2	11.7
4	12.8	38.2	21.3	12.4
5	16.5	57.7	33.7	11.9
6	15.8	39.7	28.2	12.8
7	17.0	44.8	29.6	10.5

Transformed data $\log_{10}$:

x	$y = \log_{10}$ (grams added before subject responds)			
1	1.20	1.54	1.43	1.09
2	1.27	1.59	1.61	0.98
3	1.09	1.68	1.51	1.07
4	1.11	1.58	1.33	1.09
5	1.22	1.76	1.53	1.08
6	1.20	1.60	1.45	1.11
7	1.23	1.65	1.47	1.02

$\sum x = 28$ $\sum y = 8.32$	11.40	10.33	7.44
$\sum xy$ 33.36	46.03	41.14	29.82
$\sum x^2 = 140$ $(\sum x)^2 = 784$	$n = 7$		

7F. Difficulties with Direct Assessment

Besides the question of supplementary variability treated in the next section, the major difficulties with direct assessment of variability, as just described, are these:

a) it may not be feasible to calculate meaningful results from such small amounts of data as properly chosen groups would provide, or

b) although the results would be meaningful, they would be so severely biased as to make their use unwise.

None of these difficulties arises in the examples of the last section, because the results there considered were all linear combinations of the observations with fixed weights. Thus, had we analyzed the same Johnson–Tsao data without separating the four subjects, our first act would have been to form arithmetic means over subjects, and our resulting regression coefficient would have been the arithmetic mean of those for the four individuals. Similar results hold for the particular types of interpenetrating subsampling described in Section 7E. So long as we have linearity with fixed weights, everything is simple.

In Chapter 8, we explain how to handle the more complex cases where difficulties (a) and (b) arise.

Beyond this sort of difficulty, the most prominent problem arises in the presence of two or more separately isolated measures of variability, all of which should contribute to a proper error term. Suppose that we have conducted a study of student reaction to world news in each of 20 schools, widely spread across the country, in each of 10 years. School-to-school differences, embodying regional differences as they do, are surely likely to be substantial and certainly reflect an important kind of variability. Year-to-year changes in the impact of world news cannot be neglected.

Happily, the effects of both of these major sources of uncertainty can be assessed within the data; we would not like to have had to assess either one purely as a matter of judgment. We must face the question of how to use two error terms at the same time. (Accounts of what to do are harder to find than they should be. Those that can be found refer to analysis-of-variance calculations rather than to grand means, but are routinely translatable. For formulas, see Scheffé (1959). For a worked example illustrating the arithmetic, see Cooper (1969). For a rather more complicated example, see Anderson (1947).)

It does not suffice to say that we have 200 groups, each made up of one school for one year, and that we need only use the variation among results for these groups to assess the variability of overall results.

7G. Supplementary Uncertainty and its Combination with Internal Uncertainty

We want to be ready to make allowance for the effects of systematic error and of sources of variability excluded from our assessment of internal uncertainty.

If our observations are confined to one city, city-to-city variation is not revealed in our data and cannot contribute to any assessment of internal uncertainty. (If only two cities are involved, we have already seen that it may be wise to leave city-to-city variation to judgment and include its effects in supplementary uncertainty.) Similar statements can apply to years, regions, and many other aspects of our data.

Besides the variations that might have been evident from more extensive data, some deviations are intrinsic in the way that the data were gathered. The instruments used, whether paper-and-pencil tests, questionnaires, or mercury-in-glass thermometers, are subject to imperfections of calibration and to responsiveness to other variables than the one they seek to measure.

As a homely example, consider the market-research analyst who plans to ask his respondents to perform an extra task beyond the initial questioning. In the pilot work, he finds that the interviewers report no difficulty about persuading the respondents to do the extra task. Indeed, of the respondents thus far approached, only 2 of 50 have failed to respond to either the questionnaire or the task. How shall he suppose matters are going to work out in the actual survey? He can be helped if he knows, for example, that, in his sort of survey, easy questionnaires alone lead to 15% nonresponse even with several callbacks. From his 96% indication, he is reduced at once to 85%. Next, he must consider a further discount for the extra task, but how much may depend heavily upon the enticements offered for the respondent's cooperation. At any rate, he should probably not be surprised by at least a further 15% loss.

(If the response rate is important, the analyst may design the pilot study with randomly chosen respondents and superimpose different incentives to discover what response rates they yield.)

Such sources of supplementary uncertainty should not be neglected. The fact that they have to be assessed by judgment, sometimes tempered by data from other sources, is no excuse for pretending they do not exist. Nor is the evidence that they are usually underestimated (even by physicists assessing their most fundamental constants; see DuMond and Cohen, 1958) any reason not to try to do as well with them as considered judgment permits.

How are we to do all this? As something wholly separate from our assessment of internal uncertainty? Or as something to be combined with the latter? The writers would like to be able to combine assessments of supplementary uncertainties and systematic errors with those of the internal uncertainties. In the end, all are matters of judgment. The investigator may find it worthwhile to communicate the internal uncertainty as well as the combined uncertainty; when the combining is something that only others can do, it may be better merely to communicate the separate components.

How is combination done in practice?

It is easy to expand an estimated standard deviation to an estimated root-mean-square error. One squares the standard deviation and adds on the square of the bias, and then takes the square root. Sometimes one can reduce an actual sample size to an effective sample size.

We can relatively easily combine supplementary uncertainties with results expressed as statements of confidence intervals. Suppose that we wish to add a "cookie cutter" type of uncertainty, such as "anything between −4% and +1% could be present". If our confidence interval already runs from 62% to 70%, we have only to move our cutting points outward to 58% and 71%. Similarly, if we want to make a significance test, we may wish to move the null hypothesis first one way and then the other, by a corresponding amount. If we wish to add a "distributional" type of uncertainty, such as "the systematic error tends to follow a normal distribution with average −1% and a standard deviation 3%", we will have to paste this onto our model and adjust the results somewhat more subtly. If the confidence interval is already based on a standard error of 2%, as in our 62% to 70% example, we could add the variances $2^2 + 3^2 = 13$, and get a pooled estimate of 3.6. The new center is $68\% - 1\% = 67\%$. Our final "confidence limits adjusted for supplementary uncertainty" would be about 67% ± 7.2%, or an interval from 59.8% to 74.2%.

If the supplementary uncertainty to be assessed by judgment is appreciable compared to the variability to be expected in repetitions of the overall study, then precisely what our internal "probability statements" are about is no longer important. Here, it is essential to describe *both* combined *and* internal uncertainty. Readers and hearers are entitled to see what their own judgments would lead to. When we wish to communicate an overall uncertainty, we ought to do so, and our use of probability statements should be kept flexible enough to enable us to do so.

However such details are to be handled, investigators and writers, and their statistical advisors, have a continuing and serious obligation to plan to assess supplementary uncertainties every time they assess an internal uncertainty. The results may come out in terms of "words of warning", rather than in terms of numbers. Words are often an acceptable minimum, but all of us should try to do better wherever we can. We owe supplementary and overall assessments of uncertainty to our readers, and to the researchers that come after us, even when we have made a direct assessment of internal uncertainty.

Although good methods of assessing supplementary uncertainties seem deeply bound to the subject matter of the analysis, extensive discussions by subject-matter experts may help statisticians find further methods of broad applicability. The area needs development.

In closing, let us emphasize that, for very different reasons, both the tightly controlled laboratory study, and the large study that has had to have play in its methods before it can be done at all, have special need for the appraisal of supplementary uncertainty.

Summary: Hunting out the Real Uncertainty

All results involve two kinds of uncertainty: internal uncertainty, uncertainty that can be assessed from the data, and external uncertainty, uncertainty that cannot.

$\sigma/\sqrt{n}$ may not be the correct standard error when n observations, of apparent variance σ^2, contribute equally to an arithmetic mean, and the same can be said for $\sqrt{pq/n}$.

A large, well-structured body of data offers a variety of measures of internal uncertainty based on comparisons among more or less closely related portions of the data. These measures are often of differing size. The choice among them (i) can be an important matter, and (ii) can be thought about quite rationally.

We can use equivalent subsamples, sometimes identified after the fact, as a basis for assessing internal uncertainty.

By using fewer comparisons and thus fewer degrees of freedom, we incur a cost, in increased variability of our assessment, when assessing internal uncertainty.

The major difficulties with equivalent subsample assessment of uncertainty (possible lack of definition, probable bias) are associated with small subsamples.

Assessing internal uncertainty when a result summarizes over two or more kinds of classification is not a trivial matter; in particular it will often neither suffice to add up the uncertainties exposed by the classifications separately, nor to treat the combination of classifications as a single classification.

It is appropriate to assess supplementary (external) uncertainty by informed judgment, and to combine internal and external uncertainties into a single figure, doing both as carefully as we reasonably can.

We frequently need to assess supplementary uncertainty, whether the data be experimental or observational.

References

Anderson, R. L. (1947). "Use of variance components in the analysis of hog prices in two markets." *J. Amer. Stat. Assoc.*, **42**, 612–634.

Cooper, B. E. (1969). *Statistics for Experimentalists.* Elmsford, N.Y.: Pergamon Press, Inc.; p. 167.

DuMond, J. W. M., and E. R. Cohen (1958). "Fundamental constants of atomic physics." In E. U. Condon and H. Odishaw (Eds.), *Handbook of physics.* New York: McGraw-Hill (LC: 57–6387); pp. 7–143 to 7–173.

Johnson, P. O. (1949). *Statistical methods in research.* New York: Prentice-Hall; p. 299.

Johnson, P. O., and F. Tsao (1944). "Factorial design in the determination of differential limen values." *Psychometrika*, **9**, 107–144.

Scheffé, H. (1959). *The Analysis of Variance.* New York: John Wiley and Sons; p. 247.

Shewhart, W. A. (1931). *Economic Control of Quality of Manufactured Product.* New York: Van Nostrand; p. 361.

Chapter 8/A Method of Direct Assessment

Chapter index on next page

8A. The Jackknife

For statistics more complicated than weighted averages, we are likely to find difficulties in assessing stability, even when we have moderately large amounts of data. Thus, for example, in fitting a multiple regression on k independent variables, one needs at least $k + 1$ data points, and not many people would be pleased to work with so few. Consequently, if a substantial body of data is needed in each group, the number of possible groups of data available for direct appraisal of variability by the standard method described in Section 7E may be severely restricted. Second, many statistics based on small samples give biased estimates; typically the leading term in the bias is proportional to $1/n$, where n is the sample size. Consequently, the mean of results based on several subsamples is likely to be more biased than is a single result based on all the data, at least to the extent that the individual samples are small. A method with wide application, intended to ameliorate these problems, is the jackknife.

The name "jackknife" is intended to suggest the broad usefulness of a technique as a substitute for specialized tools that may not be available, just as the Boy Scout's trusty tool serves so variedly. The jackknife offers ways to set sensible confidence limits in complex situations. The basic idea is to assess the effect of each of the groups into which the data have been divided, not by the result for that group alone, which we used in Section 7E, but rather through the effect upon the body of data that results from *omitting* that group.

An illustration will speed understanding. For a simple mean of five numbers, any single value can be easily expressed as the weighted difference of two means, the mean of all five values and the mean of the four other than the selected value. Thus, for example, for the values 3, 5, 7, 10, 15, we can represent 7 by

$$5\left(\frac{3 + 5 + 7 + 10 + 15}{5}\right) - 4\left(\frac{3 + 5 + 10 + 15}{4}\right) = 7.$$

This result is not only trivial to prove, but appears to have trivial consequences. For means of equally weighted numbers, the consequences are indeed trivial. But when we deal with more complex statistics, the analogous computation does not give us the same result as we get by applying "the same" calculation to the individual pieces. Instead, it gives us something much more useful. In particular, as we see later, such complex statistics might even be regression equations rather than mere numbers.

The two bases of the jackknife are that we make the desired calculation for all the data, and then, after dividing the data into groups, we make the calculation for each of the slightly reduced bodies of data obtained by leaving out just one of the groups.

Let, then, $y_{(j)}$ be the result of making the complex calculation on the portion of the sample that omits the jth subgroup, that is, making it on a pool of $(k - 1)$ subgroups. Let y_{all} be the corresponding result for the entire sample, and define *pseudo-values* by

$$y_{*j} = ky_{all} - (k - 1)y_{(j)}, \qquad j = 1, 2, \ldots, k. \tag{1}$$

These pseudo-values now play the role originally played in Section 7E by the results of making the computation for each group separately. Note that, as in the example with five numbers, when the calculation reduces to forming a mean using equal weights, we have $y_{*j} = y_j$, where y_j is the result for the jth piece alone. Accordingly, for simple means, the use of the jackknife technique reduces to the technique of Section 7E, as we would hope.

The final accuracy required for the y_{*j} is just about what would be needed for the y_j. Because of the multiplications by k and by $k - 1$, which may be large, one usually needs to compute y_{all} and the $y_{(j)}$ to more decimals than would be needed if they were to be used directly. Accordingly, the y_{*j} are particularly sensitive to computational errors or rounding changes in y_{all} and $y_{(j)}$, although their sensitivity to data variability is ordinarily, if anything, a little less than that of the y_j they replace.

The key idea is that, in a wide variety of problems, the pseudo-values can be used to set approximate confidence limits, using Student's t, as if they were the results of applying some complex calculation to each of k independent pieces of data. The words "as if" are vital here; Student's t still performs well in many circumstances where the y_{*j} deviate substantially from independence.

The jackknifed value y_*, which is our best single result, and an estimate s_*^2 of its variance are thus given by:

$$y_* = \frac{1}{k}(y_{*1} + \cdots + y_{*k}),$$

$$s^2 = \frac{\sum y_{*j}^2 - \frac{1}{k}(\sum y_{*j})^2}{k - 1},$$

$$s_*^2 = s^2/k.$$

(When the $y_{(j)}$ are rounded, or otherwise quantized, after or during calculation, Tukey (unpublished manuscript) has suggested the conservative practice of increasing s_*^2 by $k^2\tau^2$, where τ^2 may often be taken as the variance of a uniform distribution over the rounding or quantizing interval. Thus, if we rounded $y_{(j)}$ to 3 decimals, the corresponding displacement would have a uniform distribution over a range of 0.001, that is, from $x - 0.0005$ to

$x + 0.0005$, where x is the rounded value. Applying the usual formula $L^2/12$ for the variance of a rectangular distribution of length L, we would have $\tau^2 = (0.001)^2/12 = 0.00000008$. Unless k is large, or s_*^2 is very small, we need pay little attention to the $k^2\tau^2$ term for so small a τ^2. It costs little to compare $k^2\tau^2$ with s_*^2.)

Adjusting degrees of freedom

Difficulties with the jackknife seem to arise most frequently from two causes:

◇ excessively straggling tails,

◇ discreteness of the values produced.

No one has found a good solution for excessive straggling that applies in any circumstances or for any method of analysis; if we are willing to make a corresponding change in what we are estimating, the methods discussed in Chapter 10 may suffice to handle the difficulty. If we must stay with a more classical estimate, for example, that pointed to by the arithmetic mean, no one has found anything to help.

We can do something about discreteness. Let us look at a natural and very extreme case. Suppose that we jackknife the median of a sample of even size, say $k = 2m$. If we delete any observation in the upper half of these $2m$ values, the median of what is left will be the mth value, counting up from the lowest. If we delete any observation in the lower half, the median will be the $(m + 1)$st value in the original sample.

Accordingly the $y_{(j)}$ have but two different values, taking on each m times, and the same holds for the pseudo-values. The stability of s_*^2 is not greater than that corresponding to a squared difference between 2 values. This would lead us to use only 1 degree of freedom for t's involving such an s.

Can we obtain a more stable s_*? By jackknifing in groups, to be discussed in Section 8C, we can arrange to have more than 2 different pseudo-values appear. This can be very valuable, since the only advantage to using individuals in ordinary situations is to preserve degrees of freedom, something that cannot actually be done in this example.

Clearly a rule of thumb for reducing degrees of freedom is needed. The following simple rule should be helpful:

a)   Count the number of different numbers appearing as pseudo-values, subtract one, and use the result as degrees of freedom.

This rule is to be used only when the sameness of the pseudo-values arises from the definition of the computation, as in the case of a median or a range, and not when it arises from the nature of the basic observations or the conduct of our arithmetic. Specifically:

b) If slight changes in the basic observations—as when values that by their nature are either 0 or 1 are made -0.001, $+0.002$, 0.997, or 1.004— would make two pseudo-values different, they should *not* be considered "the same" in applying rule (a). Example: % of successes in binomial observations.

c) If carrying more decimals in the computation would have made two pseudo-values different, they should *not* be considered "the same" in applying rule (a).

APPENDIX TO 8A: Combinations and Re-expressions

The results of data analysis are not always single numbers. When we deal with several numerical results, say y, z, v, and w, we can use a single choice of k pieces and go through the jackknife computation separately for each result. Having matched sets of pseudo-values $(y_{*j}, z_{*j}, v_{*j}, w_{*j})$ for each j, we can easily find a set of k illustrative values for any combination or function of these results, forming, for example,

$$\frac{y_{*j} + z_{*j}}{v_{*j} + w_{*j}}$$

to tell us something about the estimand corresponding to

$$\frac{y + z}{v + w}.$$

Similarly, we consider the log y_{*j} as telling us about the estimand corresponding to log y. This procedure extends to combinations that depend on an auxiliary variable or variables, as in

$$ye^{-zt}$$

or in

$$y \cdot x_1 + z \cdot x_2 + v \cdot x_3 + w \cdot x_4,$$

for we can consider such illustrative functions as

$$y_{*j}e^{-z_{*j}t}$$

and

$$y_{*j} \cdot x_1 + z_{*j} \cdot x_2 + v_{*j} \cdot x_3 + w_{*j} \cdot x_4.$$

In thinking about such questions of combination and transformation, we need to bear in mind that the order of doing things will generally matter. Thus, for example, we will almost always have

$$(\log y)_{*j} \neq \log (y_{*j}),$$

although these two expressions are likely to be rather similar. We must, by definition, have

$$(\log y)_{(j)} = \log (y_{(j)}).$$

This identity, however, does not help, because

$$(\log y)_{*j} = k \cdot (\log y)_{\text{all}} - (k - 1)(\log y)_{(j)}$$
$$= k \cdot \log (y_{\text{all}}) - (k - 1) \log (y_{(j)})$$
$$= k \cdot \log (y_{\text{all}}) - (k - 1) \log \left(\frac{k \cdot y_{\text{all}} - y_{*j}}{k - 1}\right),$$

which need not equal $\log (y_{*j})$. When, for example, $k = 2$, $y_{\text{all}} = 4$, and $y_{*j} = 3$, the displayed expression becomes $2 \log 4 - \log 5 = \log 3.2$, which is not the same as $\log 3$. Two corollaries are worth noting:

⋄There may be some advantage in jackknifing one expression of a given result rather than another (as when we jackknife $\log y$ or y^2 instead of y).

⋄If we deal with a *linear* combination of results, as in

$$y - 3z + 2v,$$

or in

$$y + zt + vt^2 + wt^3,$$

where t is an auxiliary variable, or in

$$y \cdot x_1 + z \cdot x_2 + v \cdot x_3 + w \cdot x_4,$$

where x_1, x_2, x_3, and x_4 are auxiliary variables (regression variables), the order of operation will *not* matter, so that jackknifing y, z, v, w separately, using the same pieces, is enough to deal with all combinations. Consequently when considering many possible sets of x's, jackknifing the coefficients is economical.

We know little about which choices of expression tend to polish up the behavior of the jackknife. What evidence we have suggests that:

⋄It is very desirable to *avoid* situations where the sampling distribution of the quantity jackknifed has an abrupt terminus or where the possible values of its estimand are restricted to an interval or half-line. For example, if one jackknifes estimates of probabilities, he might find a few final results negative or greater than 1. One possible approach would be to jackknife the logit, $\log [p/(1 - p)]$, and then form the antilogit of the final result. This would keep the numbers in bounds.

⋄It is desirable to *avoid* sampling distributions with one or more straggling tails.

⋄It is probably desirable to *avoid* markedly unsymmetrical sampling distributions.

In summary, we can use the jackknifing of several numerical results to tell us about any combination of these results. Our conclusions will usually differ somewhat from those reached by jackknifing that combination directly. This offers us choices that can sometimes allow us to improve our conclusions. If we deal with linear combinations, as, for example, in the relation of multiple regression equations to the coefficients that appear in them, these differences disappear.

The last two paragraphs of Chapter 2 may now be profitably reread, with "jackknifing" substituted for "cross-validation".

8B. Examples with Individuals

We plan four examples. The first is a toy to get ideas in mind and deals with inference about the standard deviation from a skewed and straggling-tailed distribution. The second example is rather clean and simple, the problem it treats is nonstandard, and theory for treating it is not generally available. On the other hand, it is not a problem where the jackknife does its best work, but without it . . . ? The third deals with a small survey example where it is convenient to use groups, each consisting of several units. The fourth treats a complicated multiple-response problem in some detail. It offers more than one illustration of the power of the method in a problem where indication would otherwise have been the most we would have hoped to seek.

Example (**first of 4**). *Confidence limits on a standard deviation.* A sample from a distribution produced the 11 values

$$0.1, \quad 0.1, \quad 0.1, \quad 0.4, \quad 0.5, \quad 1.0, \quad 1.1, \quad 1.3, \quad 1.9, \quad 1.9, \quad 4.7.$$

There is no reason to suppose that the distribution is normal and some reason to suppose it is not. Set confidence limits on the standard deviation σ.

First solution. Since the data are few, let us use each measurement as a group of size 1. Call the measurements $x_1, x_2, \ldots, x_{11}$. Since each group is of size 1, we could not compute a sample standard deviation for a group. We compute standard deviations first for all the measurements together—that is, for all the groups together:

$$y_{\text{all}} = \sqrt{\sum(x_i - \bar{x})^2/10},$$

shown at the head of column 3 of Exhibit **1.** Then, leaving the jth measurement out, that is, leaving the jth group out, we compute

$$y_{(j)} = \sqrt{\sum(x_i - \bar{x}_{(j)})^2/9},$$

where $\bar{x}_{(j)}$ is the mean of the 10 x's that exclude the jth, and the summation in $y_{(j)}$ omits $(x_j - \bar{x}_{(j)})^2$. We compute this for each of the 11 measurements (groups). These values are also shown in the third column of Exhibit 1. From these we then compute pseudo-values

$$y_{*j} = 11y_{\text{all}} - 10y_{(j)},$$

shown in the fourth column of the table.

Our estimate of σ is y_*, the average of these pseudo-values, which turns out to be about 1.49. The details are shown in Exhibit 1, together with the 2/3 confidence interval for σ: from 0.85 to 2.13, and the 95% confidence interval: from 0.10 to 2.88. Since we happen to secretly know that these data came from an exponential distribution with mean and standard deviation both unity, we need not be displeased with these limits.

Second solution. We are not compelled to jackknife s itself, as we just did. We might, for example, jackknife log s instead. Thinking about this possibility, we are encouraged to do just this, because the sampling distribution of log s is

Exhibit **1** of Chapter 8

Jackknife for a sample standard deviation, first solution (y_{all} = sample standard deviation = 1.34347)

$\lfloor j \rfloor$	$\lfloor x_j \rfloor$	Standard deviations omitting x_j $y_{(j)}$	$\lfloor y_{*j} = 11y_{\text{all}} - 10y_{(j)} \rfloor$
1	.1	1.36382	1.1400
2	.1	1.36382	1.1400
3	.1	1.36382	1.1400
4	.4	1.38888	.8894
5	.5	1.39539	.8243
6	1.0	1.41457	.6325
7	1.1	1.41578	.6204
8	1.3	1.41563	.6219
9	1.9	1.39427	.8355
10	1.9	1.39427	.8355
11	4.7	.70742	7.7040
	13.1		$y_* = 1.4894$

$s_* = 0.6244$

Two-sided confidence limits on σ:

2/3: $1.4894 \pm |t_{10}|_{2/3}s_* = 1.4894 \pm 1.02(0.6244)$, or the interval from 0.85 to 2.13.

95%: $1.4894 \pm |t_{10}|_{.95}s_* = 1.4894 \pm 2.23(0.6244)$,   or the interval from 0.10 to 2.88.

known to be better behaved than that for *s*. Specifically, the distribution of the logarithm of *s* is usually more nearly symmetrical and has less straggling tails than the sampling distribution of *s*. The logarithm might be more biased; but unbiasedness is not of major importance, particularly in view of the bias-reducing feature of the jackknife.

The details and results are shown in Exhibit **2.** We have introduced *Y*'s for the logged values of the *y*'s from Exhibit 1. The 2/3 confidence interval for σ now runs from 0.94 to 3.29, while the 95% interval runs from 0.44 to 6.93.

Unsatisfactory solutions. By way of contrast, let us display some solutions which treat the data as if they came from a normal distribution. The overall s^2 is 1.805; a chi-square table gives the following % points for 10 degrees of freedom:

	2.5%	$\frac{1}{6}$	$\frac{1}{2}$	$\frac{5}{6}$	97.5%
Values	3.25	5.78	9.34	14.15	20.48

Exhibit **2** of Chapter 8

Jackknifing of $\log_{10} s$ for the data treated in Exhibit 1 ($Y_{all} = \log_{10} y_{all} = 0.12823$)

j	$Y_{(j)} = \log_{10} y_{(j)}$	$Y_{*j} = 11 \log_{10} y_{all} - 10 \log_{10} y_{(j)}$
1	.13476	.06293
2	.13476	.06293
3	.13476	.06293
4	.14266	−.01607
5	.14470	−.03647
6	.15062	−.09567
7	.15100	−.09947
8	.15095	−.09897
9	.14435	−.03297
10	.14435	−.03297
11	−.15032	2.91373

$$Y_* = 0.24454 = \text{Mean}$$

$s_*^2 = 0.071605, \quad s_* = 0.2676$

2/3 limits for $\log_{10} \sigma$: $0.2445 \pm 1.02(0.2676)$, or the interval from −0.028 to 0.517

95% limits for $\log_{10} \sigma$: $0.2445 \pm 2.23(0.2676)$, or the interval from −0.352 to 0.841

2/3 confidence interval for σ: from 0.94 to 3.29

95% confidence interval for σ: from 0.44 to 6.93

Accordingly the usual symmetric 95% confidence limits for σ^2 are

$$\frac{1.805}{20.48/10} = 0.88 \quad \text{and} \quad \frac{1.805}{3.25/10} = 5.55,$$

while the 2/3 limits are, similarly, 1.28 and 3.12. The corresponding confidence intervals for σ run from 0.94 to 2.36 and from 1.13 to 1.77. These limits may be optimistically short.

Similarly, using the range $w = 4.7 - 0.1 = 4.6$ and % points for w/σ in normal samples of 11, namely:

	2.5%	$\frac{1}{6}$	$\frac{1}{2}$	$\frac{5}{6}$	97.5%
Values	1.78	2.41	3.12	3.93	4.86

we find intervals for σ from 0.95 to 2.58 at 95% and from 1.17 to 1.91 at 2/3 confidence.

Comments. The four sets of solutions compare as follows:

Source	2/3 limits	95% limits
jackknifing s	$0.85 \leq \sigma \leq 2.13$	$0.10 \leq \sigma \leq 2.88$
jackknifing log s	$0.94 \leq \sigma \leq 3.29$	$0.44 \leq \sigma \leq 6.93$
s^2/σ^2	$1.13 \leq \sigma \leq 1.77$	$0.94 \leq \sigma \leq 2.36$
w/σ	$1.17 \leq \sigma \leq 1.91$	$0.95 \leq \sigma \leq 2.58$

The comparison between the two sets of jackknife limits shows fair similarity for the 2/3 limits and considerable difference for the 95% limits. A moment's reflection on the fact that the lower limit for σ based on jackknifing log s cannot be negative, while that based on jackknifing s can, offers one reason for preferring to jackknife log s. Nevertheless, jackknifing s itself did nicely in this example. Jackknifing $\sqrt{s}$ or $\sqrt[3]{s^2}$ is another possibility.

For the exponential distribution that we actually sampled, Var $s \approx 2(\sigma^2/n)$, whereas, for the normal distribution, Var $s \approx \frac{1}{2}(\sigma^2/n)$. The ratio of these quantities measures the relative variabilities of sample standard deviations drawn from the different distributions. Consequently, the variability in variance terms for the kind of sample we are dealing with is four times that given by normal theory. We might expect, then, a factor of roughly $\sqrt{4} = 2$ in the lengths of the confidence intervals. Consequently, normal theory directly applied cannot possibly give nearly valid confidence limits, and we must discard the last two solutions as unsatisfactory.

Example (second of 4). Estimating the 10% point of a pool of populations. Each individual in a universe has associated with it a population of measurements. For each of a sample of 11 individuals, a group of 5 measurements is taken. Exhibit **3** shows the measurements (hypothetical) arranged from greatest to least for each individual. Estimate the upper 10% point of the total population of measurements.

Solution. A variety of ways could be used to estimate the 10% point. One method pools the groups, estimates that the space between a pair of measurements contains $1/(n + 1)$ of the total distribution, and interpolates to 10%. For example, 5 measurements divide a population into 6 parts, each of which contains, on the average, 1/6 of the probability. With only 5 measurements, however, the upper 10% point is hard to estimate, because the estimate falls naturally into the block above the largest observation, which is often infinitely long. Here is another example where the pooling inherent in combining all pieces but one helps.

Our numerical work can be simplified by collecting, as in Exhibit **4,** a few of the largest observations. First, let us estimate the 10% point for the full sample of 55 points. Fifty-five points yield 56 blocks, and we want to include $56/10 = 5.6$ blocks, counting from the top. In Exhibit 4, we want y_{all} to be 0.6 of the way from the 5th to the 6th measurement. Thus,

$$5.172 - 5.137 = 0.035, \qquad 0.6 \times 0.035 = 0.021, \qquad 5.172 - 0.021$$
$$= 5.151 = y_{all}.$$

Exhibit **3** of Chapter 8

Measurements for 11 individuals

					Individual					
1	2	3	4	5	6	7	8	9	10	11
6.880	4.660	6.950	4.756	4.411	4.257	2.642	12.541	8.404	3.262	3.286
5.172	4.522	3.948	3.792	4.357	3.572	2.276	4.081	5.137	2.874	2.858
3.598	3.403	3.062	2.458	3.571	1.809	2.007	3.853	3.172	2.120	2.787
3.034	3.211	2.906	.412	2.983	1.801	1.922	.364	1.432	1.456	2.752
.628	.070	.482	−.458	1.825	1.480	1.588	−2.945	−1.415	.780	2.047

Exhibit **4** of Chapter 8

Largest 8 measurements ranked and identified

Rank	Size	Individual
1st	12.541	8
2nd	8.404	9
3rd	6.950	3
4th	6.880	1
5th	5.172	1
6th	5.137	9
7th	4.756	4
8th	4.660	2

When the first individual is omitted, we have to compute a $y_{(1)}$ based on 50 points or 51 blocks. We thus want to count down $51/10 = 5.1$ blocks. Omitting individual 1 eliminates two measurements, 6.880 and 5.172, from Exhibit 4, because both measurements come from that group. We interpolate between the 5th and 6th remaining measurements, finding

$$4.756 - 4.660 = 0.096, \qquad 0.1(0.096) = 0.0096 \approx 0.010,$$

and

$$y_{(1)} = 4.756 - 0.010 = 4.746.$$

Exhibit 5 shows the resulting $y_{(j)}$, y_{*j}, y_*, s_*, and the 95% confidence limits, where 2 degrees of freedom have been used since only 3 different values arise. Note that the estimate $y_* = 5.874$ is considerably larger than that obtained directly from $y_{all} = 5.151$. In this particular problem, we secretly know that the 10% point is 5.773, because the populations associated with individuals were constructed to be normal with $\mu = 2$, 3, or 4 with probability 1/3 and independently $\sigma = 1$, 2, 3 with probability 1/3. Thus, we had nine

Exhibit 5 of Chapter 8

Jackknife quantities

$|y_{all}|$	5.151	Pseudo-values
$y_{(1)}$	4.746	$y_{*1} = 9.201$ ($= 11(5.151) - 10(4.746)$)
$y_{(2)}$	5.168	$y_{*2} = 4.981$
$y_{(3)}$	5.099	$y_{*3} = 5.671$
$y_{(4)}$	5.168	$y_{*4} = 4.981$
$y_{(5)}$	5.168	$y_{*5} = 4.981$
$y_{(6)}$	5.168	$y_{*6} = 4.981$
$y_{(7)}$	5.168	$y_{*7} = 4.981$
$y_{(8)}$	5.099	$y_{*8} = 5.671$
$y_{(9)}$	4.746	$y_{*9} = 9.201$
$y_{(10)}$	5.168	$y_{*10} = 4.981$
$y_{(11)}$	5.168	$y_{*11} = 4.981$
Total	55.866	Total $= 64.611$
Total/11	5.0787	Total/11 $= 5.874 = y_*$

Check: $y_* = 11(5.151) - 10(5.0787) = 5.874$

$$s^2 = 2.78024$$

$$s_*^2 = 2.78024/11 = 0.25275$$

$$s_* = 0.503$$

$$y_* \pm |t_2|_{.95} s_* = 5.874 \pm 4.30(0.503), \quad \text{or 95% confidence interval from 3.71 to 8.04}$$

different equally likely normal populations. We chose one at random with replacement for each of the 11 individuals.

This example is one for which the jackknife is not especially well suited, for it deals rather repeatedly with single order statistics. The repetitions of the value 4.981 in Exhibit 5 are symptomatic of the difficulty. Generally speaking, the variations of maxima and minima and of ranges depend heavily on the exact shape of the underlying distribution. Accordingly, it is probable that robustness of validity cannot be had for any confidence procedure concerning their values. Even in such circumstances, the jackknife is often as good a procedure as we have available. An approximate idea of uncertainty is better than none.

8C. Jackknife Using Groups: Ratio Estimation for a Sample Survey

In practice, we usually divide our data less completely, working with and comparing groups made up of more than one individual or case.

Example (third of 4). *Ratio estimate.* In expounding the use of ratio estimates, Cochran (1953, p. 113; 1963, p. 156) gives sizes (number of inhabitants) in 1920 and 1930 for each city in a random sample of 49 drawn from a population of 196 large U.S. cities. Exhibit **6** repeats his values, and totals them, first by 7's and then overall. For computation on such an example by paper, pencil, and book of tables, it is worth noting that, so far, no extra work is involved, since these short additions are so much easier to check than the adding for all 49 cases would be.

Exhibit **6** of Chapter 8

Population of 49 large cities in 1920 (x) and 1930 (y) in thousands of people.

1st 7		2nd 7		3rd 7		4th 7		5th 7		6th 7		7th 7		Summary	
x	y	x	y	x	y	x	y	x	y	x	y	x	y	x	y
76	80	120	115	60	57	44	58	38	52	71	79	36	46	(751)	(915)
138	143	61	69	46	65	77	89	136	139	256	288	161	232	(977)	(1122)
67	67	387	459	2	50	64	63	116	130	43	61	74	93	(965)	(1243)
29	50	93	104	507	634	64	77	46	53	25	57	45	53	(385)	(553)
381	464	172	183	179	260	56	142	243	291	94	85	36	54	(696)	(881)
23	48	78	106	121	113	40	60	87	105	43	50	50	58	(830)	(937)
37	63	66	86	50	64	40	64	30	111	298	317	48	75	(450)	(611)
Totals 751	915	977	1122	965	1243	385	553	696	881	830	937	450	611	5054	6262

The formula for the ratio estimate of the 1930 population total is

$$\frac{(1930 \text{ sample total})}{(1920 \text{ sample total})} \times (1920 \text{ population total}),$$

so that the logarithm of the estimated 1930 population total is given by

$$\log\left(\begin{array}{c}1930 \text{ sample} \\ \text{total}\end{array}\right) - \log\left(\begin{array}{c}1920 \text{ sample} \\ \text{total}\end{array}\right) + \log\left(\begin{array}{c}1920 \text{ population} \\ \text{total}\end{array}\right).$$

Consequently, we find it natural to work with, and jackknife,

$$z = \log(1930 \text{ sample total}) - \log(1920 \text{ sample total}),$$

since this choice minimizes the number of multiplications and divisions.

Further computation is shown in Exhibit **7,** where, in the "All" column, the numbers 5054 and 6262 come direct from the previous exhibit and, in the "$i = 1$" column, the numbers $4303 = 5054 - 751$ and $5347 = 6262 - 915$ are the results of omitting the first seven cities; and so on for the other columns. Five-place logarithms have obviously given more than sufficient precision, so that the pseudo-values of z are conveniently rounded to 3 decimals. To be able to continue easily with hand calculation, an arbitrary central value of 0.100 was subtracted from each z_{*i} and the result multiplied by 1000. These working values are used for the calculation of numerical values for

$$95\% \text{ limits} = \text{mean} \pm \text{allowance}.$$

Exhibit **8** gives all the remaining details. The resulting point estimate is 28,300, about 100 lower than the unjackknifed estimate. (Since the correct 1930 total is 29,351, the automatic bias adjustment, which is effective only on the average, did not help in this instance.) The limits on this estimate are ordinarily somewhat wider than we would get if we had used each city as a separate group, since

$$|t_6|_{.95} = 2.447, \qquad |t_{47}|_{.95} = 2.012.$$

The standard error found here was $\pm.0125$ in logarithmic units, which converts to about ±830 in total (antilog $0.0125 \approx 1.0292, \quad 0.0292 \times 28,300 \approx 830$). The agreement of this value, on only 6 degrees of freedom, with Cochran's value of 604 (1953, p. 119; 1963, p. 163) is fair.

As we promised in Section 2B, we shall now develop a method that will allow the investigator who is comparing two forms of a projective test to do better than indication.

The experiment must involve a number of subjects; we want to generalize to a large class of similar subjects. We need only divide the subjects into a suitable number of groups, and jackknife the whole calculation, obtaining a jackknifed estimate, and jackknifed confidence limits for the difference of the averaged reliability coefficients according to the two scorings. While a moder-

Exhibit **7** of Chapter 8

Details of jackknifing the ratio estimate based on Exhibit 6

	all	$i = 1$	$i = 2$	$i = 3$	$i = 4$	$i = 5$	$i = 6$	$i = 7$
$x_{(i)}$ (1920 sample)	5054	4303	4077	4089	4669	4358	4224	4604
$\log x_{(i)}$	3.70364	3.63377	3.61034	3.61162	3.66922	3.63929	3.62572	3.66314
$y_{(i)}$ (1930 sample)	6262	5347	5140	5019	5709	5381	5325	5651
$\log y_{(i)}$	3.79671	3.72811	3.71096	3.70062	3.75656	3.73086	3.72632	3.75213
$z_{(i)} = \log [y_{(i)}/x_{(i)}]$	.09307	.09434	.10062	.08900	.08734	.09157	.10060	.08899
$z_{*i} = 7z_{all} - 6z_{(i)}$	—	.08545	.04777	.11749	.12745	.10207	.04789	.11755
rounded z_{*i}	—	.085	.048	.117	.127	.102	.048	.118
1000 (rounded z_{*i} − 0.100)	—	-15	-52	17	27	2	-52	18

Sum = −55; −55/7 = −7.9 = mean = z_* (in working units)

Sum sq. = 6979; 6979 − (−55)²/7 = 6547 = sum sq. deviations

$$\frac{6547}{6 \times 7} \approx 156 = s_*^2$$

$\sqrt{156} \approx 12.5 = s_*$ (in working units)

$|t_{6|.95}| = 2.447,$ $(12.5)(2.447) = 30.6 = $ allowance (in working units)

ate amount of computation is involved, the application of the jackknife is routine.

The ability to work with groups as well as individuals is a crucial advantage of the jackknife, not only as a way of keeping computation to a reasonable volume but, more importantly, as a way of ensuring the use of an appropriately large error term. In particular, the way the sample was drawn controls the proper assessment of stability of any survey result. If the units were drawn in clusters, the correct error term involves cluster-to-cluster variation, and we must be sure that each piece is made up of one or more whole clusters. If the units were stratified and stratum sizes were known, stratum-to-stratum variation must be excluded from the error term, a condition that can sometimes be ensured by the choice of pieces (and always by the choice of basic computation).

Exhibit **8** of Chapter 8

Final computations for the ratio estimate (Base data: 1920 total = 22,919; log (1920 total) = 4.360; log total = log (1920 total) + log ratio)

Quantity considered		Results found	
In words	In formulas	Good estimate	95% confidence intervals
Working units	$1000 (z_* - 0.100)$	about −7.9	−38.5 to 22.7
log ratio	z_*	about 0.092	0.062 to 0.123
log total	$\log (1920 \text{ total}) + z_*$	about 4.452	4.422 to 4.483
total	antilog (log total)	about 28,300	26,000 to 30,400

8D. A More Complex Example

We now turn to an example of jackknifing where we can offer no other bearable approach. The similarity between the "leave-one-out" form of cross-validation discussed at the end of Section 2F and the jackknife has, without doubt, occurred to the reader. The example we discuss next involves both the cross-validation issue and the complex stability issue. Thus, it should come as no surprise that we are going to use both a "leave-one-out" (that is, jackknife) assessment of stability and "leave-one-out" cross-validation. Indeed, as we shall see when we come to further discussion, there are questions about this example where it is natural to apply "leave-one-out" techniques not merely "two deep" but "three deep".

Example* (fourth of 4). *Discrimination. This example deals with an authorship problem. Alexander Hamilton and James Madison wrote during the same period about similar political matters, and their personal histories were also similar. One way of trying to ascribe authorship of certain of their writings is based upon the rates with which each used high-frequency words. Much more extensive and effective studies of this problem have been made (Mosteller and Wallace, 1964, 1963) than we are about to make. But since investigators frequently find themselves involved in problems with the same basic difficulty— many variables, few data—we set forth here, in a way that clearly reveals the jackknife methodology of assessing variability, a new small study of this problem.

We shall try to discriminate between some writings by Hamilton and some by Madison on the basis of the five words they used most frequently. Then we shall see how well the method works. An advantage of choosing the five most frequent words, from the point of view of the example, is that their choice is not based upon any prior estimate of whether these words are good or poor at separating these authors' writings, an advantage for simplification, though not necessarily for discrimination. The decision to use 5 words is arbitrary: we wanted an example complicated enough to imitate reality, but modest enough in size that neither we nor the reader need make a career of it. Another advantage is that, because one more occurrence, or one fewer, changes these rates so slightly, the rates of use of such high-frequency words should behave smoothly, making all standard techniques for measurement data sure to be effective.

We chose 11 papers known to have been written by each author, mainly from among the *Federalist* papers. These particular 22 papers were chosen, because among the 100 or so papers we had available, their lengths were nearest to 2500 words, running from about 2200 to about 2800. For some purposes, it would have been better to have chosen randomly. For convenience in applying the jackknife, each Hamilton paper was randomly paired with a Madison paper. Perhaps we could have paired them more meaningfully by order of publication, but we did not. The number $k = 11$ was chosen partly because it is one more than the round number 10, and we frequently need to multiply or divide by $k - 1$. Also 10 is only about twice as big as 5, the number of variables chosen for analysis, and one of our purposes is to illustrate the variability that may occur in a study of several weakly discriminating variables when we have only a modest set of data available for establishing the technique of making distinctions. In our discussion, we work only with papers whose authorship is known, though the method of approach illustrated should produce a means of distinguishing unknown papers as well.

We did the pairing to save calculation. We could as well have removed one paper at a time and done 22 calculations. The pairing has no intrinsic merit and may be distracting if not regarded as an economy measure.

The standard device that seems most reasonable here is the linear discriminant function. When we have two classes of individuals measured on variables $x_1, x_2, \ldots, x_k$, the linear discriminant function is a linear function

$$\hat{y} = A(b_1x_1 + b_2x_2 + \cdots + b_kx_k) + B.$$

The coefficients b_i for the x's are chosen so as to separate the observed sample values of the two classes of individuals as widely as we can, considering the internal variation within the two classes. The numbers A and B are merely scale factors chosen for convenience, either of the investigator or of computation.

When convenient, as it is here, we may regard each Hamilton paper as having a y-value of 1, and each Madison paper as having a y-value of 0. The corresponding discriminant function is just the linear function obtained by fitting a multiple-regression equation with the values of the dependent variable y (to be forecast) assigned 0 or 1 according as Madison or Hamilton is the author. As a consequence, the free coefficients A and B are automatically so chosen that the value $\hat{y}$ forecast by the discriminant function averages 1 for those Hamilton papers in the subset for which a particular discriminant function is constructed, while the average of $\hat{y}$ for the corresponding Madison papers is 0. We shall—arbitrarily, but naturally—decide that a discriminant score of over 0.5 is a Hamilton indicator, one of less than 0.5 a Madison indicator. This decision simplifies our work by making it unneccessary for us to estimate the optimum point of division for the scores—and may weaken our discrimination.

Exhibit **9** shows the rates per thousand words for the five high-frequency words *and* (x_1), *in* (x_2), *of* (x_3), *the* (x_4), and *to* (x_5) in each of the 22 papers. The numbers assigned to the papers are those assigned in Mosteller and Wallace (1964, pp. 12–14, 269–270). Exhibit 9 also shows the sums of squares

Exhibit **9** of Chapter 8

Rates per thousand for the 5 words for each of the Hamilton and Madison papers used (sums of squares and the cross-products for each author are also given)

A) Hamilton papers

Number	Group j	and 1	in 2	of 3	the 4	to 5	Total
60	1	16.1	35.3	63.9	98.3	38.4	252.0
69	2	32.2	24.5	78.2	110.0	31.4	276.3
36	3	24.3	23.5	64.7	90.8	42.3	245.6
73	4	18.0	27.2	59.6	86.8	35.9	227.5
26	5	20.6	26.9	61.4	83.6	39.5	232.0
7	6	21.8	17.4	73.1	90.4	35.6	238.3
112	7	27.9	23.1	61.9	85.4	41.3	239.6
11	8	28.5	26.1	71.3	74.5	33.3	233.7
35	9	28.9	20.9	56.9	82.7	44.9	234.3
34	10	21.3	25.0	60.4	82.2	47.7	236.6
66	11	18.5	30.7	72.7	109.3	36.6	267.8
Totals		258.1	280.6	724.1	994.0	426.9	

Exhibit **9** of Chapter 8 (continued)

B) Madison papers

Number	Group j	Words, i and 1	in 2	of 3	the 4	to 5	Total
40	1	31.6	19.9	54.8	93.8	38.6	238.7
37	2	37.3	23.3	56.8	84.2	31.0	232.6
133	3	21.2	17.5	58.2	97:6	39.9	234.4
14	4	27.9	19.1	55.8	93.1	33.5	229.4
122	5	40.7	9.3	59.0	71.5	33.6	214.1
39	6	24.4	27.9	60.0	115.3	34.8	262.4
46	7	27.7	17.7	61.1	115.3	32.7	254.5
44	8	28.1	22.3	57.0	110.9	29.7	248.0
47	9	30.6	23.6	68.3	118.6	23.2	264.3
42	10	33.9	21.8	64.9	93.7	33.6	247.9
132	11	23.3	31.4	34.8	94.3	49.6	233.4
Totals		326.7	233.8	630.7	1088.3	380.2	

C) Hamilton: Sums of squares and of cross-products of deviations

Words	1	2	3	4	5
1	275.985				
2	−139.756	226.069			
3	95.181	−12.473	471.102		
4	−67.655	181.644	455.111	1267.465	
5	−31.396	−34.801	−260.843	−227.696	244.069

D) Madison: Sums of squares and of cross-products of deviations

Words	1	2	3	4	5
1	351.920				
2	−173.050	334.287			
3	169.590	−212.312	719.265		
4	−514.910	381.708	364.715	2146.385	
5	−173.710	109.502	−479.175	−315.635	442.865

E) Pooled sums of squares and of cross-products of deviations

Words	1	2	3	4	5
1	627.905				
2	−312.806	560.356			
3	264.771	−224.785	1190.367		
4	−582.565	563.352	819.826	3413.850	
5	−205.106	74.701	−740.018	−543.331	686.934

F) Hamilton sum minus Madison sum

	−68.6	46.8	93.4	−94.3	46.7

of deviations and the sums of cross-products of deviations of these quantities for Hamilton and Madison (deviations taken from respective means). The pooled sums of squares and of cross-products of deviations are used to obtain the coefficients of the discriminant function D_{all}. The sums of columns x_i from Exhibit 9 for Madison are subtracted from the corresponding sums for Hamilton, and these mean differences are recorded in Exhibit 9. Let us first study the variability of the discriminant function in terms of the variability of its individual coefficients. Exhibit **10** shows, in its top portion, the coefficients for the five variables and the constant term—first when all 22 papers are used for the fitting and then when each group, here each matched pair, of Hamilton and Madison papers is omitted, one at a time, and the fitting executed for the other 20. Here the whole discriminant function is jackknifed column by column—that is, coefficient by coefficient—computing

$$11(\text{coefficient for all}) - 10(\text{coefficient with } j\text{th pair omitted}).$$

For example, the pseudo-coefficient for x_3 when we omit the fourth pair is given as (more decimals retained, both here and in the actual computation, than in Exhibit 10):

$$11(0.0526442) - 10(0.0563169) = 0.015917.$$

Exhibit **10** of Chapter 8

Original discriminant function D_{all}, and the 11 discriminants, $D_{(j)}$, constructed by omitting each pair of Hamilton and Madison papers in turn: from these are constructed the 11 pseudo-discriminants d_{*j} and their average D_*, which are also given

A) Original discriminant function D_{all} and 11 discriminants $D_{(j)}$

	Coefficient of					Constant
	x_1	x_2	x_3	x_4	x_5	term
D_{all}	−0.01902	0.02851	0.05264	−0.01642	0.04056	−2.83668
$D_{(1)}$	−0.01904	0.03032	0.05295	−0.01660	0.04120	−2.89257
$D_{(2)}$	−0.02747	0.02559	0.04532	−0.01964	0.03163	−1.47660
$D_{(3)}$	−0.02884	0.01467	0.04928	−0.01479	0.03962	−2.14043
$D_{(4)}$	−0.00716	0.02874	0.05631	−0.01248	0.05243	−4.21632
$D_{(5)}$	−0.01790	0.02789	0.05348	−0.01642	0.04172	−2.94733
$D_{(6)}$	−0.01695	0.03151	0.05182	−0.01455	0.04145	−3.10996
$D_{(7)}$	−0.02053	0.03063	0.05166	−0.01757	0.03681	−2.55988
$D_{(8)}$	−0.01648	0.03338	0.05660	−0.01979	0.04184	−2.99265
$D_{(9)}$	−0.02350	0.02910	0.05002	−0.01670	0.03074	−2.20700
$D_{(10)}$	−0.01093	0.03047	0.05406	−0.01559	0.04523	−3.40123
$D_{(11)}$	−0.01983	0.02953	0.05521	−0.01616	0.04188	−3.06431

➡

This result appears in the second portion of Exhibit 10 as the fourth value in the third column. Entries have been truncated to 5 decimals. These 11 new discriminant functions, together with the 12th formed by averaging them, are the pseudo-discriminants and the jackknifed discriminant, respectively. Note that whole functions are being jackknifed, not just values of functions, not just coefficients in functions, something we can do because the coefficients appear linearly in the values.

Let us look at D_*, the jackknifed discriminant function, and the standard errors of its coefficients, calculated according to the formula of Section 8A.

$$D_* = -3.0141 - 0.0195x_1 + 0.0301x_2 + 0.0547x_3 - 0.0167x_4 + 0.0420x_5$$

Standard errors
(jackknifed s_{b_i})   0.0193   0.0149   0.0105   0.00645   0.0181

Critical
ratio $|b_i|/s_{b_i}$ 1.0   2.0   5.2   2.6   2.3

These results suggest that the only coefficient seriously different from zero might be the third, which multiplies the frequency for *of*, a suggestion we shall not follow up here.

Exhibit **10** of Chapter 8 (continued)

B) Pseudo-discriminants D_{*j} and their average D_*

| | Coefficient of | | | | | Constant |
	x_1	x_2	x_3	x_4	x_5	term
D_{*1}	−0.01886	0.01041	0.04956	−0.01463	0.03422	−2.27783
D_{*2}	0.06548	0.05770	0.12584	0.01582	0.12988	−16.43747
D_{*3}	0.07914	0.16695	0.08621	−0.03268	0.04996	−9.79921
D_{*4}	−0.13765	0.02623	0.01591	−0.05584	−0.07806	10.95966
D_{*5}	−0.03020	0.03477	0.04425	−0.01639	0.02902	−1.73025
D_{*6}	−0.03967	−0.00149	0.06081	−0.03513	0.03166	−0.10390
D_{*7}	−0.00389	0.00736	0.06242	−0.00490	0.07810	−5.60468
D_{*8}	−0.04443	−0.02015	0.01308	0.01731	0.02776	−1.27696
D_{*9}	0.02582	0.02262	0.07885	−0.01364	0.13878	−9.13354
D_{*10}	−0.09988	0.00897	0.03843	−0.02470	−0.00606	2.80880
D_{*11}	−0.01090	0.01831	0.02697	−0.01905	0.02736	−0.56044
D_*	−0.01955	0.03015	0.05476	−0.01671	0.04205	−3.01416

When we apply the discriminants D_{all} and D_* to the 22 Hamilton and Madison papers used to construct these functions, we get the results shown in Exhibit **11,** which have been, arbitrarily, truncated to three decimals. First we note that the results for D_{all} and D_* are very similar. Second, when we use the value 0.5 as a cutoff, so that higher discriminant function scores are regarded as Hamilton forecasts and lower ones as Madison forecasts, we note that all 22 papers are correctly classified by both D_{all} and D_*.

Surprisingly good? Hard to be sure, since we used these same papers to select this discriminant. Cross-validation is needed.

Exhibit **11** of Chapter 8

Value of discriminants D_{all} and D_* when applied to each of the 22 papers

Group of papers	D_{all}		D_*	
	H	M	H	M
1	1.170	0.039	1.206	0.024
2	0.833	−0.017	0.859	−0.033
3	1.001	0.338	1.023	0.333
4	0.764	−0.055	0.777	−0.075
5	1.000	−0.051	1.020	−0.080
6	1.052	0.171	1.073	0.172
7	0.822	−0.209	0.836	−0.227
8	1.246	−0.351	1.275	−0.374
9	0.668	−0.157	0.673	−0.167
10	1.235	0.380	1.263	0.381
11	1.203	−0.089	1.243	−0.107
Average	0.999	0.000	1.023	−0.014

8E. Cross-validation in the Example

So far the jackknife has provided us with an honest estimate of the variability of the coefficients. Let us try next to get an honest estimate of the ability of the final discriminant to sort out Hamilton and Madison writings. This is a new and major task.

To do just this, without any assessment of stability, we would have to cross-validate. With 11 pairs and 5 variables, we clearly dare not divide the data in halves. But, as suggested in Section 2F, we could set aside each pair in turn and apply the discriminant based on the remaining 10 pairs to the set-aside pair. To do this, we need exactly the discriminants we have just calculated for a different purpose. When we apply each $D_{(i)}$ to the rates for the corresponding papers by Hamilton and by Madison, H_i and M_i, we find the values set out in the left side of Exhibit **12**. There is now one misclassification, paper M_3, whose value of 0.510 just barely assigns it to Hamilton.

Actually, the behavior of these discriminants, involving 5 coefficients fitted to 10 pairs of observations, is surprisingly good. The indicated separation of Hamilton and Madison subgroup means amounts to $0.975 - 0.015 = 0.960$. This is a surprisingly large value, but one whose stability we have not, as yet, introduced a way to estimate. (If 22 papers have been divided into 11 and 11 *at random*, the average indicated separation of such means would be zero because the signs of the differences would tend to be $+$ and $-$ at random.)

The spread of each group around its observed mean is not small; the sample standard deviation falls between 0.28 and 0.31 for both Hamilton and Madison. This time, not only do we fail to have an indication of the stability of the result, but we are not at all sure that it is not misleading. For instance, all the data enter both into the value of $D_{(3)}$ at H_3 and the value of $D_{(7)}$ at H_7.

Exhibit **12** of Chapter 8

Results of applying discriminant functions $D_{(i)}$ of Exhibit 11 to the omitted papers, thus cross-validating the discriminants.

i	$D_{(i)}$ applied to		|Shift* from D_{all}|	
	H_i	M_i	H_i	M_i
1	1.205	−.044	+.035	−.005
2	.642	−.004	−.191	−.013
3	1.025	.510†	+.024	−.172
4	.592	−.130	−.172	+.075
5	.993	−.034	−.007	−.017
6	1.018	.230	−.034	−.059
7	.792	−.253	−.030	+.044
8	1.363	−.438	+.117	+.087
9	.567	−.091	−.101	−.066
10	1.269	.459	+.034	−.079
11	1.256	−.124	+.053	+.035
Average	.975	+.015	−.024	−.015

* + means better discrimination; − means worse.
† This paper misclassified.

Accordingly we have no assurance that these values are not correlated in some way that makes the average square of their difference substantially different from the sum of the corresponding variances.

By taking the mean-squared deviations of the Hamilton papers about 1, and those of Madison about 0 (instead of about the observed means), we may obtain indications that, at the price of combining assessment of spread and shrinkage, are not exposed to this difficulty, since they are *sums* of terms each of which involves only a single value. (In this example, the numerical values are almost the same.) These measures are thus legitimate indications, but they still lack any measure of stability.

By omitting one pair at a time, then, we have had our choice:

1. estimated stability for overall discriminant function (no estimate of quality of performance), as studied in Section 8D;

2. estimated performance by cross-validation (no estimate of stability for this estimate), as studied in the present section.

8F. Two Simultaneous Uses of "Leave out One"

If both quality of performance and stability are to be estimated, we must combine cross-validation and jackknifing. This means dropping out one pair for each purpose. Accordingly we must find discriminant functions based upon sets of $9 = 11 - 2$ pairs of papers. We denote, indifferently, the discriminant obtained when both pair i and pair j are omitted by

$$D_{(i)}(j) = D_{(i)(j)} = D_{(j)}(i).$$

What we are going to do, from either point of view (that of the jackknifing or that of the cross-validation), is to set one pair aside and redo the whole analysis with the remaining 10 (instead of 11) pairs of papers.

With pair i set aside for cross-validation, the jackknife leads to pseudo-discriminants

$$D_{*j}(i) = 10D_{all}(i) - 9D_{(j)}(i) = 10D_{(i)} - 9D_{(i)(j)}.$$

For each i there are ten $D_{*j}(i)$, and one D_{*i} which is the average of the $D_{*j}(i)$. No one of these discriminant functions used papers H_i and M_i in its formation. Therefore, when we apply the discriminants to H_i and M_i, we get $2 \times (10 + 1) = 22$ honest cross-validations.

Exhibit **13** gives the results, cut to three decimals. (Two decimals would serve most purposes.) Look first at group 1 in Exhibit 13. The first entry 1.250 is gotten from $D_{*2}(1)$ applied to H_1. Note that all Hamilton values exceed 0.5, though the 10th, gotten from $D_{*11}(1)$, had a close call at 0.535. Note that one Madison value crosses 0.5—we label this a mistake by the pseudo-discriminants. For these two papers, then, the error rate by the pseudo-discriminants is 5%. For all papers the error rate by pseudo-discriminants is about 16%.

Exhibit **13** of Chapter 8

Results of the applications of pseudo-discriminant functions $D_{*j}(i)$ to the two papers of group i, and the average value of the result for each paper, which are the values of the jackknifed discriminant function $D_*(i)$, for the papers H_i and M_i not entering into their construction

j	Group 1		Group 2		Group 3		Group 4		Group 5		Group 6	
	H	M	H	M	H	M	H	M	H	M	H	M
1	—	—	0.492	−0.124	1.705	−0.388	−0.439	−0.716	0.582	−0.558	1.263	−0.701
2	1.250	0.167	—	—	0.855	0.409	−0.309	−1.170	0.811	−0.852	−0.703	0.062
3	1.723	−0.431	2.062	1.234	—	—	0.788	−0.127	2.329	−0.003	2.026	0.138
4	2.598	−0.354	−0.535	0.180	0.976	0.882	—	—	0.823	−0.019	1.127	0.206
5	1.446	−0.008	0.577	−0.213	0.958	0.429	1.283	0.133	—	—	1.021	−0.004
6	0.785	−0.126	0.146	−0.234	0.773	−0.036	0.925	−0.007	1.107	0.778	—	—
7	1.058	0.428	1.463	−0.044	1.358	0.635	0.435	−0.030	1.018	0.498	1.329	0.418
8	0.738	0.395	0.213	−0.465	0.852	1.077	0.937	0.500	0.544	−0.479	0.761	0.904
9	1.014	0.588	1.612	0.253	1.794	0.936	0.156	−0.301	1.197	0.938	1.314	0.039
10	1.383	−0.657	−0.957	−1.049	0.486	0.499	1.143	−0.211	1.036	−1.343	1.119	−0.084
11	0.535	0.102	0.976	−0.174	0.789	0.447	0.559	0.076	0.841	0.493	1.394	−0.348
Average	1.253	0.010	0.605	−0.063	1.055	0.494	0.548	−0.185	1.029	−0.054	1.065	0.063
$s^2 = 10s_*^2$	0.353	0.164	0.903	0.338	0.180	0.188	0.347	0.218	0.253	0.558	0.492	0.141

Exhibit **13** of Chapter 8 (continued)

j	Group 7 H	Group 7 M	Group 8 H	Group 8 M	Group 9 H	Group 9 M	Group 10 H	Group 10 M	Group 11 H	Group 11 M
1	1.301	0.222	1.402	−0.656	0.828	0.844	1.532	0.653	2.646	−2.106
2	0.886	−1.769	2.487	−1.165	0.402	−0.190	0.795	0.789	1.342	−0.467
3	0.681	−0.448	2.593	0.111	0.151	0.227	1.429	0.085	1.462	0.160
4	0.793	−0.215	1.167	−0.327	0.680	−0.086	1.261	0.222	1.206	0.091
5	0.810	−0.562	1.604	−0.742	0.711	−0.791	1.385	0.303	0.904	−2.121
6	0.661	−0.130	0.732	0.082	1.164	−0.157	1.548	0.671	1.464	0.372
7	—	—	0.969	−0.429	0.167	0.832	0.790	0.156	1.064	1.676
8	0.322	1.312	—	—	0.567	−0.151	2.284	0.785	1.270	1.836
9	1.634	−0.324	1.598	−0.636	—	—	1.020	0.318	1.117	−0.993
10	0.243	−0.519	0.798	−0.571	0.034	−0.468	—	—	0.862	−0.991
11	0.752	−0.225	0.696	−0.108	1.092	−0.651	0.943	0.401	—	—
Average	0.808	−0.265	1.405	−0.444	0.580	−0.059	1.299	0.438	1.334	−0.254
$s^2 = 10s*$	0.170	0.578	0.469	0.158	0.152	0.309	0.203	0.070	0.255	1.870

For each paper, we would like to know how firmly a decision in favor of Hamilton or Madison is made. Let us conceive of an infinite number of pairs of Hamilton and Madison papers, all different from pair i, on which a discriminant function $D_*(i)$ might be based, and regard the 10 pairs actually used as a sample from this infinite population. The results of applying the infinite-population $D_*(i)$ to H_i and M_i would be μ_{Hi} and μ_{Mi}, the true values for these papers. These are the values toward which our jackknifed answers are aimed. The jackknifed values for these papers, say $\bar{y}_{Hi}$ and $\bar{y}_{Mi}$ for the ith pair, combined with the variability of these jackknifed values, assessed from the individual pseudo-values by way of s_*^2, can be used to set confidence limits on the μ_{Hi} and μ_{Mi} at any chosen level. Insofar as these limits do not include 0.5, we have clear evidence, at that level, for Hamilton or for Madison. Naturally, then, we want to know how many standard errors each $\bar{y}_{Hi}$ or $\bar{y}_{Mi}$ is from the cutoff 0.5.

In Exhibit **14,** which includes values of s_*^2, we summarize the results for the 22 papers.

Note that 8 of Hamilton's papers and 8 of Madison's are beyond two standard errors from the cutoff, 1 Madison is at 1.7, while the other 5 have only relatively narrow margins.

We now have cross-validated estimates, combined with an assessment of their stability. Our results, though much more useful, are still far from perfect. We have assessed stability, climbing to the second step of the staircase, but, since we have not assessed the stability of this stability, we have not reached the third step. It is not surprising, in view of the small number of papers considered, that we have also failed to gain any appreciable information about the shape of the distribution of the μ_{Hi} (or of the μ_{Mi}).

Exhibit **14** of Chapter 8

Departures of observed means from 0.5, using s_* as the unit

Group	$\lfloor\bar{y}_{Hi} - 0.5\rfloor$	$\lfloor s_*\rfloor$	$\lfloor(\bar{y}_{Hi} - 0.5)/s_*\rfloor$	$\lfloor 0.5 - \bar{y}_{Mi}\rfloor$	$\lfloor s_*\rfloor$	$\lfloor(0.5 - \bar{y}_{Mi})/s_*\rfloor$
		H			M	
1	.753	.188	4.0	.490	.128	3.8
2	.105	.300	0.3	.563	.184	3.0
3	.555	.134	4.1	.006	.137	0.0
4	.048	.186	0.3	.685	.148	4.6
5	.529	.159	3.3	.554	.236	2.3
6	.565	.222	2.5	.437	.119	3.7
7	.308	.130	2.4	.765	.240	3.2
8	.905	.217	4.2	.944	.126	7.5
9	.080	.123	0.6	.559	.176	3.2
10	.799	.142	5.6	.062	.084	0.7
11	.834	.160	5.2	.754	.432	1.7

8G. Dispersion of the μ's

But we can learn a little more. If we had an infinite supply of papers to fix the discriminants, dropping one pair would not change the discriminants, or their coefficients. Accordingly, the μ_{Hi} must average 1, and the μ_{Mi} must average 0, since these constraints apply to the discriminant values before one pair is dropped. Thus, on the basis of quantities like

$$\sum (\bar{y}_{Hi} - 1)^2 = 1.011,$$

we could construct rough variance-component estimates of the standard deviation of the distribution (over i) of μ_{Hi}, the value that would be associated with Hamilton paper i if we had infinite material for constructing the discriminant function.

On the average, over papers and discriminants, $(\bar{y}_{Hi} - 1)^2$ will equal $\sigma_\mu^2 + \sigma^2$, where σ_μ^2 is the variance of the μ_{Hi}, and σ^2 the variability of assessing any single paper with a discriminant function based on only a finite number of pairs of other papers. We estimate σ^2 by s_*^2. We estimate the population average value of $(\bar{y}_{Hi} - 1)^2$ directly, pooling the results for the two authors, and then estimate σ_μ^2 by subtraction.

We find the average sum of squares of deviations (from 1 or 0, respectively, for Madison and Hamilton) to be $\frac{1}{2}(1.011 + 0.865) = 0.938$. To estimate the population average value of $(\bar{y}_{Hi} - 1)^2$, we divide by 11 (instead of the 10 that would have been appropriate if we had fitted a mean), finding 0.0853, thus confirming the possibly biased prediction (in Section 8E) of about 0.30 for the standard deviations.

For the 22 values of s_*^2 we get a total of 0.8364 and an average of 0.0380, which estimates σ^2. Subtracting gives $0.0853 - 0.0380 = 0.0473$ as the estimate of σ_μ^2. Consequently we estimate σ_μ to be about 0.22. For both Madison and Hamilton, the mean μ appears to be a shade more than two standard deviations from the cutoff of 0.5. Thus, on Gaussian theory and for many other distributions, roughly 2.5% of each author's papers would be incorrectly assigned if we had only these 5 words on which to base the discrimination but infinite material to determine the discriminant function used. (Our estimate of σ_μ is far from precise, being worth perhaps a dozen degrees of freedom.)

There is now some interest in comparing the mean differences and standard deviations associated with different sorts of discriminants and the Hamilton and Madison subgroups. We have found these estimates:

	Separation of group means	Standard deviation of distribution	Ratio
Discriminant based on 10 other pairs	0.96	0.29	3.3
Discriminant based on ∞ other pairs	1.00	0.22	4.5

where again we are dealing with indications of unknown stability. The improvement from basing the discriminant on many more pairs of papers appears to be rather less than one might have supposed, although the indicated gain is substantial.

Taken separately and as a whole, these results give us a picture of the strength and weakness of the proposed discriminant function when it is put into practice. Being able to carry it out required a high-speed computer and double use of "leave-one-out" procedures, one a jackknife, one a cross-validation.

Again, let us emphasize that the whole effort here is designed to illustrate the jackknife in a complicated problem with modest amounts of data, rather than to illustrate the full nature of the *Federalist* problem with its crucial problems of word selection.

8H. Further Discussion of the Example

The question of the stability of the estimate of σ_μ might be worth investigating. We have here a number calculated by a definite procedure; we seek to assess its stability. We could jackknife again. We could drop out one more pair, calculate $11 \times 10 \times 9/6 = 165$ discriminant functions, one for each set of 8 pairs of papers, and work back to 11 more jackknifed estimates of σ_μ. Combining these in another grand jackknife, with the one we already have, gives us what we sought: an estimate of σ_μ, *and* an estimate of the stability of this estimate. What other technique offers even a crude approximation to an assessment of the stability of an estimate of σ_μ?

As still a further step, consider the person who has read the sentence (toward the end of Section 8D) suggesting that one rate, that for *of*, is doing all the work, and who asks: what evidence do these data give about the superiority or inferiority of the 5-variable discriminant function to the use of *of* alone?

To answer such a question thoroughly requires double cross-validation (see Section 2F), rather than single. We ought to study the question's answer on other data than those that suggested it.

But if this cannot be done, and we must peer into the same 22 papers again, there is no difficulty in attaining a single cross-validation with assessed stability. We have only to turn to each set of 9 pairs and determine two discriminant functions, one based upon all 5 words (already done), one based upon *of* alone (very easy). Turning to each omitted pair in turn, we can calculate, for each paper,

1. the behavior of the first discriminant,

2. the behavior of the second discriminant,

3. the difference in behavior, an indicator of which is better.

The jackknifing can then proceed on quantities of type (3), producing, for each paper, both an estimate of improvement, and an assessment of the stability of this estimate. If our original question is really about how the two discriminants would compare if each could be based on a very large population of papers, this would give a relevant and useful answer.

Further references on the jackknife are Quenouille (1956), Tukey (1958), Durbin (1959), Mickey (1959), Kendall and Stuart (1961), Brillinger (1964), Miller (1964), Robson and Whitlock (1964), Jones (1965), Gray and Adams (1972), Gray, Watkins, and Adams (1972), Egman, Meyers, and Bendel (1973), Frawley (1974), Jones (1974), Miller (1974a), 45 references in a review by Miller (1974b), and Wainer and Thissen (1975). There have also been generalizations to the jackknife (see Gray and Shucany, 1972).

Summary: The Jackknife

The jackknife is designed to do any of many jobs fairly well, but it should be remembered that a special method may do better for a specific kind of analysis.

We use the jackknife both for estimation and to estimate variance.

We can learn which pieces it is natural and reasonable to divide a given body of data into.

We can calculate pseudo-values, jackknifed value, and s_*^2, and go on to set (confidence) limits on the corresponding estimand.

The major difficulties with the elementary jackknife arise (i) when the distributions of interest have excessively straggling tails, and/or (ii) when only a few different numbers appear as pseudo-values. (Occasionally the methods of Chapter 10, to come, will let us deal with difficulty (i). Rearranging the size and mutual relation of the pieces involved will often deal with (ii).)

Where jackknifing of various values comprising an overall result proceeds in a parallel way, we can frequently think of the jackknife as applying to a whole function or a whole situation (rather than to one class of "equivalent numbers" at a time).

It is desirable to *avoid* (in jackknifing) sampling distributions with (i) abrupt ends and (ii) one or more straggling tails, and it is probably desirable to avoid those that are strongly unsymmetrical (often avoidable by re-expression).

We can use the jackknife to provide estimates of the uncertainties attached to the performance of a discriminant function, as well as those of its coefficients, particularly by combining leave-out-one jackknifing and leave-out-one cross-validation. (The same would apply to results of many other analytical techniques, as well as to discriminant functions.)

We could, by bringing in a third leave-out-one operation, specifically another jackknife, add, to the two assessments just mentioned, an assessment of the uncertainties in our assessed quality of performance for the discriminant function. (This requires, overall, a "leave-out-three" technique.)

We followed through the jackknife calculations in a variety of small examples and in one of greater complexity.

References

Brillinger, D. R. (1964). "The asymptotic behavior of Tukey's general method of setting approximate confidence limits (the jackknife) when applied to maximum likelihood estimates." *Rev. Int. Statist. Inst.*, **32**, 202–206.

Cochran, W. G. (1953), *Sampling techniques.* New York: Wiley. (2nd ed., 1963.)

Durbin, J. (1959). "A note on the application of Quenouille's method of bias reduction to the estimation of ratios." *Biometrika,* **46,** 477–480.

Egman, R. K., C. E. Meyers, and R. Bendel (1973). "New methods for test selection and reliability assessment, using stepwise multiple regression and jackknifing." *Educ. Psychol. Meas.,* **33,** 883–894.

Frawley, W. H. (1974). "364: Using the jackknife in testing dose responses in proportions near zero or one—revisited." *Biometrics,* **30,** 539–545.

Gray, H. L., and J. E. Adams (1972). "Jackknifing stochastic processes." *Texas J. Sci.,* **23,** 559.

Gray, H. L., and W. R. Shucany (1972). *The generalized jackknife statistic.* New York: Marcel Dekker, Inc.

Gray, H. L., T. A. Watkins, and J. E. Adams (1972). "Jackknife statistic, its extensions, and its relation to e_n-transformations." *Ann. Math. Stat.,* **43,** 1–30.

Jones, H. L. (1965). "The jackknife method." In *Proc. IBM Scientific Computing Symposium on Statistics,* October 21–23, 1963. White Plains, New York: IBM Data Processing Division; pp. 185–201.

———— (1974). "Jackknife estimation of functions of stratum means." *Biometrika,* **61,** 343–348.

Kendall, M. G., and A. Stuart (1961). *The advanced theory of statistics.* Vol. 2: Inference and relationship. London: Charles Griffin & Company; pp. 5–7.

Mickey, M. R. (1959). "Some finite population unbiased ratio and regression estimators." *J. Amer. Statist. Assoc.,* **54,** 594–612.

Miller, R. G., Jr. (1964). "A trustworthy jackknife." *Ann. math. Statist.,* **35,** 1594–1605.

———— (1974a). "A unbalanced jackknife." *Ann. of Stat.,* **2,** 880–891.

———— (1974b). "The jackknife—a review." *Biometrika,* **61,** 1–15. (45 references)

Mosteller, F., and D. L. Wallace (1963). "Inference in an authorship problem." *J. Amer. Statist. Assoc.,* **58,** 275–309.

———— (1964). *Inference and disputed authorship: The Federalist.* Reading, Mass.: Addison-Wesley.

Quenouille, M. H. (1956). "Notes on bias in estimation." *Biometrika,* **43,** 353–360.

Robson, D. S., and J. H. Whitlock (1964). "Estimation of a truncation point." *Biometrika,* **51,** 33–39.

Tukey, J. W. (1958). "Bias and confidence in not-quite large samples." Abstract in *Ann. math. Statist.,* **29,** 614.

———— (unpublished). "Data analysis and behavioral science." Princeton University and Bell Telephone Laboratories.

Wainer, H., and D. Thissen (1975). "When jackknifing fails (or does it?)." *Psychometrika,* **40,** 113–114.

Chapter 9/Two-and More-Way Tables

A large and important class of analyses starts from two or more "handles" on the data. Here we start with the two-way case and regard the data as responses, so to speak, to two factors. The simplest case arises when the two handles can be grasped separately. Here the handles are expressed by the location of each response in a row and a column—and thus by the "values" of the two factors. We attend briefly also to the case of three or more handles that can be grasped separately and thought of as rows, columns, and layers—by the "values" of three or more factors. (We can, if we are careful, still write out the data on a single sheet of paper in such a way as to make its structure clear.) The extension to larger numbers of handles that can be grasped separately is not difficult.

Let us suppose that the fit in a two-factor analysis has the form

$$\text{fit} = \text{common} \quad \text{PLUS} \quad \text{row} \quad \text{PLUS} \quad \text{column}$$

where "common" stands for a value applied to every cell, "row" stands for a value depending only upon the cell's row, and "column" stands for a value depending only upon the cell's column. As always, we have

$$\text{residual} = \text{data} \quad \text{MINUS} \quad \text{fit}$$

where "data" is the value of the response in the cell. Thus we can write exactly

$$\text{data} = \text{common} \quad \text{PLUS} \quad \text{row} \quad \text{PLUS} \quad \text{column} \quad \text{PLUS} \quad \text{residual}.$$

We also take one short step beyond this, but for still other fits, we refer the reader to *EDA*, Chapters 11 and 12, and to McNeil and Tukey (1975).

9A. PLUS Analyses

Let us start with a one-way example. *Climatography of the U.S.* (#60–44, *Climatography of the States*: Maryland; page 9) reports the monthly mean temperatures Fahrenheit in Washington, D.C., for January to July to be

Jan.	Feb.	March	April	May	June	July
36.2	37.1	45.3	54.4	64.7	73.4	77.3

Their median is 54.4. Their mean is 388.4/7 = 55.5. We can now write down the data and a summary as either

    36.2   37.1   45.3   54.4   64.7   73.4   77.3 | 54.4

or

    36.2   37.1   45.3   54.4   64.7   73.4   77.3 | 55.5

with the form

data values | summary

where we have used a single vertical bar to cut off our summary from the values it summarizes.

The next step is to use the summary as a fit, and change the data values into residuals, by subtracting the fit from them. Since the relation of the two parts is now changed—they now have to be added together to reproduce the data—we use a double line to cut them off. For our median and mean examples, we get

    −18.2   −17.3   −9.1   0   10.3   19.0   22.9 ‖ 54.4

or

    −19.3   −18.4   −10.2   −1.1   9.2   17.9   21.8 ‖ 55.5

accordingly as we take out the median or the mean.

We may, if we wish, take out any other number, whether or not it is a well-defined summary. Thus, we might like to take out 55, or 50 (as a round number), obtaining

    −18.8   −17.9   −9.7   −0.6   9.7   18.4   22.3 ‖ 55

or

    −13.8   −12.9   −4.7   4.4   14.7   23.4   27.3 ‖ 50

respectively.

In each illustration, we have separated the given values into a sum of two contributions

data = fit   PLUS   residual

or, as we could write, in these especially simple problems,

data = common   PLUS   residual

We want next to do the same thing, apportion the data into a sum of more than two terms, when we have two handles—two factors whose values or names describe each of the numbers—each of the responses—to be analyzed. We turn to such an example.

Two-factor descriptions. If, for each of three East Coast places, we start anew with mean monthly temperatures from January through July, we can do what we just did separately for each place, as shown in Panels A and B of Exhibit **1**. Once we have done this, we see, in much more detail, how the data behave. Looking at Panel A, it is clear that it is warmer in July than in January, and warmer in Laredo (Texas) than in Caribou (Maine). We can now see, in Panel B, that the change from January to July (in °F) is larger in Caribou than in Laredo, and that Washington falls in between.

We have taken a place effect out of each row—we can go ahead and take a month effect out of each column. Panel C shows the result of writing down the medians of the columns, Panel D that of taking them out.

We now have a

two-way analysis

where we have broken down the data as

common PLUS row PLUS column PLUS residual,

expressing each value as the sum of four terms. For January in Caribou, for the analysis of Panel D, this gives

$$(54.4) + (-19.7) + (-18.3) + (-7.7) = 8.7 \qquad \text{(check!)}$$

The sum gives the original measurement of 8.7, as it should. For May in Washington, Panel D gives

$$(54.4) + (0.0) + (10.3) + (0.0) = 64.7 \qquad \text{(check!)}$$

and it adds to 64.7 as it should.

Just as we had many choices in a one-way analysis—taking out whatever value we want to remove—so in the two-way analysis we have many, many choices—taking out whatever row effects (here place effects) *and* whatever column effects (here month effects) we wish.

Panel E shows an alternative analysis of the same data, taking out effects that are multiples of 5°F. The "residuals" are not so small, but the terms are especially easy to talk about and summarize. (Note that, for Caribou in January, we have $(55) + (-20) + (-20) + (-6.3)$, which still adds up to 8.7, just as it should.)

We plan to use the pattern

	row
residuals	effects
column effects	common

as a standard effective way to break down a two-way table. We emphasize such points as these:

1. The breakdown does, carefully and quantitatively, what we often try to do qualitatively, when we "eyeball" the table.

Exhibit **1** of Chapter 9

A two-way example: median monthly temperatures. Medians are used for the fit in Panels A through D.

A) The DATA—and some medians: data | medians

	Jan.	Feb.	March	April	May	June	July	
Caribou	8.7	9.8	21.7	34.7	48.5	58.4	64.0	34.7
Washington	36.2	37.1	45.3	54.4	64.7	73.4	77.3	54.4
Laredo	57.6	61.9	68.4	75.9	81.2	85.8	87.7	75.9

B) The same ANALYZED ONE WAY: residuals ‖ medians

Caribou	−26.0	−24.9	−13.0	0	13.8	23.7	29.3 ‖	34.7
Washington	−18.2	−17.3	−9.1	0	10.3	19.0	22.9 ‖	54.4
Laredo	−18.3	−14.0	−7.5	0	5.3	9.9	11.8 ‖	75.9

C) And some medians the OTHER WAY: residuals ‖ medians for rows

							median of column residuals ‖	grand median
Caribou	−26.0	−24.9	−13.0	0	13.8	23.7	29.3 ‖	34.7
Washington	−18.2	−17.3	−9.1	0	10.3	19.0	22.9 ‖	54.4
Laredo	−18.3	−14.0	−7.5	0	5.3	9.9	11.8 ‖	75.9
	−18.3	−17.3	−9.1	0	10.3	19.0	22.9 ‖	54.4

D) And with these TAKEN OUT also: residuals ‖ row terms (residual row medians)

							column terms (medians of column residuals) ‖	common (median of row medians)
Caribou	−7.7	−7.6	−3.9	0	3.5	4.7	6.4 ‖	−19.7
Washington	.1	0	0	0	0	0	0 ‖	0.0
Laredo	0	3.3	1.6	0	−5.0	−9.1	−11.1 ‖	21.5
	−18.3	−17.3	−9.1	0	10.3	19.0	22.9 ‖	54.4

E) An ALTERNATIVE ANALYSIS

Caribou	−6.3	−10.2	−3.3	−.3	3.5	3.4	4.0 ‖	−20
Washington	1.2	−2.9	.3	−.6	−.3	−1.6	−2.7 ‖	0
Laredo	2.6	1.9	3.4	.9	−3.8	−9.2	−12.3 ‖	20
	−20	−15	−10	0	10	20	25 ‖	55

2. Not only does the breakdown usually do it better, but it exposes the residuals to our view—now we can "eyeball" them to see what else is going on—moreover, we can, and usually should, subject them to further quantitative analysis.

3. We now have more numbers in our table than we started with, a general phenomenon, essential if we are to both analyze and preserve the original data. At some stage we may replace either all residuals or each of several groups of residuals by summary descriptions, thus decreasing the number of numbers we attend to, but we should do this only after we have looked hard at the residuals.

4. The approach offers great freedom of choice of analysis. Standard versions have important uses, but we may choose the analysis that helps us most.

When we deal with any such breakdown, we can always call the parts "contributions," or "terms"—neutral words. Sometimes, especially when we have been careful in making the fit, we call them "effects". To use "effects" gives notice that we are trying to go beyond describing the data, trying to estimate as best we can something that lies behind the particular data before us, possibly even arguing for causation, or suggesting that we have in mind a structural model that the fitting is being used to estimate.

9B. Looking at Two-Way PLUS Analyses

Given the two-way PLUS analysis just presented, we need to answer

◇what is the fit like?

◇what are the residuals like?

We will soon see how answering the first helps answer the second.
Consider the alternative analysis of Panel E in Exhibit 1, namely

(Residuals omitted)							‖	−20
							‖	0
							‖	20
−20	−15	−10	0	10	20	25	‖	55

The same fit can be written in many ways, including:

(Residuals omitted)							‖	35
							‖	55
							‖	75
−20	−15	−10	0	10	20	25	‖	Not used, or 0, whichever interpretation is preferred

which shows the fit for any month–place combination as the sum of only two terms, one determined by the place (here, 35, 55, or 75) and one determined by the month (here, −20 to +25).

Let us start a plot with a month-axis and a place-axis. The beginning looks like Exhibit **2**. Because we have used the same scale (in °F) on the two axes, we can now look at the effects of changing place or of changing month (the differences in place contribution or in month contribution) and see them as pictorial sizes, not just as numbers.

We could therefore enter the values of the residuals on the picture of Exhibit 2, as shown in Exhibit **3**. You may well be able to see the general features of the table of residuals by studying these numbers. Probably you can bring them out better by coding them and making them more visual. Just what coding seems best may vary with the problem, but 5 to 7 classes seems a good number. We have made a stem-and-leaf of the residuals and found the median, −0.3, and the difference between the upper and lower 25% points, $I = 5.2$.

Exhibit **2** of Chapter 9

Starting to picture the fit from Panel E of Exhibit 1

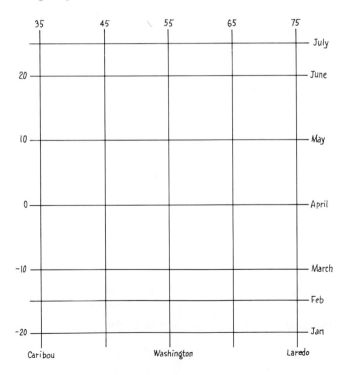

We chose, as a central interval, $M \pm \frac{1}{4}I$, then, as the next interval, items falling between there and $M + \frac{1}{2}I$ (beyond the central interval and within the hinges), next $M + I$, and beyond that. We also assigned the following symbols

	$\begin{matrix}M - I \\ \text{to} \\ M - \frac{1}{2}I\end{matrix}$	$\begin{matrix}M - \frac{1}{2}I \\ \text{to} \\ M - \frac{1}{4}I\end{matrix}$			$\begin{matrix}M + \frac{1}{4}I \\ \text{to} \\ M + \frac{1}{2}I\end{matrix}$	$\begin{matrix}M + \frac{1}{2}I \\ \text{to} \\ M + I\end{matrix}$	$\begin{matrix}\text{Above} \\ M + I\end{matrix}$
Below $M - I$			$M - \frac{1}{4}I$ to $M + \frac{1}{4}I$				
$\odot$	$\bigcirc$	$\circ$	$\bullet$		$+$	$+$	$\#$

The symbols are probably satisfactory for other occasions, but the intervals chosen might often be varied.

When we use these cutoffs, we get Exhibit **4**. It shows strikingly, if Exhibit 3 did not, that opposite corners of the rectangle have residuals of the same sign

Exhibit **3** of Chapter 9

Picture of the fit with residuals from Panel E of Exhibit 1

	35	45	55	65	75	
	4.0		-2.7		-12.3	July
20	3.4		-1.6		-9.2	June
10	3.5		-.3		-3.8	May
0	-.3		-.6		.9	April
-10	-3.3		.3		3.4	March
	-10.2		-2.9		1.9	Feb.
-20	-6.3		1.2		2.6	Jan.
	Caribou		Washington		Laredo	

and that the middle of the table has been fit rather closely. This is a fairly common pattern for residuals after fitting row PLUS column PLUS common.

Exhibit **4** of Chapter 9

Residuals (of Exhibit 3) coded

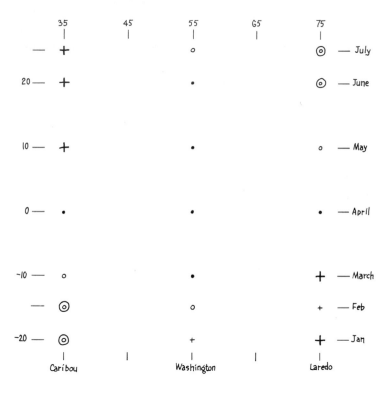

9C. Taking Advantage of Levels

Our fit is still of the form

common   PLUS   row   PLUS   column

(where we have made the common zero). Let us consider all combinations of place and month that give a chosen value to the fit, say 55°F. As Exhibit **5** shows, these combinations fall exactly on a diagonal straight line.

This leads us to:

1. Draw a selection of constant-fit lines for other constants as well as for 55°F;

2. Put in +'s for the actual combinations of places and months;

3. Look at the result—see how it would be simplified by being turned through 45°—and draw it so;

4. Pull back the constant-fit lines to the sides of the new picture;

5. Replace the many crosses by the intersections of a smaller number of lines, one for each row and one for each column.

Exhibit **5** of Chapter 9

Exhibit 2 with 55°F line added.

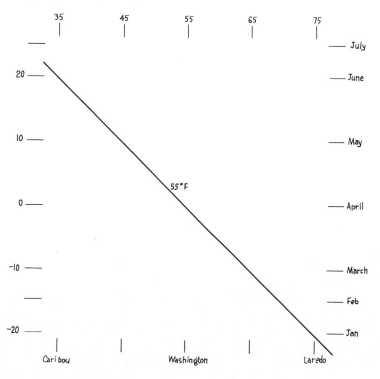

Exhibit **6** shows the results after steps (1) and (2), and Exhibit **7** after all five are completed.

The final picture differs from many graphs in that we do not try to interpret horizontal position; only *vertical* position, which shows the fitted value, is of direct importance to us.

Residuals. Now that the fit has been conveniently arrayed, it offers a natural place to put the residual for each combination of row and column—in our example, each combination of place and month. Exhibit **8** shows the residuals left by our alternative fit, each located at the place where the corresponding place and month lines in Exhibit 7 cross. We have now put the residuals at sensible places, and we get a good visual feel from them.

Exhibit **6** of Chapter 9

Exhibit 5 with more constant-fit lines and the data points individually marked.

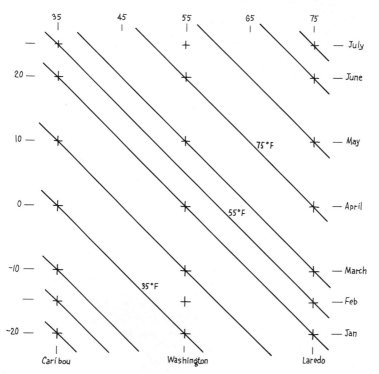

Some may prefer to see these residuals coded in a different way. We can code our values and show them on a figure like Exhibit 7. Let us use the same coding and intervals as in Exhibit 4, except that, when we turn the picture of the fit through 45°, we turn the crosses also:

⊙   ◯   ∘   •   ×   ✕   ※

These symbols, applied to Exhibit 7, appear in Exhibit **9**.

Exhibit **7** of Chapter 9

Exhibit 6 with added changes: (3) turned through 45°, (4) the constant lines shown at edges, (5) the 3 × 7 = 21 crosses replaced by 3 + 7 = 10 lines, one for each place and for each month.

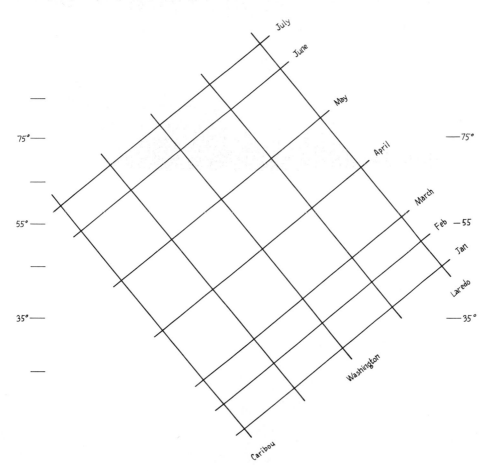

Because this coding shows our residuals to be highly structured, in that
alternate corners have alternate signs,
we need to carry the fitting further.

Exhibit **8** of Chapter 9

Exhibit 7 with the residuals plotted vertically

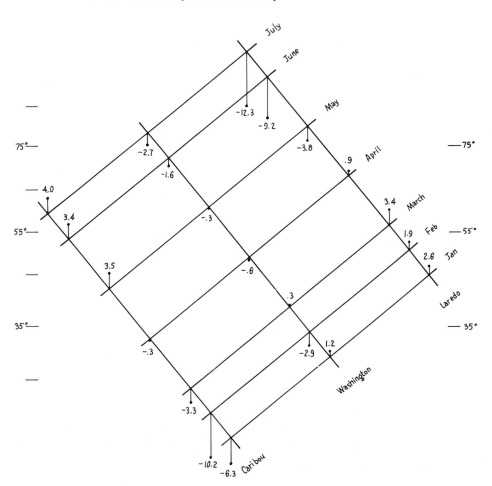

In thinking over this display of residuals, bear in mind that (1) these data will eventually be seen to be very well fitted rather simply, and (2) we started on these data knowing a lot about the effects of places and months (Laredo hot, Caribou cool; January to July warming, most rapidly near April)—we

Exhibit **9** of Chapter 9

Exhibit 7 with coded residuals

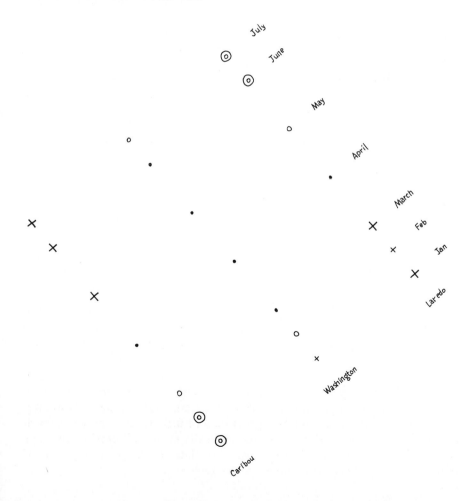

could have put places and months in order without seeing the data, and thus gone a long way toward Exhibit 9. For most sets of data we will not know so much and we may need the plot of the two-way fit, as in Exhibit 7, to organize a picture of the residuals. Vertical position lets us see easily what happens if we change months or change places or both. We have a good hold on the entire behavior of the fit.

   We did this for one fit to one set of data. We can do it for any fit of the form

   (common   PLUS   row)   PLUS   column

to any set of data, since we can convert to

   (common   PLUS   row)   PLUS   column

and do exactly what we have done in this example.

9D. Polishing Additive Fits

When we use means to carry out the fitting in a two-way table, the effects for rows and columns can be found in one "iteration." Further iterations leave the results unchanged. When we use other kinds of averages, such as medians, the first iteration may not be adequate. In this section, we illustrate this point. Along the way, we encourage the reader to compute residuals and examine them as a routine action in data analysis. In exemplifying the iterative approach and attention to residuals, we consider how means and medians treat residuals in a table with an outlier and how they treat a table with an empty cell.

   Although it may sometimes seem like extra computing trouble, routine good practice requires the calculation and analysis of residuals,

   ◇to appreciate the fit itself,

   ◇to look for peculiarities and interpretations,

   ◇to make comparisons with other analyses,

   ◇to check for blunders.

   For exploratory data analysis, we recommend starting with medians, but, of course, other averages such as the mean may be especially appropriate in a particular problem. Similarly, re-expressions such as logarithms may improve the analysis.

The basic approach. Up to this point in analyzing two-way tables, we have taken out row effects, then taken out column effects, and stopped. Ordinarily this one full step of iteration will do a great deal of fitting, but it may not finish the job, perhaps because the method used basically requires iteration, perhaps because of rounding problems when we calculate to a given number of decimals, or perhaps because of holes in the table.

For the steps in an iteration, we use single lines to separate data at one stage from summaries. Then we use double lines to separate the summaries from the residuals. Thus, the sequence would go:

First	Second	Third	Fourth	Fifth
original data in block	row or column summary	residuals in block	column or row summary	new residuals

and so on. When we use rows first and then columns, the physical layout for two-way tables is convenient if we move snakewise left to right, down, left, down, right, as in this pattern:

First	Second	Third
original block	row summary	residuals from rows

		Fourth
		column summary

Seventh	Sixth	Fifth
row residuals	row summary	new residuals from columns

Eighth . . .

To illustrate this method, Exhibit **10** shows an example whose median fit requires more than one iteration to be complete. Recall that the original example in Exhibit 1 was completed after a single cycle. It will be simplest to think of it as another set of monthly temperature data for the same three places. Although with the previous explanation, Exhibit 10 largely spells itself out, it may be worth mentioning the steps and details.

First, start with the temperature data |

Second, get row medians ‖

Third, subtract row medians from cells in their row

Fourth, get column medians

Fifth, subtract column medians from residuals in columns |

Sixth, get row medians and use check mark ($\checkmark$) if median is zero ‖

Exhibit 10 of Chapter 9

Median polish of the fit. First table in Panel A is reexpressed temperature by month (column) and by place (row). (These are the values for d = 36 in the exponential re-expression of Exhibit 21, Panel G.)

A) THE POLISH

28.9	29.3	33.3	39.5	49.1	58.6	65.2	39.5
40.4	40.9	46.5	54.4	66.1	79.0	85.9	54.4
57.7	62.6	71.2	83.3	93.5	103.7	108.3	83.3

-10.6	-10.2	-6.2	0	9.6	19.1	25.7	39.5
-14.0	-13.5	-7.9	0	11.7	24.6	31.5	54.4
-25.6	-20.7	-12.1	0	10.2	20.4	25.0	83.3
-14.0	-13.5	-7.9	0	10.2	20.4	25.7	-.7

3.4	3.3	1.7	0	-.6	-1.3	0	✓
0	0	0	0	1.5	4.2	5.8	✓
-11.6	-7.2	-4.2	0	0	0	0	-.7

3.4	3.3	1.7	0	-.6	-1.3	0	✓
0	0	0	0	1.5	4.2	5.8	✓
-10.9	-6.5	-3.5	.7	.7	.7	0	-.7

3.4	3.3	1.7	0	-1.3	-2.0	0	✓
0	0	0	0	.8	3.5	5.8	✓
-10.9	-6.5	-3.5	.7	0	0	0	✓

B) PUTTING TOGETHER the PLACE FIT

		gives		or	
39.5	0	39.5	-14.9	‖	54.4
54.4	0	54.4	0	‖	54.4
83.3	-.7	82.6	28.2	‖	54.4

C) PUTTING TOGETHER the MONTH FIT

-14.0	-13.5	-7.9	0	10.2	20.4	25.7
0	0	0	0	.7	.7	0

gives

-14.0	-13.5	-7.9	0	10.9	21.1	25.7

Seventh, get row residuals
_____

Eighth, get column medians, use check marks
===

Ninth, get column residuals |

Tenth, get row medians and *stop*,

because row medians are all checks and step 9 made all column medians checks. This completes Panel A.

In Exhibit 10 we introduce the idea of a half-step. We call finding (computing and writing down) a row (or column) median and then finding the residuals a half-step. Thus, pairs of instructions listed above represent half-steps, i.e., the second and third, the fourth and fifth, and so on. For Panel B, we assemble the place (city) fit by adding the corresponding terms for each iteration, here the first and third half-steps. The third row was the only one changed in third half-step. We could stop there, or we could break up these row medians into the sum of two parts; the first is a median of row medians, and the second is the row effects. To do this, we first take out the median of the row medians, 54.4. That median plays the role of a grand mean, of a common term, and of a baseline value.

The pieces of the assembly of the monthly fit for Panel C are a little harder to find because they don't line up as neatly as the row effects do; only May and June were affected in the fourth half-step.

In our example, we want four half-steps. Our experience is that four to eight half-steps are ordinarily enough. Since arithmetic errors in this iterative process are easy to make and not self-correcting, it is well, whenever reasonable, to have a computer do this work and print it out. Although it is possible to get slow drifts that could seem to be asking for many half-steps, it may not be wise to follow them to the bitter end. For a complete table, one may want to pick some number of half-steps, 4, 5, or 6, say, and quit after that. (A table with holes may require several more half-steps.)

Fitting by means. We can use the same approach with means—or with any other kind of summary. Exhibit **11** shows first mean polish and then median polish for another set of temperature data. We make two half-steps before quiescence (nearest 0.1°F) using means, and four half-steps using medians.

This result is in agreement with the theoretical demonstration that analysis by means requires only two half-steps. It may surprise the reader to note that sometimes analysis by means takes longer than two half-steps. The reason that this "contradiction" occurs is that our theoretical demonstration applies to arithmetic using exact values—infinite decimal places or exact fractions. It does not apply when we calculate to a fixed number of decimals and round our mean accordingly. Although we rounded in Exhibit 11, analysis by means still required only two half-steps, as analysis by means often will.

Exhibit **11** of Chapter 9

Mean polish and median polish. (These are the values for *d* = 70 in the exponential re-expression of Exhibit 21, Panel C.)

A) MEAN POLISH

21.1	21.7	28.5	37.4	48.8	58.5	64.6	40.1
38.5	39.2	45.9	54.4	65.4	76.0	81.3	57.2
57.6	62.2	69.8	79.4	86.8	93.7	96.7	78.0

−19.0	−18.4	−11.6	−2.7	8.7	18.4	24.5
−18.7	−18.0	−11.3	−2.8	8.2	18.8	24.1
−20.4	−15.8	−8.2	1.4	8.8	15.7	18.7
−19.4	−17.4	−10.4	−1.4	8.6	17.6	22.4

.4	−1.0	−1.2	−1.3	.1	.8	2.1
.7	−.6	−.9	−1.4	−.4	1.2	1.7
−1.0	1.6	2.2	2.8	.2	−1.9	−3.7

B) MEDIAN POLISH

21.1	21.7	28.5	37.4	48.8	58.5	64.6	37.4
38.5	39.2	45.9	54.4	65.4	76.0	81.3	54.4
57.6	62.3	69.8	79.4	86.8	93.7	96.7	79.4

−16.3	−15.7	−8.9	0	11.4	21.1	27.2
−15.9	−15.2	−8.5	0	11.0	21.6	26.9
−21.8	−17.1	−9.6	0	7.4	14.3	17.3
−16.3	−15.7	−8.9	0	11.0	21.1	26.9

0	0	0	.4	0	.3	0
0	.1	−.4	−.4	.1	−.4	.4
−1.9	2.2	2.9	0	−3.2	−6.0	−3.6

0	0	.4	0	.7	−.4
−.1	0	−.4	.1	0	
2.1	2.9	0	−3.2	−5.6	

Exhibit 11's purpose is to allow the comparison of the two sets of residuals. If we look at the final residuals of Panel A (analyzed by means), we see an undistinguished mess, in which 13 of the 21 residuals are greater than or equal to 1 in absolute value. Only with the utmost concentration might we see informative patterns among these residuals. The mean spreads the deviations around, thus making the sum of their squares small, as we illustrate again later in this section.

The final residuals of the analysis by medians, as we have already noticed, are much more interpretable: Only 6 of the 20 residuals are greater than 1 in absolute value, and all 6 of these, together with one zero, make up the 7 residuals for Laredo. Thus we learn that Caribou and Washington are in very good agreement, and we should start thinking about Laredo.

If we draw the line at a size of ± 0.5, we still have 7/21 for the analysis by medians, but now count 17/21 for the analysis by means.

Exhibit **12** shows the stem-and-leaf displays for the two sets of final residuals from Exhibit 11, compared with each other and with the final

Exhibit **12** of Chapter 9

Stem-and-leaf displays for three analyses of the same data (two analyses are from Exhibit 11)

	Analysis by means	Analysis by medians	Analysis by midrange
6		6	6
5		5	5
4		4	4
3		3 | 6	3
2	128	2 | 09	2 | 5414
1	276	1	1 | 37
0	41872	0 | 004070001	0 | 0666
−0	694	−0 | 104400	−0 | 356
−1	023409	−1 | 9	−1 | 7053
−2		−2	−2 | 1535
−3	7	−3 | 2	−3
−4		−4	−4
−5		−5 | 6	−5
−6		−6	−6

	Analysis by means	Analysis by medians	Analysis by midrange
High	2.8	3.6	2.5
Low	−3.7	−5.6	−2.5
Range	6.5	9.2	5.0
Sum of squares	51.04	71.57	58.57
Sum of absolute values	27.2	21.3	30.5

residuals obtained by taking midranges (mean of the highest and lowest) as chosen summaries. In addition to the stem-and-leaf displays, we also give three measures of spread: range, sum of squares, and sum of absolute values. Each of the three analyses beats the other two according to exactly one of these criteria, just as they should.

Unusual values. To continue our discussion of means and medians, we look at their comparative performance in a very pure case—in a two-way table that has all values identical except exactly one. We do this in a 3×7 table. We lose nothing in the illustration by making all values 0 except the center one and by making it 21 (so it is divisible by 3 and 7). Putting the odd value in the center promotes symmetry and visual contrast but makes no other difference.

0	0	0	0	0	0	0
0	0	0	21	0	0	0
0	0	0	0	0	0	0

1. *Median analysis.* The medians of all rows and columns are zero and the original table is also the residual table. (It is not the speediness of this analysis that recommends it, for that criterion would leave all tables untouched.)

2. *Mean analysis.* Here we find

0	0	0	0	0	0	0	$\sqrt{}$		0	0	0	0	0	0	0
0	0	0	21	0	0	0	3		−3	−3	−3	18	−3	−3	−3
0	0	0	0	0	0	0	$\sqrt{}$		0	0	0	0	0	0	0

$$ -1 \quad -1 \quad -1 \quad 6 \quad -1 \quad -1 \quad -1 $$

$\sqrt{}$	1	1	1	−6	1	1	1
$\sqrt{}$	−2	−2	−2	12	−2	−2	−2
$\sqrt{}$	1	1	1	−6	1	1	1

Thus what started as a table well summarized as 21 values, 20 of one kind and 1 of another, winds up with 4 kinds of values, distributing the deviations all over the table.

One might say, of course, that, if one had such a table, the fact would be recognized and means would not be used. Such a plan for recognizing outliers is more easily requested than executed. When the data are many and complicated, situations like the one revealed so starkly in the example may be well camouflaged, and a great deal of sophisticated scanning may be required to discover outliers.

Another position would be that part of the purpose of an analysis is to reallocate the deviations and that the pattern of residuals produced by the analysis by means is one satisfactory analysis. And an analysis that doesn't reallocate variation at all may not be appealing.

Our purpose here is not so much to settle this sort of argument as to make clearer how the two methods respond to outliers, the median tending to leave them as spikes, the mean tending to reduce them and spread the deviations around. The mean analysis has a final panel of residuals where absolute values sum to 48. Thus, using the mean analysis led to more than doubling the total absolute deviation (48 versus 21), at the same time that the sum of squared residuals was reduced from $441 = (21)^2$ to $252 = (15.9)^2$. If we were sure that we wanted to minimize the sum of squares, we would use the mean; if we want to minimize the sum of absolute values, we would gain much more, in this instance, by using the median ($48/21 = 2.3$ but $21/15.9 = 1.3$).

The example suggests that it takes only a few large outliers in a table to allow an analysis by means to conceal what would otherwise be a rather easily recognized structure—structure on a scale intermediate between that of the outliers and that of the usual irregularities.

Holes. Sometimes tables have empty cells, cells with no observation, or sometimes we wish to set an observation aside and see what would have been estimated in its absence. When we do the two-way iteration with some empty cells, we treat each row or column according to the number of cells it has present. We take the summary for the filled cells. Usually this situation leads to more iterations.

We analyze Exhibit **13** both with and without the entry 1035, using both medians and means, four analyses in all. Panel B of Exhibit 13 shows the mean coping with the outlier by spreading error over the table, and the median leaving the spike pretty well in place. When the outlier cell is removed, Panel C shows that the residuals for the two methods are much more nearly comparable, though the median has 10 residuals smaller than 10 in absolute value and

Exhibit **13** of Chapter 9

Analysis of a set of data on wheat, ratio of dry to wet grain (× 1000), by medians and means, with and without one cell value, 1035, present.

A) Original data

Block	None	Early	Middle	Late
1	718	732	734	792
2	725	781	725	716
3	704	1035	763	758
4	726	765	738	781

➡

the mean has only 6. Of course, this is a feature of the median—to give more very small and more very large residuals than the mean does.

Even more important are the changes of column (treatment) effect due to omission of the anomalous value. These are 0, −12, 0, 14 for median fitting, but 17, −51, 17, 17 for mean fitting. This shows how much less—sum of squared changes of 313 instead of 3468—the one anomalous value perturbs the column (treatment) effects when medians are used.

Fitting by means and medians together. We have been comparing the virtues of mean, median (and midrange) analyses. Sometimes a combination of methods may be used, as the following example illustrates.

Example. Seasonal adjustment. Exhibit **14** shows the logarithm of the indices of department-store sales, by months, from 1941 through 1945, with the mean of 1935 through 1939 set at 100. The entries under the years in the left half of

Exhibit **13** (continued)

B) With cell present

	Medians							Means				
−1	−38	1	33	−3			18	−78	12	48	−18
7	12	−7	−42	−4			32	−22	10	−21	−25
−30	250	15	−16	12			−67	154	−30	−57	53
1	−11	−1	16	3			18	−53	8	29	−10
−27	24	−13	13	749			−44	66	−22	0	762

C) With one cell (value = 1035) omitted

	Medians							Means				
−1	−26	1	19	0			1	−27	−5	31	−1
7	24	−7	−56	−1			15	29	−7	−38	−8
0	□	45	0	−15			−16	□	21	−6	2
0	0	−2	1	7			0	−3	−10	11	8
−27	12	−13	27	746			−27	15	−5	17	745

S) SOURCE

W. G. Cochran (1947). "Some consequences when the assumptions for the analysis of variance are not satisfied." **Biometrics, 3,** p. 27. With permission from the author and from the Biometric Society.

Exhibit 14 are actually $1000 \log_{10}(\text{index}/100)$. (We use logarithms because financial movements are most often multiplicative rather than additive in their effects. Considerable experience encourages using the logarithm, even though the fit to raw values would be about as satisfactory in this example.) Returning to Exhibit 14, we observe that the numbers are growing larger as the years go by. To see the degree of periodicity by years, we rank the values for months within each year in Exhibit **15**, and compute the median rank for each month and the range of that month's ranks.

For six of the months, the range of the ranks is 2 or less, October, November, and December being especially notable for the stability of their high ranks. January and July compete for the lowest sales. February, March, and August have the largest ranges. Notice especially the sharp break between each December and the following January. This means that we cannot expect to fit a smooth curve across years to data that have not been seasonally adjusted.

To fit the original log data of Exhibit 14, we first remove the 5-year median for each month and take the residuals in each row, as shown in the righthand side of Exhibit 14. The regression model we plan to fit is, except for

Exhibit **14** of Chapter 9

Removing the median monthly effect from the log values of the department-store sales index

Months	1941	1942	1943	1944	1945	Median*	\multicolumn{5}{c}{Residual$_1$ = log y − median}
	\multicolumn{5}{c}{log y}						
Jan.	−41	93	111	140	193	111	−152	−18	0	29	82
Feb.	−13	68	193	152	233	152	−165	−84	41	0	81
Mar.	45	146	158	230	328	158	−113	−12	0	72	170
Apr.	114	149	215	238	243	215	−101	−66	0	23	28
May	121	124	193	250	262	193	−72	−69	0	57	69
June	93	93	190	212	270	190	−97	−97	0	22	80
July	4	29	104	152	215	104	−100	−75	0	48	111
Aug.	117	104	146	196	225	146	−29	−42	0	50	79
Sept.	179	207	241	292	320	241	−62	−34	0	51	79
Oct.	140	230	272	320	364	272	−132	−42	0	48	92
Nov.	199	274	330	394	438	330	−131	−56	0	64	108
Dec.	364	418	438	507	548	438	−74	−20	0	69	110
Mean residual in column							−102	−51	3	44	91

* Median over five years for each month.

error terms,

$$y = \text{constant} + \text{month } i + \text{year } j,$$

where month i and year j stand for the month and year effect, i = Jan., Feb., ..., Dec., and j = 1941, ..., 1945. When we remove a median for each month, as we do in Exhibit 14, we are taking out an estimate of the "month effect." (We might alternatively have taken out a mean for each month.) What is left can be used to estimate an average effect for each year separately.

The reader will observe that the average residuals by year, given at the foot of the righthand side of Exhibit 14, are nearly linear in the years. It is tempting then to fit a straight line to the five points. Arguing against this is our knowledge that recessions and prosperous times make economic series jump around in nonlinear ways. And so, although a straight line might be useful for some purposes, in general we will want to take out an effect for each year, or possibly fit some more complicated curve, or use some smoothing device. Therefore, we choose to remove the column means from the first residuals and to examine them further. (Means instead of medians were taken only for variety.)

In Exhibit **16**, the months for the five years are assigned the numbers $t = 1, 2, ..., 60$. The year effects from the bottom of Exhibit 14 are subtracted from the first residuals given in the righthand side of that exhibit, to produce the final residuals shown in Exhibit 16. These have been smoothed using

Exhibit **15** of Chapter 9

Ranks of months within years for the original entries of Exhibit 14, together with the median rank and range of ranks for each month

Months	1941	1942	1943	1944	1945	Median	Range
			Ranks of months within years				
Jan.	1	$3\frac{1}{2}$	2	1	1	1	$2\frac{1}{2}$
Feb.	2	2	$6\frac{1}{2}$	$2\frac{1}{2}$	4	$2\frac{1}{2}$	$4\frac{1}{2}$
Mar.	4	7	4	6	9	6	5
Apr.	6	8	8	7	5	7	3
May	8	6	$6\frac{1}{2}$	8	6	$6\frac{1}{2}$	2
June	5	$3\frac{1}{2}$	5	5	7	5	$3\frac{1}{2}$
July	3	1	1	$2\frac{1}{2}$	2	2	2
Aug.	7	5	3	4	3	4	4
Sept.	10	9	9	9	8	9	2
Oct.	9	10	10	10	10	10	1
Nov.	11	11	11	11	11	11	0
Dec.	12	12	12	12	12	12	0

Exhibit **16** of Chapter 9

Residuals₁, in sequence instead of in tabular form (where residuals₁ = log y − median from Exhibit 14) and both raw and smoothed values of residual₂ = residual₁ − column means of residuals₁.

t	Resid₁	Resid₂*	Smoothed Resid₂	
1	-152	-50		
2	-165	-63	-50	
3	-113	-11		
4	-101	1		
5	-72	30	5	
6	-97	5		
7	-100	2	5	
8	-29	73	40	
9	-62	40		
10	-132	-30	-29	
11	-131	-29		
12	-74	28		
13	-18	33	28	
14	-84	-33	33	28
15	-12	39	-15	
16	-66	-15		
17	-69	-18		
18	-97	-46	-24	
19	-75	-24		
20	-42	9		
21	-34	17	9	
22	-42	9		
23	-56	-5	9	
24	-20	31	-3	

t	Resid₁	Resid₂	Smoothed Resid₂	
25	0	-3	31	-3
26	41	38	-3	
27	0	-3		
28	0	-3		
29	0	-3		
30	0	-3		
31	0	-3		
32	0	-3		
33	0	-3		
34	0	-3		
35	0	-3		
36	0	-3		
37	29	-15		
38	0	-44	-15	
39	72	28	-21	-15
40	23	-21	13	-21
41	57	13	-21	4
42	22	-22	4	
43	48	4		
44	50	6		
45	51	7	6	
46	48	4	7	
47	64	20		
48	69	25	20	

t	Resid₁	Resid₂	Smoothed Resid₂
49	82	-9	-9
50	81	-10	-9
51	170	79	-10
52	28	-63	-22
53	69	-22	
54	80	-11	
55	111	20	-11
56	79	-12	
57	79	-12	
58	92	1	
59	108	17	
60	110	19	

* resid₂ = resid₁ − column means of resid₁.

running medians of length three. Then the final smoothed residuals have been plotted against t, the number of the month, in Exhibit **17**.

The outstanding pattern of Exhibit 17 is that of small variability in the middle years and of larger variability for t in the first two or last two years. We know, from both Exhibits 14 and 15, that the middle year was fitted extremely closely, because for eleven of the twelve months, the median value was in the middle year. Inevitably, then, the more extreme years will fail to fit as well. Why the first two years should produce more residual variability after smoothing than the last two is not apparent, but the graph strongly suggests this. Conceivably, during the early war years, people could buy more or less at will, while during the later years they bought what was available.

9E. Fitting One More Constant

Although our PLUS fit to the table of mean monthly temperatures (3 places $\times$ 7 months) helped our understanding, we concluded it did not go far enough. The smallest further step we can take would be to fit one more constant. And what constant should it be?

Exhibit **17** of Chapter 9

Smoothed residuals$_2$ from Exhibit 16, where residuals$_2$ = residuals$_1$ − column means of residuals$_1$.

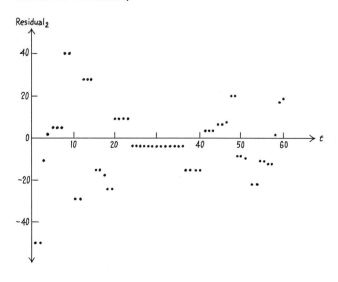

To bring in a multiplicative effect, we propose to add terms proportional to the product of the row and column effects. We write them as

$$\text{constant}\,\frac{(\text{row})(\text{column})}{\text{common}},$$

where the "constant" is the constant we spoke of fitting. This move makes our fit

$$\text{common}\ \ \text{PLUS}\ \ \text{row}\ \ \text{PLUS}\ \ \text{column}\ \ \text{PLUS}\ \ \text{constant}\,\frac{(\text{row})\,(\text{column})}{\text{common}}.$$

To execute the fit, we can recall that, for a given cell of the table,

first residual = data   MINUS   (common   PLUS   row   PLUS   column).

We plan to use the new term constant (row)(column)/common to fit these first residuals. Excluding the constant, we can call the remaining part of the new term a comparison value, where

$$\text{comparison value} = \frac{(\text{row})(\text{column})}{\text{common}},$$

the quantities on the right being the ones we have already computed. We can exhibit this relation by making a graph—plotting first residuals against comparison values—or by going to some more formal calculation, even a least-squares line. Exhibit **18**, Panel A, shows the calculation of comparison values for the alternative analysis given in Panel E of Exhibit 1, and the left side of

Exhibit **18** of Chapter 9

Calculation of comparison values for the alternative fit of Panel E of Exhibit 1 and its use to array the first residuals.

A) Calculation of COMPARISON VALUES

		Month						
		Jan.	Feb.	March	Apr.	May	June	July
Column		−20	−15	−10	0	10	20	25
Column/common		−.364	−.273	−.182	0	.182	.364	.454
	Row							
Comparison values	−20:	−7.28	−5.46	−3.64	0	3.64	7.28	9.08
= (row)(column)	0:	0	0	0	0	0	0	0
common	20:	7.28	5.46	3.64	0	−3.64	−7.28	−9.08

→

Panel B shows the results of ordering the pairs of comparison values and first residuals on the basis of their comparison values. The similar arrangement of positive and negative values for the pairs indicates the dependence of comparison values and first residuals. For each group of equal comparison values, the medians of the corresponding first residuals are written on the right side of Exhibit 18, Panel B. Clearly the dependence is strong.

Exhibit **19** shows the corresponding diagnostic plot, with comparison value as abscissa and first residual as ordinate. As is shown there, a line of slope +1.00 seems a reasonable fit, if we decide to stop with a line.

Panels A and B of Exhibit **20** show an analysis in which

$$1.00 \frac{(\text{row})(\text{col})}{\text{common}}$$

has also been fitted, by subtracting it from the first residuals of Exhibit 1, Panel E. We see (1) that the sizes of the new residuals are considerably smaller, and (2) that their signs are heavily negative (16 out of 21). The stem-and-leaf display of the new residuals in Panel C suggests that an even better fit (in the sense of reducing largest residuals) would be found by subtracting −1.0 (or even −2.0) from the new residuals and adding −1.0 (or −2.0) to the common, giving the fit

$$54 \text{ PLUS column } \text{PLUS } \text{row } \text{PLUS } \frac{(\text{row}) (\text{column})}{55}$$

or

$$53 \text{ PLUS column } \text{PLUS } \text{row } \text{PLUS } \frac{(\text{row}) (\text{column})}{55}.$$

Exhibit **18** of Chapter 9 (continued)

B) COMPARISON VALUE—FIRST RESIDUAL pairs ORDERED according to their COMPARISON VALUES

Comp. value	First resid.		Comp. value	First resid.		Comp. value	Median first resid.
9.08	4.0		0	−0.6		9.08	4.0
7.28	2.6		0	−1.6		7.28	3.0
7.28	3.4		0	−2.7		5.46	1.9
5.46	1.9		0	−2.9		3.64	3.4
3.64	3.4		−3.64	−3.8		0	−0.3
3.64	3.5		−3.64	−3.3		−3.64	−3.6
0	1.2		−5.46	−10.2		−5.46	−10.2
0	0.9		−7.28	−9.2		−7.28	−7.8
0	0.3		−7.28	−6.3		−9.08	−12.3
0	−0.3		−9.08	−12.3			
0	−0.3						

For the latter, the largest residuals would be about 3.2 in absolute value, in contrast to 5.1 (in Exhibit 20) or 12.3 (in Exhibit 1). Fitting one more constant has considerably improved the adequacy of our partial description.

Additive and multiplicative fitting. When we make this further fit, the values of the constant multiplying the comparison values can often be given a helpful interpretation (see *EDA*, Chapter 12). This particular extended additive fit can be equally well described as a multiplicative fit, because

$$\text{common} \quad \text{PLUS} \quad \text{row} \quad \text{PLUS} \quad \text{column} \quad \text{PLUS} \quad \frac{(\text{row})\,(\text{column})}{(\text{common})} \qquad (*)$$

is identical to

$$\text{common}\left(1 + \frac{\text{row}}{\text{common}}\right)\left(1 + \frac{\text{column}}{\text{common}}\right), \qquad (**)$$

as we see by multiplying (**) out. Here the quantities in parentheses could be renamed "row*" and "column*" since the value in the first parenthesis depends on only the row and the second on only the column, giving

$$\text{common TIMES row* TIMES column*.}$$

Exhibit **19** of Chapter 9

Diagnostic plot of first residuals against comparison values for data of Exhibit 18, Panel B. Circled point indicates two observations.

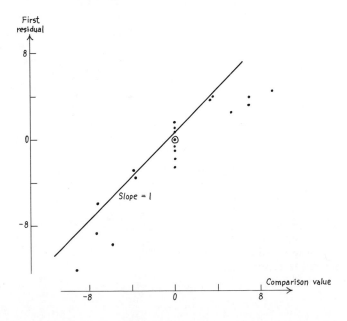

9F. Using Re-expression

In this section we illustrate the use of re-expression to make a fit involving one additional constant. This fit is tuned to a particular set of data, with no implication that the particular constant appearing in the re-expression would be good for another set of similar data. (This is quite different from the choice between raw values and log values, where a good choice for one set of data may well be a good choice for similar sets of data.) As in other cases of fitting a single constant, we feel free to fit "any old value." If we have no better way to go, we proceed by trying a variety of values for the constant and then fixing on the one that seems best. (When we tinker with the expression of the response, we shall use matched re-expressions as introduced in Section 5E.)

Exhibit **20** of Chapter 9

Fitting the extra constant to the first residuals in Exhibit 1, Panel E.

A) The PATTERN

	New residuals	Row contribution
$1.0 \times \dfrac{(row)(column)}{common}$		
	Column contribution	Common contribution

B) RESULTS

	Jan.	Feb.	March	Apr.	May	June	July	
Caribou	1.0	−4.7	.3	−.3	−.1	−3.9	−5.1	−20
Washington	1.2	−2.9	.3	−.6	−.3	−1.6	−2.7	0
Laredo	−4.7	−3.6	−.2	.9	−.2	−1.9	−3.2	20
Caribou	−7.3	−5.5	−3.6	0	3.6	7.3	9.1	
Washington	0	0	0	0	0	0	0	
Laredo	7.3	5.5	3.6	0	−3.6	−7.3	−9.1	
	−20	−15	−10	0	10	20	25	55

C) STEM-and-LEAF DISPLAY for NEW RESIDUALS

```
 1 | 20
 0 | 933
-0 | 122336        Median   −0.6
-1 | 69            Upper hinge   −0.1
-2 | 79            Lower hinge   −3.2
-3 | 269           Interquartile range  3.1
-4 | 77
-5 | 1
```

Here we combine the use of re-expression of response with the use of two-way analysis. The most used re-expressions are logarithms of amounts or counts, which, in view of

$$\log ((\text{common})(\text{row}^*)(\text{column}^*)) \equiv \log \text{common} + \log \text{row}^* + \log \text{column}^*,$$

converts situations adapted to multiplicative fitting into situations adapted to additive fitting.

Effective re-expression of the response by a logarithm corresponds to a slope of plus one on our diagnostic plot (Exhibit 19). As a consequence, we may take other positive slopes as suggesting re-expressions in the general direction of the use of logs. The slopes for square and cube roots go in the same direction, but are less vigorous than logs. The slopes for reciprocals and reciprocal roots go in the same direction but are more vigorous. But what if the slope of the diagnostic plot is negative—something less common but possible? Or what if the responses to be re-expressed take negative values as well as positive ones, as our Fahrenheit temperature example in Exhibit 1 could, and would if we extended our places some hundreds of miles north?

In many of the cases, where the slope is negative and the responses are routinely nonnegative, we can effectively re-express our response by squaring it—or by taking some other power greater than one. For potentially negative responses, however, squaring is unlikely to be satisfactory—since two values of y, one positive and one negative, correspond to each single value of y^2 (except zero). Confusion of values leads almost inevitably to confusion of result.

The simplest re-expressions that apply equally to negative and positive values, and whose plots, like that of y^2, are hollow upward, are probably the exponentials, introduced in Section 5A:

$$y \rightarrow e^{cy} = e^{y/d}$$

whose matched form—matched at y_0—corresponds to replacing y by

$$de^{(y-y_0)/d} + (y_0 - d),$$

which is the same as

$$(de^{-y_0/d})e^{y/d} + (y_0 - d).$$

Thus, if in our temperature example of Exhibit 1, Panel A, we choose $y_0 = 55$ (close to the center of the data, which was 54.4) and $d = 80$, we would use

$$(80e^{-55/80})e^{y/80} + (55 - 80) = 40.23e^{y/80} - 25,$$

something that can be evaluated in a few key strokes on a hand-held calculator.

Exhibit **21** shows the results of making a variety of such exponential re-expressions on our East Coast temperature data of Exhibit 1, Panel A. Note that, as d changes, there is a gradual change of the fitted values and of the residuals.

The chosen values of d are either multiples of 10 or values for which $120/d$ are simple, or both. The number 120 is not actually used in the

Exhibit **21** of Chapter 9

Re-expressed values (on the left) of the data of Exhibit 1, Panel A, and a two-way median analysis (on the right) for each of eight exponential re-expressions, no re-expression (raw values) and, for comparison, re-expression by squares. (All re-expressions matched at and near 55.)

A) No re-expression—raw values $d = \infty$, $120/d = 0$

8.7	9.8	21.7	34.7	48.5	58.4	64.0
36.2	37.1	45.3	54.4	64.7	73.4	77.3
57.6	61.9	68.4	75.9	81.2	85.8	87.7

−7.7	−7.6	−3.9	0	3.5	4.7	6.4	−19.7
.1	0	0	0	0	0	0	0
0	3.3	1.6	0	−5.0	−9.1	−11.1	21.5
−18.3	−17.3	−9.1	0	10.3	19.0	22.9	54.4

B) Re-expression for $d = 80$, $120/d = 1.5$

19.8	20.5	27.8	37.1	48.8	58.5	64.5
38.2	39.0	45.9	54.4	65.3	75.7	80.7
57.6	62.2	69.6	78.9	86.0	92.6	95.4

0	−1.2	−.8	0	.8	.1	1.1	−17.3
1.1	0	0	0	0	0	0	0
−.2	2.5	3.0	3.8	0	−3.8	−6.0	20.7
−17.3	−15.4	−8.5	0	10.9	21.3	26.3	54.4

C) Re-expression for $d = 70$, $120/d = 1.7143$

21.1	21.7	28.5	37.4	48.8	58.5	64.6
38.5	39.2	45.9	54.4	65.4	76.0	81.3
57.6	62.3	69.8	79.4	86.8	93.7	96.7

0	−.1	0	0	.4	0	.7	−17.4
0	0	0	−.4	−.4	.1	0	0
−1.9	2.1	2.9	3.6	0	−3.2	−5.6	21.0
−16.3	−15.6	−8.9	0	11.0	21.1	26.5	54.8

D) Re-expression for $d = 60$, $120/d = 2$

22.7	23.2	29.4	37.8	48.8	58.5	64.7
38.9	39.5	46.0	54.4	65.5	76.5	82.0
57.7	62.3	70.0	80.0	87.9	95.3	98.5

.4	0	0	0	0	0	0	−16.6
0	−.3	0	0	.1	1.4	.7	0
−3.7	0	1.5	3.1	0	−2.3	−5.3	22.5
−15.5	−14.6	−8.4	0	11.0	20.7	26.9	54.4

Exhibit 21 of Chapter 9 (continued)

E) Re-expression for $d = 48$, $120/d = 2.5$

25.3	25.7	31.0	38.4	48.9	58.5	64.9
39.4	40.1	46.2	54.4	65.7	77.4	83.4
57.7	62.4	70.5	81.2	89.9	98.2	101.9

1.9	1.6	.2	0	-.8	0	0	-16.0
0	0	-.6	0	0	2.9	2.5	0
-5.4	-1.4	0	3.1	.5	0	-2.7	23.7
-15.0	-14.3	-7.6	0	11.3	20.1	26.5	54.4

F) Re-expression for $d = 40$, $120/d = 3$

27.6	27.9	32.4	39.1	49.0	58.5	65.1
40.0	40.6	46.4	54.4	66.0	78.4	84.9
57.7	62.5	70.9	82.4	92.0	101.4	105.6

2.9	2.6	1.3	0	-1.7	-2.4	0	-15.3
0	0	0	0	0	2.2	4.5	0
-7.5	-3.3	-.7	2.8	.8	0	0	25.2
-14.4	-13.8	-8.0	0	11.6	21.8	26.0	54.4

G) Re-expression for $d = 36$, $120/d = 3.33$

28.9	29.3	33.3	39.5	49.1	58.6	65.2
40.4	40.9	46.5	54.4	66.1	79.0	85.9
57.7	62.6	71.2	83.3	93.5	103.7	108.3

3.4	3.3	1.7	0	-1.3	-2.0	0	-14.9
0	0	0	0	.8	3.5	5.8	0
-10.9	-6.5	-3.5	.7	0	0	0	28.2
-14.0	-13.5	-7.9	0	10.9	21.1	25.7	54.4

H) Re-expression for $d = 30$, $120/d = 4$

31.4	31.6	34.9	40.2	49.2	58.6	65.5
41.0	41.5	46.7	54.4	66.5	80.4	88.1
57.7	62.8	71.9	85.2	96.8	108.8	114.2

4.6	4.3	2.4	0	-2.6	-5.2	0	-14.2
0	0	0	0	.5	2.4	-3.7	0
-14.1	-9.5	-5.6	0	0	0	4.7	30.8
-13.4	-12.9	-7.7	0	11.6	23.6	29.0	54.4

Exhibit 21 of Chapter 9 (continued)

I) Notes on Panels A through H

1. The order of columns is January to July.
2. The order of rows is Caribou (top), Washington, Laredo (bottom).

J) Letter-value displays for residuals—selected d's

d = 70 (spr)			
M11		0	
H6	.1	-.1	.2
E3½	1.4	-1.2	2.6
2	2.9	-3.2	6.1
1	3.6	-5.6	9.2

d = 60 (spr)			
M11		0	
H6	.1	0	.1
E3½	1.0	-1.3	2.3
2	1.5	-3.7	5.2
1	3.1	-5.3	8.4

d = 40 (spr)			
M11		0	
H6	1.3	0	1.3
E3½	2.7	-2.0	4.7
2	2.9	-3.3	6.2
1	4.5	-7.5	12.0

4.8	4.4	.1	0	-.2	0	0	-16.0
0	0	-1.5	0	.5	2.0	1.1	0
-4.0	0	0	3.2	0	-2.4	-5.6	22.3
-15.0	-14.4	-6.7	0	10.7	20.1	26.3	54.4

K) Re-expression by squares

28.2	28.4	31.8	38.4	48.9	58.5	64.7
39.4	40.0	46.2	54.4	65.6	76.5	81.8
57.7	62.3	70.0	79.9	87.4	94.4	97.4

L) Note on Panel K

The re-expression used for squares follows the general formula of Section 5E, namely,

$$\frac{A}{p}\left(\frac{y}{A}\right)^p + \left(1 - \frac{1}{p}\right)A$$

which, for $A = 55$ and $p = 2$, becomes

$$\frac{1}{110}y^2 + \frac{55}{2}.$$

re-expressions or calculations of Exhibit 21. The advantage to using 120 and having $120/d$ simple is that this makes plotting against the reciprocal of d easy, where we expect—and in this instance confirm—simpler behavior in terms of (any multiple of) the reciprocal of d than in terms of (any multiple of) d itself, particularly since large values of d correspond to no re-expression, to using the raw values with which we started (as can be verified by taking limits).

When we look at the selected letter-value displays for residuals from fits with selected d's in Panel J, it is clear—since we used matched re-expressions—that $d = 60$ is better than either $d = 120$ or $d = 40$. What value of d is best? To approach an answer, Exhibit **22** plots the H-spreads, E-spreads, 2-spreads (2 in from each end), and 1-spreads (ranges) for all available values of d. In every case we see a minimum in the spread for d near 60–70, near 60 for the three outer spreads, and near 70 for the H-spread, which appears to depend much more irregularly on d.

Why do we have a minimum just here? Turning back to the residuals, which were, for $d = 60$,

Caribou	.4	0	0	0	0	0	0
Washington	0	−.3	0	0	.1	1.4	.7
Laredo	−3.7	0	1.5	3.1	0	−2.3	−5.3

and, for $d = 70$,

Caribou	0	−.1	0	0	.4	0	.7
Washington	0	0	0	−.4	−.4	.1	0
Laredo	−1.9	2.1	2.9	3.6	0	−3.2	−5.6

we see that the residuals for Caribou and Washington are very small, only tenths of a degree, while those for Laredo are not. However, looking up and down Exhibit 21, we see no closer agreement than for $d = 60$ and $d = 70$ between Laredo and either Washington or Caribou. Thus we conclude that the best exponential re-expression

◇ has d close to 60 or 70.

◇ brings Washington and Caribou into close agreement.

◇ does reasonably well in bringing Laredo into agreement with the other two, but is not as successful.

Since it is easy to argue that Laredo, though not far from the Gulf of Mexico, is not genuinely "East Coast," it would be natural for us to next examine a larger variety of genuine East Coast places and see how well $d = 70$ does.

Note that in such an exponential re-expression, we are again fitting

one more constant

although this time the constant is in the re-expression of the response, rather than in the fit itself (as it was in Section 9E, when the constant times a comparison value was added to the fit).

9G. Three- and More-Way Analyses

If we wish to analyze tables with more than two ways, the arithmetic inevitably gets heavier. The general technique is much like that of the two-way attack. For a three-way analysis, let us call the ways A, B, and C. Then the paired variables are AB, BC, and CA. Let us begin with AB. For each AB pair, there is a line of entries differing only in C. We take a half-step, finding and removing a summary for every such line. Then we turn to the BC pairs, and take a half-step for the residuals in the lines where only A changes. Then we turn to the CA pairs—and continue after that with AB again.

The approach is illustrated and treated more extensively in *EDA*, Chapter 13.

Exhibit **22** of Chapter 9

Dependence on *d* of various spreads of residuals. The spreads are calculated from the values of $de^{(y-y_0)/d} + (y_0 - d)$, the re-expressions of the mean monthly temperatures in Exhibit 21.

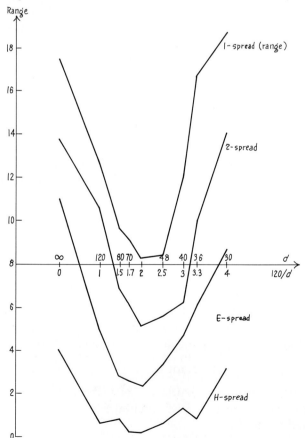

Summary: Two-Way Tables of Responses

Analyses of tables of responses that come out in the form: common PLUS row PLUS column PLUS residual offer a general pattern within which there can be important details of fitting.

These analyses extend what we do when we "eyeball" a table of responses.

Having more numbers in a full analysis than in the data is a normal phenomenon, often necessary for a careful examination of what appears to be going on; such an examination does not bar further summarization.

Starting from scratch (or from some other analysis), successive removal of medians, alternately by rows and by columns (median polish), produces common PLUS row PLUS column fits.

Such analysis by medians leaves unusual perturbations in the data much more clearly delineated in the residuals than does analysis by means, which tends to smear the consequences of isolated perturbations quite widely.

Such an analysis by medians works whether the table is complete or whether a certain number of values are either missing or chosen to be set aside for a particular analysis.

Any row-PLUS-column-PLUS-common fit can be graphically displayed in terms of two families of parallel lines, with the vertical coordinate of the intersections giving the values of the fit.

Coded symbols, showing both their signs and rough sizes, let us make effective pictures of residuals.

The appearance of matching signs of residuals in opposite corners, of either a well-rearranged numerical table or a coded plot of residuals, gives a basis for trying a specific sort of further fitting—fitting one more constant.

Fits of the form

$$\text{common} \quad \text{PLUS} \quad \text{row} \quad \text{PLUS} \quad \text{column} \quad \text{PLUS} \quad 1.0\,\frac{(\text{row})\,(\text{col})}{(\text{common})}$$

can also be written as

$$\text{common TIMES row* TIMES column*.}$$

Plotting residuals against comparison values allows us to choose the "constant" in a term

$$\text{PLUS (constant)}\,\frac{(\text{row})(\text{col})}{(\text{common})}$$

to be added to our previous common-PLUS-row-PLUS-column fit.

We can often re-express y as a convenient alternative to fitting the one more constant. To see which re-expression allows common-PLUS-row-PLUS-column to fit very well, we can try varied re-expressions of the responses.

The qualities of such different fits may be compared by using re-expressions matched near the center of the values of our responses, and looking at the behavior of various simple measures of spread for residuals.

References

McNeil, D. R., and J. W. Tukey (1975). "Higher-order diagnosis of two-way tables, illustrated on two sets of demographic empirical distributions." *Biometrics*, **31,** 487–510.

EDA = Tukey, J. W. (1977). *Exploratory Data Analysis*. Reading, Mass.: Addison-Wesley.

Chapter 10/Robust and Resistant Measures of Location and Scale

Chapter index on next page

In this chapter we press beyond the ideas of robust and resistant methods given in Chapter 1 and introduce a number of ideas lightly. We referred to them in Chapter 9 and will return to them in more detail in Sections 14G, 14H, and 14I.

10A. Resistance

Resistance is a property we like summary statistics to have. If changing a small part of the body of data, perhaps drastically, can change the value of the summary substantially, the summary is not resistant. Conversely, if a change of a small part of the data, no matter what part or how substantially, fails to change the summary substantially, the summary is said to be *resistant*.

The arithmetic mean is the prototype of a nonresistant summary. If, in

$$\frac{1 + 2 + 2 + 3 + \cdots + 23}{101} = 9.58,$$

we change the second measurement of 2 to 101002, the arithmetic mean will change to

$$\frac{1 + 2 + 101002 + 3 + \cdots + 23}{101} = 1009.58.$$

Changing 1/101st of the data has changed the mean tremendously.

The median is the prototype of a simple resistant summary. In our example of 101 values, suppose we have

$$\begin{aligned}
50 \text{ values} &\quad \{1, 2, 2, 3, \ldots, 8, 9\} \\
1 \text{ value} &\quad \{9\} \\
50 \text{ values} &\quad \{9.5, 10, \ldots, 23\}
\end{aligned}$$

where the median is 9. Then changing the second 2 to 101002 changes the body of data to

$$\begin{aligned}
50 \text{ values} &\quad \{1, 2, 3, \ldots, 8, 9, 9\} \\
1 \text{ value} &\quad \{9.5\} \\
50 \text{ values} &\quad \{10, \ldots, 23, 101002\}
\end{aligned}$$

with median 9.5—not far from 9.

No change of a single value could move the median any further than this. Clearly the median is resistant.

Other kinds of summaries are also resistant. Many important ones are defined implicitly and have to be calculated by iteration, as in the following example, where iteration is required because we cannot compute y^* until we know the w's and we cannot compute the w's until we know y^*:

$$y^* = \frac{\sum w_i y_i}{\sum w_i}$$

where

$$w_i = \begin{cases} \left(1 - \left(\dfrac{y_i - y^*}{cS}\right)^2\right)^2 & \text{when } \left(\dfrac{y_i - y^*}{cS}\right)^2 < 1, \\ 0, & \text{otherwise,} \end{cases}$$

and

$$S = \text{median } \{|y_i - y^*|\}$$

or, perhaps,

$$S = \tfrac{1}{2}(\text{spread between hinges}),$$

and c is a constant, often taken as 9 or 6. (It may help to think of S as estimating roughly $\frac{2}{3}\sigma$, so that, with $c = 6$, we allow the residuals to count up to about 4σ, where the spread is being measured more by the middle observations than by the large ones.) The estimate thus defined is often called a *biweight estimate*. (See Chapter 14 for more about such estimates.)

10B. Robustness

Why not be satisfied with the median? Why squander the considerable amount of arithmetic needed to calculate a more complex resistant estimate of location like the *biweight*?

We are concerned with efficiency—with using a summary that is thorough in extracting from bodies of data most of what we can learn about the aspect being summarized.

How can we measure thoroughness of extraction? We must

◇ decide how to measure effectiveness, and

◇ decide in what body of data the extraction will take place.

Except for small samples from distributions with very stretched tails, it is usually satisfactory to measure effectiveness in terms of the variance of the summary (assuming that, on the average, the summary comes close to assessing what we would like assessed—otherwise special circumstances, such as special loss functions, might rule). Using variances, a natural measure of efficiency is

$$\text{Efficiency} = \frac{\text{lowest variance feasible}}{\text{actual variance}}.$$

We usually report efficiency as a percentage. What then should we think about 90% efficiency? The answers are that

◇ it is very good.

◇ this summary's behavior could be distinguished from that of a 100% efficient summary.

◇ but it would be a massive job to do this in practice.

Gains of a single percent of efficiency are essentially never detectable in practice.

What we want is

robustness of efficiency.

We want to have high efficiency in a variety of situations, rather than in any one situation. Our attention naturally shifts to the poorest efficiency in a reasonable collection of situations. If this is high, we can rightfully feel that we have a good summary.

10C. Robust and Resistant Estimates of Location

How do the three estimates of location mentioned above survive the new test of robustness of efficiency? Exhibit 1 has the story.

The conclusion is that, except for quite small samples, the biweight estimate has all the desired properties. (For the smallest samples, say those of count three, four, and five, we might do better with the median.)

Exhibit **1** of Chapter 10

Resistance and robustness of efficiency of some location estimators

Estimator	Sample size	Resistant?	Gaussian efficiency	Stretched-tail efficiency	Robustness of efficiency
Arithmetic mean	Small	No	100%	Poor	Poor
	Large	No	100%	Very poor	Very poor
Median	Small	Yes	High	Higher	High
	Large	Yes	62%	Higher	Moderate
Biweight with $c = 6$ or 9	Small	Reasonably	Highish	Higher	Highish
	Large	Yes	>90%	>90%	High

In practice, then, we tend to:

◇ use the median in exploration, and in other circumstances where moderate efficiency in a wide variety of situations is enough.

◇ use the biweight, or one of its relatives, when high performance is needed.

◇ use the arithmetic mean only after careful study: when traditions or meaningfulness in the field of exploration of application require it; when the cost of redoing computer programs is exorbitant; when better confidence or significance techniques are available only for means; when the convenience of linearity is required; or when the type of data has short tails and no outliers, as we explain next.

Sometimes distributions tend to have tails tighter than or about like a normal distribution, rather than more straggling. Under these circumstances, means are more efficient than biweights, and, for very short tails, special location statistics can be developed that give more weight to the far-out observations than to those in the center of the sample. The uniform distribution has especially short tails for a continuously spread-out distribution. For samples from the uniform, theory tells us that the average of the two extreme observations uses all the sample information. This illustrates how, for short-tailed distributions, the extreme observations should get more weight.

10D. Robust Estimates of Scale

Here we do not know as much as we would like. Either the

$$\text{MAD} = \text{Median Absolute Deviation} = \text{median } |y_i - y^*|,$$

where y^* is a resistant estimate of location, or the interquartile range I (distance between the hinges or 25% points),

$$I = \text{H-spread} = \text{upper hinge} - \text{lower hinge},$$

is a resistant estimate of scale (of spread). We recognize these two as barely robust of efficiency.

Ranges, and range-based estimates of scale, are perhaps a shade less robust than $s = $ (sample variance)$^{1/2}$. Neither is competitive with MAD or I.

An alternative. David Lax (1975) has appraised a rather complicated statistic that does more than twice as well as MAD in some fairly realistic situations. Let

$$\overset{|}{y} = \text{median of } y\text{'s},$$

$$u_i = \frac{y_i - \overset{|}{y}}{9(\text{MAD})}$$

Then he uses a measure of scale derived from the asymptotic variance of a biweight, namely,

$$\frac{n\sum' (y - \overset{\shortmid}{y})^2(1 - u^2)^4}{[\sum' (1 - u^2)(1 - 5u^2)]^2},$$

where $\sum'$ indicates summation for $u^2 \le 1$ only. Roughly speaking, the u's define weights. When u^2 is small, the weights are about equal, the denominator reduces to n, and the whole expression becomes $\sum (y - \overset{\shortmid}{y})^2/n$, something that looks like a reasonable variance estimate.

A modification that reduces to

$$s^2 = \frac{\sum (y - \overset{\shortmid}{y})^2}{(n - 1)},$$

and is known to perform somewhat better, is

$$ns_{bi}^2 = \frac{n\sum' (y - \overset{\shortmid}{y})^2(1 - u^2)^4}{[\sum' (1 - u^2)(1 - 5u^2)][-1 + \sum' (1 - u^2)(1 - 5u^2)]}.$$

Note here that $(1 - u^2)^4 = w^2$, where w was introduced in Section 10A. Inevitably other estimates will be developed.

10E. Robust and Resistant Intervals

The word "robust" has been used in statistics with many different meanings—it will probably be used with many more. Until about 1970, it usually meant robustness of validity—that what was nominally 5% was in fact 5% for a wide variety of situations. This is separate from the meaning here—separate but not necessarily contradictory.

Indeed, Alan Gross (1976) has shown us that we can have both robustness of efficiency and robustness of validity in a simple procedure for setting confidence limits for centers of symmetric situations. In later work, Gross tells us, he has shown that one can do essentially as well with an interval based upon a biweight. A modified procedure uses, for $\hat{y}$, the biweight estimate, namely

$$u_i = \frac{y_i - \hat{y}}{9(\text{MAD})}$$

and then takes, as the estimated variance of $\hat{y}$, the (asymptotically correct) expression (cf. Section 10D):

$$\text{vâr } \hat{y} = s_{bi}^2 = \frac{\sum' (y - \hat{y})^2(1 - u^2)^4}{[\sum' (1 - u^2)(1 - 5u^2)][-1 + \sum' (1 - u^2)(1 - 5u^2)]}.$$

(Whether we use the biweight $\hat{y}$ or the median $\overset{\shortmid}{y}$ is not important.) Then use Student's t with 0.7 times the usual number of degrees of freedom, $\nu =$

$0.7(n - 1)$, to give the interval

$$\hat{y} \pm t_\nu \sqrt{\text{vâr } \hat{y}}.$$

This works well for $n \geq 8$.

10F. Resistant and Robust Regression

For the iterative use of biweights in regression, see Sections 14G, 14H, and 141.

10G. Multiple-Component Data

When we have multiple-component data—often called vectors—we are likely to want:

⬦ resistant-and-robust location estimates—which we can get by applying biweights to each component separately.

⬦ resistant-and-robust substitutes for variances, covariances, and correlations.

How are we to find the latter?

Classically, the most frequent use for calculated estimates of variances and covariances was to compute linear regressions. We now turn this around, and use resistant-and-robust linear regressions as a tool in calculating resistant-and-robust estimates of variances and covariances.

The fundamental fact about variances and covariances is their *quadratic nature*, best shown forth by the identities they satisfy, such as

$$\text{var} (fu + gv) = f^2 \text{ var} (u) + 2fg \text{ cov} (u, v) + g^2 \text{ var} (v),$$

and, more generally,

$$\text{cov} (fu + gv, ju + kv) = fj \text{ var} (u) + (fk + gj) \text{ cov} (u, v) + gk \text{ var} (v),$$

and not at all by how they are calculated. Our usual estimates v̂ar and ĉov of variances and covariances, of course, satisfy the same identities.

If we are to define estimates of analogs of variances and covariances, we want them to satisfy the same identities. Since they will not be quadratic functions of the data, there is only one easy way to satisfy the identities: First, define estimates for special components, and then use the identities to extend the definition.

Suppose, then, that we are willing to begin with specially chosen components x, $y - bx$, $z - cy - dx$, ..., and that we are willing

1. to take zero as the estimate of the analog of covariance, which we will denote ĉob rather than ĉov, between each pair of these special components; and

2. to take the square of a resistant estimate of scale as the analog of the variance, which we will denote b̂ar rather than v̂ar, for each of these special components.

We use b for robust, since r (= "rar") would be far from euphonious—note, too, that "b" sounds a lot like "v".

The identities teach us, in particular, that:

$$\hat{b}ar\,(y) \equiv \hat{b}ar\,((y - bx) + bx)$$
$$= \hat{b}ar\,(y - bx) + b^2\,\hat{b}ar\,(x),$$

$$\hat{c}ob\,(x, y) \equiv \hat{c}ob\,(x, (y - bx) + bx)$$
$$= b\,\hat{b}ar\,(x),$$

$$\hat{b}ar\,(z) \equiv \hat{b}ar\,(z - cy - dx) + c^2\,\hat{b}ar\,(y - bx) + (d + bc)^2\,\hat{b}ar\,(x);$$

$$\hat{b}ar\,(2y + 3z) \equiv 4\,\hat{b}ar\,y + 6\,\hat{c}ob\,(y, z) + 9\,\hat{b}ar\,z$$
$$= 9\,\hat{b}ar\,(z - cy - dx) + (9c^2 + 6c + 4)\,\hat{b}ar\,(y - bx)$$
$$+ (9(d + bc)^2 + 6(d + bc)b + 4b^2)\,\hat{b}ar\,(x),$$

so that $\hat{b}ar$ and $\hat{c}ob$ are defined for all linear combinations of x, y, z, . . . Thus it remains only for us to:

1. choose the constants b, c, d, . . .

2. decide which resistant-and-robust estimates of scale are to define $\hat{b}ar\,(x)$, $\hat{b}ar\,(y - bx)$, $\hat{b}ar\,(z - cy - dx)$, . . .

The simple choices seem to be quite satisfactory for the present, namely:

1. $y - bx$, $z - cy - dx$, . . . shall be the residuals from our conventional resistant-and-robust regression of y on x, and of z on x and y, and

2. we will take

$$\hat{b}ar\,(x) = ns_{bi}^2 \quad \text{(for the values of } x\text{)},$$
$$\hat{b}ar\,(y - bx) = ns_{bi}^2 \quad \text{(for the value of } y - bx\text{)},$$

and so on, where the ns_{bi}^2 are the biweight variances defined in Section 10D.

Other robust variables might be chosen, such as the square of the median deviation from the median, or even the square of the difference between two order statistics.

A comment. We have chosen to keep the property of resistance for $\hat{b}ar$ and $\hat{c}ob$, namely that:

⋄changing a small part of the data, no matter how much, makes only a small change in the results.

We have chosen to give up—in part because we do not know how to keep both this and resistance—a property that some might have thought equally important, namely that:

⋄changing the order of the components, say from $(x, y, z, . . .)$ to

$(z, x, y, \ldots)$, or replacing one set of components by another linearly equivalent to it, say from $(x, y, z, \ldots)$ to $(x + y, x - y, 2y - z, \ldots)$, does not change the results.

We are acting as if invariance were nice, but can be dispensed with, while resistance and robustness of efficiency cannot be spared. And that is just what we think and recommend.

Alternatives. If we have only three components, they can be written down in only six orders. Those who wish to have the order of their components not matter could start from each of the six permutations, go all the way to the b̂ars and ĉobs of the original components—and then combine the six sets of values, taking either the arithmetic mean or some robust summary. With 10 components there are far too many orders to do anything like all of them, but one could do the calculations for a random sample of 10 or 20 of the permutations (cp. Moses and Oakford, 1963). While the means—or the summaries—of the 10 or 20 results would not be free of all effects of where we started, the corresponding spreads would show, by their sizes, just about how much effect where we started has on the results. We call attention to these possibilities without recommending their use.

(Optional) Invariance under permutation, or an estimate of how much permutation matters, may not satisfy a few. They would like invariance under all rotations, or even under all affine transformations. If they were willing to be satisfied with an estimate of how much difference rotation, or even affine transformation, matters, they can get this by drawing a random sample of rotations, or of affine transformations, applying each to the given form of the data, carrying through the calculations to identify special components, and then establishing b̂ars and ĉobs for the original data as found, starting from each transformed set of initial components. Again, the spreads among corresponding results would show how large are the differences that arise from the original choice of components.

Consequences. Now that we are not going to use the estimates of analogs of variance and covariance to calculate linear regressions, one of their most frequent uses (deplored by some) will be to calculate estimated analogs of correlation coefficients, for example:

$$\hat{\text{corr}}(x, y) \equiv \frac{\hat{\text{cob}}(x, y)}{\sqrt{\hat{\text{bar}}(x)\,\hat{\text{bar}}(y)}}$$

$$= \frac{b\,\hat{\text{bar}}(x)}{\sqrt{\hat{\text{bar}}(x)[\hat{\text{bar}}(y - bx) + b^2\,\hat{\text{bar}}(x)]}}$$

$$= \pm \frac{1}{\sqrt{1 + \left(\dfrac{\hat{\text{bar}}(y - bx)}{b^2\,\hat{\text{bar}}(x)}\right)}},$$

where the sign in the last form is that of b.

Another frequent use will be in choosing linear combinations of our original components—principal components, for instance—or in assessing properties of such newly chosen components—canonical correlations, for instance. In such uses, we may find it important to protect ourselves from the consequences of giving up invariance. There is a simple way to do this:

1. Calculate b̂ar and ĉob values as above.

2. Use these as if they were v̂ar and ĉov values, to calculate the new components.

3. Return to step (1), using the new components instead of the original components.

4. Repeat step (2), finding, usually, minor modifications of what we found at step (2).

5. Use the result.

This plan does not eliminate all the consequences of loss of invariance, but it does protect us from most of them.

Example 1. Andrews case. Exhibit **2** gives 11 x and y pairs and the corresponding values of v̂ar, ĉov, b̂ar, and ĉob. The conventional correlation is

$$\hat{r}(\text{classical}) = -0.7333.$$

The resistant and robust analog is

$$\hat{\text{corr}}\,(x, y) = 0.99992.$$

We see how great a difference doing things resistantly can make.

Example 2. Ten-in-a-box. Exhibit **3** gives 17 (x, y, z) triples, and goes on to find:

i)   as column 6, the residuals of y for a resistant straight-line fit to x;

ii) as column 8, the residuals of z for a resistant straight-line fit to x;

iii) as column 9, the residuals of column 8 for a resistant straight-line fit to column 6; and

iv) as column 10, the residuals of x for a resistant fit of a constant.

The values of s_{bi}^2 and ns_{bi}^2 can then be calculated for columns 6, 9, and 10, since we know that the biweight fit to each of these columns is zero (to a more than adequate approximation).

We have now calculated both the b, c, and d in x, $y - bx$, $z - cy - dx$ and the ns_{bi}^2 for these three quantities (which are also their b̂ars). We go on to calculate the complete table of b̂ars and ĉobs and, for comparison, the complete table of v̂ars and ĉovs. The two tables show no trace of resemblance.

Exhibit **2** of Chapter 10

Andrews case; resistant correlation example

A) The DATA

x	y	$y - x$	(4)	$x - 5$	$x - 4.64$
.02	.04	.02	.0093	−4.98	−4.62
.99	1.03	.04	.0349	−4.01	−3.65
2.01	1.97	−.04	−.0391	−2.99	−2.63
2.98	2.96	−.02	−.0135	−2.02	−1.66
4.03	3.97	−.06	−.0474	−.97	−.61
5.01	4.98	−.03	−.0117	.01	.37
6.05	6.07	.02	.0443	1.05	1.41
6.98	7.03	.05	.0797	1.98	2.34
8.07	8.00	−.07	−.0340	3.01	3.37
9.03	8.96	−.07	−.0284	4.03	4.12
25.00	−25.00	−50.00	−49.8658	20.00	20.36
H		.02	.0221	2.50	2.86
H		−.05	−.0366	−2.50	−2.14
S		.035	.0294	2.50	2.50
cS		.315	.2641	22.50	22.50

B) RESULTS of VARIOUS COMPUTATIONS

Biweight fit of $y - x$ to x: $0.0108 - 0.0058x$

Column (4): $(y - x) - (0.0108 - 0.0058x) = y - 0.0108 - 0.9942x$

s_{bi}^2 (for col. (4), using $cS = 0.2641$) $= 0.00019136$

ns_{bi}^2 (for same) $= 0.0021050 = \hat{b}ar\,(y - bx)$

Biweight fit of $x - 5$ to a constant: 0.36

Column (5): $(x - 5) - (-0.36) = x - 4.64$

s_{bi}^2 (for $x - 4.64$, using $cS = 22.5$) $= 1.13587$

ns_{bi}^2 (for same) $= 12.4945 = \hat{b}ar\,x$

$$\hat{c}orr = \left\{ 1 + \frac{.0021050}{(0.9942)^2 (12.4945)} \right\}^{-1/2} = 0.99992$$

$$\hat{r}(\text{classical}) = \frac{(-46.7004)}{\sqrt{(46.4410)(87.3414)}} = -0.7333$$

Exhibit **3** of Chapter 10

Ten-in-a-box example, perturbed form

A) DATA, RESIDUALS, SPREADS

x	y	z	$y-x$	(5)	(6)	$z-x$	(7)	(8)	(9)	(10)
1001	999	-1001	-2	-.817	-.752	-2002	-2000.564	-2000.522	-2000.514	1000.883
999	-1001	-1000	-2002	-2000.820	-2000.820	-1999	-1997.566	-1997.524	-1997.252	998.883
15	14	33	-1	-1.060	-1.060	18	18.219	18.258	18.266	14.883
-7	-6	-17	1	-.929	.927	-10	-9.808	-9.769	-9.762	-7.117
1000	1001	999	1	2.182	2.247	-1	.435	.477	.484	999.883
-1001	1001	-1000	2002	2000.658	2000.590	1	-.035	.001	-.255	-1001.017
22	21	23	-1	-1.052	-1.052	1	1.227	1.266	1.274	21.883
12	10	11	-2	-2.064	-2.065	-1	-.785	-.746	-.738	11.883
1	-1	0	-2	-2.078	-2.078	-1	-.789	-.760	-.751	.883
-10	-11	-12	-1	-1.094	-1.094	-2	-1.812	-1.773	-1.765	-10.117
-23	-22	-21	1	.892	.889	2	2.172	2.111	2.218	-23.117
1001	-1000	999	-2001	-1999.817	-1999.752	-2	-.564	-.523	-.250	1000.883
-1001	-999	-1000	2	.658	.590	1	-.035	.001	.009	-1001.117
6	8	16	2	1.928	1.927	10	10.208	10.247	10.254	5.883
-155	-14	-33	1	.902	.900	-18	-17.818	-17.779	-17.772	-15.117
-1000	1001	999	2001	1999.660	1999.592	1999	1997.964	1998.022	1997.746	-1000.117
-1001	-999	1000	2	.658	.590	2001	1999.963	2000.001	2000.009	-1001.117
H 22			2	.902	.902	2	2.170	2.111	2.218	21.883
H -23			-2	-1.092	-1.094	-2	-1.814	-1.773	-1.765	-23.117
S 22.5			2	.997	.997	2	1.992	1.992	1.992	22.5
cS 202.5			18	8.973	8.973	18	17.928	17.928	17.928	202.5
s_{bi}^2					.1799				1.0033	25.583
ns_{bi}^2					3.0577				17.0556	434.914

Exhibit **3** of Chapter 10 (continued)

B) STRAIGHT-LINE FITS, COLUMNS OF PANEL A

Biweight fit of $y - x$ to x: $.079 - .001261x$
Column (5): $y - x - (.079 - .001261x) = y - .079 - .998739x$
Biweight fit of (5) to x: $.001 - .000067x$
Column (6): column (5) $- (.001 - .000067x) = y - .080 - .998672x$

Biweight fit of $z - x$ to x: $-.200 - .001235x$
Column (7): $z - x - (-0.200 - .001235x) = z + .200 - .998765x$
Biweight fit of (7) to x: $-0.039 - .000003x$
Column (8): column (7) $- (-.039 - .000003x) = z + .239 - .998762x$

Biweight fit of (8) to (6): $-.007 + .000132$ (column 6)
Column (9): column (8) $-(-.007 + .000132 (y$ $.080 - .998672x))$
           $= z + .246 - .000132y - .998630x$

Biweight fit of x to a constant: $.117$
Column (10): $x - .117$

C) CALCULATION OF $\hat{B}$AR and $\hat{C}$OB

Special combinations

	$\hat{b}$ar
x	434.914
$y - bx$	3.0577
$z - cy - dx$	17.0556

where

$$b = .998672$$
$$c = .000132$$
$$d = .998630$$

so that

$\hat{b}$ar $y = 3.0577 + (.998672)^2(434.914) = 436.758$
$\hat{c}$ob $(x, y) = (.998672)(434.914) = 434.336$
$\hat{b}$ar $z = 17.0556 + (.000132)^2(3.0577) + (.998672^*)^2(434.914) = 450.815$
$\hat{c}$ob $(x, z) = (.998672^*)(434.914) = 434.336$
$\hat{c}$ob $(y, z) = (.00132)(3.0577) + (.998672)(.998672^*)(434.914) = 433.760$

* These are $d + bc = .998672$ and not $b = .998672$

D) COMPARISON OF $\hat{V}$AR/$\hat{C}$OV and $\hat{B}$AR/$\hat{C}$OB

$\hat{V}$AR AND $\hat{C}$OV				$\hat{B}$AR AND $\hat{C}$OB		
500362.43	−81.49	−97.70	x	434.914	434.336	434.336
	500221.61	332.59	y		436.758	433.760
		500217.57	z			450.815

How can the $\hat{bar}$–$\hat{cob}$ table be so different from the $\hat{var}$–$\hat{cov}$ table? Essentially because our example has been carefully loaded so that the eight outer points, which come very close to the (±1000, ±1000, ±1000) corners of a large cube, will dominate the $\hat{var}$–$\hat{cov}$ entries and have very little effect on the $\hat{bar}$–$\hat{cob}$ entries. When they dominate, the variances are large—close to half a million—and the shape appears very nearly spherical (or cubical). When the corners of the large cube drop away, the $\hat{bar}$s are small—around 435—and the shape corresponds to a flattened cigar, to an ellipsoid with semiaxes of, roughly, 20.9, 4.1, and 1.8. The latter shape is determined almost entirely by the locations of the nine inner points. (The sizes are enhanced over what the nine points alone would have given, because each ns_{bi}^2 is increased when a point gets very little—or no—weight—when it appears so large as to deserve "skipping". This happens when the size of $y - \hat{y}$ approaches or exceeds cS.)

Comment

A Roman-numeral jack-in-the-box is about to pop out at you! Plotting—or thinking hard—about the unperturbed points, as in Exhibit **4,** for the ten-in-a-box example, as given in Exhibit 3, could persuade the ten-in-a-box configuration to show itself to the reader almost as suddenly.

Exhibit **4** of Chapter 10

Ten-in-a-box example, unperturbed form

x	y	z
1000	1000	−1000
1000	−1000	−1000
14	14	32
−7	−7	−16
1000	1000	1000
−1000	1000	−1000
22	22	22
11	11	11
0	0	0
−11	−11	−11
−22	−22	−22
1000	−1000	1000
−1000	−1000	−1000
7	7	16
−14	−14	−32
−1000	1000	1000
−1000	−1000	1000

Example 3. *Alternative solution to the Andrews case.* Let us return to the first example, and open the resistant analysis of the Andrews case from another start, taking $y + x$ as a residual, rather than $y - x$. Exhibit 5 has the first few steps of the arithmetic leading to a slope of

$$-1.0023$$

(which is close, compared with the -1.0056 of the classical analysis). We thus see that, for this set of data, we can use the same resistant straight-line fitting procedure to obtain two quite different fits, one with slope 0.9942 (Exhibit 2) and one with slope -1.0023 (Exhibit 5).

It would often be claimed that it was regrettable that a single fitting procedure would give either of two such different answers, and that it ought to give only one. However, if we plot the data of Exhibit 2, even roughly, it is

Exhibit **5** of Chapter 10

Another look at the Andrews configuration

x	y	y + x	(4)	(5)	(6)
.02	.04	.06	−7.50	−8.06	−8.118
.99	1.03	2.02	−5.54	−6.10	−6.155
2.01	1.97	3.98	−3.58	−4.14	−4.193
2.98	2.96	5.94	−1.62	−2.18	−2.230
4.03	3.97	8.00	.44	−.12	−.169
5.01	4.98	9.99	2.43	1.87	1.824
6.05	6.07	12.12	4.56	4.00	3.956
6.98	7.03	14.01	6.45	5.89	5.848
8.01	8.00	16.01	8.45	7.89	7.850
9.03	8.96	17.99	10.43	9.87	9.833
25.00	−25.00	0	−7.56	−8.12	−8.121
H		13.06	5.50	4.94	4.902
H		3.00	−4.56	−5.12	−5.174
S		5.03	5.03	5.03	5.038
cS		45.27	45.27	45.27	45.34

Biweight of $y + x$: 7.56
Column (4): $y + x - 7.56$

Biweight of column (4) (using $cS = 45.27$): .56
Column (5): $(y + x - 7.56) - .56 = y + x - 8.12$

Biweight fit of column (5) on x (using $cS = 45.2$): .058 + .0023x
Column (6): $(y + x - 8.12) - (.058 - .0023x) = y - 1.0023x - 8.062$

clear that both fits are sensible, and that either might be the one we need in a specific situation. Which is which depends on our attitude toward the "odd point" $(25, -25)$. What we really need, in such situations, is to be presented with both fits and be told "think hard, and see if you can choose".

Regrettably, no one has yet designed a computer program to find all plausible distinct solutions and tell us to choose.

10H. Closing Comment

We have now seen that resistant and robust techniques are already available for a wide variety of questions. Further extensions—and more polished techniques for use in place of today's suggestions—are by now inevitable. But enough is already known to make resistant and robust techniques widely usable.

Summary: Resistant and Robust Techniques

Both resistance and robustness of efficiency are concepts important to the practice of data analysis, as well as to its theory.

We can do well using the median for exploration and the biweight for careful work, reserving the arithmetic mean for very special situations.

If we have reason to believe that tails *less* straggling than those of a Gaussian ("normal") distribution are a solid possibility, the choice of an estimate is a new question, not discussed here.

The classical estimates of spread (of scale) are not very robust of efficiency, with the MAD (median absolute deviation from the median) or the H-spread (or the interquartile range) among the best of an inferior lot.

A somewhat more complex measure of spread (of scale), denoted "ns_{bi}^2" in this book, is robust of efficiency and highly resistant.

We can, using a method of Gross's, combine biweight and s_{bi}^2 to obtain confidence intervals that are resistant, robust of efficiency, and robust of validity.

If we want analogs of variances and covariances for multiple-component data, we can follow a three-step path:

1. Find components we are willing to treat as orthogonal (obtained as residuals from successive robust-and-resistant regressions).

2. Assess a spread for each (we suggest using ns_{bi}^2).

3. Apply the usual identities for vars and covs to extend the definition of $\hat{b}$ar and $\hat{c}$ob to all linear combinations (of either the original or the new components).

We could assess the extent of the dependence of the resulting $\hat{b}$ars and $\hat{c}$obs on the choice of components from which to start.

When the purpose of finding substitutes for $\hat{v}$ars and $\hat{c}$ovs is to pick out certain newly chosen components, we may repeat the whole $\hat{b}$ar–$\hat{c}$ob process, starting from the first set of newly chosen components as if they were the original components. This will greatly reduce any serious dependence of the final answers on how we chose the original components.

References

Gross, A. M. (1976). "Confidence-interval robustness with long-tailed symmetric distribution," *J. Amer. Statist. Assoc.*, **71,** 1976, 409–416.

Lax, D. A. (1975). "An interim report of a Monte Carlo study of robust estimators of width." *Technical Report No. 93*, (Series 2). Department of Statistics, Princeton University.

Moses, L. E., and R. V. Oakford (1963). *Tables of Random Permutations.* Stanford, California: Stanford University Press.

Chapter 11/Standardizing for Comparison

When responses can be expressed as rates or proportions, we want to compare the effects of various treatments. Once the overall rates are computed and compared, we usually ask for more information. For example, if we find that one form of teaching makes students spell better than another, the skeptic immediately asks whether the groups tested were comparable. Similarly, if a new chemical for preventing wound infection in abdominal operations shows a lower rate of infection than the old treatment, we want to make sure that the patients getting the new treatment were not better off initially.

The use of randomized controlled trials goes a long way toward providing such controls. Even so, chance occasionally plays sad tricks, and it may be useful to make adjustments for this. Of more importance are adjustments to equate the performance of groups in observational or poorly controlled studies. Comparisons frequently can be improved by using one of the methods of standardization described in this chapter.

Although we do not illustrate this here, essentially similar problems arise and essentially similar solutions are sometimes useful when dealing with means or medians of measurements instead of with rates or proportions.

11A. The Simplest Case

Example 1. Two treatments, dichotomous populations. Let us begin with the problem in its simplest form: Two treatments are applied, one to one group and one to the other; then the groups are split into two strata, those Easy and those Hard to achieve success with, as determined by a criterion *other than* the present treatments. The report of the overall investigation is that Treatment I got 60% success and Treatment II got 44% success (Panel A of Exhibit **1**). The field of vital statistics would call these numbers "crude success rates." They are crude because they do not consider the comparability of the groups taking the different treatments, because they essentially ignore the additional information provided by the classification into strata.

In our example, let us suppose that the numbers came out as in Panel A of Exhibit 1. The crude success rate is computed for Treatment I, for example, thus:

$$\frac{0.7 \times 800 + 0.2 \times 200}{1000} = \frac{600}{1000} = 0.60,$$

or 60%.

We note an alarming thing. Although the crude rate is higher in Treatment I, Treatment II performs better on both kinds of people, the Easy and the

Hard (see Panel C). In this example, the comparison of the crude success rates is misleading about the relative value of the treatments. Treatment II was almost forced to have a lower crude success rate because it was applied to a much larger proportion of people who are hard to get success with.

Once this is pointed out, we can consider what to do instead. Naturally, the set of four percentages of success gives us a good summary of the situation, and knowing them, we can answer various questions. Four possible questions are:

1. What would the comparison of rates be if both treatments were to be applied to a population with the composition of Treatment I's group?

2. Same question, about a population like Treatment II's group?

3. What about the average population? (Half that of Treatment I, half that of Treatment II.)

	Numbers exposed to Treatment	
	I	II
Easy	450	450
Hard	550	550

Exhibit **1** of Chapter 11

Calculation of crude success rates, two treatments, two strata.

A) Number of successes

Stratum	Treatment	
	I	II
Easy	560	80
Hard	40	360
Total	600	440

B) Numbers exposed

Easy	800	100
Hard	200	900
Total	1000	1000

C) Success rates

Easy	70%	80%
Hard	20%	40%
Crude rate	60%	44%

4. What about an arbitrary population?

| | Percentage of successes | | |
	Treatment I	Treatment II	Diff. (II − I)
Answers			
Question 1	60	72	12
Question 2	25	44	19
Question 3	42.5	58	15.5

Question 4. When we come to question 4, Panel C shows that the difference is somewhere between 10 percent in favor of Treatment II (80 − 70) and 20 percent in favor of Treatment II (40 − 20), depending on the mixture of those Easy and Hard to succeed with. When all are Easy, Treatment II wins by 10 percent; when all are Hard, Treatment II wins by 20 percent. A 50–50 mixture of Easy and Hard gives 15 percent (60 − 45), just midway between.

Which of all these populations should we choose? The answer depends on our purpose, of course, which might be merely reviewing possibilities as we have just done, to see how sensitive the results are to the change in mix. If so, we have done enough already. But we may want to report for a specific situation.

We would then make our best estimate of the kind of population we were expecting to have to treat, and perform the calculation on that basis, with perhaps some consideration given to deviations from that chosen population. The chosen population is called the *standard population;* that is, it is the mix for which the comparison is standardized. Any mix can be chosen as the standard. Now consider our second example.

Example 2. Crossing percentages. Let us suppose that the rates of success for the Easy and Hard groups are as follows:

| | Successes with treatment | |
	I	II
Easy	70%	80%
Hard	40%	10%

Now when we have all Easy subjects, Treatment II wins by 10%, but when they are all Hard, Treatment I wins by 30%, and therefore it makes quite a difference to us what the standard population is. In Example 1, we merely missed the difference a bit, but now if we can use only one treatment, our choice will depend upon the mixture that exists in the population we plan to treat. Above 75% Easy we choose Treatment II, and below that we choose Treatment I.

Of course, if one can identify the subject's class beforehand and give the more successful treatment for that class, we are in the best of all possible worlds. In these cheery circumstances, the interest in standard populations may well wane.

Needless to say, this approach can be generalized to handle several treatments and to handle many classes. The method is usually called **direct standardization**.

The results of direct standardization are often computed—just as we have done so far—with no attention to the indicated variability, but we may wish some indication of variability. In question 1 above, for Example 1, we have

$$p_{Std} = 0.8 p_{Easy} + 0.2 p_{Hard}$$

as our computed success rate. Treating 0.8 and 0.2 as fixed, and using

$$\hat{var} \, p = \frac{pq}{N},$$

where $q = 1 - p$ and N is the group size, and letting u and v be the observed proportions, we have

$$\hat{var} \, (0.8u + 0.2v) = 0.64 \, \hat{var} \, u + 0.32 \, \hat{cov} \, (u, v) + 0.04 \, \hat{var} \, v,$$

in which we can calculate the appropriate variances.

Let $p_{Std,I}$ be the observed proportion of success when the standard population sizes (0.8 Easy and 0.2 Hard) are applied to the percentages of success from Treatment I. Similarly, $p_{Std,II}$ is the observed proportion of success when the standard population (same mixture) is applied to the percentages of success of Treatment II. Then $p_{Std,I} - p_{Std,II}$ is the difference in the observed proportion, and we want to compute its variance. We get, assuming zero correlation between results for the Easy and Hard strata:

$$\hat{var} \, p_{Std,I} = 0.64 \frac{(0.70)(0.30)}{800} + 0.04 \frac{(0.20)(0.80)}{200} = 0.000200 = (0.014)^2$$
$$= (1.4\%)^2;$$

$$\hat{var} \, p_{Std,II} = 0.64 \frac{(0.80)(0.20)}{100} + 0.04 \frac{(0.40)(0.60)}{900} = 0.001035 = (0.032)^2$$
$$= (3.2\%)^2;$$

$$\hat{var} \, (p_{Std,I} - p_{Std,II}) = 0.64 \left(\frac{(0.70)(0.30)}{800} + \frac{(0.80)(0.20)}{100} \right)$$
$$+ 0.04 \left(\frac{(0.20)(0.80)}{200} + \frac{(0.40)(0.60)}{900} \right)$$
$$= 0.001235 = (0.035)^2 = (3.5\%)^2;$$

or, more simply, since there is no correlation across treatments,

$$\hat{var} \, (p_{Std,I} - p_{Std,II}) = (1.4\%)^2 + (3.2\%)^2 = (3.5\%)^2.$$

Thus, our difference of 12% (under Question 1) has a standard error of 3.5% and can be regarded as certain of sign and as known to within about a factor of two.

11B. Direct Standardization

Let us suppose that there are J treatments and K strata. In the standard population, we have the proportions W_k, $k = 1, 2, \ldots, K$, which might well be thought of as weights, so that $\sum W_k = 1$. And we have success rates corresponding to the jth treatment in the kth stratum P_{jk}, where $j = 1, 2, \ldots, J$. In Example 1, the strata were Easy and Hard and the P_{jk} appear in the bottom panel of Exhibit 1.

Then D_j, the directly standardized success rate for Treatment j, is given by

$$D_j = \frac{\sum_{k=1}^{K} W_k P_{jk}}{\sum_{k=1}^{K} W_k},$$

which simplifies, when $\sum W_k = 1$, to

$$D_j = \sum_{k=1}^{K} W_k P_{jk}.$$

Example 3. Death rates in Maine and South Carolina. Theodore D. Woolsey, in his description of adjusted death rates, gives a beautiful example where the crude death rates produce a misleading result in comparing the death rates for Maine and South Carolina in 1930. Exhibit **2** gives the basic data. The point of the example is similar to that of our Example 1: South Carolina had a higher death rate than Maine in all but one age class, and even there they were nearly equal. Nevertheless, the crude death rate gave South Carolina the better overall showing. The reason was that Maine's population was generally much older than that of South Carolina. Maine had a crude death rate of 1390.8 per 100,000, while South Carolina had only 1288.8 per 100,000.

To get a better comparison of the overall death rates, Woolsey did not choose the age distribution of either of the states. Instead, he took as his standard population, the age distribution of the whole United States, and applied the death rates for each age interval—in Maine, $P_{\mathrm{ME}k}$, and in South Carolina, $P_{\mathrm{SC}k}$—to this standard population, which he represented as 1 million people spread across the age groups. The state corresponds to a treatment, the age group to a stratum. His calculation is shown in Exhibit **3**.

11C. Precision of Directly Standardized Values

We want some idea of the possible precision of the directly standardized numbers of deaths. We may not need to be as precise as can be. So let us develop both a rough calculation (to be defined below) and a crude calculation, as well as a relatively precise one.

Exhibit 2 of Chapter 11

Age-specific death rates and populations, Maine and South Carolina, 1930 (adjusted rates and indices)

Age (in years)	Maine		South Carolina		Percent distribution of population	
	Specific rates (per 100,000 population)[1]	Population[1]	Specific rates (per 100,000 population)[1]	Population[1]	Maine	South Carolina
0–4	2,056	75,037	2,392	205,076	9.4	11.8
5–9	186	79,727	185	240,750	10.0	13.9
10–14	140	74,061	184	222,808	9.3	12.8
15–19	223	68,683	426	211,345	8.6	12.2
20–24	370	60,575	645	166,354	7.6	9.6
25–34	391	105,723	871	219,327	13.3	12.6
35–44	545	101,192	1,242	191,349	12.7	11.0
45–54	1,085	90,346	1,994	143,509	11.3	8.3
55–64	2,036	72,478	3,313	80,491	9.1	4.6
65–74	5,219	46,614	6,147	40,441	5.8	2.3
75+	13,645	22,396	14,136	16,723	2.8	1.0
Total	...	796,832	...	1,738,173	99.9	100.1
Crude death rate (per 100,000 population)	1,390.8	...	1,288.8	...	...	...

[1] Deaths and populations of unknown age excluded.

S) SOURCE

T. D. Woolsey (1959). Chapter 4, page 67, in F. E. Linder and R. D. Grove, *Vital Statistics Rates in the United States, 1900–1940.* National Office of Vital Statistics, U.S. Government Printing Office, Washington, D.C., 1959.

Exhibit 3 of Chapter 11

Work sheet showing adjustment by the direct method of the death rates for Maine and South Carolina, 1930

Age (in years)	Standard million for the United States in 1940	Maine				South Carolina			
		Population[1] in 1930	Deaths[1] in 1930	Specific rates (per 100,000 population)	Expected deaths in standard million	Population[1] in 1930	Deaths[1] in 1930	Specific rates (per 100,000 population)	Expected deaths in standard million
	(1)	(2)	(3)	(4) = (3) ÷ (2)	(5) = (4) × (1)	(6)	(7)	(8) = (7) ÷ (6)	(9) = (8) × (1)
0–4	80,100	75,037	1,543	2,056	1,647	205,076	4,905	2,392	1,916
5–9	81,100	79,727	148	186	151	240,750	446	185	150
10–14	89,200	74,061	104	140	125	222,808	410	184	164
15–19	93,700	68,683	153	223	209	211,345	901	426	399
20–24	88,000	60,575	224	370	326	166,354	1,073	645	568
25–34	162,100	105,723	413	391	634	219,327	1,910	871	1,412
35–44	139,200	101,192	552	545	759	191,349	2,377	1,242	1,729
45–54	117,800	90,346	980	1,085	1,278	143,509	2,862	1,994	2,349
55–64	80,300	72,478	1,476	2,036	1,635	80,491	2,667	3,313	2,660
65–74	48,400	46,614	2,433	5,219	2,526	40,441	2,486	6,147	2,975
75+	20,100	22,396	3,056	13,645	2,743	16,723	2,364	14,136	2,841
Total	1,000,000	796,832	11,082	...	12,033	1,738,173	22,401	...	17,163

[1] Deaths and populations of unknown age excluded.

$$D_{ME} = \frac{12,033}{1,000,000} \times 1,000 = 12.03; \qquad D_{SC} = \frac{17,163}{1,000,000} \times 1,000 = 17.16; \qquad D_{SC}/D_{ME} = \frac{17.16}{12.03} = 1.43$$

Notes

1. The formula is set up for rates per thousand population. If proportions are preferred, then divide D_{ME} and D_{SC} by 1000.
2. The table includes the calculation of the rates already presented in Exhibit 2.

S) SOURCE

T. D. Woolsey (1959). Chapter 4, page 67, in F. E. Linder and R. D. Grove, Vital Statistics Rates in the United States, 1900–1940. National Office of Vital Statistics, U.S. Government Printing Office, Washington, D.C.

Let N be the sample size and p the proportion dying. Then, assuming binomial variation and taking advantage of p being small, so that $q = 1 - p$ is near 1, we have

$$\text{var (Actual deaths)} = Npq \approx Np \approx \text{Actual deaths.}$$

Method 1. *Crude binominal.* For South Carolina, there are 22,401 actual deaths. This estimates both the mean Np and the variance Npq, because the latter is close to Np.

For Maine the corresponding estimate of mean and crude variance is 11,082, the actual deaths.

Method 2. *Standardized binomial (rough standard error).* If D is the number of deaths, let KD be the standardized deaths, regarding K as a constant. Then the mean is KNp and the variance $K^2Npq \approx K^2Np$. We can take

$$K = \frac{\text{Standardized deaths}}{\text{Actual deaths}} \approx \frac{\text{Standardized deaths}}{Np}.$$

Thus, an alternative, improved approximation for the variance of standard deaths is

$$\frac{(\text{Standardized deaths})^2}{\text{Actual deaths}}, \qquad \text{since } q \approx 1.$$

To take still more care, we would multiply by q.

For South Carolina, the estimate with q set at 1, is

$$\frac{(17,163)^2}{22,401} = 13,150.$$

For Maine, the corresponding estimate is

$$\frac{(12,033)^2}{11,082} = 13,066.$$

Using these estimates for the variances, the standardized difference,

South Carolina MINUS Maine $= 17,163 - 12,033 = 5,130,$

the number of excess deaths for South Carolina, gets variance (for this estimate of excess deaths)

$$13,150 + 13,066 = 26,216 = (162)^2.$$

Thus we see that the difference is close to 30 times its approximate standard error. We are unlikely, in such a case, to need a better standard error, but we shall go ahead, as an illustration of the method.

Method 3. Stratified binomials. To improve our approximate calculation, we need only apply the

$$\frac{(\text{Standardized deaths})^2}{\text{Actual deaths}}$$

approximation to each stratum (age group, here) separately, starting with, for South Carolina,

$$\frac{(1916)^2}{4905} + \frac{(150)^2}{446} + \frac{(164)^2}{410} + \cdots$$

or

$$748.4 + 50.4 + 65.6 + 176.7 + 300.7 + 1043.8$$
$$+ 1257.7 + 1928.0 + 2653.0 + 3560.2 + 3414.2 = 15{,}198.7,$$

and, for Maine,

$$1758.0 + 154.1 + 150.2 + 285.5 + 474.4 + 973.3$$
$$+ 1043.6 + 1666.6 + 1811.1 + 2622.6 + 2462.1 = 13{,}401.5,$$

so that our rough approximation for the variance of the difference is now

$$15{,}198.7 + 13{,}401.5 = 28{,}600.2 = (169)^2,$$

less than 5% greater than the rougher approximation.

Allowing for q. To improve the approximation still further, we must remember the q in Npq and use

$$\frac{(\text{Standardized deaths})^2}{\text{Actual deaths}}(1 - \text{probability of death})$$

for each stratum, starting with

$$\frac{(1916)^2}{4905}(1 - 0.02392) = 730.5 \quad \text{and} \quad \frac{(150)^2}{446}(1 - 0.00185) = 50.3$$

for South Carolina, and reaching

$$730.5 + 50.3 + 65.5 + 175.9 + 298.8 + 1034.7$$
$$+ 1242.1 + 1889.6 + 2565.1 + 3341.4 + 2931.6 = 14{,}325.5$$

for that state, and for Maine

$$1721.9 + 153.8 + 150.0 + 284.9 + 472.6 + 969.5$$
$$+ 1037.9 + 1648.5 + 1774.2 + 2485.7 + 2126.1 = 12{,}825.1,$$

giving, for the estimated variance of the difference,

$$14325.5 + 12825.1 = 27150.6 = (165)^2.$$

Comparison of Approximations

In the example, it matters little whether we take the standard error of the difference of 5130 standardized deaths as ±162 or ±169 or ±165, particularly since sources of error beyond binomial variation usually need to be considered. For example, year-to-year variation in the South Carolina–Maine difference might dwarf the variation we have carefully calculated. In more extreme circumstances, the difference between the various approximations and the full stratified Npq form can be more important.

When we deal with small numbers of deaths or small numbers of survivors, we may need to go a little further. In the most extreme case—when either deaths or survivors are zero—where pq, as usually calculated, vanishes so that, if we are not more careful, we assess a zero estimated variance—we may especially want to enlarge the estimated variance because we are confident it is larger than 0. We have no way to do anything exactly although, in more complicated problems, Sutherland, Holland, and Fienberg (1974) have ways to offer some help. We offer here a method that does fit in with the rest of our treatment of variables. But this consistency does not, so far as we know, guarantee that our correction is nearer the true value than was the original raw estimate.

The proposed adjustment is to add $\frac{1}{6}$ to the count of every category, having

$$p^* = \frac{\text{actual deaths} + \frac{1}{6}}{\text{cell count} + \frac{1}{3}} \quad \text{and} \quad q^* = \frac{\text{actual survivors} + \frac{1}{6}}{\text{cell count} + \frac{1}{3}}$$

for each cell for which an estimated variance would be required, thus leading to the use of

Estimated variance

$$= \left(\frac{\text{standard population}}{\text{\# exposed}}\right)^2 (\text{actual deaths} + \tfrac{1}{6}) \left(1 - \frac{\text{actual deaths} + \frac{1}{6}}{\text{cell count} + \frac{1}{3}}\right),$$

for the contribution from each cell in which there are no deaths.

11D. Difficulties with Direct Standardization

We began as if there were no uncertainty about the P's and went on to calculate some uncertainties. In Example 3, the death rates at each age, called age-specific rates, are fairly well determined. But often our rates are based upon the investigation under discussion, and then it is a common experience that some cells (treatment–stratum pairs) may have small numbers of cases and so the P's may be estimated with considerable uncertainty. When this sort of uncertainty is combined with a substantial weight, then the result of the entire calculation may be unduly sensitive to the specific outcomes.

Example 4. Ill-determined rates. Let us give an example where such an effect is easy to demonstrate. If we work with death rates associated with an elective operation done under two different treatments, we might have three classes of patients, those in Excellent health, those in Fair health, and those in Poor health. The distribution of patients might fall as in Exhibit **4**.

Let us note that the one patient in cell [Poor × Treatment I] died. Had she survived, the cell rate would have been $0/1 = 0$, the death rate/1000 would have gone down from 26.48 to 1.47, and Treatment I would have looked very favorable for the standard population. Instead of looking about 6 times as bad, it would have looked 3 times as good as Treatment II.

Exhibit **4** of Chapter 11

Table with heavily weighted unreliable cell

A) Deaths

Stratum	Treatment I	Treatment II
Excellent	10	4
Fair	9	20
Poor	1	2

B) Numbers exposed (cell counts)

Excellent	10,000	5,000
Fair	3,000	4,000
Poor	1	20

C) Rates/1000

			Standard population	Expected deaths I	Expected deaths II
Excellent	1	0.8	14,500	14.5	11.6
Fair	3	5	5,000	15.0	25.0
Poor	1000	100	500	500.0	50.0
Totals			20,000	529.5	86.6
Rate/1000				26.48	4.33

Had the Poor–Treatment I patient survived:

Rate/1000				1.48	4.33

How can this be fixed? Some things just can't be fixed. In the example, we can, of course, investigate to try to find out whether the lone patient who died suffered some sort of horror story, was *doomed* from the start, and possibly was out at the "hopeless" edge of the category Poor. But so, of course, may have been the deaths in this category for the other treatment. Such troubles require careful consideration and cannot all be provided for by technical statistical devices.

What we are illustrating is that, when basically small probabilities accidentally produce high estimates, possibly by factors of 10 or 100, and when they are given substantial weight by the standard population, then the direct method of standardizing can give rather wild results. Consequently, it is important to review any direct standardizing calculation for this sort of difficulty. One grave danger is that the number of cells may be very large and that a computer is going to compute the final result without displaying any intermediate calculations; then one can be badly fooled. Whenever a direct standardization gives a wild value, the possibility should be reviewed that this was caused by a small probability having been heavily weighted. Whenever such a wild value doesn't occur, we may wonder whether it should have. In either case, it will help us to know estimates of the relevant standard deviations.

If we estimate the standard error by the revised formula, using p^* and q^*, we find, for the expected deaths for the Poor–Treatment I cell, the following:

$$p^* = \frac{7/6}{4/3} = \frac{7}{8} \quad \text{and} \quad q^* = \frac{1/6}{4/3} = \frac{1}{8},$$

$$\text{Estimated variance} = \frac{(500)^2}{1} \frac{7}{6}\left(1 - \frac{7}{8}\right) = (191)^2,$$

which contributes ± 9.55 to the estimated standard deviation of the 26.48 rate, which may be enough to warn us. (The difference is only 2.3 times its standard error.)

If we had had a computer program that reported (a) estimated variance of total expected deaths and (b) largest contribution from a single cell, we would have been adequately warned, since these figures would have been

$$36505 = (191)^2 \quad \text{and} \quad 36458 = (191)^2.$$

Here,

$$36505 = \frac{(500)^2}{1}\left(\frac{7}{6}\right)\left(1 - \frac{7}{8}\right) + \left(\frac{5000}{3000}\right)^2 9\tfrac{1}{6}\left(1 - \frac{9\tfrac{1}{6}}{3000\tfrac{1}{3}}\right)$$

$$+ \left(\frac{14500}{10000}\right)^2 (10\tfrac{1}{6})\left(1 - \frac{10\tfrac{1}{6}}{10000\tfrac{1}{3}}\right)$$

$$= 36458.33 + 25.39 + 21.35.$$

(Had we added $\frac{1}{2}$, as is sometimes done, instead of $\frac{1}{6}$, we would have found

$$p^* = \frac{\frac{3}{2}}{2} = \frac{3}{4} \quad \text{and} \quad q^* = 1 - p^* = \frac{1}{4},$$

$$\text{Estimated variance} = \frac{(500)^2}{1} \left(\tfrac{3}{2}\right)\left(1 - \tfrac{3}{4}\right) = (306)^2,$$

increasing the estimated contribution to the standard error by almost a factor of $\sqrt{3}$. The difference between using $\frac{1}{6}$ and $\frac{1}{2}$, though sometimes large, rarely matters. If we should have been adding 0.01 instead of either of these, the estimated variance would have been $(49)^2$ and that might matter.)

None of these remarks or calculations is intended to deny the basic uncertainty created in this technique by having one cell with one observation, which then is given a large weight. It could, of course, have been even worse—no observation at all.

This feature of direct standardization may have been largely responsible for the development of another method of standardization, to which we now turn.

11E. Indirect Standardizing

For direct standardizing, we had to reach outside the system for a standard population, or invent it as an average of some sort from the two populations we had at hand. We applied the specific rates associated with each treatment to the standard population. In indirect standardization, we go the other way round. We reach out for a set of standard specific rates and apply them to the population associated with each treatment, and see how the resulting frequency of success compares with the crude rate for the population. Then we compare the two ratios. Fleiss (1973) offers a good discussion of the dangers and anomalies associated with direct and indirect standardization, as well as medical and birth- and death-rate examples.

Where shall we get our standard rates for *indirect* standardization? In a problem like the Maine–South Carolina comparison that we treated in Example 3, Woolsey chose the specific death rates for the given age groups in the United States as a whole. This uses the U.S. experience as a baseline, just as he used the sizes of U.S. national age groups as the standard population in the direct attack.

Sometimes in direct standardization, people choose the population formed by pooling the populations from the treatments. The corresponding choice for indirect standardization would be to use as standard rates some weighted combination of the treatment rates for the several treatments, and, of course, one attractive set of weights for that purpose is "weighted proportional to count of stratum." Using these weights is equivalent to using the rates pooled across treatments. Let us return to the data of Example 1 and carry through with this particular weighting.

Example 5. Indirect standardization applied to Example 1 in Exhibit 5. First, we compute the standard rates for each stratum separately. Then we apply the standard rates to the treatment counts in the two strata, for each treatment separately. This gives us the reference percent of success for each treatment. (These numbers are not to be compared with one another! They are to be compared with their own crude success rates.)

Exhibit **5** of Chapter 11

Indirect standardizing for Example 1

A) Numbers of successes

Stratum	Treatment I	II	Totals
Easy	560	80	640
Hard	40	360	400
Total	600	440	

B) Numbers exposed

Easy	800	100	900
Hard	200	900	1100
Total	1,000	1,000	

C) Success rates

			Standard rates
Easy	70%	80%	71.1% $(= \frac{640}{900})$
Hard	20%	40%	36.4% $(= \frac{400}{1100})$
Crude rate	60%	44%	

Applying standard rates to:

Reference success rates

$$\text{Treatment I:} \quad \frac{71.1 \times 800 + 36.4 \times 200}{1,000} = 64.2\%$$

$$\text{Treatment II:} \quad \frac{71.1 \times 100 + 36.4 \times 900}{1,000} = 39.9\%$$

$$\text{Standard success ratio} = \frac{\text{crude rate}}{\text{reference rate}}$$

$$\text{Treatment I:} \quad \frac{60\%}{64.2\%} = 0.93; \quad \text{Treatment II:} \quad \frac{44\%}{39.9\%} = 1.10$$

The results suggest that Treatment II has a relatively higher success rate than does Treatment I. Relative to what? Treatment II is 10% favorable compared to the standard, and Treatment I is 7% unfavorable compared to its standard.

The principles of the rough approximation to the (binomial) variance, used earlier for direct standardization, are easily applicable here. The most direct application, **treating n_{Std} as known**, would be to compute either

$$\frac{n_{Obs} - n_{Std}}{\sqrt{n_{Std}}} \quad \text{or} \quad \sqrt{4n_{Obs} + 2} - \sqrt{4n_{Std} + 1}$$

as an approximately standard normal deviate. Actually, except for $n_{Obs} = 0$, $\sqrt{4n_{Obs} + 2}$ is a good enough approximation to $\sqrt{n_{Obs}} + \sqrt{n_{Obs} + 1}$, which is approximately standard normal [see Freeman and Tukey (1950)]. For $n_{Obs} = 0$, replace $\sqrt{4 \cdot 0 + 2}$ by 1. In our example, we have 600 successes $= n_{Obs}$ and 642 standard successes $= n_{Std}$, and

$$Z_I = \frac{600 - 642}{\sqrt{642}} = \frac{-42}{25.3} = -1.66 \quad \text{or} \quad = \sqrt{2402} - \sqrt{2569}$$

$$= 49.01 - 50.68 = -1.67.$$

The minus sign reflects the failure of the observed number of successes to come up to the standard.

For Treatment II, we have

$$Z_{II} = \frac{n_{Obs} - n_{Std}}{\sqrt{n_{Std}}} = \frac{440 - 399}{\sqrt{399}} = \frac{41}{20.0} = 2.05 \quad \text{or} \quad \sqrt{1762} - \sqrt{1597} = 2.02.$$

Then the quantity we are concerned about is

$$\frac{Z_I - Z_{II}}{\sqrt{2}} = \frac{-1.65 - 2.05}{1.414} = -2.62 \quad \text{or} \quad \frac{-1.67 - 2.02}{\sqrt{2}} = -2.61,$$

which suggests that Treatments I and II do perform differently, because the final quantity, if they were identical, would be a unit-variance, approximately normal random variable.

Example 6. Another set of standard rates. The standard rates need not be the weighted averages; instead we might have weighted by treatment totals, which would give, as standard rates,

Easy 75% ($= [(1000 \times 0.70) + (1000 \times 0.80)]/2000)$)   and   Hard 30%.

Applying these rates gives

Treatment I: 66%   Treatment II: 34.5%

Ratios: $\dfrac{\text{crude}}{\text{standard}}$ $\dfrac{600}{660} = 0.91$ $\dfrac{440}{345} = 1.28$

and again Treatment II shows the higher rate.

Discussion

The net difference in Example 6 is:

$$(-9\%) - (28\%) = -37\%,$$

instead of

$$(-7\%) - (10\%) = -17\%$$

(as in Example 5). (And even $(Z_{I} - {'}Z_{II})/\sqrt{2}$ is nearly double its previous value.) Without regard to the assessment of uncertainty, how are we to think about these two answers, both negative but considerably different in size? If we do have strong external evidence about standard rates, we should use that evidence. What if we do not? Clearly, the first thing to do is to understand, if we can, why the strata were so unbalanced for the two treatments. Sometimes this can tell us something about which standardized rates to use. But what if this does not decide for us either?

The most extreme standard rates that are "consistent" with the data are as follows, and lead to the following ratios:

| | | Ratio: $\dfrac{\text{crude successes}}{\text{standard successes}}$ | | |
Easy	Hard	I	II	II − I
70%	20%	1.00	1.76	0.76
70%	40%	0.94	1.02	0.08
80%	20%	0.88	1.69	0.81
80%	40%	0.83	1.00	0.17

All these agree that Treatment I seems, on balance, to have a lower death rate than Treatment II—in this regard, any of these is much better than the crude comparison, which gave the wrong sign. They do differ quite considerably among themselves, however, reminding us that it often matters appreciably which standard rates we use.

This uncertainty is over and beyond the uncertainty assessed above, which treats the standard rates as known.

Often the unbalance is not nearly so great as here. Accordingly, the exact choice of standard rates is often not so important. In some studies, we can find reasons for a choice of standard rate. In the others, one plausible policy might be to use the median rate across all treatments (including control treatments). In our example, of course, this would again lead to the 75% and 30% that we got by using treatment totals for weights.

Example 7. Indirect approach with an ill-determined rate. Let us apply the indirect method to Example 4, which had one seriously ill-determined rate. Let us arbitrarily use the pooled counts to get standard rates, as shown in Exhibit **6**.

Exhibit **6** of Chapter 11

Indirect standardization with an ill-determined rate

A) Numbers of successes

| | Treatment | | |
Stratum	I	II	Total
Excellent	10	4	14
Fair	9	20	29
Poor	1	2	3
Total	20	26	

B) Numbers exposed

Excellent	10,000	5,000	15,000
Fair	3,000	4,000	7,000
Poor	1	20	21

C) Rates/1,000

			Standard rates
Excellent	1	0.8	$\dfrac{14}{15,000}$
Fair	3	5	$\dfrac{29}{7,000}$
Poor	1,000	100	$\dfrac{3}{21} = \dfrac{1}{7}$

D) Standard successes

Excellent	9.33	4.67
Fair	12.43	16.57
Poor	.14	2.86
Total	21.90	24.10

Ratios: $\dfrac{\text{Crude successes}}{\text{Standard successes}}$:

0.913	1.08

Rough strength:

$$\left.\begin{array}{l} \sqrt{4(20) + 2} - \sqrt{4(21.90) + 1} = \ 9.06 - 9.41 = -0.35 \\ \sqrt{4(26) + 2} - \sqrt{4(24.10) + 1} = 10.30 - 9.87 = \ \ \ 0.43 \end{array}\right\} \quad \dfrac{-0.35 - 0.43}{\sqrt{2}} = -0.55$$

If we had 10 times the data, the final result would be multiplied by about $\sqrt{10} \approx 3.16$, and, if we had 100 times the data by about 10. (Thus we would need vastly more data to have a reasonable chance of reaching significance.)

This approach suggests that Treatment II has the higher death rate but that we can't be sure. The approach suppresses the effect of the single case in the I–Poor cell, so far as indication goes, but preserves it in assessing uncertainty.

In Exhibit 6 we also look to see what 10 or 100 times as much data, divided in exactly the same way, would provide in the way of strength of evidence.

Discussion

We mention two awkwardnesses with the indirect approach.

1. Even if the treatments have equal rates for each group, if the groups have different proportions, then the indirect method will produce a difference in overall success rates, usually slight, but annoying.

2. Woodworth offers an example to show that, when one compares 3 or more treatments, it is possible for one treatment to have higher rates in each group than a second treatment and yet wind up with a lower standardized rate. This seems more distressing than the previous difficulty.

Exhibit **7** shows Woodworth's example. We note that, in Stratum 1, Treatment I has a higher success rate than Treatment II, and the same is true in Stratum 2. But the indirectly standardized rates for Treatments I and II are 1.02 and 1.08, respectively. Thus, Treatment I, which starts out with a *relatively* higher success rate of about 10 or 11%, winds up with about a 6% lower rate. Woodworth reports that, if all comparisons are made by pairs, this effect does not occur.

This example deserves some further attention. We face a quite inconsistent situation: Two treatments give higher rates in Stratum 1, while the third shows a higher rate in Stratum 2. If we analyze the table of rates in the manner of Chapter 9, we find:

$$
\begin{array}{ccc|c}
0 & 0 & -22 & 10 \\
0 & 0 & 22 & -10 \\
\hline
10 & 0 & -10 & 100
\end{array}
$$

for (1000 times) the rates themselves, and

$$
\begin{array}{ccc|c}
0 & 0 & -10 & 4 \\
0 & -1 & 10 & -4 \\
\hline
4 & 0 & 5 & -100
\end{array}
$$

for (100 times) their logarithms. Either table shouts for an explanation as to why Treatment III is different.

It is probably even more important to understand this difference between Treatment III and the other two, than to make the comparison of Treatments I and II come out right.

Exhibit **7** of Chapter 11

Woodworth's example of reversal of rates through indirect standardization

A) Occurrences

Stratum	Treatment			Total
	I	II	III	
1	12	99	39	150
2	90	9	51	150
Totals	102	108	90	300

B) Numbers exposed

1	100	900	500	1,500
2	900	100	500	1,500
Totals	1,000	1,000	1,000	

C) Rates/1,000

				Standard rates
1	.120	.110	.078	.1
2	.100	.090	.102	.1

D) Standard occurrences

1	10	90	50
2	90	10	50
Totals	100	100	100

Ratios: $\dfrac{\text{Crude occurrences}}{\text{Standard occurrences}}$:

1.02	1.08	0.90

Differences between crude and standard occurrences, in standard deviation units, using $\sqrt{4n_{\text{Obs}} + 2} - \sqrt{4n_{\text{Std}} + 1}$:

.22	.81	−1.00

S) SOURCE

Woodworth, G. G. (1971). *Standardized Mortality Comparisons: Sketch of a Review*. Unpublished manuscript. Used with permission of the author.

Had we used the median rule in this example, our standard rates would have been 0.115 and 0.095, so that standardizing would have produced a correct sign for the comparison of Treatments I and II. This works often but, sad to say, not always. (As we can see if we make up a new example, replacing Treatment III by three Treatments, say IIIa, IIIb, IIIc that seem to behave exactly the same, with the same total counts as for Treatment III. In such a case, the two-way analysis would shout even more loudly.) There is no substitute for looking at the two-way analysis of the individual rates!

Once again the basic message is that

1. One number cannot honestly do the work of many.
2. We should look as carefully as we can at the basic tables.
3. Sometimes we are, nevertheless, forced to summarize statistics in a single number, or find it desirable to do so.

11F. Adjustment for Broad Categories

We have so far treated our strata as they stood. We need to be concerned about the possibility that our strata are wider than we wish.

Let us go back to Example 1 (in Exhibit 5, in particular), where we have only two strata: Easy and Hard. While we may not be able—either from lack of records or lack of insight—to classify the cases more finely, we can hardly doubt that some Easy cases are easier than others—or that some Hard cases are harder than others.

Whatever mechanism determined that Treatment I should be used on 80% Easy cases and Treatment II on 90% Hard ones is unlikely to be blind to shades of ease or hardness. If we could cut the data into four strata, rather than two, we might expect something like Exhibit **8**. The difference between treatments looks a little greater than it did in Exhibit 5. If we calculate the indirectly standardized data, they, too, will now look more different. Having four strata instead of two can lead to an appreciably greater correction. What are the morals of this story?

First and foremost, we must not feel that standardization for the strata that we happen to have on hand has eliminated the influence on whatever variable underlies our stratification. We have gotten rid of some, but not all, of that variable's effects. We must be prepared for a need for further adjustment.

Second, we need to seek out a plausible method of further adjustment. Such a method may be useful, even if its formal development calls on unlikely assumptions. Its use helps, even if it gives us only an order-of-magnitude indication of the size of the further adjustments required. (And we can usually do better than just this.)

Adjusting for effects of broad categories. When we adjust for broad categories, as in the "Easy"- and "Hard"-to-cure example in Section 11A, we think of these two categories as broad categories on a continuum representing degree of difficulty of success. And we think of the individuals as spread along this continuum. In the example, under Treatment I, 80% were easy to cure, 20% hard. And so we think of the distribution as in Exhibit **9.** The cutting point of the distribution splits the easy cases from the hard ones. The areas under the curve to the left and right equal the percentages in the two groups, 80% and 20%, respectively. If we have such a difficulty scale and such a distribution, then, in principle, we can compute the center of gravity for the easy cases and for the hard ones. In Exhibit 9, these are represented by the small fulcrums ▲ beneath the horizontal axis. These centers of gravity can be regarded as measuring the average difficulty of successful treatment in the two groups. Don't worry for a moment about where that scale comes from or where the distribution comes from. We supply it later.

Next, we have a corresponding picture for Treatment II of the example, 10% easy, 90% hard, as shown in Exhibit **10**. Again we have centers of gravity for the two groups.

Exhibit **8** of Chapter 11

A hypothetical further breakdown of Exhibit 5.

A) Successes

Stratum	Treatment I	Treatment II	Total
Very easy	360	20	380
Semi-easy	200	60	260
Semi-hard	35	250	285
Very hard	5	110	115

B) Numbers exposed

	I	II	Total
Very easy	400	20	420
Semi-easy	400	80	480
Semi-hard	150	500	650
Very hard	50	400	450

C) Rates

	I	II	Total
Very easy	90%	100%	90.5%
Semi-easy	50%	75%	54.2%
Semi-hard	23.3%	50%	43.8%
Very hard	10%	27.5%	25.6%

We want to see how the rates of success (bottom panel of Exhibit 1) relate to the centers of gravity for the two treatments and two classes of individuals treated, and so we form Exhibit **11**, where the ×'s show rates of success plotted against centers of gravity for the four groups. An important finding is that the two Easy groups are not matched on the difficulty scale. Similarly, neither are the two Hard groups. Our purpose is to adjust for these inequities in difficulty.

To do this, we need to pick an average difficulty and adjust the success rates to this average difficulty. We might choose the mean or median (if there are several treatments or several groups), possibly weighted by numbers of cases. If the lines connecting the points in Exhibit 11 are approximately parallel, it will make little difference which average we choose. For the moment, let us adjust to the average difficulty for the Easy groups and the average difficulty for the Hard groups. The geometry is shown in Exhibit **12**, where the abscissas of the open circles represent the average standard difficulty for the Easy groups and for the Hard ones.

Exhibit **9** of Chapter 11

An illustrative distribution of individuals along the scale of difficulty, with 80% easy to cure, 20% hard to cure, as in Treatment I in Section 11A. The ▲'s represent centers of gravity for the two groups.

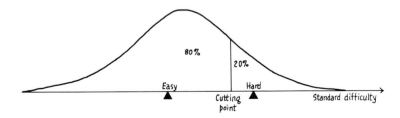

Exhibit **10** of Chapter 11

Hypothetical distribution of difficulty for the individuals undergoing Treatment II, from Section 11A.

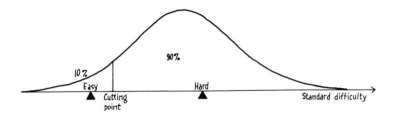

To get a final summary of the difference in success rates for the two treatments, adjusted for difficulty, we might average the differences for the Easy and Hard groups, perhaps weighting by the numbers in the original groups, or by the numbers expected in a population to be treated.

With one omission, we have given the overall plan for the two-treatment, two-category problem. The omission is the distribution and its association with the scale of difficulty. Although many distributions might be chosen, we would like one with an easily computed center of gravity for segments. It should not matter a great deal just what distribution we choose, provided we take a distribution whose shape roughly agrees with our ideas of the distribution of difficulty in a population. The logistic recommends itself. It has an attractive shape, symmetric with a single mode that is not unduly peaked. Its formulas are convenient and we have a table that helps us compute the centers of gravity (Exhibit 11 of Section 5F). The idea of scoring grades by way of centers of gravity has already been presented in Section 5H, together with the needed formulas.

Exhibit **11** of Chapter 11

Plot of success rates against centers of gravity (for two Easy and two Hard groups).

The ×'s show rates of success plotted against centers of gravity for the four groups. We observe that the Easy groups are not matched for difficulty, nor are the Hard groups.

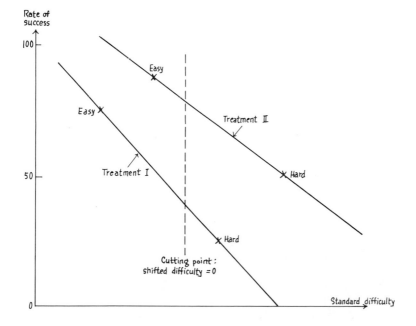

Exhibit **12** of Chapter 11.

Plot of average standard difficulty versus success rates for Easy and Hard groups.

The abscissas of the open circles represent the average standard difficulty for the Easy groups and for the Hard ones. The arrows show the movement required to adjust the success rates for each of the four groups. The lengths of the vertical line segments represent the estimated difference in success rates for the two treatments after being adjusted for difficulty.

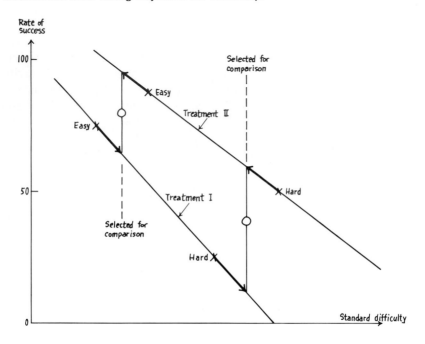

Exhibit **13** of Chapter 11.

The standard logistic density function. The abscissa is log $[p/(1 - p)]$, where the cumulative is p.

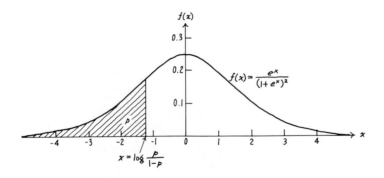

The logistic. The standard logistic distribution has a probability density with the general shape shown in Exhibit **13**. It is symmetric about zero. The defining property is that the abscissa at the point where the area to the left is p has the value

$$x = d = \log \frac{p}{1 - p}.$$

We note that when $p = \frac{1}{2}$, $d = 0$, we use d as a mnemonic for standard difficulty.

We choose the logistic distribution and translations of it to represent distributions of degrees of difficulty in treatment populations. We will want the cutting points moved to zero, and so we shall have to translate the distributions by the distance from $\log[p/(1 - p)]$ to 0; or, equivalently, we measure the distance of the center of gravity from $\log[p/(1 - p)]$.

The general idea is that the difficulty of treating a population successfully can be approximately represented by a bell-shaped curve whose center moves to the right as the population becomes harder to treat successfully.

To get an idea of the scale of the standard logistic, the central interval containing 95% of the density runs from -3.23 to 3.23.

Finally, we need the center of gravity for such an interval. From Section 5H we have that, if the upper cumulative level is B and the lower is A, the center of gravity of the interval is

$$CG = \frac{[B \log B + (1 - B) \log (1 - B)] - [A \log A + (1 - A) \log (1 - A)]}{B - A}.$$

When we have only two classes, one runs from 0 to p, the other from p to 1. Substituting $A = 0$, $B = p$ in the formula, we get the center of gravity of this interval for the standard logistic as

$$0 \text{ to } p: \quad CG = \frac{p \log p + (1 - p) \log (1 - p)}{p} = \frac{\phi(p)}{p}.$$

Then, letting $A = p$, $B = 1$, gives, for the other segment,

$$p \text{ to } 1: \quad CG = -\frac{p \log p + (1 - p) \log (1 - p)}{1 - p} = \frac{-\phi(p)}{1 - p} = \frac{\phi(1 - p)}{1 - p},$$

where $\phi(p)$ is tabled in Exhibit 11 of Section 5F. Recall that to translate the distributions, we measure the distance of these centers of gravity from the cutting point

$$\log \frac{p}{1 - p},$$

thus effectively translating the cutting point to zero. We now carry out, as a numerical example, the example given at the beginning of the chapter.

The basic data we need are shown in Exhibit **14**.

Numerical example. Let us carry out the calculations first for Treatment I. For the Easy group, $p = 0.8$, and so the cutting point on the standard logistic is

$$\log_e \frac{0.8}{1 - 0.8} = \log_e 4 = 1.386.$$

To compute the two centers of gravity, we read, for $p = 0.8$ from Exhibit 11 of Section 5F, that $p \log_e p + (1 - p) \log_e (1 - p)$ is -0.5004. The centers of gravity for the standard logistic then are

$$0 \text{ to } 0.8: \quad CG = \frac{-0.5004}{0.8} = -0.6255 \approx -0.626.$$

$$0.8 \text{ to } 1: \quad CG = \frac{-(-0.5004)}{0.2} = +2.502.$$

We want to find the positions of the centers of gravity when the cutting point is translated to zero, and so we must subtract 1.386 from each, to get:

Treatment I. Results for shifted distribution, zero at $p = 0.8$:

Easy	0 to 0.8	$-2.012 = (-0.626 - 1.386)$
Hard	0.8 to 1	$1.116 = (2.502 - 1.386)$
Cutting point		$0 = (1.386 - 1.386).$

The corresponding calculations for Treatment II, where 10% were in the Easy category and 90% in Hard, are, using the numerator -0.3251 from Exhibit 11 of Section 5F to get the cutting point -2.187 and the centers of gravity: -3.250 and 0.361.

Exhibit **14** of Chapter 11

Crude success rates: two treatments, two strata (repeated from Panels B and C of Exhibit 1).

A) Numbers exposed

	Treatment	
Stratum	I	II
Easy	800	100
Hard	200	900
Total	1,000	1,000

B) Success rates/1000

	Treatment	
Stratum	I	II
Easy	70%	80%
Hard	20%	40%
Crude rate	60%	44%

Treatment II. Results for shifted distribution, zero at $p = 0.1$:

Easy	0 to 0.1	$-1.063 = (-3.250 + 2.187)$
Hard	0.1 to 1	$2.548 = (0.361 + 2.187)$
Cutting point		$0 = (-2.187 + 2.187), \log(0.1/0.9) = -2.187.$

The average difficulty for the two groups in each category is:

Easy groups:	$\frac{1}{2}(-2.012 - 1.063) = -1.538,$
Hard groups:	$\frac{1}{2}(1.116 + 2.548) = 1.832.$

We can read the adjusted numbers off the graph shown in Exhibit **15**, or interpolate after we get the slopes of the lines, as follows.

Exhibit **15** of Chapter 11

Success levels after adjustment for broad categories (indicated by heavy arrows).

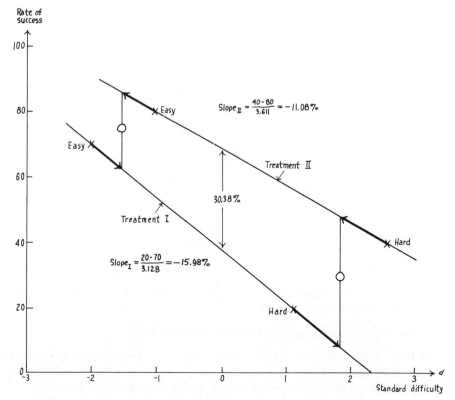

Adjusted values of success rates in percents

For the Easy groups:

Treatment I $70 - 15.98(-1.538 - (-2.012)) = 62.4\%$
Treatment II $80 - 11.08(-1.538 - (-1.063)) = 85.3\%$

For the Hard groups:

Treatment I $20 - 15.98(1.832 - 1.116) = 8.6\%$
Treatment II $40 - 11.08(1.832 - 2.548) = 47.9\%$

The gains, Treatment II minus Treatment I, are then

Easy $85.3 - 62.4 = 22.9$
Hard $47.9 - 8.6 = \underline{39.3}$
Average gain 31.1

Our calculation of the average gain going from Treatment I to Treatment II is 31.1%, compared with the unadjusted crude increase 16% = 60% − 44% and the simply adjusted value of $\frac{1}{2}(10\% + 20\%) = 15\%$.

The example brings out the point that adjusting for broad categories can make a substantial difference.

Standard population. If we wanted to take as the standard population 45% Easy and 55% Hard—where "Easy" is a compromise between "Easy as given Treatment I" and "Easy as given Treatment II" and "Hard" is a similar compromise—then we could weight the adjusted differences 0.45 and 0.55, and we would get

$$0.45(22.9) + 0.55(39.3) = 31.9\%,$$

compared with the original values (simple adjustment):

$$0.45(10) + 0.55(20) = 15.5\%.$$

Comment

The adjustment we have made would be exact (for huge samples) if:

1. the percent success depends linearly on the CG of the corresponding difficulty distribution, and

2. the difficulty distributions have the same spread.

The first of these assumptions is exact only when percent success is linear in standard difficulty. Since only approximate good behavior is needed to make the adjustment, this issue seems not to be so important for practice.

Acting as if the difficulty distributions have the same spread seems rather more likely to cause trouble. We can avoid this by calculating %'s of success *at the cutting point*, which only requires us to proceed as though % success is linear in the standard difficulty CG's. For our example, at the cut between Easy and Hard, this gives:

For Treatment I: $\dfrac{1.116}{3.128}(70\%) + \dfrac{2.012}{3.128}(20\%) = 37.84\%$,

For Treatment II: $\dfrac{2.548}{3.611}(80\%) + \dfrac{1.063}{3.611}(40\%) = 68.22\%$

(where $3.128 = 1.116 + 2.012$ and $3.611 = 2.568 + 1.063$). This difference of 30.38%, quite close to the difference obtained in the other way above, is pictured in Exhibit 15 as the difference between Treatment I and Treatment II at the point where the standard difficulty equals zero for both treatments.

11G. More than Two Broad Categories

The case treated in the last section is actually the most difficult one—just because there is *only one cutting point*. Once we have two or more cutting points, our problem separates into two quite different parts:

◇All but two categories are internal categories, bounded by two cutting points. These we can tackle quite directly and effectively.

◇Two categories are external categories, which often involve only small fractions of the cases, and hence, for a reason described below, do not cause great difficulty.

While we have more alternatives to consider, working with more than two categories is actually much more robust, and its results are correspondingly more satisfactory.

Let us recapitulate some crucial aspects of the overall situation:

◇cases are of varying "difficulty".

◇ skilled judgment has divided them into three or more strata on the basis of difficulty.

◇we believe that difficulty actually varies continuously, so that it is reasonable to treat the strata we actually have as if there were a continuous scale of difficulty that had been cut in a few places.

◇one way to put numbers on difficulty—without any attention to observed results—is to assume a shape for some or all of each difficulty distribution. The logistic shape is convenient.

◇For internal categories, as we shall see, the shape chosen is not very important.

◇For external categories, as we shall see, the logistic shape is often neutral in behavior.

Underlying these facts are two questions. (1) How well do we expect the logistic scale to work? If the logistic scale works well enough, (2) how do we think percent success should be related to it? We turn now to these points.

If only a small amount of selectivity has crept in, the two difficulty distributions (on the unknown but most relevant scale) will be fairly similar. In this case, a way of attaching numbers to difficulties that makes one distribution reasonably closely logistic is likely to do the same for the other also. Conversely, if there is a great difference in difficulty for the populations, given the different treatments, so great that the overlap is small, the results of making each apparently exactly logistic can conflict only in this short overlap, so we can again make both nearly enough logistic on a single scale. If there is any trouble, it will arise, between these two extremes, for cases combining largish difference with largish overlap.

The question of relation of response to numerical difficulty is taken up below.

Internal Categories and CG'S

Saying at least something about the center of gravity (CG) of an internal category is easy and is not heavily dependent on our choice of standard distribution. Let us think for a bit about categories 30% wide. If such a category extends from 35% to 65% on a scale of qualitative difficulty, we will have little hesitation in putting its CG at its midpoint, especially since this will be exact for every symmetrical distribution of quantitative difficulty.

If the category extends from 40% to 70%, or from 50% to 80%, we will have little doubt that the CG is toward the 40%-point or 50%-point end from the midpoint. This has to happen for every symmetrical single-peaked distribution. Thus the quality of location of the CG—left of mid, close to mid, or right of mid—is very clear.

Trial of a variety of alternative distribution shapes—Gaussian, logistic, even Cauchy—shows that the numerical location of the CG is rather closely the same for internal categories, so that any one of these will do and we might as well use the easiest.

Now we investigate the second question. Estimating the rate at which percent success varies with CG location is neither so easy nor so crucial as locating the CG is. Suppose only that our correction, for a given category, is always in the right direction and between 0.5 and 1.5 of the right size. Then, after correction,

$$|\text{remaining error}| < \tfrac{1}{2}|\text{initial error}|$$

and we have made real progress.

Let us plan to adjust to category midpoints. To get the direction right, we need only be right about the quality of location of the CG and the direction of change of percent success with difficulty. These are just exactly the simplest things to deal with. We can get them right almost all the time. Accordingly, we do not need high precision in either numerical location of CG or slope of percent success against quantitative difficulty. So let us look at the arithmetic.

For the example given in Exhibit 8, when we have four categories and two treatments, Exhibit **16** shows the calculations leading to the centers of gravity of the segments. They are calculated using our usual formula with the help of the table of Exhibit 11 of Section 5F.

Exhibit **16** of Chapter 11

Calculations of centers of gravity for the grades of Exhibit 8

A) Calculations of the Location of Centers of Gravity

Grade	Cut	$\phi(p)$*	$P - p$	Standard center of gravity	Diff.	Value** of log $(p/(1-p))$	Translated (extremes)	Adjust for ***
	0%	0		Treatment I				
Very easy			.4	−1.682			−1.276	−.134
	40%	−.6730			2.114	−.406		
Semieasy			.4	.432		(.490)		−.058
	80%	−.5004			1.581	1.386		
Semihard			.15	2.013		(2.165)		−.152
	95%	−.1985			1.957	2.944		
Very hard			.05	3.970			1.026	−.125
	100%	0						
	0%	0		Treatment II				
Very easy			.02	−4.900			−1.008	+.134
	2%	−.0980			2.061	−3.893		
Semieasy			.08	−2.839		(−3.045)		+.206
	10%	−.3251			2.143	−2.197		
Semihard			.50	−.696		(−.896)		+.200
	60%	−.6730			2.378	.405		
Very hard			.40	1.682			1.276	+.125
	100%	0						

Median

Very easy	−1.142 ± .134
Semieasy	
Semihard	
Very hard	1.151 ± .125

* $\phi(p) = p \log_e p + (1 - p) \log_e (1 - p)$.
** Values in ( ) are midway between adjacent values of log $(p/(1 - p))$.
*** For extreme categories: values of standard center of gravity MINUS median of same (across treatments). For internal categories: values of std. CG MINUS midvalue (in parenthesis).

As an example, let us run across the Semieasy line for Treatment I. That 80% of the area is to one side of the cut is determined from the middle panel of Exhibit 8. Exhibit 11 of this chapter gives −0.5004 for 80%. We need the difference, −0.5004 − (−0.6730), divided by 0.4 to get the center of gravity 0.432, as referred to the standard logistic. The column of differences comes in handy later in getting the slopes.

From the bottom panel of Exhibit 8, we copy the rates of success onto Panel B of Exhibit 16. Then we can plot success rates against difficulty as measured by centers of gravity, as is done in Exhibit **17** for Treatment I. The figure is of some help because it reminds us which chord we are interpolating on and in which direction.

For Treatment I and the Semieasy category, we have these standard difficulties:

$$\begin{array}{lll} \text{One end:} & -0.406 & \\ \text{CG:} & 0.432 & \text{mid } 0.490 \qquad \text{adjust by } 0.058 \\ \text{Other end:} & 1.386 & \end{array}$$

where $0.490 = \frac{1}{2}(-0.406 + 1.386)$, and the direction of $+0.058$ is very clear.

Exhibit **16** of Chapter 11 (continued)

B) Percentage of Success by Treatment

Rate of success and slopes

Pooled percentages	Category	Treatment		Diff.	Adjusted percentage		Diff.
		I	II		I	II	
21%	Very easy	90%	100%	10%	87.5	101.6	14.1*
	Slope	**−18.92**	**−12.13**				
24%	Semi-easy	50%	75%	25%	49.0	77.4	28.4
	Slope	**−16.89**	**−11.67**				
32.5%	Semi-hard	23.3%	50%	26.7%	20.7	52.3	31.6
	Slope	**−6.80**	**−9.46**				
22.5%	Very hard	10%	27.5%	17.5%	9.2	28.7	19.5
			Weighted average	20.7		Weighted average	24.4

*Note that only 2% of those receiving Treatment I fall in this line.

Since the CG of the Semihard category falls at 2.013, the adjustment is by

$$\frac{0.058}{2.013 - 0.432} = 0.0367$$

of the $50\% - 23.3\% = 26.7\%$ difference in success. This adjustment gives (Treatment I) success at Semieasy midpoint $= 50\% - 0.98\% = 49.02\%$, recorded in Panel B of Exhibit 16 and shown by the circle in Exhibit 17. For Treatment II, a similar adjustment gives $75\% + 2.39\% = 77.39\%$.

We can write the first adjustment as:

($+0.058$ of standard difficulty)(16.9% per unit of standard difficulty),

where

$$16.9\% = \frac{50\% - 23.3\%}{2.013 - 0.432},$$

the slope of the dotted line in Exhibit 17. Here the $+0.058$ would change very little for different assumed distributions—which matter over only this one interval so far as $+0.058$ goes, while the 16.9 would change somewhat, not

Exhibit **17** of Chapter 11

Plot of rates of success for Treatment I against centers of gravity measuring difficulty.

Vertical bars indicate the midpoints of the internal intervals, but medians (over the two treatments) of the standard CG for external treatments. Circles show the adjusted values of the success rate.

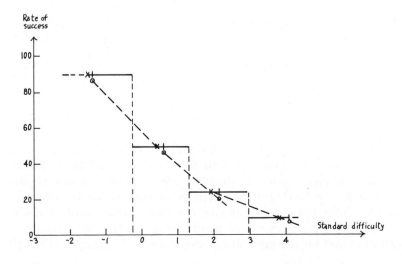

greatly, for different assumed distributions—which now matter over only two categories (here Semieasy and Semihard). We would be surprised by anything that deviated as much as either 0.05 and 12%, giving a 0.6% adjustment, or 0.065 and 20%, giving a 1.3% adjustment. An adjustment of 0.98% is not large, but it will help us a lot. That is what we want from a procedure for adjusting for broad groups.

We are attracted to this type of adjustment for every internal category because it has no need for approximate equality of spread for the distribution of standard difficulties for different treatments—and because it requires reasonable behavior only within one category, or within two adjacent ones.

Because the external categories present a slightly different situation (to be discussed subsequently), we cannot rely much on the slope between the internal and external categories to calculate the adjusted percentage for the internal group, in this case, the Semihard group. Instead, it is safer to use the slope between the internal groups to adjust internal percentages. This slope is also pictured in Exhibit 17.

External Categories

External categories may still give us difficulty, however. We are not very well tied down when we look at only one cut. The best we can hope to do, usually, is to work with two cuts by concentrating on two categories, the extreme category and the adjacent category.

Things are a little better than one might fear, however. The distance of the CG of an extreme category from the cut that defines that category behaves in a way that is often quite helpful, as Exhibit **18** suggests. We see that, for extreme categories that contain no more than 10% of the cases, the CG is just about 1.0 unit outside the defining cut.

Thus, if the spreads of the tails of the distributions of quantitative difficulties are the same for various treatments, and all extreme categories are small, all CG's will be about 1.0 unit from the defining cuts, and cannot differ much from one another, so that the required adjustment will be small.

This is a special property of distributions, like the logistic, that have nearly exponential tails. This sort of "neutral" behavior, leading to small adjustments when all extreme categories are small, seems a safe choice when—as is almost always true—we are uncertain about exact shapes of distribution of quantitative difficulty.

Sometimes, as in our standard example, the extreme categories are not all small. In these cases we can apply either of the methods used above in the two-category case—the case where all categories are extreme. We can use the less robust approach and find adjusted percents of success for the extreme categories working from the nearest cut, or we can, rather more robustly, interpolate to that cut. (It is sometimes handy to give adjusted percents of success for every cut and for the midpoint of every internal category, although we shall not present them.) We notice that we got an adjusted value of over 100% in one cell.

Finally, we need a standard set of weights. We chose the pooled counts in the groups for the two treatments, and they appear in the leftmost column of Panel B of Exhibit 16. Using these weights, we compute the weighted average of the differences in success rates for the raw percentages and the adjusted ones. There is a difference of 20.7 in the raw and of 24.4 in the adjusted. This is a smaller adjustment than we got for the two-group case, but it is still substantial. Exhibit **19** summarizes the differences (Treatment II minus Treatment I) in the success rates already calculated for the two- and four-group examples, as well as the one-group example, the example that ignores the classification into easy and hard and yields only the crude success rate. The group adjusted weighted average entry in the two-group example was discussed in Section 11F in the paragraph on standard population.

Direct standardization. The sort of adjustment we have just been making corresponds to direct standardization, in that we have tried to calculate a success rate appropriate to each treatment (for each category) and have then

Exhibit **18** of Chapter 11

Separation of CG from defining cut (in standard difficulty) for extreme categories

Content of extreme category	Separation of CG from cut
99%	4.652
98%	3.991
95%	3.153
90%	2.557
80%	2.012
60%	1.627
40%	1.277
20%	1.116
10%	1.054
5%	1.026
2%	1.008
1%	1.005

looked at differences and, finally, combined the differences with weights. Had we combined the adjusted success rates with weights and then taken differences, an arithmetically equivalent procedure, we would have been proceeding strictly parallel to direct adjustment, except for the allowance for category breadth.

Summary: Standardizing for Comparison

Responses expressed as rates or proportions can often be more properly compared after standardization for some background variable.

We can directly standardize rates or proportions for background variables.

We can calculate, using various compromises between reduced effort and quality of approximation, estimated standard errors for such directly standardized rates.

Ill-determined rates—perhaps in only one cell—can cause trouble for direct standardization and we need (i) a warning bell when this is likely to occur, and (ii) a method of standardization that avoids this difficulty.

Exhibit **19** of Chapter 11

Differences (Treatment II minus Treatment I) in percentage of success.

("Standardized only" here means that inequalities in difficulty among Easy groups and among Hard groups have not been taken into account; "Broad-group adjusted" means that they have been.)

One group example:

44% − 60% = −16%

Two-group example

	Stand'ized only	Broad-group adjusted
Average	15%	31.1%
Weighted average	15.5%	31.9%

Four-group example

	Stand'ized only	Broad-group adjusted
Average	19.8%	23.4%
Weighted average	20.7%	24.4%

Standard error calculations—perhaps overall, but better in terms of the largest contributions from one or a few cells—warn of the presence of ill-determined rates and the need for a different approach.

Indirect standardization is available whenever it is more appropriate than direct standardization.

Indirect standardization offers advantages and has a reasonable logic and straightforward calculations.

Standardization for a background variable given in broad catetories is never enough to eliminate all the biases associated with that background variable, and often requires supplementation.

To adjust further the results of standardization for broad categories, we can use one technique for end categories (hence only this technique is needed when but two categories are used for the background variable) and another for intermediate categories.

The logistic distribution offers a convenient way to assign centroids to broad categories, a way that generally reflects the true differences in center of gravity for the parts of the populations (for specific treatments) to be compared.

References

Fleiss, J. L. (1973). *Statistical Methods for Rates and Proportions.* New York: Wiley and Sons. Chapter 13.

Freeman, M. F., and J. W. Tukey (1950). "Transformations related to the angular and the square root." *Ann. Math. Stat.*, **21**, 607–611.

Sutherland, M., P. Holland, and S. E. Fienberg (1974). "Combining Bayes and frequency approaches to estimate a multinomial parameter." In S. E. Fienberg and A. Zellner (Eds.), *Studies in Bayesian Economics and Statistics.* Amsterdam: North-Holland; p. 585–617. Or see Y. M. M. Bishop, S. E. Fienberg, and P. Holland (1975). *Discrete Multivariate Analysis.* Cambridge, Mass.: MIT Press; pp. 429–433.

Chapter 12/Regression for Fitting

Introduction

The ABC's of a subject are supposed to be simple, but the ACD's of the relationship between two or more variables—the meanings of *Association*, *Causation*, and *Dependence*, and especially the distinctions among them—are complex. Therefore, after beginning with explanations, distinctions, and definitions, we illustrate them by a variety of examples, some where actual numbers are available, and some where we all know enough to be confident about what would be found if the numbers were gathered.

French dictation test. Let us consider a test of children's ability to write French text correctly from dictation. For simplicity we may assume that the spoken French is standardized by using a tape recorder, that the scoring has been made uniform, and that the scorers can decipher all the handwritings involved. Let y be the score on this dictation test, and let x be the weight of the child. What is the relation of y and x?

In addition to sounding a bit facetious, with nothing more said, this question is almost meaningless.

Widely varying ages. Are we considering children over a substantial range of ages, say 5 to 15, or are we considering only children of age exactly 15 years, plus or minus a few weeks? If all ages are mixed together, heavier children will in general be older and thus, at least where French is taught or spoken, more likely to do well. For such a mixed group there is likely to be a strong positive relation between weight x and dictation score y.

Nearly constant age. If all children are 15 years old, plus or minus a few weeks, there are other questions. If nothing else, there is the difference between girls and boys, which is likely to show the lighter children performing better (since, at this age, girls are better in languages).

A mix. But what if we mix 15-year-olds from different countries where different amounts of French are taught before age 15, for example, in France, Holland, and the U.S.A? If the French are lighter than the Dutch who, in turn, are lighter than the Americans, we would get a strong negative relation between weight and French dictation.

These illustrations show us that to discuss the relation of x and y we need to specify the circumstances—perhaps even a specific population of people or

259

instances that would be sampled. With this need in mind, we can turn to some concepts and their definitions.

Some Statistical Concepts

Association. The weakest concept is *association*—do the values of x and y seem to be paired in a somewhat related way in the population (presence of association), or do they seem to be paired in a completely unrelated way (lack of association)? We have discussed a few ways that weight and French dictation could appear associated, sometimes positively with y increasing as x increases (different ages) and sometimes negatively with y decreasing as x increases (girls vs. boys, children from three different countries mixed).

Independence. To illustrate independence, if we had a population of 15-year-old girls, all of whom had been equally exposed to the teaching of French and all of whom came from homes where their food had been equally nourishing, we might have a population where weight and score on French dictation were not associated, where their numbers seemed to behave independently. To get this to happen exactly rather than approximately, using the full mathematical idea of *statistical independence* might be hard. (Strictly, X and Y are independent random variables if and only if

$$\Pr(X \le a, Y \le b) = \Pr(X \le a)\Pr(Y \le b)$$

for all a and b; and similar definitions hold for discrete unordered variables.)

When, more generally, we find an association between weight and French dictation, do we believe that increased weight "causes" French dictation to be better or worse (no matter which direction the relation may be)? Not if we understand the situation as we have just described it. We would be willing to accept amount of training, native ability, sex, and amount of French spoken at home as possible causes, perhaps even age, but we would be unlikely to accept weight as causing better or worse scores at French dictation. Why?

Causation. Two or three sorts of ideas usually are required to support the notion of "cause":

1. Consistency.

   ◇that when other things are equal in the population we examine, the relation between x and y is consistent across populations in direction— perhaps even in amount.

2. Responsiveness.

   ◇that if we can intervene and change x for some individuals, their y's will respond accordingly.

3. A mechanism.

◇that there is a mechanism, which someone might sometime understand, through which the "cause" is related, often step by step, with the "effect"— the sort of mechanism where, at each step, it would be natural to say "this causes that".

Of course, none of these applies to the relation between weight and performance in French dictation.

Of these three, only one—consistency—can be confirmed by observation alone. We might look at a variety of different populations and see whether the relationship between x and y is consistent in direction or in direction and amount.

The second—responsiveness—can be confirmed by experiment, if an experiment is feasible. If we can intervene and change x, we can see if y then changes. Occasionally natural experiments as well as manmade ones offer some evidence here, the natural ones being almost always less valuable. (Beware especially of "natural experiments" where no change has been made.)

The third—the mechanism—can be confirmed only by constructing a detailed mechanism, and supporting the correspondence between each step in the mechanism and that in the process under study.

Causation, though often our major concern, is usually not settled by statistical arguments (indeed, is usually not settled at all in social problems), though statistics has many inputs that can help. Causation is made more plausible by considerations drawn from data, most effectively perhaps when three essentials are provided:

◇a clear and consistent association between x and y;

◇a showing that either there are no plausible common causes of x and y or that the quantitative relationships between the plausible common causes and x and y are inadequate to explain an observed clear and consistent association. (The frequent existence of partial causes, as in the classical nature– nurture and heredity–environment arguments, can haunt attempts to make such showings. Clearly both heredity and environment cause stature in men and women, but how much of the effect to assign to each—or whether such an assignment based on a single number makes sense—is a difficulty.) An example of common-cause creating is that inflation increases both interest rates and prices; another example, the Industrial Revolution produced a larger population and hence both more scotch whisky imported into New York and more ministers there as well!

◇a showing that it is unreasonable for y to cause x, which is often far from easy. (In our example, suppose fond parents give more candy to children who do well in French—and even send those who do poorly to bed without their supper!)

In these chapters, then, our concern has to be with the presence, absence, or amount of association. We should, and will, leave the question of causation aside. We have to be careful with our language and give repeated warnings about what we are not doing.

Dependence. It is not enough to straighten out association and causation. We must be clearer about that abused word "dependence" and its relatives. When we say "*y* depends on *x*", sometimes we intend *exclusive dependence*, meaning that, if *x* is given, then the value of *y* follows, usually because the existence of a law is implied. In mathematics if *y* is the area of a circle and *r* the radius, then $y = \pi r^2$ illustrates this exclusive dependence. (This use is found in mathematics courses, and is frequently used in the more exact parts of the physical sciences.)

At other times, "*y* depends on *x*" means *failure of independence*, usually in the sense of "other things being equal" as in "the temperature of the hot water depends on how far the faucet is from the heater." Clearly it may depend on other things such as the setting of the heater and the building's temperature as well, to say nothing of whether the long pipe between is on a cold outside wall or a warm inside one. These are two quite different ideas of dependence, and the use of one word for both has often led to troublesome, if not dangerous, confusion.

Then there are the mathematical usages "dependent variable" and "independent variable". These have been extremely effective in producing confusion when dealing with data. We will try to avoid them here entirely.

12A. The Two Meanings of Regression

Regression methods bring out relations between variables, especially between variables whose relation is imperfect in that we do not have one *y* for each *x*. To choose physical variables as examples of imperfect relations, we could cite the relations in man between weight and height, or between weight and height and girth. Such inquiries were common in physical science long before the name "regression" emerged from Galton's studies of inheritance in biology. In his particular example, Galton noted that tall fathers had tall sons, but not as tall on the average as the fathers. Similarly, short fathers had short sons, but not as short as the fathers. These tendencies of the average characteristics of selected groups in the next generation to move toward the average of the population, rather than reproduce the averages of their parents, Galton called regression—regression toward the mean. We mention this bit of history because without it the name regression is rather a puzzle.

The first meaning of regression: column (local) averages. We study the more general ideas of regression and correlation. What is regression? To begin, suppose that we have two variables, illustrated by height *x* and weight *y* for a large population of men. Then for each small interval of *x* (say, of length an inch or a centimeter), we have a distribution of weights *y*. We could compute a summary of these weights for that interval. The summary might be the arithmetic mean, the median, or even the geometric mean. Imagine the chosen summary weight being computed for each successive one-inch interval from, say, 5' 2" to 6' 4". Then the points $(x_i, \bar{y}_i)$, where x_i is the center of the ith height interval and $\bar{y}_i$ is the average weight for that interval, will likely fall close to a curve that could summarize them, possibly close to a straight line. Such a smooth curve approximates the regression curve called the regression of y on x. We give a more mathematical description later.

Example. Age at notable contribution (y) versus age at death (x). Lehman (1953) has explored the distribution of ages when people make notable contributions for ten fields of endeavor. To avoid exaggerating the effects of youth on creativity, since early death precludes further contributions, he grouped his data according to age at death of the person making a contribution. Exhibit 1 gives, for each class interval of age of death, the distribution of age of making contributions. It parallels the height–weight description above with age of death being x, held approximately fixed, and y being the age at contribution.

Exhibit **1** of Chapter 12

Percentages of notable contributions produced in each decade of life, among 980 individuals who died at various ages. Total percentage of notable contributions for each longevity group is 100. In all, 1,540 notable contributions.

Age at which contribution made	$x = $ Age at time of death							
	Under 50	50–59	60–64	65–69	70–74	75–79	80–84	85+
Under 20	5		1		2			
20–29	32	23	17	8	15	10	12	8
30–39	50	39	32	38	36	28	32	29
40–49	14	28	27	28	28	27	28	26
50–59		9	20	16	13	20	15	22
60–69			4	10	6	10	9	9
70–79						4	3	3
80–89							1	2
Totals*	101	99	101	100	100	99	100	99
Medians**	32.7	36.8	40.2	41.4	36.4	44.3	42.1	44.8

* Totals may not add to 100 because of rounding.
** Interpolated linearly. Thus the died-under-50 category, where 37% contributed before age 30 and 87% before age 40, interpolates to 50%, the median, by age 32.7.

S) SOURCE

Lehman, H. C. (1953). *Age and Achievement,* Princeton University Press; p. 317. Copyright 1953 by the American Philosophical Society. Reprinted by permission of Princeton University Press.

Once we have such a table, we want to boil it down. Why? To clarify or accent the relation between the specific values of one variable and the corresponding values of the other. A drastic summary might note that the modal age-interval for contribution for every death interval provided in the table is 30–39. These seem to be highly productive years.

If we want an average, we might compute the mean or median for each column. Then to get a regression we might plot these averages against the midpoints of the class intervals for age at death. (It is not immediately clear what age at death to give to the died-under-50 class or the lived-past-85 class.) If we use the median for age at contribution, we will thereby evade the problem of what age to assign to contributions under age 20.

The medians are shown at the bottoms of the columns of Exhibit 1. Then Exhibit **2** shows a plot of the points, together with a fitted line and an equation for it. Apparently each additional 5 years of life adds about 1 year ($\approx 5(0.19)$) to the median age at contribution for people dying between ages 50 and 85. The two endpoints have arbitrarily been given ages at death of 45 and $87\frac{1}{2}$, respectively.

Formal definition of first meaning of regression. In the mathematical case where for each value of x there is a distribution of Y, with density $f(y \mid x)$ (read f of y given x), the regression of y on x is defined as follows. For each x, compute the mean value of Y for that x, namely

$$\bar{y}(x) = \int_{-\infty}^{\infty} y f(y \mid x) \, dy.$$

Then the function defined by the set of ordered pairs $(x, \bar{y}(x))$ is called the regression of y on x. We have followed the custom, in this case, of choosing the mean as the average for defining the curve (we could have used the median, for example).

Example. Mean. Suppose that the density function is

$$f(y \mid x) = \frac{2y}{x^2}, \qquad 0 \le y \le x \le 1.$$

For a given value of $x, 0 \le x \le 1$, the mean is

$$\bar{y}(x) = \frac{2}{x^2} \int_0^x y \cdot y \, dy$$

$$= \frac{2}{x^2} \left(\frac{x^3}{3} \right) = \frac{2}{3} x.$$

And so $\bar{y}$ is a linear function of x, passing through the origin.

Let us look at the median using the same conditional distribution of y.

Example. Median. The median $y_{med}(x)$ is obtained by finding what value splits the density in half. Thus we require

$$\frac{2}{x^2} \int_0^{y_{med}(x)} y \, dy = \frac{1}{2},$$

or

$$\frac{2}{x^2} \frac{[y_{med}(x)]^2}{2} = \frac{1}{2}.$$

This yields

$$y_{med}(x) = \frac{1}{\sqrt{2}} x,$$

and so in this example, using the median as the average also gives a straight line through the origin, but with a slightly different slope, about 0.71 instead of 0.67.

Sometimes for each y there is also a distribution of X, with density $g(x \mid y)$, and then we can define $\bar{x}(y) = \int x g(x \mid y) \, dx$, and the regression of X on y is given by the set of ordered pairs $(\bar{x}(y), y)$.

Exhibit **2** of Chapter 12

Regression of median age of contribution on age at death; line is based on middle 6 points equally weighted; x values of two endpoints arbitrarily assigned to 45 and $87\frac{1}{2}$, respectively.

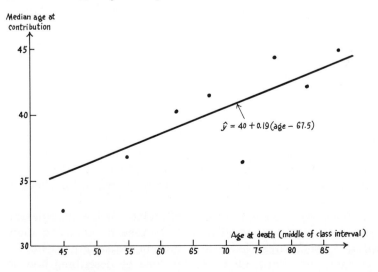

In practical work, we ordinarily do not have continuous populations with known functional forms. But the data may be very extensive. When they are, we can break one of the variables into small intervals and compute averages for each interval and, without severe assumptions about the shape of the curve, essentially get a regression curve, as we did in the age-of-innovation example.

What the regression curve does is give a grand summary for the averages for the distributions corresponding to the set of x's. We could go further and compute several different regression curves corresponding to the various percentage points of the distributions and thus get a more complete picture of the set. Ordinarily this is not done, and so regression often gives a rather incomplete picture. Just as the mean gives an incomplete picture of a single distribution, so the regression curve gives a corresponding incomplete picture for a set of distributions. This was the first meaning of regression.

When the data are more sparse, we may find that sampling variation makes it impractical to get a reliable regression curve in the simple averaging manner described.

The second meaning of regression: fitting a function. One device occasionally used is to introduce a smoothing procedure, applying it either to the column summaries or to the original y's (once these have been ordered in terms of increasing x). For an introduction to such methods, see Sections 3F and 3G, and *EDA*, Chapters 7 and 16. Using that approach gives a curve that may be smooth, but not necessarily of any simple functional form. Sometimes such a result is good enough in itself; sometimes it serves to suggest a simple functional form that we can then fit.

By force when the data is still sparser, by suggestion when we have looked at and thought about smoothed results, by lack of thought in very many other circumstances, we usually come to a second approach. We assume a shape for the curve—linear, quadratic, logarithmic, or whatever—and fit the curve by some statistical method, often least-squares.

In doing this, we do *not* even pretend that the resulting curve has the shape of the regression curve that would arise if we had unlimited data—only that we have an approximation.

Since conditions leading to this second approach to regression—forced fitting of a functional form—arise so often, we tend to forget the first and more fundamental meaning discussed above—curve connecting averages of column distributions. The second approach considerably extends the usefulness of regression methods, for we often have comparatively modest amounts of data, rather than the hundreds or thousands of (x, y) pairs often needed to make narrow columns work well, or the rather smaller number required for effective smoothing.

We ordinarily choose for the curve a form with relatively few parameters. And we have to choose how to fit it. (This can be done by choosing some criterion for fitting—least-squares, least first powers, least pth powers— whichever we can stand and afford. Or it can be done by describing how to recognize a fit in terms of the residuals that it leaves. It can also be done by describing the steps we are to take in constructing a fit—or by some combination of two or three of these approaches.)

By fitting a simple shape, one may be able to detect the need for complications. For example, when we fit to data the horizontal line $y = c$ where c is a constant, we may at once see the need for some slope to the line. Or, having fitted a sloping straight line, we may then see the need for curvature.

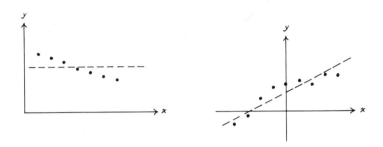

Such insights may come more easily by looking at plots of residuals $(y - \text{fitted } y)$. They may also come as a result of smoothing such plots, either by formal methods or by passing a smooth curve by eye through the scatter of points.

More Than One Carrier

Up to this point, we have mainly, but not entirely, emphasized regressing one y variable (response) against one x variable, or as we shall call it, a carrier. But the whole discussion can be extended to the case of more than one carrier (other than the constant). The move to two carriers (beyond the constant) is an important step, and it has the advantage that the geometry can be kept to three dimensions, leaving us feeling that we can have a strong intuitive grip on the situation. Once we are comfortable with two nonconstant carriers, the next important move is to a large but fixed number of carriers. A third move is to treat several, but an undecided number. Both choosing the set of carriers from which a final subset is to be drawn and choosing that subset can be most disconcerting processes.

12B. Purposes of Regression

A simple, important use of regression is:

1. to get a summary, a use we illustrated in Section 12A with the data on age at contribution.

That same example illustrated another use,

2. to set aside the effect of a variable that might confuse the issue—there, age at death.

3. Contributions to attempts at causal analysis are a popular use for regression, and our example illustrated that as well. Although age alone does not cause contributions, death nearly always stops them.

4. Sometimes, as a corollary to item 3, we want to measure the size of the effect through a regression coefficient, as we did in the age-at-contribution example. As we mention in Chapter 13, this use is fraught with difficulties when there are multiple causes and when various noncausal variables are associated with other causal ones.

5. An extreme instance of the causal approach occurs when we use it to try to discover a mathematical or empirical law.

   A more common use for regression is:

6. for prediction, as when we use information from several weather stations to predict the probability of rain at a particular area several hours hence, or use information about composition of the population and measures of pollution in various cities to forecast the death rate from a class of diseases believed caused by such pollution.

Such predictive uses are widespread and diversified. Sometimes, indeed, a predictive use may be intended to be causative as in item 4—if we reduce a particular sort of pollution, we estimate the effect on the death rate (a possibly misleading use); or the regression equation may be used to estimate the anticipated death rate in a new place and thus to see whether that place is better off than "average", as in item 2, an "other things being equal" prediction.

We can try to fit

$$y = f(x), \quad \text{where } f \text{ is empirically determined,}$$
$$y = bx,$$
$$y = a + bx,$$
$$y = a + bx + cx^2,$$
$$y = a + bx + ct,$$

and their generalizations to more carriers (more coefficients, more variables, or both) for such purposes. The problems we need to solve can be different in the different cases.

Regression as exclusion. Sometimes we make a fit, perhaps by a quadratic, say

$$y = a + bx + cx^2$$

with a general purpose,

7. of getting x "out of the way." We may know that x affects y, perhaps substantially, and be curious as to whether t affects y, too. (Here we use "affect" as a shorthand for "is associated with, possibly, but not certainly, through a causal mechanism." We will also use the "effects of x" in a similar way.)

For example, we might like to remove the effects of years of education from measures of performance on knowledge of current events, and then see what is left to be explained by frequency of media readership.

One approach would take the effects of x out of y and see if what remains is associated with t. To try to do this, we might first fit

$$y = a + bx + cx^2$$

and then continue with $y_{\cdot x}$, where

$$y_{\cdot x} = y - a - bx - cx^2$$

is a residual after fitting x, a quantity freer of the effects of x than y was.

In other circumstances, we can estimate, for each x, the distribution of y given x and then replace y by a measure of where y stands in the distribution of y given x, such as the number of standard deviations above the mean. This is, of course, easier when x takes on only a few values, as when x is sex (male or female), or a few class intervals, as in our age-at-contribution example. In this class of problem, we compare y with what we would predict knowing x, or knowing the values of several variables (x's).

Those who are primarily concerned with getting x out of the way need not interpret or give meaning to the coefficients, formulas, and tables they use to do this. They should look hard at what carriers are used to see that they are reasonable—not an easy task—but the coefficients need not have meaning in themselves. If an investigation shows that the score, y, on French dictation is related to age x by

$$y \approx 50 + 7(x - 10) - 0.5(x - 10)^2,$$

we do not expect the coefficients 50 (of 1), 7 (of $x - 10$), and -0.5 (of $(x - 10)^2$) to be the same in other investigations. Others may choose other curves. If we want to brush x out of the way while carrying this investigation further, we can use

$$y - 50 - 7(x - 10) + 0.5(x - 10)^2$$

without prejudice to what we might do some other time.

More than this, we would like a good fit, without regard to whether it is the best kind of fit. In this example, some other function of age might be better than a polynomial. Thus, if French is not taught below age 7, the time dependence of y on x must be very close to $y = 0$ (if 0 means no skill) for all $x < 7$, something that a polynomial finds very hard to approximate. But if our study includes no children under 9, the quadratic may be quite good enough.

And in general. Such a fit is a practical tool. A simple expression is a fairly good description. It may also be the right way to express a relationship.

In the next chapter, we will look at a question of intermediate difficulty: If we accept the form of dependence, what ought we to make of the coefficients? The purpose there—getting coefficients that can be interpreted—is quite different from the purposes in the rest of the present chapter: getting a good fit, either for getting the x's out of the way or for predicting y from the x's.

These last two purposes sound almost like versions of the same thing. They are similar, but some differences can matter.

Let us look a little harder at a concrete prediction problem. Consider a large state university, so large that for most applicants the decision on admission has to depend on a formula that forecasts degree of success. How should the variables and the coefficients in such a formula be chosen? Clearly the prediction needs to be at least partly successful.

Should father's occupation be an x? If it is, what should its coefficient be? If it isn't available, then there is no choice. If it is, moral and political considerations may control whether it is used. (At a smaller institution, one that prides itself on a balanced student body, x may be used for quite other reasons than prediction of performance.)

Within the constraints of availability (and cost) and moral and political principles, probably some set of x's can be used. We will want to combine them into our prediction formula, but we ought not to care much about the values of the coefficients. What is important is the fit—what sort of y do we predict for these x's? And how close will prediction come to performance? Many different formulas will do about equally well. We would like one of them—it doesn't matter much which. (If we were trying to understand what mattered in college success, we would have to take a very different attitude.)

A word more about the political side. When the formula is chosen, since many formulas will be about equally good at forecasting, it would be well to note whether the chosen one has any bad quirks that will lead to hilarious newspaper columns or to the appearance of injustice for reasons that might be anticipated from its form or values. One must assume that the formula will become public property and will be criticized, at best, by neutral and, more likely, by unsympathetic observers. Consequently, if it has features such as that the outcome on one true–false spelling item is worth more than a B in two years' study of the calculus, it is going to be hard to explain that it wouldn't matter much which formula was used, however true that might be. Consequently, among the many formulas that might be used, it would be well to choose one that is reasonably presentable and defensible if other people choose to make interpretations.

In this chapter, we treat the easier problems of finding a good fit: something usable either for getting x out of the way or for predicting y from x, but something whose structure and coefficients are not—or at least not yet—to be taken seriously.

12C. Graphical Fitting by Stages

Let us suppose that we wish to fit y with x_1 and x_2 in the form $\beta_1 x_1 + \beta_2 x_2$. Although we can fit

$$y = \beta_1 x_1 + \beta_2 x_2$$

directly, we may improve our understanding by a more sequential approach. In this second approach we represent the fit not in terms of x_1 and x_2, but in terms of x_1 and x_2-adjusted-for-x_1, which we label $x_{2;1}$. Specifically, we construct

$$x_{2;1} = x_2 - d_{2;1} x_1,$$

where $d_{2;1}$ is chosen to free x_2 of x_1, at least approximately. Specifically $d_{2;1}$ is the slope of a line through the origin in the x_1-vs.-x_2 plane, chosen to fit the scatter of points well, especially to predict x_2 from x_1. Then $x_{2;1}$ is the set of residuals of x_2 based on the forecast $d_{2;1} x_1$. In this section we do not use least squares explicitly to get $d_{2;1}$ but elsewhere we may. We are doing for x_2 what we often do to free y of x when we have only y and x.

The overall plan is first to fit y to $\gamma_1 x_1$ and get

$$y = \hat{\gamma}_1 x_1 + \text{residual}.$$

Once this is done we compute residuals

$$y_{;1} = y - \hat{\gamma}_1 x_1,$$

where the subscript ;1 reminds us that we are computing a residual for y and the 1 indicates that we have, in a specific sense, removed the effect of x_1. Second, we use these residuals to fit

$$y_{;1} = \hat{\beta}_2 x_{2;1} + \text{residual}.$$

That is, we take the part of y that was free of x_1 and relate it to the part of x_2 that is free of x_1. Then we assemble the pieces to see that

$$y = \hat{\gamma}_1 x_1 + y_{;1}$$

is fitted by

$$\hat{\gamma}_1 x_1 + \hat{\beta}_2 x_{2;1}.$$

If we wish, this can be converted to a form that is identically equivalent, where

$$y \text{ is fitted by } \hat{\beta}_1 x_1 + \hat{\beta}_2 x_2$$

provided we set

$$\hat{\beta}_1 = \hat{\gamma}_1 - d_{2;1}\hat{\beta}_2.$$

We have illustrated the point that removing the effects of variables can be done one variable at a time.

Here the removals symbolized by the semicolon in the subscript may be formal and carefully executed or they may be very informal. In this section, we are rather informal, fixing our coefficients graphically, but carefully, in two stages.

Exhibit **3** gives the U.S.A. data for a simple example where

$$y = \text{total derived employment} - 65317 \quad \text{(in 1000 jobs)}$$
$$x_1 = \text{GNP price deflator} - 101.7 \quad \text{(as a \%)}$$
$$x_2 = \text{gross national product} - 387,698 \quad \text{(in \$1000)}$$

for the years 1947 to 1962. (The subtracted constants 65317, 101.7, and

Exhibit **3** of Chapter 12

Employment, price deflator, and gross national product by years.

Year	Employment, y	Price deflator, x_1	Gross national product, x_2
1947	−4,994	−18.7	−153,409
1948	−4,195	−13.2	−128,272
1949	−5,146	−13.5	−129,644
1950	−4,130	−12.2	−103,099
1951	−2,096	−5.5	−58,723
1952	−1,678	−3.6	−40,699
1953	−328	−2.7	−22,313
1954	−1,556	−1.7	− 24,586
1955	702	−.5	9,771
1956	2,540	2.9	31,482
1957	2,852	6.7	55,071
1958	1,196	9.1	56,848
1959	3,338	10.9	95,006
1960	4,247	12.5	114,903
1961	4,014	14.0	130,475
1962	5,234	15.2	167,196

Note. The precision given is that of the original sources.

S) SOURCE:

A. E. Beaton, D. E. Rubin, and J. L. Barone (1976). "The acceptability of regression solutions: another look at computational accuracy." *J. of Amer. Statist. Assoc.,* 71, 158–168.

Quoted from J. W. Longley (1967). "An appraisal of least-squares for the electronic computer from the point of view of the user." *J. of Amer. Statist. Assoc., 62, 819–841.*

387,698 used in these definitions are, to the accuracy given, the means over this sequence of years.)

Exhibits **4** and **5** show the two steps of fitting by a constant times x_1. The first, found by eye, gives a trial slope of +300. We use a prime to indicate a trial value. And so $y'_{;1}$, $x'_{2;1}$, and so on, are trial residuals. From the first line of Exhibit 3 we get

$$y'_{;1} = -4{,}994 - 300(-18.7) = 616.$$

A further fit of x_1 to $y'_{;1} = y - 300x_1$, shown in Exhibit 5, then finds a slope of 5, giving a final slope of 305 and a $y_{;1}$ of $y - 305x_1$. Then for 1947

$$y_{;1} = -4{,}994 - 305(-18.7) \approx 710$$

from the values in Exhibit 3. (The arithmetic is shown in Exhibit **6**.)

Exhibit **4** of Chapter 12

First try at regression of y on x_1.

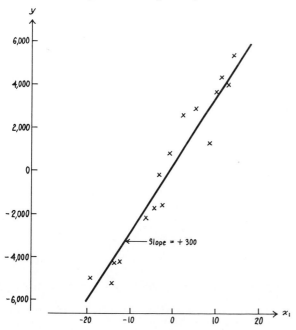

Exhibit **5** of Chapter 12

Second try at regression of $y'_{;1}$ on x_1.

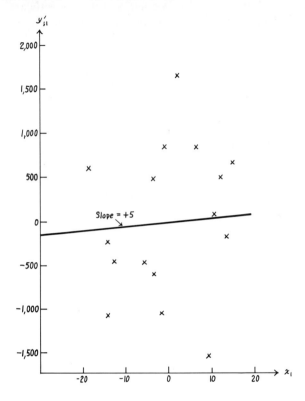

Exhibit **6** of Chapter 12

First tries and fitted values of regressions (noted at bottom of exhibit), using the values from Exhibit 3.

Year	x_1	First tries		Fitted values		First try,
		$y'_{;1}$	$x'_{2;1}$	$y_{;1}$	$x_{2;1}$	$y'_{;12}$
1947	−18.7	616	33,591	710	1,801	584
1948	−13.2	−235	3,728	−169	−18,712	1,141
1949	−13.5	−1,096	5,356	−1,028	−17,594	204
1950	−12.2	−470	18,901	−409	−1,839	−280
1951	−5.5	−446	−3,723	−418	−13,073	497
1952	−3.6	−598	−4,699	−580	−10,819	177
1953	−2.7	482	4,687	496	97	489
1954	−1.7	−1,046	−7,586	−1,038	−10,476	−305
1955	−.5	852	14,771	854	13,921	−120
1956	2.9	1,670	2,482	1,656	7,412	1,137
1957	6.7	842	−11,929	808	−539	846
1958	9.1	−1,534	−34,152	−1,580	−18,682	−272
1959	10.9	68	−13,994	14	4,536	−304
1960	12.5	497	−10,097	434	11,153	−347
1961	14.0	−186	−9,525	−256	14,275	−1,255
1962	15.2	674	15,196	598	41,036	−2,275

Notes.

$$y'_{;1} = y - 300x_1,$$
$$y_{;1} = y'_{;1} - 5x_1 = y - 305x_1,$$
$$x'_{2;1} = x_2 - 10,000x_1,$$
$$x_{2;1} = x'_{2;1} + 1700x_1 = x_2 - 8300x_1$$
$$y'_{;12} = y_{;1} - 0.07x_{2;1}$$

Exhibits **7** and **8** go on to do a similar thing for x_2 on x_1. This time the second term is 17% of the first. (Arithmetic is also shown in Exhibit 6.) Exhibit 3 shows for 1947

$$x'_{2;1} = -153,409 - 10,000x_1 = 33,591,$$

and then

$$x_{2;1} = 33,591 + 1,700x_1 = 1801.$$

Exhibit **7** of Chapter 12

First try at regression of x_2 on x_1.

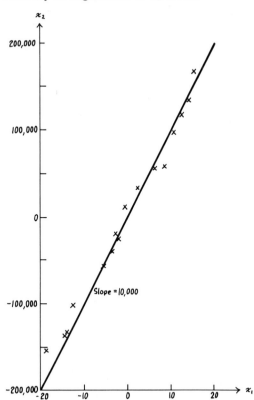

We are ready to go ahead with the final step, the regression of $y_{;1}$ on $x_{2;1}$. Exhibit **9** has the first step. It is not always easy to judge what would be a good regression coefficient; we decided to try 0.07. The next exhibit, Exhibit **10,** shows that this may have been too much, but we decided to stop here, in view of the generally horizontal nature of all but two of the sixteen points, namely those for the last two years, 1961 and 1962. (Some analysts might have come back by a slope of as much as -0.015 or -0.02.)

Exhibit **8** of Chapter 12

Second try at regression of $x'_{2;1}$ against x_1.

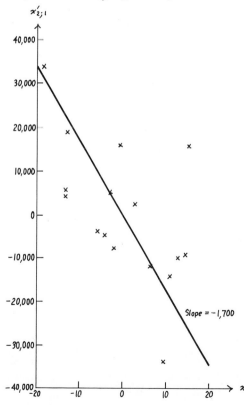

In summary, then we have found the following rough fits:

$$y \text{ by } 305x_1, \qquad \text{so that } y_{;1} = y - 305x_1;$$
$$x_2 \text{ by } 8300x_1, \qquad \text{so that } x_{2;1} = x_2 - 8300x_1$$
$$y_{;1} \text{ by } .07x_{2;1}, \qquad \text{so that } y_{;12} = y_{;1} - 0.07x_{2;1},$$

where $y_{;12}$ means residuals of y after removing both x_1 and $x_{2;1}$, or, more briefly, x_1 and x_2. Combining, we find

$$
\begin{aligned}
y_{;12} &= y_{;1} - 0.07x_{2;1} \\
&= (y - 305x_1) - 0.07(x_2 - 8300x_1) \\
&= y + 276x_1 - 0.07x_2,
\end{aligned}
$$

Exhibit **9** of Chapter 12

First try at the regression of $y_{;1}$ against $x_{2;1}$.

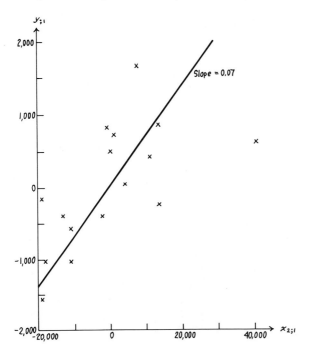

so that our fit to y is

$$-276x_1 + 0.07x_2.$$

Had we come back by 0.01 in the slope of $y'_{;12}$ on $x_{2;1}$, our fit would have been

$$-199x_1 + 0.06x_2,$$

showing how much change in coefficients can sometimes be associated with a very small adjustment in fit (only one point, that for 1962, is moved more than ±200, out of a trend amounting to 10,000).

Exhibit **10** of Chapter 12

Basis for a second new try at the regression of $y'_{;12}$ against $x_{2;1}$.

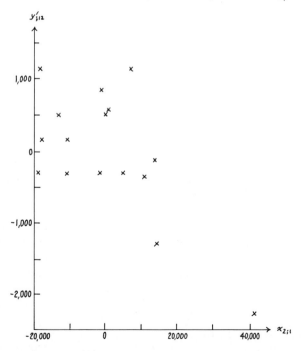

12D. Collinearity

Many opportunities for trouble arise in regression problems. One well known difficulty goes under the heading of collinearity. In the simplest case, the difficulty arises from measuring essentially the same quantity under different names, and then trying to use the several measures to get a regression relation between all these nearly-equivalent carriers and the response variable. Essentially, the idea is that we get into trouble when we try to treat one piece of information as if it were several pieces. This inevitably leads to arbitrariness about the allocation of the weights to be given the several pieces. This situation can arise very readily in social science work and in economics, when we have many variables that may enter a regression relation and clumps of these variables measure much the same thing. We illustrate this in the church-membership example at the close of this section.

Example. Chest and heart measurements. A biostatistician had two measurements of chest size in boys about 12 years old taken a few months apart, as well as a direct measure of heart size. He decided to use each boy's two chest measures to develop a regression equation for estimating heart size. His equation was something like the following, where H = heart size, and C_1 and C_2 are the first and second chest measurements,

$$H = 10 + 8C_1 + 3C_2. \tag{1}$$

His equation was based on least squares from one group of boys. He had an additional set of data taken on boys of the same age in the same part of the country, and when he fitted his equation for the second group, he got something like

$$H = 10 + 5C_1 + 6C_2. \tag{2}$$

Since the two groups of boys were expected to be quite comparable and in both groups the number was substantial, he was bothered by the large differences between the coefficients of the C's in the two equations. Which equation should he believe?

What is going on here is that, for each boy, C_1 and C_2 are nearly identical, differing because of slight growth, and possibly differing more because of measurement error. Thus C_1 and C_2 might well have been averaged, producing a single value for chest measurement $\bar{C}$, leading to a regression equation of about

$$H = 10 + 11\bar{C}. \tag{3}$$

Note that the coefficient 11 is the sum of the coefficients of C_1 and C_2, and that both Eqs. (1) and (2) had coefficients of C's adding to 11 (the sums of these coefficients were approximately, but not exactly, identical in the actual example).

Geometrical clarification. From a geometric point of view, in the three-dimensional space of (C_1, C_2, H), the points are strung out almost along a straight line. This straight line is being asked to determine a plane. Insofar as the points are on the line, this is an impossible task. When the points are close to the line, it is an uncertain one; a slight change in the positioning of a point can tip the plane just like a seesaw.

Estimation. Even though the two estimates of H,

$$10 + 8C_1 + 3C_2 \qquad \text{and} \qquad 10 + 5C_1 + 6C_2,$$

were very different in the coefficients of the C's, their estimates of H were, in the application, very near one another. From the point of view of the geometric explanation, the reason is that any point (C_1, C_2) had to be close to the $C_1 = C_2$ line in the (C_1, C_2) plane. That point, projected vertically to the regression plane, then, will be close to the original line of points in the three-dimensional space for a great variety of planes through the line. All these planes will then give much the same fit or prediction for the point.

All told then, when two reliable, identically distributed measures of the same quantity are used as predictors in a multiple regression equation, we find:

a)   the sum of their regression coefficients is relatively stable, and

b)   the prediction of the dependent variable is much the same because of the stability mentioned in (a).

Naturally the very special case we have just treated has implications for other problems, or we would not have brought it up. Whenever variables used for prediction are highly correlated with one another, or when two or more are estimating essentially the same thing, the sizes of the corresponding coefficients are likely to be uncertain indicators of the "importance" of the predictor. We must therefore anticipate that, at the very least, it is difficult to tell what the effect might be on the response variable if we were able to change the value of one of the predictor variables. Note that, in these high-correlation situations, we cannot change the value of one carrier substantially without changing the other at the same time.

Example. Church membership. If one thinks of number M of individual church members as the consequence of a church's activity, using salaries L as a measure of labor (actually salaries plus 10% of the value of its parsonages), and of the value of its buildings less debt as a measure of capital, K, then an economist might wish to relate these by

$$M = 10^{\beta_0} L^{\beta_L} K^{\beta_K}.$$

Taking logarithms gives

$$\log M = \beta_0 + \beta_L \log L + \beta_K \log K.$$

Exhibit **11** gives the logarithms of M, L, and K for one state from each region for the Methodist Episcopal Church, according to the 1936 religious census. Before fitting the proposed regression, the plot of $\log L$ against $\log K$ was made, as shown in Exhibit **12.** The plot shows a nearly linear relation with slope near 1 between $\log L$ and $\log K$, giving warning that the chest-measurement sort of collinearity has struck again. This means that β_L and β_K would be poorly determined by the data. Still, estimating $\log M$ can be readily done from $\log L$, $\log K$, or a combination. Let us plot $\log M$ against the sum of $\log L$ and $\log K$.

Exhibit **13** shows the result of the plot. Again the relation is close and near-linear with slope close to $\frac{1}{2}$.

Exhibit **11** of Chapter 12

Logarithms of membership, salaries, and capital for one state from each region in the Methodist Episcopal Church in 1936.

State	Membership	Salaries	Capital
Maine	4.30	5.33	6.20
New York	5.48	6.47	7.60
Ohio	5.58	6.34	7.44
Minnesota	4.87	5.75	6.77
Delaware	4.41	5.30	6.33
Kentucky	4.38	5.19	6.13
Arkansas	3.62	4.49	5.45
Montana	4.12	5.05	5.93
Washington	4.61	5.44	6.33
Median	4.41	5.33	6.33
7th minus 3rd	0.57	0.56	0.64

S) SOURCE.

G. Mosteller, personal communication, drawn from *Religious Bodies: 1936,* Vol. II, Part 2, Bureau of the Census, United States Department of Commerce, U.S. Government Printing Office; pp. 1086–1096.

Exhibit **12** of Chapter 12

Plot of logarithm of salaries against logarithm of value of property (less debt) for the Methodist Episcopal Church, 1936, in one state from each of nine geographical regions.

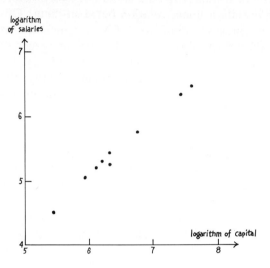

Exhibit **13** of Chapter 12

Plot of logarithm of membership against (logarithm of salaries) + (logarithm of value of property) for the Methodist Episcopal Church, 1936, in nine states.

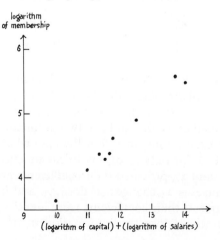

12E. Linear Dependence, Exact and Approximate

When we have a set of carriers and plan to use the regression on them for prediction, we frequently call the carriers *predictors*.

Let us push the collinearity idea further and deal with predictors that are functionally related, with no error. In a most extreme situation, two or more predictors may be measured with exactly a linear relation between them. This could happen if a number of percentages adding to 100% are included as separate variables (such as compositions of alloys in metals, or percentages of families having different kinds of housing). Or we might have redundancy, for example, for variables having to do with a child in a completed family: the child's birth order B, the family size N, and number of younger children, Y. These yield the exact relation

$$B + Y = N.$$

Other examples of special interest to social and clinical psychologists are certain tests whose parts are scored in such a way as to add up to a constant, as in the Allport–Vernon–Lindzey *Study of Values*.

How do such dependencies affect the fitting of multiple-regression equations? They create difficulties equivalent to those that arise from trying to divide by zero in ordinary arithmetic. Let us take the simplest case, where the variables x_1 and x_2 add to a constant k, $x_1 + x_2 = k$, or where one is a linear function of the other, $x_2 = bx_1 + c$.

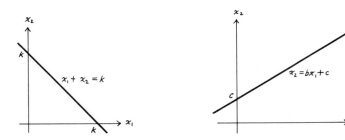

Let x_1 and x_2 be predictors for the variable y. Then the observed points in three dimensions (x_{1i}, x_{2i}, y_i) are all in the plane perpendicular to the (x_1, x_2) plane through the line relating x_1 and x_2. This goes to the extreme suggested in the chest–heart example. The graphs tell us that we do not have two predictors, only one. We can choose a variety of representations for it. For the case where $x_1 + x_2 = k$, we could get along with only x_1 or only x_2 or any linear function of x_1 or of x_2. Indeed, we can combine x_1 and x_2, provided their coefficients are not identical: $ax_1 + bx_2 + c$, $a \neq b$, because as x_1 changes so does x_2, and by the same amount but with the opposite sign. If their coefficients were identical, the function would be constant and we would lose the information in both predictors.

In the second illustration, we can also use any linear function of x_1 and x_2, provided it is not of the form $a(x_2 - bx_1) + d$, because the latter would be a constant.

The only easy remedy, then, for the forms of exact collinearities discussed in this subsection is to abandon the attempt to use the two predictors separately. Instead, replace them by a single carrier. This will fix the arithmetic—and may or may not help us with the prediction problem.

Which single carrier we use is not determined by anything discussed here. Convenience or relations with further carriers might determine the answer for regression as prediction. If we are trying for regression as measurement or regression as a causal indicator, we are likely to be in deep trouble because there the coefficients are usually required to have meaning.

The simplest polynomial; a nearly linear relation. One form of near collinearity generated by functional relations arises when we try to fit a regression involving a linear function of 1, x, and x^2—to fit a quadratic $a + bx + cx^2$ to a set of data.

Since x and x^2 are closely related, we may have computational difficulties leading to the need for many places of accuracy to keep any of the precision we actually have. In computing a mean, we are used to the idea of picking a central value, obtaining deviations from it, averaging these deviations, and then recapturing the mean. *Example*: Average 13879, 13881, 13864. It is convenient to choose 13870, say, as an arbitrary value, obtain deviations, 9, 11, -6, sum to get 14 and the average $4\frac{2}{3}$, and a final average $13874\frac{2}{3}$. People who do the calculation in their heads usually use some form of this approach. The main point is that the original 5-digit numbers led to 1- and 2-digit deviations, and the arithmetic was easier.

In multiple regression, especially with highly correlated variables, a related device—similar to choosing a central value—can be even more helpful in keeping numbers of decimal places down, and is often even more needed. Fitting $a + bx + cx^2$ is equivalent to fitting

$$a* + b*x + c(x^2 - A - Bx)$$

for any A and B, where $a = a* - Ac$ and $b = b* - Bc$, and the coefficient, c, of the quadratic term is unchanged. If an x_0 were centrally located in the data, then $(x - x_0)^2$ would be small compared with x^2. Also

$$(x - x_0)^2 = x^2 - 2x_0x + x_0^2$$

has the form

$$x^2 - Bx - A$$

with $A = -x_0^2$ and $B = 2x_0$. Thus it could be advantageous to fit

$$a* + b*x + c(x - x_0)^2$$

or even

$$a** + b*(x - x_0) + c(x - x_0)^2.$$

In addition to the point about keeping down the arithmetic, this form tells us more about the uncertainty of the estimates of the coefficients. If all x's are close to x_0, the carrier $(x - x_0)^2$ is very small in comparison with $x - x_0$ or 1, and so, as we next explain, we cannot expect to estimate c with much numerical precision.

As discussed in Chapter 14, the denominator of the variance of $\hat{c}$ will be the sum of the squares of what is left after $(x - x_0)^2$ has been regressed on 1 (important) and $x - x_0$ (usually less important) (that is, we fit $\gamma_0 + \gamma_1(x - x_0)$ to $(x - x_0)^2$ and get the sum of the squared residuals). So we are in somewhat more trouble than the smallness of $\sum (x - x_0)^2$ suggests. This doesn't matter as much for our present "good fit" aim, since the values of $(x - x_0)^2$ we substitute in are also likely to be small compared to 1 and $x - x_0$. (Extrapolation is another matter; there parabolic fits are far more dangerous than straight-line fits because the contribution from the quadratic term can swamp the rest of the fit.)

Returning to practical arithmetic, we note that it is often important to avoid trying to fit

$$a + bx + cx^2$$

instead of

$$a** + b*(x - x_0) + c(x - x_0)^2.$$

A few "canned" computer programs fit the second form automatically and then reconvert and report a, b, and c. If yours does not, users may get badly misleading results, unless they have changed to a good set of carriers in advance.

Even for what appear to be long intervals of x, the correlation of x and x^2 may be considerable. For example, if x is uniformly distributed over some interval $(0, A)$, then the correlation between x and x^2 is 0.97 for any choice of A. Many would associate this degree of relation with the idea of a practically perfect linear relation. For this situation the variance of our coefficients is increased by a factor of about

$$\frac{1}{1 - r^2} = 12$$

over what it would be if we sought to fit either x or x^2 alone. And so again we find ourselves close to an exact linear relation.

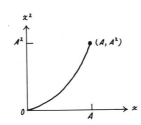

In spite of all these troubles, it may nevertheless be useful to detect and remove a quadratic term from a scatter diagram.

As one would expect, the computational difficulties and the uncertainty in the coefficients become even more critical when fitting a cubic or higher-degree polynomial to any fund of data.

Near-linear and even exact dependencies are also common between variables with different-seeming names, although we have used x and x^2 as an easy example. The examples at the beginning of the section illustrate this.

12F. Keeping Out What is Imprecisely Measured—Regression as Exclusion

(Before reading this section, the reader may wish to read the introduction and the first section of Chapter 14.)

To do a fair job of excluding the effect of a variable (the seventh purpose of regression, mentioned in Section 12B) that cannot be assessed directly, but is related to something we can measure, requires additional data and techniques. An example is maturity in children, which is related to chronological age. Essentially, we need at least two measures of such a latent variable to get a handle on its effect.

The case of no deviations. Let us review, first, our usual approach. When x is measured precisely and we want to take its linear contribution out of y, we can do this by looking at either the residual

$$y - bx$$

or the residual

$$y - a - bx,$$

where $a + bx$ is as good a linear fit to y as we can get. So long as x measures exactly what we want to eliminate, keeping x out is exactly like predicting from x.

Because the estimate b will not exactly coincide with the true β, our elimination of x's linear contribution will not be perfect. We can write

$$y - bx = (y - \beta x) - (b - \beta)x,$$

or

$$y - a - bx = (y - \alpha - \beta x) - (a - \alpha) - (b - \beta)x,$$

where $y - \beta x$ (or $y - \alpha - \beta x$) is what we wish we had. The deviation $-(b - \beta)x$ is unbiased on the average because $(b - \beta)x$ behaves, from one data set to another, just as $b - \beta$ does. If we do a good job of fitting, $(b - \beta)x$ is sometimes positive and sometimes negative.

If other contributions from x are important, we may want to go to additional terms or forms of x and use, say,

$$y - bx - cx^2,$$

where $bx + cx^2$ is a better fit to y.

The customary situation. But what if x is a useful measure of what we want to keep out, but is not an exact measure? Perhaps we are studying children and wish to keep out—adjust for—the effect of maturity. We are likely to have available chronological age, which tells us a lot about maturity, but not the whole story. We have no difficulty using chronological age as an x in prediction, but how should we use it in adjustment?

More generally, then, we have an x which measures a z only approximately; how can we use it to remove the effect of z from y? It turns out that we at least need an extra variable u, called an instrumental variable, which also measures z, like the score on a maturity test.

Suppose that we have an *instrumental variable u*, meaning by this:

1. that u is also associated with z,

2. that the residual of u, after subtracting its true regression on z, is uncorrelated (has zero covariance) with the residuals of x and y.

More precisely, the statistical model is

$$x = \alpha_x + \beta_x z + \text{residual}_x,$$

$$y = \alpha_y + \beta_y z + \text{residual}_y,$$

$$u = \alpha_u + \beta_u z + \text{residual}_u,$$

$\beta_x, \beta_y, \beta_u$ are regression coefficients of z,

$$\text{cov}\{\text{residual}_x, \text{residual}_u\} = 0,$$

$$\text{cov}\{\text{residual}_y, \text{residual}_u\} = 0.$$

We might try to use a linear function of either u or x to eliminate z from y. If we want to exclude z from y by subtracting a multiple of x, since we have to remove the term $\beta_y z$ from y, we need to multiply x by β_y/β_x so that the x's $\beta_x z$ term will also contribute $\beta_y z$. We get

$$y - \frac{\beta_y}{\beta_x} x = \beta_y z + \text{residual}_y - \frac{\beta_y}{\beta_x} \beta_x z - \frac{\beta_y}{\beta_x} \text{residual}_x + C$$

$$= \text{residual}_y - \frac{\beta_y}{\beta_x} \text{residual}_x + C$$

where $C = \alpha_y - (\beta_y/\beta_x)\alpha_x$. (The value of C is not important in this discussion.) To do this, we need an estimate of β_y/β_x, something we cannot get directly, because β_y and β_x are regression coefficients on z, and we do not know z.

In this ideal situation, the covariances of x and y with u are

$$\text{cov}\{x, u\} = \text{cov}\{\beta_x z, \beta_u z\} = \beta_x \beta_u \text{ var}\{z\},$$

$$\text{cov}\{y, u\} = \text{cov}\{\beta_y z, \beta_u z\} = \beta_y \beta_u \text{ var}\{z\},$$

as we see by substituting for x, y, and u in the first equalities and noticing that, in both lines, three of the four resulting covariances vanish. Accordingly, from the least-squares definition for the population, the regression coefficients of x and y on u are

$$\beta_{xu} = \frac{\text{cov}\{x, u\}}{\text{var}\{u\}} = \beta_x \cdot \frac{\beta_u \text{ var}\{z\}}{\text{var}\{u\}}$$

and

$$\beta_{yu} = \frac{\text{cov}\{y, u\}}{\text{var}\{u\}} = \beta_y \cdot \frac{\beta_u \text{ var}\{z\}}{\text{var}\{u\}},$$

so that

$$\frac{\beta_{yu}}{\beta_{xu}} = \frac{\beta_y}{\beta_x}.$$

Now we can estimate β_{yu} by b_{yu}, since we know both y and u, and β_{xu} by b_{xu}. This lets us take

$$\frac{b_{yu}}{b_{xu}}$$

as a reasonable estimate of β_y/β_x and

$$y - \frac{b_{yu}}{b_{xu}} x$$

as a reasonably adjusted y—adjusted for z, not only for x. All this is sound so long as u satisfies the zero-covariance conditions for the residuals, and $\text{cov}(x, u) \neq 0$. (Economists might use the language that says that we have adjusted for the structural regression instead of for the predictive regression.)

The u's that we have available are often crude. If they are, we still use them rather than nothing. Even a u that takes two or three values—that divides the data into two or three groups—is quite usable if we can believe that it satisfies the zero-covariance conditions for residuals and the nonzero covariance condition.

Note that

$$\frac{b_{yu}}{b_{xu}}$$

is ordinarily larger than b_{yx} would be. To see this, suppose first that

$$\text{cov}\{\text{residual}_x, \text{residual}_y\} = 0.$$

then

$$\beta_{yx} = \frac{\text{cov}\{y, x\}}{\text{var}\{x\}} = \frac{\beta_y \beta_x \text{ var}\{z\}}{\beta_x^2 \text{ var}\{z\} + \text{var}\{\text{residual}_x\}},$$

while we can rewrite β_y/β_x as

$$\frac{\beta_y}{\beta_x} = \frac{\beta_y \beta_x \text{ var}\{z\}}{\beta_x^2 \text{ var}\{z\}},$$

a fraction with the same numerator but a smaller denominator than β_{yx}. If the covariance between the residuals of x and y does not vanish, we can still estimate β_y/β_x—though it may be even harder to find a u—but we cannot be as sure that β_y/β_x is farther from 0 than β_{yx} is.

Although we shall not prove it, the variability of

$$y - \frac{b_{yu}}{b_{xu}} x$$

will ordinarily be larger than the variability of

$$y - b_{yx}x$$

because trying to predict

$$y \quad \text{adjusted for } z$$

must be harder than trying to predict

$$y \quad \text{adjusted for } x,$$

since we know so much more about x than about z.

We now have a technology for removing the effects of a latent variable, here z, from a response variable y, provided that we have two measures that depend on the latent variable and meet other conditions. Let us apply it.

When illustrating a new statistical technology, perhaps the most helpful example is one where the structure is secretly known, so that we can see how well the method behaves and find the sources of deviations. Let us display first a set of data as it might appear in a real-life example where the structure is not known.

Example. Known structure. Exhibit **14** shows the original data for y, x, and u, and we want to remove, even if we cannot quite succeed, the effect of z where, as a special case of what was described above, we suppose that

$$y = \beta_y z + \text{residual}_y, \qquad x = \beta_x z + \text{residual}_x, \qquad u = \beta_u z + \text{residual}_u,$$

and that the residual terms have zero covariances.

Solution. Exhibit 14 shows also the calculation of $b_{yu}/b_{xu} = 2.505$. Using this result, we can compute the residuals shown in the first column of Exhibit **15**. (The true values are plotted against the estimated values in Exhibit **17**.) Note that they follow the true residual$_y$'s in a general way, though by no means perfectly.

Discussion

We secretly know exactly the structure of this example because we built it to fit the model perfectly. We chose 10 values of z ($= -5, -4, -3, -2, -1, 1, 2, 3, 4, 5$)—(*Note.* Zero happens not to be present)—and $\beta_y = 5$, $\beta_x = 2$, $\beta_u = 3$. We chose $\alpha_x, \alpha_y, \alpha_u = 0$. The residual terms were chosen to be independently normally distributed with means 0 and standard deviations 2, 1, and 3 for y, x,

Exhibit **14** of Chapter 12

Data for removal of z effect from y, where u is the instrumental variable.

y	x	u
−26	−8	−16
−21	−7	−14
−14	−7	−8
−5	−4	−2
−8	−2	−2
7	3	3
8	5	2
13	6	11
16	8	9
25	11	14
Totals −5	5	−3

$$\sum yu = 1522$$
$$\sum xu = 605$$
$$\frac{b_{yu}}{b_{xu}} = \frac{152.2 - (-.5)(-.3)}{60.5 - (.5)(-.3)} = 2.505$$

Exhibit **15** of Chapter 12

Estimated and true residuals.

Estimated residual$_y$ $y - 2.505x$	True residual$_y$
−6.0	−1
−3.5	−1
3.5	1
5.0	5
−3.0	−3
−.5	2
−4.5	−2
−1.5	−2
−4.0	−4
−2.6	0
Totals −17.1	−5

and u, respectively. (*Note.* These are not standard deviations of y, x, and u, but of their residual terms.) Using random normal deviates, 10 values of residual terms were found for each variable, and, for convenience, these were rounded to the nearest integer. The actual values and components leading to the y's, x's, and u's are shown in Exhibit **16.** In real life only the y, x, and u columns would be available.

Exhibit **16** of Chapter 12

Construction of the y, x, u values

z	$\beta_y z$	Residual$_y$	y	$\beta_x z$	Residual$_x$	x	$\beta_u z$	Residual$_u$	u
−5	−25	−1	−26	−10	2	−8	−15	−1	−16
−4	−20	−1	−21	−8	1	−7	−12	−2	−14
−3	−15	1	−14	−6	−1	−7	−9	1	−8
−2	−10	5	−5	−4	0	−4	−6	4	−2
−1	−5	−3	−8	−2	0	−2	−3	1	−2
1	5	2	7	2	1	3	3	0	3
2	10	−2	8	4	1	5	6	−4	2
3	15	−2	13	6	0	6	9	2	11
4	20	−4	16	8	0	8	12	−3	9
5	25	0	25	10	1	11	15	−1	14

Exhibit **17** of Chapter 12

Plotted from values of Exhibit 15.

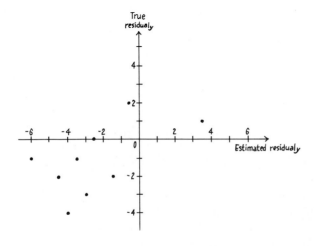

Estimating the residuals. Our estimate of β_y/β_x is 2.505, extremely—and unusually—close to the true value 2.5. Nevertheless, we cannot retrieve the residual$_y$'s exactly because, even with the ratio almost exactly right, x is still not z. Since x is not z, the errors in such estimates of the residual of y will have size about $2.505 \times$ residual$_x$, as algebra will show. In the example, the residual$_x$'s are all ± 2, ± 1, or 0, and so the error in estimating the residual y is about ± 5, ± 2.5, or 0. The only residual$_x$ of 2 is the first one, and that gives an error of 5, to one decimal place. In addition to this, had the estimate of the ratio β_y/β_x been farther from the true value, the deviations from the true residuals would have had a larger contribution from that error in slope.

Some readers may have noted that we might also have used x as the instrumental variable for the regression of y on z, estimating β_y/β_u by b_{yu}/b_{xu}. If we had chosen to do that, our estimate of residual$_y$ would be

$$y - (b_{yx}/b_{xu})u.$$

Since β_y/β_u is 5/3, the new estimated residual$_y$'s would be in error by roughly 5/3 residual$_u$.

Both methods could be used, and some weighting scheme could combine them to make different estimates of the residual$_y$'s. We do not pursue that line of discussion here.

★12G. Which Straight Line? (Optional.)

Suppose, for the moment, that we have just one carrier x—and that y is well fitted by a straight line

$$y \equiv A + Bx$$

in x. Does this mean—if there is a "right" straight line—that we have the "right" straight line? Perhaps—and perhaps not.

Suppose that $Y(y)$ is a function of y that increases when y increases; then

$$Y(y) \equiv Y(A + Bx),$$

so that

$$Y(y) \equiv A^* + B^*X(x),$$

where

$$X(x) = \frac{Y(A + Bx) - A^*}{B^*},$$

which is a function of x alone, once the constants A, B, A^*, and B^* are known. Thus, any such Y has a whole family of X's (A^* and B^* are at our choice) such that $Y(y)$ is well fitted by a straight line in $X(x)$. How then are we to pick out the right straight line?

If there appears to be a single choice for which the two re-expressions— one of y and one of x—both seem simple, perhaps we can believe that this is the right straight line. (A nice special case arises when the first-aid rules lead to a straight line.) But what if there is no such guidance? Or if two or three pairs of relatively simple re-expressions all seem to fit relatively well? (If this seems unlikely, recall that if $y = bx$ then $\log y = \log b + \log x$, and that if $\log y = \log b + k \cdot \log x$, then $y = bx^k$.)

It is easy to say, especially for a physical scientist, that theory should tell us what to choose. But if theory says that $y = bx$, it also says that $\log y = \log b + \log x$. Theory is no help to a single set of data.

Sometimes parallel cases are sufficient to settle a choice. If

$$y = a_i + b_i x$$

in the ith case, where the a_i and b_i vary rather unpredictably, then having

$$\log(y - a_i) = \log b_i + \log x$$

with different values of a_i, is far from being attractive. After all, $\log(y - a_i)$ is a different function of y for each a_i. We have lost consistency of description by "taking logs" when this means something different for each parallel case. Making parallel cases come out parallel can be a good reason for choosing (y, x) instead of (Y, X).

Beyond this, only one other criterion sees some general use and has some arguments of plausibility in its form. This criterion says choose (y, x) or (Y, X) so that the spread of the data around the straight line (or around the more general fit) is roughly constant.

By this criterion, for instance, if we can fit $y = bx$, and note that the spread of y's for a given x is roughly proportional to that given x (and hence to the center of the y-distribution for that x), we find ourselves urged to go over to

$$\log y = \log b + \log x$$

for which the residuals will be more nearly of constant width. This choice depends on the behavior of imperfections; we can only do it if the fit is not perfect.

Why might this criterion be plausible? First, because it makes the situation easier to explain and think about. One can say that 90% or 95% limits on a single new observation are about so-and-so wide, without having to bother about x. One can look at a residual and assess its plausible relevance—without bothering about x.

Second, we often fit y against x and then go on to analyze the residuals, perhaps by regression on a further carrier—usually more usefully, of course, by regression on the residuals of the new carrier on x. In this process, it helps if the variables used at the second or later stage are at least moderately compatible with the residuals; a good standard for both is roughly constant variance when displayed against x.

Without other dominating considerations, in choosing between (y, x) and (Y, X), we appear to have two good criteria, which may apply:

⬦ making parallel situations become well analyzed in parallel ways,

⬦ making the imperfections of fit of approximately constant size.

We cannot expect both of these to be available for the same data. When they are both available, they often agree; when they disagree, follow the first of the two.

12H. Using Subsamples

Regression is applied to varying amounts of data—as few as 3 or 4 points (data sets) and, perhaps, as many as 5 million. In any ordinary circumstances, starting one's regressions with more than 1000 points is almost certainly bad practice, and starting with more than 200 can well be. Regression should ordinarily be a flexible process. Residuals ought to be looked at in several ways. Starting in on an oversized collection of data sets severely discourages flexibility.

If we are lucky enough to have 10,000 data points, it is almost certain that we should begin our work on them by preparing subsamples, perhaps a subsample of 1000 and a sub-sample of 100. Perhaps a subsample of 2500, a sub-sample of 625, and a sub-sub-subsample of 160 or so. In either case, we would then begin the analysis by working with the smallest subsample until we had gone as far as we could, and then move up to the next subsample.

A simple consequence is that large data files should be so stored that it is easy to extract subsamples of various sizes. When the order of data sets is unimportant, the easiest way to arrange for this may be to pull successive subsamples out, putting each ahead of the sample or subsample from which it is drawn.

Taking one from every 2, 4, 8, 16 or, more generally, 2^k (integer $k > 0$) successive values is easily accomplished, using uniform random numbers to pick the one value. At worst, such a random stratified sample will be almost as good as a simple random sample. At best, as when the original file is quite structured, it will be much better. If we need more careful sampling, perhaps taking explicit account of some stratification of the file, C. T. Fan, Mervin Muller and Ivan Rezucha (1962) have prepared programs that make such sampling easy on IBM equipment.

Another use of subsamples, of course, is to assess the stability of answers, either directly or via the jackknife (see Chapter 8).

Summary: Regression

We have recognized the difficulties of establishing causation, the pressures for belief in establishment of causation that needs for decisions impose upon us, and the consequent dangers of serious error when established association is wrongly interpreted as causation.

We have reviewed the different meanings of dependence (and of related words) and the advantages of avoiding the use of such words.

"Regression" has two quite different meanings, one a curve of typical values of y for (nearly) fixed x (or, more generally, a surface of such values) and the other a choice of a collection of possible fits distinguished by values of constants and a fitting of these constants.

Every regression (according to either definition) is an incomplete description. Thus, though our regression is often extremely convenient or extremely useful, it has no claim either to begin in "the right form" or to be telling us the whole story.

Regression has several different purposes, including summarization, measurement (of the coefficients in a specified regression), exclusion (of the effects of interfering variables), and prediction—and what is good practice for one purpose need not be good practice for another.

Near or exact collinearity, either of two carriers or of several, (i) is something we must be prepared for, (ii) is something that can destroy (one or both of) precision of estimation and simplicity of interpretation of coefficients while leaving the quality of fit high, (iii) is something that has caused many misinterpretations and error, and (iv) is something that we can learn to deal with.

Polynomials—even quadratic ones—are likely instances of near collinearity (unless we are careful).

A good fit of the form $y = a + bx$ may, or may not, mean that this is the "right" straight line.

We can fit regressions by stages, removing each carrier in turn—to an amount assessed by either graphical or arithmetical techniques—not only from the response but also from each other carrier not so far fitted.

We replace bundles of closely associated (nearly collinear) carriers by either a single composite or by a single composite together with the residuals of the individual carriers after regression on that composite (or together with only those residuals that appear enough larger than measurement error to be likely to be useful).

In regression as exclusion, we can do better than use regression coefficients appropriate for prediction (we need to use larger coefficients, specifically structural ones).

★Rough constancy of size of residual is a guide in choosing among alternative good fits by straight lines (where the alternatives involve re-expressions, sometimes of both x and y). Consistent description of parallel sets of data is a better guide.

To deal effectively with any significantly large set of data, plan to subsample and to work first on the smallest sample that is reasonable for what we are doing, stepping up (by a factor of 2, 3, 5, or 10) to successively larger bodies of data as our analysis progresses.

References

Allport, G. W., P. E. Vernon, and G. Lindzey (3rd ed., 1960). *Study of Values.* Boston: Houghton Mifflin.

EDA = Tukey, J. W. (1977). *Exploratory Data Analysis.* Reading, Mass.: Addison-Wesley.

Fan, C. T., M. E. Muller, and I. Rezucha (1962). "Development of sampling plans by using sequential (item-by-item) selection techniques and digital computers." *J. of Amer. Statist. Assoc.*, **57,** 387–402.

Riley, M. W., and A. Foner (1968). *Aging and Society*, Vol. 1. New York: Russell-Sage Foundation. p. 437. (Shows Exhibit 1, page 263.)

Chapter 13/Woes of Regression Coefficients

We know that regression coefficients can suffer from serious difficulties. Let us look systematically at what can be done about them. We discuss what troubles arise, how they happen, and to what extent we can prepare for them, avoid them, or become warned of their presence.

Coefficients that we can apply in new situations may be the most useful kind of knowledge we can try to gain or reasonably hope to have (except in a few highly structured situations).

13A. Meaning of Coefficients in Multiple Regression

Form of the variable. As a first problem in the meaning of a regression coefficient, we consider a quadratic function, which can be thought of as having structure $\beta_0 \cdot 1 + \beta_1 x_1 + \beta_2 x_2$, where $x_2 \equiv x_1^2$. We consider different ways of writing the basic variable x (dropping the subscript) and their effect upon the size of the coefficients.

If

$$117 - 3x + 2x^2$$

is a good fit for $-2 \leq x \leq 5$, what meaning do we give x's coefficient, -3? Each of the following expressions is numerically identical with $117 - 3x + 2x^2$, namely:

$$115 + x + 2(x - 1)^2$$
$$109 + 5x + 2(x - 2)^2$$
$$117 + x + 2x(x - 2)$$
$$117 + 5x + 2x(x - 4).$$

Whatever interpretation we can properly give to -3, the coefficient of x in our original formula, there must be a parallel interpretation that we can give to 1 or 5, each the coefficient of x in two of our later formulas, because all these formulas produce identical results.

If we had started with

$$12 - 3x_1 + 5x_2$$

and gone to

$$12 - 8x_1 + 5x_2^*$$

where $x_2^* = x_1 + x_2$, we might be tempted to talk about the coefficient "with

x_1 held constant," because the remaining coefficients in the two expressions have the same value. But recalling that x_2 might be x^2, note that for positive x's we cannot hold x^2 constant and vary x very much. This cannot be the answer.

The only general fact is that the coefficient of any one carrier—in our example, x—depends on which other carriers are offered for fitting at the same time—in this example, 1 and x^2, or 1 and $(x - 1)^2$, or 1 and $x(x - 2)$, or 1 and $x(x - 4)$.

As the differences, in our simple example, among

$$-3 = \text{coefficient of } x \text{ when } 1 \text{ and } x^2 \text{ are also offered}$$

and

$$+1 = \text{coefficient of } x \text{ when } 1 \text{ and } (x - 1)^2 \text{ are also offered,}$$

$$+5 = \text{coefficient of } x \text{ when } 1 \text{ and } (x - 2)^2 \text{ are also offered,}$$

emphasize, a coefficient in a multiple regression—either in a theory or in a fit—depends on MORE than just:

⋄the set of data and the method of fitting.

⋄the carrier it multiplies.

It also depends on:

⋄what else is offered as part of the fit.

Subsets of variables. Part of the difficulty of giving meaning to regression coefficients arises because the coefficients themselves change depending upon which variables are *present* in the regression. This could be regarded as an extension of the points made above. Since we frequently do not know which ones to put in, this leaves us in an awkward spot. To drive this point home, let us illustrate with an easily verified numerical example.

Example. Plane through the origin. Let us suppose that, unknown to us, it is exactly true that

$$y = \beta_1 x_1 + \beta_2 x_2 + \beta_3 x_3,$$

but that we plan to fit by least squares

$$y = \gamma_1 x_1 + \gamma_2 x_2$$

or

$$y = \delta_1 x_1 + \delta_3 x_3$$

or

$$y = \epsilon_1 x_1,$$

instead. Let us assume that $\beta_1 = 2$, $\beta_2 = 4$, $\beta_3 = 10$, and, for arithmetic simplicity, that we have complete cross-product symmetry among our variables so that, for our (x_1, x_2, x_3) triples,

$$\sum x_1^2 = \sum x_2^2 = \sum x_3^2 = 1, \qquad \sum x_1 x_2 = \sum x_1 x_3 = \sum x_2 x_3 = \tfrac{1}{2}.$$

What are γ_1, δ_1, and ϵ_1, and how do they compare with β_1?

Solution. For least-squares fitting a plane through the origin,

$$\gamma_1 = \frac{(\sum x_1 y)(\sum x_2^2) - (\sum x_2 y)(\sum x_1 x_2)}{(\sum x_1^2)(\sum x_2^2) - (\sum x_1 x_2)^2}$$

and the formula for δ_1 is the same except that x_2 is replaced by x_3 throughout. We can readily get $\sum x_i y$, $i = 1, 2, 3$, from

$$\sum x_i y = \beta_1 \sum x_1 x_i + \beta_2 \sum x_2 x_i + \beta_3 \sum x_3 x_i, \qquad i = 1, 2, 3,$$

and we find for the numerical case chosen that $\sum x_1 y = 9$, $\sum x_2 y = 10$, $\sum x_3 y = 13$.
    We then have or compute

$$\beta_1 = 2, \qquad \gamma_1 = 5\tfrac{1}{3}, \qquad \delta_1 = 3\tfrac{1}{3}.$$

And so, even in this very symmetrical problem, the coefficient of x_1 varies substantially depending on which variables are present. If we fit only $y = \epsilon_1 x_1$, the coefficient of x_1 is

$$\epsilon_1 = 9.$$

When we treat this example, as we have, as a straight problem in arithmetic, our first reaction may well be, "Whoever said there was to be a simple relation or invariance when the variables changed?" That is a fair enough comment. But in real-world problems, we do ransack the variables, keeping some and throwing others away. In the end we will have a strong urge to interpret the chosen coefficients in a physical manner, such as "If we change x_1 by a given amount, we will change y by a certain amount, and therefore public policy should be . . .". Our example shows that even in a deterministic system (the original $y = \beta_1 x_1 + \beta_2 x_2 + \beta_3 x_3$ had no error), different subsets of the variables used for regression can give substantially different coefficients for the same variable. Indeed, even the sign can be reversed from one set to another.
    We need then to speak of the coefficient of x_1

⋄ when x_2 and x_3 are also offered,

⋄ when x_2 is also offered,

⋄ when x_3 is also offered,

⋄ when nothing else is offered,

and appreciate that these four ordinarily give substantially different results.

A variety of x's. This point is so important that we make it again more algebraically.

Given several x's, the coefficients—c_1, c_1^*, c_1^{**}, c_1^{***}—of x_1 when we fit

$$c_1 x_1 + c_2 x_2 + c_3 x_3 + c_4 x_4$$
$$c_1^* x_1 + c_2^* x_2 + c_3^* x_3$$
$$c_1^{**} x_1 + c_2^{**} x_2 + c_6^{**} x_6 \tag{1}$$
$$c_1^{***} x_1 + c_7^{***} x_7$$

are usually different. Indeed, rarely will they be the same.

We are now looking at four different fits—in the sense that the set of possibilities that can be represented by $c_1^* x_1 + c_2^* x_2 + c_3^* x_3$ (for some values of c_1^*, c_2^*, c_3^*) is a very different set of possibilities from all those that can be represented by $c_1^{***} x_1 + c_7^{***} x_7$ (for all possible values of c_1^{***} and c_7^{***}). Thus there is no way to go from a fit of (c_1, c_2, c_3, c_4) without further information to a fit of (c_1^*, c_2^*, c_3^*) or of $(c_1^{**}, c_2^{**}, c_6^{**})$ or of (c_1^{***}, c_7^{***}) and no way to use a fit to one of these sets of coefficients without further information to get (c_1, c_2, c_3, c_4) or even the individual c's. We would need to know much more about the data (or population) involved to make such conversions. In our deterministic example we used all the sums of squares and cross-products.

(The case of $12 - 3x_1 + 5x_2$ and $12 - 8x_1 + 5x_2^*$, with $x_2^* = x_1 + x_2$, is a different matter, where we deal with the same fit expressed in different terms. Here, knowing that $x_2^* = x_1 + x_2$ enables us to convert one set of coefficients into the other. This sort of possibility, although often useful computationally, does not often arise in subject-matter oriented analyses.)

Stocks. Each fit is selected from a set of possible fits. In expressions (1), we gave four such sets. We call such a set a "stock" because we want to think about it as a collection of potential fits from which *one* can be selected by choosing the values of the c's, just as men's shirts can be ordered by giving neck and arm measurements. They are the set of possible fits chosen for the regression.

Any sum-of-terms stock—the only kind we discuss in this chapter—can be written

$$c_1 x_1 + c_2 x_2 + \cdots + c_k x_k$$

if we are willing to admit that any or all of $x_2, x_3, \ldots, x_k$ may also involve x_1. (Thus, $x_2 \equiv x_1^2$, $x_3 \equiv x_3^*$, $x_4 \equiv x_1^3 + x_4^*$, where x_3^* and x_4^* do not involve x_1, would define a perfectly good stock of the form $c_1 x_1 + c_2 x_2 + c_3 x_3 + c_4 x_4$.)

It is important to know what other carriers are present in addition to the carrier whose coefficient concerns us. Are the details of the form of expression of a stock important? Not so long as we can get back and forth algebraically from one form to the other. The stocks

$$c_1 x_1 + c_2 x_2 + \cdots + c_k x_k$$

and

$$c_1 x_1 + c_2^* x_2^* + \cdots + c_k^* x_k^*$$

are the same, if the smaller stocks, called costocks of x_1 (to have a convenient way of talking about them), one made up of everything of the form

$$c_2x_2 + \cdots + c_kx_k \qquad (\text{Any } c\text{'s}),$$

and the other of everything of the form

$$c_2^*x_2^* + \cdots + c_k^*x_k^* \qquad (\text{Any } c^*\text{'s}),$$

are identical. These costocks include everything in the whole stock with $c_1 = 0$.

The values of c_1 in the two fits will be the same provided the fitting criteria: (i) are the same, (ii) depend only upon residuals; for example, if both forms are fitted by least squares. Thus, in a sum-of-terms fit, what matters in fixing the coefficient of x_1 are:

◇ the set of data and the method of fitting (including any weights),

◇ the stock defined by the other terms that are also being offered.

13B. Linear Adjustment as a Mode of Description

In Section 12C, we introduced and illustrated graphically a technique for removing one variable at a time; we called it fitting by stages. In this section we develop this idea algebraically, partly

a)   because we wish to use it further,

b)   because we want to show that the technique does produce a least-squares solution, and

c)   because it emphasizes what we have just said by observing that the least-squares coefficient of each carrier is the regression of the response on this carrier linearly adjusted for this carrier's costock.

A word about notation. When we have y and several x's, $x_1, x_2, \ldots, x_k$, we have introduced notation for residuals exemplified by $y_{;12}$ and $x_{3;12}$ to indicate that x_1 and x_2 have been removed from y and x_3, respectively. When we use least squares we replace the semicolon by a period, and the least-squares residuals above are written $y_{.12}$ and $x_{.12}$. Sometimes x_1 is a constant, frequently $x_1 \equiv 1$. The example we are about to discuss has three variables y, x, and t, and a constant is also to be fitted. In the x_1 notation our variables are $x_1 = 1$, $x_2 = x$, $x_3 = t$. It will be convenient in this example to let 1, x, and t stand for themselves instead of the subscripts. Thus $y_{.1x}$ will mean that the constant and x have been removed from y by least squares.

In this discussion we describe a step-by-step procedure for "taking out" the linear effect of variables, one variable at a time. Let us begin with only three variables, y, x, and t and the constant 1—one response and three carriers, respectively. Suppose we have regressed y on 1 and x, finding the fit $y = a + bx$. Then we can compute the residuals, which in this context are

$$\text{y-adjusted-linearly-for-1-and-}x \equiv y_{\cdot 1x} \equiv y - a - bx,$$

and suppose that this adjusts y adequately for x. If then we want to ask whether the adjusted y seems to depend on t as well, we could start by plotting the adjusted y against t and looking.

What if t were rather like x? We have tried to free $y_{\cdot 1x}$ of x, so it cannot look like the part of t that behaves like x. We will get a much clearer picture if we regress t on x also, thus largely freeing t of x and finding

$$\text{t-adjusted-linearly-for-1-and-}x \equiv t_{\cdot 1x} \equiv t - c - dx.$$

Now we can plot $y_{\cdot 1x}$ against $t_{\cdot 1x}$. Suppose we find that there is a clear dependence. It is natural for us now to fit a linear regression of $y_{\cdot 1x}$ on $t_{\cdot 1x}$, say

$$y_{\cdot 1x} \sim e + ft_{\cdot 1x}.$$

What can we say about e and f when we do everything by least squares? We will show, at the end of the next chapter, that this approach gives us, in an unfamiliar form, exactly the least-squares fit of y to $\beta_1 1 + \beta_2 x + \beta_3 t$. More specifically

$$a + bx + f(t - c - dx)$$

is the least-squares fit in question, as proved in Section 14K.

The general case. We can have several x's instead of 1, as explained in Section 14K. The general result is that if, for example,

$$y \sim a + bx + ct + du + ev + fw,$$

then:

1. b is the regression coefficient of x when y, linearly adjusted for 1, t, u, v, and w, is regressed on x, linearly adjusted for 1, t, u, v, and w.

2. c is the regression coefficient of t when y, linearly adjusted for 1, x, u, v, and w, is regressed on t, linearly adjusted for 1, x, u, v, and w.

    .   .   .

    .   .   .

    .   .   .

6. a is the regression coefficient of 1 when y, linearly adjusted for x, t, u, v, and w, is regressed on 1, linearly adjusted for x, t, u, v, and w.

When we deal with a particular situation, we want to write this sort of statement out for each coefficient that we want to interpret. There is no substitute for facing the facts in detail.

When we want a statement to remember, we can condense a little, and say:

The (least-squares) coefficient of each carrier is the regression of the response, linearly adjusted for this carrier's costock, on this carrier also linearly adjusted for this carrier's costock (where the costock contains all possibilities in the whole stock for which the corresponding carrier has coefficient zero).

13C. Examples of Linear Adjustment

Using fitting methods to recover meaningful coefficients from data is one of the jobs often asked of regression. Let us look at an example where we assume the correct functional form, but where the data have involved rounding. We assume that $a + bx + cx^2$ is the correct form to fit. The true quadratic is $y = -1 + x + 0.5x^2$. When we computed the values of y for a set of x's, we rounded the term $0.5x^2$ to two decimals, rounding to the nearest even second decimal when necessary. Thus for $x = 0.9$ we replaced $0.5(0.81)$ by 0.40 instead of the correct 0.405. We are starting with values close to the correct quadratic. Let us see how nearly we recover it by regression.

Beginning with the constant. If, for computational convenience, we wanted to take out 1 first, the calculation analogous to that given in Section 13B would begin

$$y._1 \equiv y - \bar{y}, \qquad x._1 \equiv x - \bar{x}, \qquad t._1 \equiv t - \bar{t};$$

then taking out x would give

$$y._{1x} \equiv y._1 - gx._1, \qquad t._{1x} \equiv t._1 - hx._1;$$

and finally

$$y._{1xt} \equiv y._{1x} - ft._{1x};$$

and the fit would be

$$y_{\text{fitted}} \equiv \bar{y} + gx._1 + ft._{1x}.$$

We use this approach in the examples.

Example 1. Quadratic. Exhibit **1** gives data for y and x and we will take as t the quadratic x^2. We want the multiple regression of y on 1, x, and t.

Solution. From y, x, and x^2 we removed their averages 0.94, 1.2, and 1.48 to get

$$y._1 \equiv y - \bar{y}, \qquad x._1 \equiv x - \bar{x}, \qquad t._1 \equiv (x^2)._1 \equiv x^2 - \overline{x^2}.$$

(Note that $\overline{x^2}$ is the average of the squares.)

Next we adjust $y_{\cdot 1}$ and $t_{\cdot 1}$ by removing $x_{\cdot 1}$ to get

$$y_{\cdot 1x} \equiv y_{\cdot 1} - gx_{\cdot 1} = y_{\cdot 1} - 2.2x_{\cdot 1},$$

$$t_{\cdot 1x} \equiv t_{\cdot 1} - hx_{\cdot 1} = t_{\cdot 1} - 2.4x_{\cdot 1}.$$

Then at last we regress $y_{\cdot 1x}$ on $t_{\cdot 1x}$ to get

$$y_{\cdot 1xt} \equiv y_{\cdot 1x} - ft_{\cdot 1x} = y_{\cdot 1x} - .48t_{\cdot 1x}.$$

Exhibit 1 of Chapter 13

Quadratic regression using stepwise approach.

y	$y_{\cdot 1} \equiv$ $y - \bar{y}$	$y_{\cdot 1}x_{\cdot 1}$	x	$x_{\cdot 1} \equiv$ $x - \bar{x}$	$(x_{\cdot 1})^2$	$t \equiv x^2$	$t_{\cdot 1} \equiv (x^2)_{\cdot 1}$ $\equiv x^2 - \overline{x^2}$	$t_{\cdot 1}x_{\cdot 1}$
0.30	−.64	.192	0.9	−.3	.09	.81	−.67	.201
0.50	−.44	.088	1.0	−.2	.04	1.00	−.48	.096
0.70	−.24	.024	1.1	−.1	.01	1.21	−.27	.027
0.92	−.02	0	1.2	0	0	1.44	−.04	0
1.14	.20	.020	1.3	.1	.01	1.69	.21	.021
1.38	.44	.088	1.4	.2	.04	1.96	.48	.096
1.62	.68	.204	1.5	.3	.09	2.25	.77	.231
$\bar{y} = 0.94$		.616	$\bar{x} = 1.2$		.28	$\overline{x^2} = 1.48$		.672

$$g \equiv \frac{\sum y_{\cdot 1}x_{\cdot 1}}{\sum x_{\cdot 1}^2} = \frac{.616}{.28} = 2.2, \qquad h \equiv \frac{\sum t_{\cdot 1}x_{\cdot 1}}{\sum x_{\cdot 1}^2} = \frac{.672}{.28} = 2.4$$

$y_{\cdot 1}$	$gx_{\cdot 1}$	$y_{\cdot 1x} \equiv$ $y_{\cdot 1} - gx_{\cdot 1}$	$t_{\cdot 1}$	$hx_{\cdot 1}$	$t_{\cdot 1x} \equiv$ $t_{\cdot 1} - hx_{\cdot 1}$	$y_{\cdot 1x}t_{\cdot 1x}$	$t_{\cdot 1x}^2 \equiv$ $[(x^2)_{\cdot 1x}]^2$	$y_{\cdot 1xt}$
−.64	−.66	.02	−.67	−.72	.05	.0010	.0025	−.004
−.44	−.44	0	−.48	−.48	0	0	0	.000
−.24	−.22	−.02	−.27	−.24	−.03	.0006	.0009	−.006
−.02	0	−.02	−.04	0	−.04	.0008	.0016	−.001
.20	.22	−.02	.21	.24	−.03	.0006	.0009	−.006
.44	.44	0	.48	.48	0	0	0	.000
.68	.66	.02	.77	.72	.05	.0010	.0025	−.004
						.0040	.0084	

$$f \equiv \frac{\sum y_{\cdot 1x}t_{\cdot 1x}}{\sum t_{\cdot 1x}^2} = \frac{.0040}{.0084} = 0.48.$$

$$y_{\text{fitted}} = \bar{y} + gx_{\cdot 1} + f(x^2)_{\cdot 1x} = 0.94 + 2.2x_{\cdot 1} + 0.48(x^2)_{\cdot 1x}$$

$$y_{\text{fitted}} = \bar{y} - (g - fh)\bar{x} - f\overline{x^2} + (g - fh)x + fx^2 = -1.03 + 1.05x + 0.48x^2$$

That result, shown in the final column of Exhibit 1, gives $y_{.1xt}$ as zero to two decimals, when we round $0.5x^2$ to the nearest even second decimal.

The final fit can be written

$$y_{\text{fitted}} = 0.94 + 2.2x_{.1} + 0.48(x^2)_{.1x},$$

or rewritten in terms of 1, x, and x^2 as

$$y_{\text{fitted}} = -1.03 + 1.05x + 0.48x^2.$$

The result is most encouraging when we recall that we started out with $y = -1 + x + 0.5x^2$, but that we rounded the $0.5x^2$ term. We started very close to a true quadratic and came recognizably close to recovering it. We note that introducing error into one term disturbed all three coefficients slightly.

The calculations have been carried to the extent shown and the various rounding errors in the calculations have some effect. In the top panel, second column, the $y_{.1}$'s do not sum to zero as they should in theory. We observe that the final residuals add up to a negative value rather than zero as theory says they must. But the theory is for exact calculations with unlimited decimals. The magnitude of the residuals $y_{.1xt}$, as calculated in the last column of Exhibit 1, is misleading because, when we use the desired formula, we get smaller residuals, as we now show.

Let us look at the difference between the original unrounded function and the fitted function. It is

$$y_{\text{true}} - y_{\text{fitted}} = -1 + x + 0.5x^2 - (-1.03 + 1.05x + 0.48x^2)$$
$$= 0.03 - 0.05x + 0.02x^2.$$

By completing the square we have

$$0.02\left[x^2 - \frac{0.05}{0.02}x + \frac{1}{4}\left(\frac{0.05}{0.02}\right)^2\right] - \frac{(0.05)^2}{4(0.02)} + 0.03$$
$$= 0.02(x - 1.25)^2 + 0.03 - 0.03125$$
$$= 0.02(x - 1.25)^2 - 0.00125.$$

Between 0.9 and 1.5, the largest departures of x from 1.25 are at $x = 0.9$ and $x = 1.5$. At $x = 0.9$ we find the largest error is about 0.0012, and at $x = 1.5$ it is 0. Therefore, the residuals in the final column exaggerate the actual errors in the final fit if it were calculated to unlimited decimals. In the next similar example, we will be just a little more careless, and observe the consequences.

Example 2. Rounded quadratic. When the quadratic $y = 1 + x^2$ is rounded to one decimal place for $x = 0.9, 1.0, 1.1, 1.2, 1.3, 1.4, 1.5$, find what quadratic is recovered when we fit a general quadratic.

Solution. The results are laid out in Exhibit 2 in a manner completely parallel to that for Exhibit 1. The graph of Exhibit 3 exhibits the irregularities introduced by the rounding.

We have recovered

$$y_{\text{fitted}} = 0.95 - 0.02x + 1.01x^2,$$

which again gives coefficients close to those of the true function. The error $-0.05 - 0.02x + 0.01x^2 = -0.06 + 0.01(x - 1)^2$ takes values close to -0.06 for x's between 0.9 and 1.5. We are, of course, aided by the lack of any random error beyond the rounding and by the fact that the function we happen to be fitting has *exactly* the algebraic form that we are trying to retrieve. The

Exhibit 2 of Chapter 13

Recovering $y = 1 + x^2$ rounded to one decimal.

y	$y_{.1}$	$y_{.1}x_{.1}$	x	$x_{.1}$	$(x_{.1})^2$	$t = x^2$	$t_{.1}$	$t_{.1}x_{.1}$
1.8	−.67	.201	0.9	−.3	.09	.81	−.67	.201
2.0	−.47	.094	1.0	−.2	.04	1.00	−.48	.096
2.2	−.27	.027	1.1	−.1	.01	1.21	−.27	.027
2.4	−.07	0	1.2	0	0	1.44	−.04	0
2.7	.23	.023	1.3	.1	.01	1.69	.21	.021
3.0	.53	.106	1.4	.2	.04	1.96	.48	.096
3.2	.73	.219	1.5	.3	.09	2.25	.77	.231
$\bar{y} = 2.47$		.670	$\bar{x} = 1.2$		.28	$\bar{t} = 1.48$		.672

$$g = \frac{.670}{.28} = 2.4, \qquad h = 2.4$$

$y_{.1}$	$gx_{.1}$	$y_{.1x}$	$t_{.1}$	$hx_{.1}$	$t_{.1x}$	$y_{.1x}t_{.1x}$	$t^2_{.1x}$	$y_{.1xt}$
−.67	−.72	.05	−.67	−.72	−.05	.0025	.0025	0
−.47	−.48	.01	−.48	−.48	0	0	0	.01
−.27	−.24	−.03	−.27	−.24	−.03	.0009	.0009	0
−.07	0	−.07	−.04	0	−.04	.0028	.0016	−.03
.23	.24	−.01	.21	.24	−.03	.0003	.0009	.02
.53	.48	.05	.48	.48	0	0	0	.05
.74	.72	.04	.77	.72	.05	.0020	.0025	−.01
						.0085	.0084	

$$f = \frac{.0085}{.0084} = 1.01$$

$$y_{\text{fitted}} = \bar{y} + gx_{.1} + ft_{.1x} = 2.47 + 2.4x_{.1} + 1.01t_{.1x}$$

$$y_{\text{fitted}} = 0.95 - 0.02x + 1.01x^2$$

residuals are not as near zero as they were for Example 1. But we note that the residuals are smaller than the error, for example, in the constant term, which shows again that a function can fit well over a range, even though the coefficients do not match the true ones very closely. We can suppose then that even when we have the wrong form for the function, fitting may be fairly close.

Example 3. Rounding to the nearest integer. What happens when we try to recover $y = 1 + x^2$ from data at the same values of x as in Example 2 after we round y to the nearest integer?

Solution. Exhibit **4** shows the calculations, which, except for $y._{1xt}$, are essentially the same on the righthand half of the page as in the previous two tables. The final result is

$$y_{\text{fitted}} = 3.31 - 3.67x + 2.38x^2,$$

and even a motherly eye will find it very difficult to recognize $1 + x^2$ hidden on the righthand side of this equation.

Exhibit **3** of Chapter 13

Plot of y_{fitted} versus x (small x's) for data shown in Exhibit 2 and $y_{\text{true}} = 1 + x^2$ (indicated by curve).

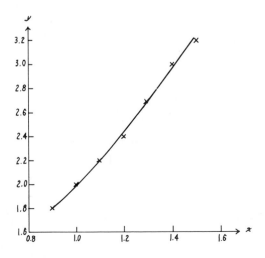

From the difference

$$y_{\text{true}} - y_{\text{fitted}} = 1 + x^2 - (3.31 - 3.67x + 2.38x^2),$$

we find that the error is now, completing the squares as before,

$$0.12 - 1.38(x - 1.33)^2,$$

whose largest absolute value, for x between 0.9 and 1.5, is

$$|0.12 - 1.38(-0.43)^2| \approx |-0.14| = 0.14.$$

Exhibit **4** of Chapter 13

Retrieving $y = 1 + x^2$ from integer rounding.

y	$y_{\cdot1}$	$y_{\cdot1}x_{\cdot1}$	x	$x_{\cdot1}$	$(x_{\cdot1})^2$	$t = x^2$	$t_{\cdot1}$	$t_{\cdot1}x_{\cdot1}$
2	−.43	.129	0.9	−.3	.09	.81	−.67	.201
2	−.43	.086	1.0	−.2	.04	1.00	−.48	.096
2	−.43	.043	1.1	−.1	.01	1.21	−.24	.027
2	−.43	0	1.2	0	0	1.44	−.04	0
3	.57	.057	1.3	.1	.01	1.69	.21	.021
3	.57	.114	1.4	.2	.04	1.96	.48	.096
3	.57	.171	1.5	.3	.09	2.25	.77	.231
$\bar{y} = 2.43$		.600	$\bar{x} = 1.2$		.28	$\bar{t} = 1.48$		.672

$$g = \frac{.600}{.28} = 2.14, \qquad h = 2.4$$

$y_{\cdot1}$	$gx_{\cdot1}$	$y_{\cdot1x}$	$t_{\cdot1}$	$hx_{\cdot1}$	$t_{\cdot1x}$	$y_{\cdot1x}t_{\cdot1x}$	$t^2_{\cdot1x}$	$y_{\cdot1xt}$
−.43	−.64	.21	−.67	−.72	.05	.0105	.0025	.09
−.43	−.43	0	−.48	−.48	0	0	0	0
−.43	−.21	−.22	−.27	−.24	−.03	.0066	.0009	−.15
−.43	0	−.43	−.04	0	−.04	.0172	.0016	−.33
.57	.21	.36	.21	.24	−.03	−.0108	.0009	.43
.57	.43	.14	.48	.48	0	0	0	.14
.57	.64	−.07	.77	.72	.05	−.0035	.0025	−.19
						.0200	.0084	

$$f = \frac{.0200}{.0084} = 2.38$$

$$y_{\text{fitted}} = \bar{y} + gx_{\cdot1} + ft_{\cdot1x} = 2.43 + 2.14x_{\cdot1} + 2.38t_{\cdot1x}$$

$$y_{\text{fitted}} = 3.31 - 3.67x + 2.38x^2$$

When we have rounded to units, moving values as much as 0.5, a maximum shift of 0.15 in the fit does not seem unreasonably large. The values of the function fit decently well in the relevant range of *x*, even though the coefficients of the fitted function are distant from those of the true function.

We might return to the beginning of this section to see how closely we recovered the coefficients in the first example, and to contrast this with how poorly we just now recovered them in the third example. The difference is striking, and we may well ask "Why such a difference?"

Let us recall that the scale of the disturbance in Example 3 was 100 times that in Example 1. To reduce the consequences of a random disturbance 100-fold would take a sample 10,000 times as big. Such a change ought to correspond to a big difference in behavior.

The general lesson here contains no surprises, but it is one we must continually keep in mind: As the data themselves become more and more degraded, the chance that the coefficients will reflect the true effects of their carriers becomes less and less *even when we have exactly the correct model*. The reader will want to consider whether Examples 1, 2, or 3 or something worse more properly represents the problems he or she ordinarily deals with.

Example **4.** *Race relations, expectations of biracial living, and previous experience.* Exhibit **5** displays the expectations of blacks and whites for the outcome of biracial living as the expectation depends on previous experience, which Herbert Hyman (1965) gives. We wish to treat the variables "race" and "previous experience" as adjustment variables for the "expectations" variable.

Exhibit **5** of Chapter 13

Previous experience, race, and expectations. (Numbers in parentheses are counts of cases.)

	t: 1 = Have previously lived in biracial area		t: 0 = Have not previously lived in biracial area	
y: Expectations	x: 1 = black	0 = white	x: 1 = black	0 = white
2 = Integration of races	11% (21)	9% (10)	5% (1)	5% (5)
1 = Accommodation of races	69% (133)	72% (83)	50% (9)	39% (35)
0 = Conflict between races	20% (39)	19% (22)	45% (8)	56% (50)
	193	115	18	90

S) SOURCE

Hyman, Herbert (1965). *Survey Design and Analysis*. New York: The Free Press; p. 291.

Consequently, we have arbitrarily assigned scores for

Variable x: Race
$$\begin{cases} \text{black} = 1 \\ \text{white} = 0 \end{cases}$$

Variable t: Previous experience
$$\begin{cases} \text{lived in biracial area} = 1 \\ \text{not lived in biracial area} = 0 \end{cases}$$

Variable y: Expectations
$$\begin{cases} \text{integration of races} = 2 \\ \text{accommodation of races} = 1 \\ \text{conflict between races} = 0 \end{cases}$$

The choice of scores for variables x and t does not matter because there are only two categories. The choice of scores for y does matter a bit, and some might prefer other spacings than the equal spacing we have chosen. If so, it will be easy to redo our analysis.

Arithmetic is simplified if we recall that the least-squares fit to an array of numbers is their mean. Furthermore, when a line can be passed through the mean of y for each x, then that line is the least-squares fit. Because we have only two values of x, we can readily compute the regression lines by taking means of y and of t for each x. The intercept will be the mean value for $x = 0$, and the slope will be the mean value at $x = 1$ minus the mean value at $x = 0$ divided by 1 [the "run" of (rise/run) is $(1 - 0)$].

Exhibit **6** shows the calculations for getting the regression equations for estimating y and t from x and tables of the residuals $y_{\cdot 1x}$ and $t_{\cdot 1x}$. The final regression equation is obtained from calculations outlined in Exhibit **7** and the result is

$$y_{\text{fitted}} = 0.514 + 0.029x + 0.371t.$$

The cell with the largest number of cases is that for $y = 1$, $x = 1$, $t = 1$, and there the final equation estimates $y_{\text{fitted}} = 0.914$ for a residual $y_{\cdot 1xt} = 1.09$. The overall swing from inexperienced whites to experienced blacks is a change in estimated score from 0.514 to 0.914, or 0.4 score points. Thus experience and race make some difference in expectations, but not a great deal. We must remember that we study here populations as they stand. The events leading to experience or lack of it may also be related to expectations. Previous expectations themselves could be part of the cause of degree of familiarity with integrated housing. Therefore we must be careful to remember that we are primarily describing a situation rather than proving cause and effect.

Example 5. *Linear constraints on variables.* We mentioned in Section 12E the questions arising from

$$B + Y = N$$

where B = birth order, Y = number of younger children, and N = family size. The investigator studying birth order has a choice between fitting

$$\text{response} = a + b_B B,$$

or

$$\text{response} = a + b_B^* B + b_Y Y,$$

or

$$\text{response} = a + b_B^{**} B + b_N N,$$

or something more complex. Anything that can be put in the form

$$\text{response} = a + b_B B + b_Y Y + b_N N$$

can be reduced, identically, to either of the two two-term forms we have just given. To interpret the coefficient of B, which measures the effect of birth

Exhibit **6** of Chapter 13

Freeing y and t of x.

A) Counts f_y

y	x 0	1	Totals
2	15	22	37
1	118	142	260
0	72	47	119
Totals	205	211	416

$\sum y f_y = $ 148   186

$\bar{y}_{x=0} = 0.722,$ $\bar{y}_{x=1} = 0.882$

$\hat{y}_x = 0.722 + 0.160x$

y	$y_{.1x}$ $x = 0$	$x = 1$
2	1.278	1.118
1	0.278	0.118
0	−0.722	−0.882

B) Counts f_t

t	x 0	1	Totals
1	115	193	308
0	90	18	108
	205	211	416

$\sum t f_t = $ 115   193

$\bar{t}_{x=0} = 0.56,$ $\bar{t}_{x=1} = 0.915$

$\hat{t}_x = 0.56 + 0.354x$

t	$t_{.1x}$ $x = 0$	$x = 1$
1	0.439	0.085
0	−0.561	−0.915

order, he or she must decide among b_B, b_B^*, and b_B^{**}. The real basis for this decision has to be the costock, which here is either

$$\{\text{all constants}\}$$

or

$$\{\text{all } a + b_Y Y\}$$

or

$$\{\text{all } a + b_N N\}.$$

Which should the investigator choose?

Exhibit **7** of Chapter 13

Freeing $y_{.1x}$ of $t_{.1x}$ for the expectations for the integration example.

y	x	t	counts f_{yxt}	$y_{.1x}$	$t_{.1x}$	$f_{yxt}y_{.1x}t_{.1x}$	$y_{.1xt}$	$\|y - \bar{y}\|$
2	1	1	21	1.118	.085	2.0	1.09	1.2
2	1	0	1	1.118	−.915	−1.0	1.46	1.2
2	0	1	10	1.278	.439	5.6	1.12	1.2
2	0	0	5	1.278	−.561	−3.6	1.49	1.2
1	1	1	133	.118	.085	1.3	.09	.2
1	1	0	9	.118	−.915	−1.0	.46	.2
1	0	1	83	.278	.439	10.1	.12	.2
1	0	0	35	.278	−.561	−5.5	.49	.2
0	1	1	39	−.882	.085	−2.9	−.91	−.8
0	1	0	8	−.882	−.915	6.5	−.54	−.8
0	0	1	22	−.722	.439	−7.0	−.88	−.8
0	0	0	50	−.722	−.561	20.3	−.51	−.8
						24.8		

$\sum f_{yxt}t_{.1x}^2 = 66.9$

$f = 24.8/66.9 = 0.371$

$y_{\text{fitted}} = 0.722 + 0.160x + 0.371t_{.1x}$

$y_{\text{fitted}} = 0.514 + 0.029x + 0.371t$

Note that the choice cannot be

$$\{\text{all } a + b_Y Y + b_N N\}$$

for any $c + dB$ can be written in this form as

$$c + dY - dN,$$

so that the costock of B covers the whole stock, leaving nothing to use to estimate a coefficient for B.

If the investigator believes that Y matters only because of B and N, he or she should accept

$$a + b_B^{**} B + b_N N.$$

Similarly, if one is convinced that N does not matter except through Y, one will accept

$$a + b_B^* B + b_Y Y.$$

In either case, we need to report (a) that this choice was a matter of judgment, and also whose judgment it was, and (b) that data with $Y = N + B$ can never shed any light on the adequacy or wisdom of this judgment.

Of course, if all the investigator wanted was a fit to the data, either of these two latter choices gives the same fit as the fit (with its indeterminacy of coefficients) of

$$a + b_B B + b_N N + b_Y Y.$$

Here no problem arises until the moment when we begin to think about the coefficients, something that is often hard to avoid.

13D. The Relative Unimportance of the Exact Carrier

We have seen that, in reporting a regression coefficient, it is not enough to report its carrier. If we are to understand what contributes to the coefficient of the carrier, we must know the carrier's costock also. It is now time to go further.

Consider two investigators. One fits, by least squares,

$$y \sim a + bx + ct + du.$$

The other fits, by least squares,

$$y \sim A + Bx^* + Ct + Du,$$

where

$$x^* = kx + \text{a linear combination of } (1, t, u).$$

How will their b's and B's compare?

If they do this for only one set of data, which we call a situation, they will have

$$b = kB,$$

as we see by substitution. If we don't know k, and we rarely do in social and economic problems, we have trouble in relating the two reports.

If they do do this in a whole variety of situations:

$$\text{situation}^{(1)}, \quad \text{situation}^{(2)}, \quad \ldots, \quad \text{situation}^{(H)},$$

finding

$$b^{(1)}, \quad b^{(2)}, \quad \ldots, \quad b^{(H)},$$

and

$$B^{(1)}, \quad B^{(2)}, \quad \ldots, \quad B^{(H)},$$

then we will have

$$b^{(i)} = kB^{(i)}$$

with the same k for all i. This means that

$$\frac{b^{(1)}}{B^{(1)}} = \frac{b^{(2)}}{B^{(2)}} = \cdots = \frac{b^{(H)}}{B^{(H)}}.$$

Thus, except for a scale factor, the b's tell us just the same things about the various situations that the B's do.

If our purpose is comparative—and it almost always is, ultimately—it doesn't matter much whether we use x or x^*. What really matter are:

1. the costock of x (or of x^*), and

2. the whole stock that is fitted.

(They were the same for 1, x, t, u as for 1, x^*, t, u.)

Except for scale factors it is exactly the costock (all of the fit that doesn't involve the coefficient of interest) and the whole stock (the collection of all fits considered possible) that matter in the result, not the specific carrier used.

13E. Proxy Phenomena

Suppose x_2 makes a serious contribution to a regression equation—to reducing the residual variance. Suppose further that x_{22} is highly correlated with x_2, so that

$$x_2 \approx ex_{22}$$

for some e. (The presence or absence of a constant term can be important, but introducing it does not add any essential new features to the discussion.) Then we can account for nearly as much variance by putting

$$b_2(ex_{22})$$

in the regression as we can by putting in

$$b_2(x_2).$$

This is an inevitable consequence of the close correlation and is not at all affected by the presence or absence of an established or believable relationship between x_2 and x_{22}.

In such a case we often say, especially when x_2 is part of the regression but x_{22} is not, that

$$x_2 \text{ is a } \textit{proxy} \text{ for } x_{22}.$$

Sometimes this is good, but often it is bad.

Channeling through a proxy. If x_2 and x_{22} are closely correlated, if x_2 is *not* related to what we are studying, if x_{22} *is* quite strongly related, if x_2 *IS* in the regression, but x_{22} is *NOT* in, then we are likely to find x_2 carrying an appreciable part of our regression. When this happens, we are tempted to believe that x_2 is "relevant," but a more appropriate interpretation would be that "x_2 appears relevant because it is a proxy for x_{22}, which I am sure ought to be relevant because . . ."

A very simple example comes from geometry. Suppose that, without knowing the actual structure of our problem, we actually are dealing with squares whose sides are $x_2 = 4, 5, 6$ (unknown to us) and we are interested in estimating circumferences which here are $y = 16, 20, 24$, respectively. If, as might happen, we had a variable "related to x_2," namely $x_2^2 \equiv x_{22} = 16, 25, 36$, respectively, then we could get the least-squares regression line through the origin of

$$y = 0.745 x_{22}$$

with circumference estimates of 12, 19, 27. (Had we allowed a constant term, we could do much better using x_{22}.) Yet $x_{22} = x_2^2$ does not even have the right dimensions for the problem we are working on. We ought to have a term linear in x_2, not a squared term. The coefficient is absorbing the dimensional problem. But x_{22} is strongly correlated with x_2 here.

We may often have to face circumstances similar to this one.

Distraction by a proxy. If, on the other hand, we include both $b_2 x_2$ and $b_{22} x_{22}$ terms in our regression, where x_2 is not related to y but x_{22} is, we have a tendency toward dilution or distraction, just as in the chest-and-heart-measurements example, Section 12D. Both

$$b_2 x_2 \quad \text{and} \quad b_{22} x_{22}$$

terms will appear in the fit, and both will be serving essentially the same purpose as far as numbers go, so that

$$b_2 + e b_{22}$$

plays the role that we would have assigned to b_2 if only we had been able to get the necessary insight and data to leave x_{22} out. As a result, b_2 may be smaller, perhaps far smaller than we would have wished, and perhaps even of the wrong sign.

Extreme distraction. A bad case of distraction occurs when x_{22} is so competitive as to reduce b_2 essentially to zero. This may not happen too often with one x_{22}, but when there are several variables—say $x_{22}, x_{23}, x_{24}, \ldots, x_{27}$—each of which is highly correlated with x_2, we are almost sure to come close to knocking b_2 out of the ring. (Again, we can even make b_2 go through zero and come out with the wrong sign.) A single x_{22} can also make b_2 come out with the proper sign, but twice too large.

More variables, in general. All the complexities we have ascribed to high correlation of two x's can arise when there is weaker linear dependence of three or more variables. Suppose, for example, that u, v, and w are independent, conceptually and statistically, and that

$$x_1 = u + v + \text{smidgen}_1,$$
$$x_2 = v + w + \text{smidgen}_2,$$
$$x_3 = w - u + \text{smidgen}_3,$$

where "smidgen$_i$" represents something quite small. Then $x_2 - x_1$ is very highly correlated with x_3, and the same sort of difficulty cannot help arising.

Example. During World War II, in investigating aiming errors made during bomber flights over Europe, one of the research organizations developed a regression equation with several carriers. Among its nine or so carriers were altitude, type of aircraft, speed of the bombing group, size of group, and the amount of fighter opposition. On physical grounds, one might expect higher altitudes and higher speeds to produce larger aiming errors. It would not be surprising if different aircraft differed in performance. What the effect of size of group might be can be argued either way. But few people will believe that additional fighter opposition would help a pilot and bombardier do a better job. Nevertheless, amount of fighter opposition appeared as a strong term in the regression equation—the more opposition, the smaller the aiming error. The effect is generally regarded as a proxy phenomenon, arising because the equation had no variable for amount of cloud cover. If clouds obscured the target, the fighters usually did not come up and the aiming errors were ordinarily very large.

13F. Sometimes x's can be "Held Constant"

We have been careful to point out—using x and $t = x^2$—that it does not generally make sense to try to interpret the coefficients of x_i in terms of what "would happen if the other x's were held constant". In this section, we try to go ahead a little, sounding a few of the most necessary warnings.

Polynomial fits. When it comes to fitting polynomials, whether as simple as

$$b_1 x + b_2 x^2$$

or as complex as

$$b_0 + b_1 x + b_2 x^2 + b_3 x^3 + b_4 x^4 + b_5 x^5,$$

it rarely pays to try to interpret coefficients. Pictures of the fits—or of the difference in two fits to two sets of data—can be very helpful, but the coefficients themselves are rarely worth a hard look.

Unrelated x's. If the x's are not closely related, either functionally or statistically, we may be able to get away with interpreting b_i as the "effect of x_i changing while the other x's keep their same values." If we want to tap expert judgment about the value of b_i, some set of words like those in quotes may be the best we can use.

In practical or policy situations, however, we need to recognize how large a difference there can be between:

1. x_i changing while the other x's are not *otherwise* disturbed or clamped, and
2. changing x_i while holding the other x's fast.

Such differences are not only possible but likely in social and economic problems, because the x's we are working with there are usually neither the most fundamental variables in the situation nor the complete set of variables. Consider the example of performance on tests of cognitive achievement as related to parents' education, socioeconomic status, and years of schooling. Note that we have no measures of innate intelligence, attention paid in school, parental or teachers' encouragement, or even hours spent on the subject matter being tested, to say nothing of physical handicaps. Our regression works so long as x's and y's together are driven by the fundamental variables acting as they had acted, at particular times and places, before we collected the data. If we interfere with their activity, we are likely to change the regression and find that the effect of the change cannot be predicted by either of the two regressions listed above.

Holding all but one fixed and changing that one, if we can indeed do this, is likely to interfere with the underlying pattern of variability and covariability, thus changing the regression. In most policy situations this danger is very real.

Multicollinear carriers. We have already explained that the only hope in dealing with sets of carriers where certain linear combinations are nearly constant is to change our coordinate system, bringing in, on the one hand, carriers whose coefficients we can assess reasonably solidly, and on the other hand carriers that we can say little about. The coefficients of the latter carriers scarcely need interpretation.

The former carriers often have a general interpretation about what must be constant; thus, looking ahead, in the wide-receiver example in Section 15E we might well be able to understand the consequences of one wide receiver being 10% bigger than another in all lengths. The coefficient of $x_{2\ldots90}$, however, would not always measure such a thing. In Section 15E, all three of $x_{123\ldots90}$, x_{567890}, and x_{7890} would increase by 10% under such a change, so that $b_{123\ldots90}$, b_{567890}, and b_{7890} would all contribute to the fitted effect of a general 10% increase in size.

13G. Experiments, Closed Systems, and Physical versus Social Sciences, with Examples

George Box has [almost] said (1966): "The only way to find out what will happen when a complex system is disturbed is to disturb the system, not merely to observe it passively." These words of caution about "natural experiments" are uncomfortably strong. Yet in today's world we see no alternative to accepting them as, if anything, too weak.

Regression is probably the most powerful technique we have for analyzing data. Correspondingly, it often seems to tell us more of what we want to know than our data possibly could provide. Such seemings are, of course, wholly misleading. Some examples of what can happen may help us all to understand Box's point, which covers these examples as well as many others.

First, suppose that what we would like to do is measure people (or items) in a population and use the regression coefficients to assess how much a unit change in a background variable (say x_1) will change a response variable (say y). Since the regression coefficient of x_1 depends upon what other variables are used in the forecast, we cannot hope to buy the information about the quantitative effect of x_1 so cheaply. These remarks do not deny the potential use of forecasting the value of y from several variables x_1, x_2, and so on, in the population *as it now exists*. What they do cast grave doubts on is the use to forecast a change in y when x_1 is changed for an individual (class, city, state, country), without verification from a controlled trial making such a change. (Strictly speaking, but unrealistically and impractically, if we want to verify what happens when only x_1 changes, the controlled trial should be made so that x_1 changes and there is no chance for the other variables to change the way they naturally would when the underlying variables are manipulated to change x_1. This sort of study may not be feasible, and it may not yield what we need to know. We ordinarily want to know what will actually happen when we change x_1.)

When such issues are raised, proponents of observational studies plus regression analysis are likely to cite the physical sciences for illustrations of the success of the method.

The idea that such regression-as-measurement methods are successful in the physical sciences is seriously misleading for a variety of reasons. First, because so many physical-science applications of regression-as-measurement are to experimental data. And second, because the relatively few useful applications that remain involve systems in which "the variables" are:

◇ few in number,

◇ well clarified, and

◇ measured with small error.

Example 1. *Heat capacity.* When an engineer fits

$$A + BT + CT^2 + DT^3$$

to data on heat capacity of a well-defined substance as a function of temperature, for example, he or she takes advantage (1) of knowledge that only temperature is involved, (2) of knowledge that temperature and only temperature is changing, (3) of many decades of work that have gone into the definition of the temperature scale, and (4) of measurement precision that is very high, compared to the changes in T. Analogs of any of these four supports, to say nothing of all four, are not common in social, economic, or medicosurgical analyses.

   Some researchers who have a keen appreciation of certain of these difficulties, try to use regression in a more qualitative way. They admit they cannot really use it to measure the size of the effects involved, but they hope that they can use it to show that some effect or other actually exists. They, too, are likely to fail—often without knowing that they have failed.

   Why might they fail? We raised this point in the discussion of the example of expectations in integrated housing, where we noted that previous expectations (nearly the response variable) may have partly caused the experience or lack of it with such housing. Let us say it more generally. If we fear that what we measure with error as x_2 may also contribute to the "causation of y" alongside what we measure with error as x_1, it is not enough to do multiple regression of y on x_1 and x_2. The coefficient of x_1 may be reliably different from zero when what we measure with error as x_1 has no effect on y whatsoever (see Tukey, 1973 and 1974).

   Let us illustrate this paragraph further.

Example 2. *Child development.* Suppose we are studying some aspect, y, of children's development as a response to some out-of-home variable, x_1, and we want to separate any response to x_1 from the varied responses to home atmosphere, either material or mental. First, we choose as appropriate a measure of home atmosphere as we can, call it x_2, and carry out multiple regression of y on x_1 and x_2. If we still find that x_1 clearly contributes to the fitting, should we believe that "even with home conditions held constant" x_1 contributes to y?

   Regrettably, we cannot be confident. First, while we may have used much wisdom and insight in choosing our x_2, only a limited number of measures of home atmosphere could be considered for measurement, so that, at best, the x_2 we choose is an imperfect measure of the aspect of home atmosphere that matters. And, second, x_1, whose effect we are probing, and x_2, which falters under the burden of carrying all relevant home atmosphere, may turn out to be correlated in the population our data comes from. Under such circumstances, a substantial coefficient for x_1 does not ensure that x_1 has an effect "with home atmosphere held constant."

(Part of Section 13G above has been adapted from Gilbert *et al.*, 1977.)

Example 3. Long-run vs. specific educational gains. At the risk of over-explaining the point, suppose that the performance on a test in some units is approximately related to the amount of education (measured in school years), x_1, one has had and to the amount of special training (measured in weeks), x_2, one has had in the following way:

$$y = 6 + x_1 + 2x_2.$$

The coefficients 1 and 2 of the carriers x_1 and x_2 might be used to estimate the average value of additional years of schooling or of additional weeks of special training. This sort of use is one that those social scientists are likely to want.

When the original observations come from an experiment, such an interpretation may be valid, but when the interpretation comes from an observational study, it is not so likely to hold. For example, let us suppose that we have looked at a population of individuals having various values of the x's and that we have related their y's to the x's as we have described here. We are justified in saying that, on the average, people with two more years of schooling score two units higher on the test, and that three more weeks of special training go with people getting 6 more units in the performance test.

What we are not so justified in saying is that if we took a random person who had, say, 9 years of schooling and gave him 2 more years, he would, on the average, gain two points on the test. This is not because of the lack of reliability in determining the equation. We are thinking of this additional schooling being given to many people. The difficulty arises because the data are not based on an experiment. It may be that the reason the people with 9 years of schooling didn't get more schooling in the first place is part of the reason that they did not score higher on the test. And if they are somehow given more schooling, we may not get the results implied by the equation. When we do an experiment, we may be able to apply the change in treatment to all the kinds of people we have in mind, or restrict our prediction to the kinds of people who are studied. A classical example is the Salk vaccine experiment where volunteers were *more* likely to contract polio than nonvolunteers. An observational study therefore would show the effectiveness of the vaccine as less than its actual worth if outcomes of volunteers all vaccinated were compared with those of nonvolunteers all unvaccinated.

In spite of this danger, the interpretation of the effect of changing the variables is very frequently made, and with little warning for the reader. The idea that increased schooling probably increases the score is likely correct, though not because of these data. (For a serious attempt to prove this see Hyman, Wright, and Reed (1975).) Our general knowledge of the world may well be worth much more than the data. In particular we often can't count much on the magnitude of the fitted regression coefficient.

In physical science, we may be just trying to get an expression for an approximate law relating y and the x's. The coefficients may also have some special meaning there, such as the acceleration of gravity.

In physical science we often are in a position where we think we know the variables that "cause" the dependent variable. For example, once we know the radius of a circle, it is the "cause" of the area of a circle. At least, knowing the radius, there is no escape from the area. But in many social problems, where the causes are both multiple and entangled, if indeed there are identifiable causes, it is hard to choose the variables.

Example 4. Poverty and birth defects. What are the variables that "cause" poverty? To get at this, one has to have some sort of view of an economic or social process. Or to take another tack, what is the cause of birth defects? We have to pick and choose among variables. One view is that it is some problem in the biological process leading to the birth of the baby. That is a good causal approach for a biologist. It might also do for a medical student until he begins to say something like "better prenatal care would have reduced the proportion of birth defects in this class of individuals." Now inadequacy of care becomes a cause. This groping for causes is a process that is endless, and one person's causes are another's consequences.

Example 5. Measuring health. When we measure health for groups of people, we might use death rate, morbidity rate (days of illness), and amount spent for food, as a few relevant variables. We could also use last year's measure of health for the same individuals to help measure this year's health. After all, what "caused" much of this year's health was the state prior to this year. If we put all these variables in the equations to forecast health, it may be hard to interpret the coefficients since all these variables are intended to estimate the same thing. For the purpose of estimating fitness, as appraised by the physician, it still may be all right to use all these variables, but we may not have much of an interpretation appropriate for any particular coefficient. Once we see that the matter has this fuzziness, we also see that, when we try to insert variables to describe the causes, we are likely to have several variables that measure much the same thing.

Example 6A. Egg data with full linear model. As a way of rounding out our discussion of the more exact parts of the sciences, begun by our discussion of quadratics, let us return to physical measurements for a further illustration. A. P. Dempster measured the diameters (long and short) of some hen's eggs and their volume V with a view to estimating the relation between these measures.

If an egg were an ellipsoid of revolution, we would expect its volume to be

$$V = kLW^2,$$

where L is the long diameter, W the diameter of the largest circular cross section (how *do* hens make eggs circular in cross section?), and k a constant, $\pi/6$. Dempster got the diameters with calipers and the volume in the same way you or Archimedes would have measured the volume of a lobster. The data appear in Exhibit **8**. Dempster fitted the equation

$$KV = cL^{\beta_2}W^{\beta_3}$$

by taking logarithms to the base 10 and then using least squares. Thus he fitted (letting $v = \log KV$, $x_1 = 1$, $x_2 = \log L$, $x_3 = \log W$, $K = 6/\pi$),

$$v = \beta_1 \cdot 1 + \beta_2 x_2 + \beta_3 x_3,$$

where $\log c = \beta_1$.

If his measurements were exact and the ellipsoid law were, too, then $\hat{\beta}_1 = 0$, $\hat{\beta}_2 = 1$, and $\hat{\beta}_3 = 2$. The actual fit he got was approximately

$$v = 0.320 + 0.728x_2 + 1.812x_3.$$

Exhibit **8** of Chapter 13

Data for the egg example: $x_2 = \log L$, $x_3 = \log W$, $v = \log(6V/\pi)$; logarithms to the base 10.

	x_2	x_3	v
1	0.7659	0.6360	2.031
2	0.7353	0.6198	1.982
3	0.7416	0.6280	1.995
4	0.7600	0.6280	2.019
5	0.7861	0.6239	2.031
6	0.7539	0.6156	1.956
7	0.7747	0.6156	2.007
8	0.7718	0.6239	1.995
9	0.7889	0.6114	1.995
10	0.7659	0.6072	1.995
11	0.7689	0.6156	1.995
12	0.7478	0.6239	2.007

S) SOURCE

Dempster, A. P. (1969). *Elements of Continuous Multivariate Analysis*. Reading, Mass: Addison-Wesley; p. 151.

Reproduced with the permission of the author and the publisher.

This result does not look very close to

$$v = 0 + 1x_2 + 2x_3,$$

but both equations fit the data fairly well, just as our experience with quadratics might lead us to expect. Because x_2 is near 0.75 and x_3 is near 0.62, the contribution of x_1 (= 1) to v is worth, on the average, about $1\frac{1}{3}x_2$ or $1\frac{1}{2}x_3$. Also, the three coefficients add up to 2.860, or nearly the correct 3. One way to get closer to 3 would be to add about half of 0.320 to the righthand side. Thus a roughly appropriate sort of trading-off is going on. This helps prediction, but leaves it devilishly hard to get the correct constants. This illustrates that even in a physical-science example with a closed system and an almost correct form of relation, the coefficients can appear to be far from the "true" values. (The values found are, however, NOT significantly different from 0, 1, and 2.) We say *almost correct* and put quotes around true because eggs are ovoid rather than elliptical in one cross section, and this might make a little difference. (Historians say that this difference between an ovoid and an ellipse held the astronomer Kepler up for several years.)

Example 6B. Egg data with ellipsoidal model. Had we set $\hat{\beta}_1$ to its true value for the ellipsoid, namely zero, we would have fitted the equation

$$\hat{v} = 0.858x_2 + 2.168x_3.$$

This result looks a bit more like

$$v = 1x_2 + 2x_3,$$

and we note that the coefficients add to 3.026, extremely close to 3.

Not all examples turn out to be contrary in their results:

Example 7. Steam consumption, weather, and operating days. Draper and Smith (1966) give an example of steam consumption in a factory.

y = number of units of steam used per month,

x_8 = average atmospheric temperature for month in degrees Fahrenheit,

x_6 = number of operating days per month.

Their fitted equation was

$$y = 9.1 - 0.0724x_8 + 0.2029x_6.$$

Note that the coefficient of x_8 is negative, which agrees with our expectation. If the outside temperature is higher, we need to generate less heat. If the factory runs more days, it is likely to use more heat, so we expect the coefficient of x_6 to be positive, as it turned out to be. If $x_8 = 60°F$ and $x_6 = 20$ days, then the estimated y is 8.8. The signs are sensible, but the sizes are uncertain.

Example 8. Yield, fertilizer, and rainfall. Wonnacott and Wonnacott (1969) give an example of yield of wheat:

$$y = \text{yield in bushels/acre,}$$

$$x = \text{fertilizer in pounds/acre,}$$

$$z = \text{rainfall in inches,}$$

which gives

$$y = 11.33 + 0.0689x + 0.6038z.$$

Both coefficients are positive, which we expect as long as we don't overdo the fertilizer and as long as the fields don't become too soggy. The values of the variables used to produce the equation were in the ranges

$$y: \quad 40 \text{ to } \quad 80$$

$$x: \quad 100 \text{ to } 700$$

$$z: \quad 32 \text{ to } \quad 37$$

But we illustrate again the contrariness of many social-science examples.

Example 9. Pupils' achievement, home, school and teacher variables. A sample of information on 20 schools from the northeast and middle Atlantic states drawn from the population of the Coleman Report is tucked into the computer. Its 5 variables are:

$y = $ verbal achievement score (6th graders)

$x_1 = $ staff salaries per pupil,

$x_2 = $ 6th grade % white collar (father),

$x_3 = $ SES (socioeconomic status),

$x_4 = $ teachers' average verbal scores,

$x_5 = $ mothers' average education (1 unit = 2 school years),

and the regression that comes out is

$$y = 19.9 - 1.79x_1 + 0.0432x_2 + 0.556x_3 + 1.11x_4 - 1.79x_5.$$

The coefficients of x_1 and x_5 are unexpected, certainly in sign and probably in magnitude.

When we have lots of variables, especially variables competing to measure the same thing, as variables 1, 2, 3, and 5 all do, it is very hard to interpret the coefficients. One way around this is to try to make one variable out of all the variables supposed to measure the same thing, here some combination of socioeconomic status and interest in schooling.

Here the SES variable x_3 is already an attempt to weight together several economic variables. It might be wise, alternatively, to use it and x_4 without the other variables.

The general position is roughly as follows:

◇ when we have several variables, all trying to measure the same thing, they are likely to be highly correlated;

◇ when this happens each is a proxy for all;

◇ if they are expressed in equivalent units, the sum of their coefficients is fairly well determined, though no individual coefficient is;

◇ the data are not going to tell us what linear combination will serve us well—in particular, they are not going to tell us which of these closely related variables is important;

◇ so we will do better by using good judgment in condensing these variables into a composite, and fitting one regression coefficient for the composite and other regression coefficients for other variables, if any. In doing this, we would give no detailed attention to the response and certainly no detailed attention to the apparent relationships of these variables to the response.

Sometimes, especially in borderline cases, we want to take this approach a stage further, as follows:

◇ pick a judgment composite of the j constituents (essentially without regard to the response or to their relationship to it),

◇ find the residuals after regressing each constituent variable on the composite—this gives us j residual variables;

◇ study these residual variables, particularly in their relation to measurement errors, known, evaluated, or guessed (if only a few percent of their values are large enough to be clearly not routine measurement error, we will have to do what we can to assess why the few are large—was it unusual measurement error, or are these unusual individuals?—only in the latter case will such almost-small residual variables be worth using);

◇ put into a regression (a) the composite and (b) those residual variables found worthy of further use.

When we do this, even the surviving residual variables will tend to be small—and their small sums of squares will be reflected in large estimated variances for their coefficients—regrettable, but inevitable, and a proper reflection of what we do not know. This sort of approach saves what information we have about which variables among this group seem important. Occasionally we are able to learn a lot this way.

Just dumping in a lot of closely correlated variables, and expecting a fit to the data to tell us, directly and simply, which one or ones are important usually expresses unjustified optimism. Any appearance this approach produces has a good chance of being misleading. It is far better to know what "these data cannot tell us" rather than erroneously to believe the results when they seem to have told us more than they actually can.

★13H. Estimated Variances Are Not Enough

An easy way to try to pass off the woes of regression coefficients is to say: "But I always calculate, and report, standard errors for my regression coefficients; surely that is enough to protect me!" We are about to learn, on the contrary that

◇in many of these problems, a standard error is no help;

◇in many others, standard errors do too much.

This "easy way out" doesn't work.

External proxies and correlation among variables measured with error. Two troublesome situations arise when:

◇a carrier is really performing as a proxy for a variable not present in the regression [example of 13E, 3 of 13G (this could fall here or below), and 4 of 13G];

◇two carriers measure with error two correlated variables of primary interest [examples 4 of 13C, 2 of 13G, 3 of 13G (could fall here or above), 5 of 13G, and 6 of 13G].

In either situation, having even an arbitrarily large amount of data does nothing to resolve the difficulty. Accordingly, standard errors, which tell us how close our finite-data numbers come to their infinite-data ideal, cannot be expected to show us the indeterminacies and uncertainties we face. To avoid these kinds of troubles, if this is possible, we require some means other than getting more data. And we need warning from something other than such measures of statistical uncertainty as standard errors.

Proxies within the fit. In a third important situation, enough data would resolve our dilemmas, so that measures like standard errors have a chance to help us. (Example 9 of 13G probably falls here.) The simplest case arises when two carriers x_1 and x_2, say, are highly correlated. This results in b_1 and b_2 being poorly determined, but some combination of them being much better determined. If x_1 and x_2 are in equivalent units, then $b_1 + b_2$ will be well determined if they point in the same way, while $b_1 - b_2$ will be well determined when they point oppositely. (Large negative correlations make exactly as much trouble as large positive ones of the same magnitude.)

If we only look at

the estimated variance of b_1

and

the estimated variance of b_2

in such a situation, we will see only two very large numbers. Unless we look

---

★ This section may be skipped on a first reading.

further, say at the estimated covariance of b_1 and b_2, we will receive no warning that at least one linear combination of b_1 and b_2 is well determined.

When we have only two carriers, looking at 1 covariance, as well as 2 variances (for the coefficients) is quite possible. With 5 carriers, though, it is 10 covariances for 5 variances, and for 10 carriers it is 45 for 10. It is desirable to have some better way of detecting what is going on.

Beaton and Tukey (1974) have proposed one way to do this, first finding and reporting the c's such that each of

$$b_1 + c_{12}b_2 + c_{13}b_3 + \cdots + c_{1k}b_k,$$

$$c_{21}b_1 + b_2 + c_{23}b_3 + \cdots + c_{2k}b_k,$$

$$\vdots \qquad\qquad \vdots$$

$$c_{k1}b_1 + c_{k2}b_2 + c_{k3}b_3 + \cdots + b_k$$

has as small an estimated variance as possible and then also reporting what that minimized estimated variance is.

If, for example, a calculation showed that the original coefficients, minimized-variance residuals, and their estimated variances are:

$\hat{C}$oefficient	$\hat{V}$ariance	$\hat{L}$inear combination	$\hat{V}$ariance
b_1	(47.2)	$b_1 - 0.11b_2 - 0.07b_3$	(44.0)
b_2	(3.0)	$-0.04b_1 + b_2 - 0.01b_3$	(2.9)
b_3	(25.3)	$-0.08b_1 - 0.23b_2 + b_3$	(21.4)

we would feel that things were almost perfect, so far as dependence interfering with the interpretation of variances is concerned.

If, on the other hand, we found the following:

b_1	(47.2)	$b_1 - 0.11b_2 + 1.37b_3$	(1.54)
b_2	(3.0)	$-0.06b_1 + b_2 - 0.03b_3$	(2.7)
b_3	(25.3)	$0.67b_1 - 0.08b_2 + b_3$	(0.47)

we would recognize that $b_1 + 1.4b_3$ is relatively very much better determined than either b_1 or b_3 (because x_1 and x_3 are closely correlated). Triple—and higher-order—dependences between the b's (which express closeness of the x_1 to planes, etc.) will also be diagnosable if this approach is used.

To do a good job of detection, we need to take matters a step further. After having found the

$$b_i + \sum{}' c_{ij}b_j$$

(where $\sum{}'$ indicates summation over all $j \neq i$), which minimizes the estimated variance, we would like to drop as many terms out of the summation as we can (readjusting the c's for the others freely, of course) without much raising the estimated variance of the result.

For such a purpose, the "backwards" techniques to be discussed in Chapter 15 seem quite suitable. How far should we go? Two standards suggest themselves, one more conservative, one more liberal, namely: If $V = $ est var $\{b_i\}$ and $V_{min} = $ est var $\{b_i + \sum' c_{ji}b_j\}$, we agree,

If $V_{min} \leq \dfrac{V - V_{min}}{K}$, \qquad to allow increase up to V_{min};

If $\dfrac{V_{min}}{K} \leq \dfrac{V - V_{min}}{K} \leq V_{min}$, \qquad to allow increase to $\dfrac{V - V_{min}}{K}$;

If $\dfrac{V - V_{min}}{K} \leq \dfrac{V_{min}}{K}$, \qquad to allow increase to $\dfrac{V_{min}}{K}$,

where $K = 10$ for a conservative rule and $K = 5$ for a liberal one. Note that the rule can be written:
    Allow an increase of up to the median of

$$\left\{ \frac{V_{min}}{K}, \frac{V - V_{min}}{K}, V_{min} \right\}.$$

This allows us a reasonable increase, both when $V_{min} \ll V$ and when $V - V_{min} \ll V$.
    In our two examples above, this might lead, in the first case, to

b_1	(47.2)	b_1	(47.2)
b_2	(3.0)	b_2	(3.0)
b_3	(25.3)	$-0.27b_2 + b_3$	(23.0)

which tells us that the only dependence of even small-to-moderate consequence involves b_3 and b_2, and, in the second case, to

b_1	(47.2)	$b_1 + 1.38b_3$	(1.97)
b_2	(3.0)	b_2	(3.0)
b_3	(25.3)	$0.71b_1 + b_3$	(0.52)

which really brings out the dependence between b_1 and b_3 and the fact that an appropriate linear combination is well-determined.

Comment

Other special cases may require appropriately matched treatment, but we have covered the major points:

◇some kinds of problems are not at all touched by standard errors;

◇in other kinds, standard errors, taken alone, can mask knowledge we have actually gained;

◇ there are steps we can take routinely to have the latter situation brought to our attention;

⋄ we know of no mechanistic way to tackle the possibility of the former situation; so far we can only recommend careful and deep thought.

Summary: Woes of Regression Coefficients

It is not enough to know what a regression coefficient multiplies; we must know what other carriers are offered. (The same carrier can appear with two or more different costocks even when one and the same stock is presented in two or more different ways.)

The coefficient of a carrier can always be judged from a simple linear regression, that of the response upon the carrier, *provided both* the response *and* the carrier have first been linearly adjusted for the carrier's costock.

If we represent the same stock in two ways by changing one carrier while keeping the others the same, the change in the fitted coefficient of the changeable carrier is by a fixed factor, so that, if we compare results for several "populations," the coefficients of the initial carrier will tell us essentially the same story as the coefficients of the changed carrier. (All this is true so long as we keep the costock the same, whether or not the other carriers are kept the same.)

Combining the last points, we recognize the *costock* of the carrier which a regression coefficient multiplies as more important than the carrier itself.

We must be prepared for one variable—or carrier—to serve as a proxy for another, and worry about the possible consequences, in particular, whether the proxy's coefficient siphons off some of the coefficient we would like to have on the proper variable, or whether a variable serves us well only because it is a proxy. (In either case, interpretation of the regression coefficient requires very considerable care.)

Two kinds of circumstances need to be distinguished. There may be a large difference between the consequences of x_i changing (i) while the other x's are not otherwise disturbed or clamped, or (ii) while the other x's are held fast. When, as so often happens, we collect data according to (i) and wish for results applicable to (ii), the discrepancy can be misleading.

Internal standard errors can do nothing to warn us about any of the difficulties considered in this chapter.

Looking at estimated variances of regression coefficients, without considering covariances, can often leave us with impressions of overlarge variability. This leads us to the following sequence:

⋄ Find the modifications $b_i + \sum' c_{ij}b_j$ (where $j \neq i$) of the regression coefficients b_i whose estimated variances are least;

⋄ Compare the variances of these modifications with those of the bare regression coefficients;

◇ Ask to what extent we can simplify the expression of these modifications, by dropping out terms and readjusting the remaining coefficients, without too greatly increasing the estimated variance of the result;

◇ Look at these remaining modifications as telling us about important dependences (in estimation) among the estimated regression coefficient.

References

Beaton, A. E., and J. W. Tukey (1974). "The fitting of power series, meaning polynomials, illustrated on band-spectroscopic data." *Technometrics*, **16**, 147–185.

Box, G. E. P. (1966). "Use and abuse of regression." *Technometrics*, **8**, 625–629.

Coleman, J. S., E. Q. Campbell, C. J. Hobson, J. McPartland, A. M. Mood, F. D. Weinfeld, R. L. York (1966). *Equality of Educational Opportunity.* 2 volumes. Washington, D.C.: Office of Education, U.S. Department of Health, Education, and Welfare, U.S Government Printing Office. [OE-38001; Superintendent of Documents Catalog No. FS 5.238:–38001.]

Draper, N., and H. Smith (1966). *Applied Regression Analysis.* New York: John Wiley and Sons, Inc.; p. 352.

Gilbert, J. P., F. Mosteller, and J. W. Tukey (1977). "Steady social progress requires quantitative evaluation to be searching," in Chapter 4 of *The Evaluation of Social Programs* (C. C. Abt, Ed.). Beverly Hills, Ca.: Sage Publications, Inc.; pp. 295–312.

Hyman, H. H., C. R. Wright, and J. S. Reed (1975). *The Enduring Effects of Education.* Chicago: University of Chicago Press.

Tukey, J. W. (1973). "The zig-zagging climb from initial observation to successful improvement." *Frontiers of Educational Measurement and Information Systems* (W. E. Coffman, Ed.). Boston: Houghton Mifflin; pp. 113–120.

Tukey, J. W. (1974). "Instead of Gauss–Markov least squares, what?" In R. P. Gupta (Ed.), *Applied Statistics.* Proceedings of a Conference at Dalhousie University, Halifax, Nova Scotia, May 2–4. North-Holland Publishing Co.; pp. 351–372.

Wonnacott, T. H., and R. J. Wonnacott (1969). *Introductory Statistics.* New York: John Wiley and Sons, Inc.; p. 255.

Chapter 14/A Class of Mechanisms for Fitting

Chapter index on next page

In exploring this chapter, the reader may want to keep in mind that it discusses basic ideas in regression without giving all details of execution. In developing these important ideas, we exploit a rather different approach and language from the usual ones. If readers are not always ready to construct the functions or weights described here, they need not be dismayed. If, instead, they assume that such constructions are possible, they can use the ideas to appreciate the subtleties of fitting functions, particularly in several variables. The ideas presented can, should, and will be used for data analysis, but in this chapter their practical and conceptual implications, not their execution, are what count.

Additivity is the hallmark, both of the most classical techniques of fitting and of the individual steps of more flexible ones: If $\hat{y}$ is the fit to some values y, and if these values had instead been z or $z + y$, with fitted values $\hat{z}$ or $\widehat{z + y}$ respectively, then it may be that

$$\widehat{z + y} = \hat{z} + \hat{y}.$$

When this happens for all y's and all z's, this fit, or this step of a more complex fit, is said to be additive.

If a fitting method is additive in this sense, it is possible to write it in terms of adding up over data sets. For if

$$y = y_{(1)} + y_{(2)} + \cdots + y_{(n)},$$

the fits must satisfy

$$\hat{y} = \hat{y}_{(1)} + \hat{y}_{(2)} + \cdots + \hat{y}_{(n)}.$$

Let us then define

$$y_{(j)}(i) = \text{Value of } y_{(j)} \quad \text{for data set } i$$

$$= \begin{cases} y(i) & \text{when } i = j, \\ 0 & \text{else.} \end{cases}$$

Then $\hat{y}_{(j)}$ depends only on y at data set i, so the fit $\hat{y}$ comes about by adding up across data sets.

Fitting processes that can be thought of as working by *add*ing up information across data sets are important:

⋄because they are simple,

⋄because they include the classical techniques,

⋄because they can provide great flexibility (particularly by taking several

steps and using the result of each step to determine the way things are added up for the next).

How general is this sort of fitting? Although not universal, it can be made nearly so. All linear least-squares fits are additive, and so is every fitting technique that can be reduced to least squares, in one step or iteratively. Both nonlinear least squares and modern techniques of resistant or robust fitting can be treated as iterations of appropriately changing linear least-squares procedures.

How should we think about these kinds of fitting? This chapter offers a treatment using the concept of "matchers" that deals mathematically and heuristically with many ideas of fitting. The matching idea applies widely to thinking about fitting, covering all the cases just mentioned, provided we are willing to change matchers whenever this is appropriate—as after every step in an iterative fit.

The generality of iteratively modified least squares can also be hinted at by mentioning that we treat least absolute deviation fitting as a special case in Section 14I.

14A. Fitting Lines—Some Through the Origin

If we fit βx to y with no constant term, ordinary least squares leads, as we verify below, to the estimate of β

$$\hat{\beta} = \frac{\sum xy}{\sum x^2}.$$

Here x is called the "carrier"—if the sum of several terms were fitted, as in $\beta x_1 + \gamma x_2 + \delta x_2^2$, then x_1, x_2, and x_2^2 would each be carriers. (Throughout the chapter we shall not bother to remind the reader that dividing by zero is illegal, but we act as if denominators are nonzero unless the matter needs special emphasis.) Hats ^ over quantities mean estimates, in this chapter some kind of least-squares estimates.

If the variance of each y is σ^2 and all covariances of the y's with each other vanish, we have

$$\text{var } \hat{\beta} = \frac{\sum x^2 \sigma^2}{(\sum x^2)^2} = \frac{\sigma^2}{\sum x^2}$$

$$= \frac{(\text{Residual variance})}{\sum x^2}.$$

This is the basic relation, and we can frequently reduce other variance results to a similar form.

If we fit $\alpha + \beta x$ to y, which means taking 1 and x as "carriers" ($\alpha \cdot 1 + \beta \cdot x$), least squares gives

$$\hat{\beta} = \frac{\sum (x - \bar{x})y}{\sum (x - \bar{x})^2}.$$

Again, if the variance of y is σ^2, and covariances of the y's vanish, we get

$$\operatorname{var} \hat{\beta} = \frac{\text{Residual variance}}{\sum (x - \bar{x})^2},$$

a form which suggests that fitting

$$\mu + \beta(x - \bar{x}),$$

where $\mu = \alpha + \beta \bar{x}$, is equivalent to fitting

$$\alpha + \beta x.$$

We now notice that, for least squares (done as if all y_i have equal variances), the presence or absence of μ or β has no influence on our estimate of the other. That is, the value of $\hat{\beta}$ in fitting

$$\beta(x - \bar{x})$$

is identical to that in fitting

$$\mu + \beta(x - \bar{x}).$$

(This is also true for β in $\alpha + \beta x$, since the estimate is the same.) Furthermore, the value of $\hat{\mu}$ in fitting μ alone, namely $\bar{y}$, is identical to that in fitting

$$\mu + \beta(x - \bar{x}),$$

though this is not true for α in $\alpha + \beta x$.

The lack of influence of the presence of μ and β on one another, together with the previous results, means that

$$\operatorname{var} \hat{\beta} = \frac{\text{Residual variance}}{\sum (x - \bar{x})^2}$$

and

$$\operatorname{var} \hat{\mu} = \operatorname{var} \{\bar{y} \mid \{x_i\}\} = \frac{\text{Residual variance}}{n},$$

where $\{x_i\}$ stands for the given set of values of x.

Note that in fitting $\mu \cdot 1$, the carrier is 1 for each x_i, and that the n in the denominator is the sum of squares of n values, each value being 1. To summarize,

◇when we fit βx, the denominator for var $\hat{\beta}$ is $\sum x^2$;

◇when we fit $\mu + \beta(x - \bar{x})$, the denominator for var $\hat{\beta}$ is $\sum (x - \bar{x})^2$, and

◇when we fit μ, the denominator for var $\hat{\mu}$ is n $(= \sum 1^2)$.

Leaning on our previous work (in Chapter 13), we see that the form $\hat{\mu} + \hat{\beta}(x - \bar{x})$ could come smoothly from taking 1 out of both y and x, thus fitting $\hat{\mu}$ and getting the residual $y_{.1}$, and then fitting $x - \bar{x}$ to $y_{.1}$. Fitting $\alpha + \beta x$ does not have this take-one-variable-out-at-a-time interpretation (unless $\bar{x} = 0$).

14B. Matching as a Way of Fitting

Given a variety of carriers,

$$x_1, \quad x_2, \quad \ldots, \quad x_k,$$

the x's now being different variables, no two identical, each taking on a set of values, let us think of fitting

$$\beta_1 x_1 + \beta_2 x_2 + \cdots + \beta_k x_k,$$

a linear multiple regression on our carriers. Some of the x_i may be functions of others. One of the x's may be a constant; for example, x_1 will often have the single value 1 for all data sets.

To get values $\hat{\beta}_1, \hat{\beta}_2, \ldots, \hat{\beta}_n$ we need some equations. Life is simplest when these are linear equations in the $\hat{\beta}$'s. Let

$$\hat{y} = \hat{\beta}_1 x_1 + \hat{\beta}_2 x_2 + \cdots + \hat{\beta}_k x_k$$

be the fit, once we have reached it.

Let us now lead up to a method of getting simple linear equations for the $\hat{\beta}$'s.

If we have a set of coefficients $\{h(i)\}$ (where the letter i corresponds to the ith set of observations, $i = 1, 2, \ldots, n$), we will call the set a *matcher*, if and only if it produces the following equality for any data and the corresponding fit

$$\sum h(i)y(i) = \sum h(i)\hat{y}(i). \qquad (*)$$

Thus the matcher-weighted sum of the observed values is required to equal the matcher-weighted sum of the fitted values of the response variable. Every such set of coefficients that satisfies (*) for a specific kind of fit we call a

matcher

for that fit.

What are some simple examples? When we fit $y = \beta x$ by ordinary least-squares, x is a matcher, for to require

$$\sum x(i)y(i) = \sum x(i)\hat{y}(i) = \sum x(i)\hat{\beta}x(i) = \hat{\beta}\sum x(i)x(i)$$

is to require

$$\sum x(i)y(i) = \hat{\beta}\sum x(i)x(i).$$

This is equivalent to the equation for $\hat{\beta}$ we used earlier

$$\hat{\beta} = \frac{\sum xy}{\sum x^2}.$$

When we fit $y = \alpha + \beta x$ or, equivalently, $y = \mu + \beta(x - \bar{x})$, then 1 (a constant), x, and $x - \bar{x}$ are all matchers. The corresponding conditions are for $\hat{\mu}$ and $\hat{\beta}$:

1 as matcher:

$$\sum 1 \cdot y(i) = \sum 1 \cdot \hat{y}(i),$$
$$n\bar{y} = n(\hat{\alpha} + \hat{\beta}\bar{x}) = n\hat{\mu};$$

x as matcher:

$$\sum x(i)y(i) = \sum x(i)\hat{y}(i) = \hat{\alpha}\sum x(i) + \hat{\beta}\sum x(i)x(i)$$
$$= \hat{\mu}\sum x(i) + \hat{\beta}\sum x(i)(x(i) - \bar{x});$$

$(x - \bar{x})$ as matcher:

$$\sum (x(i) - \bar{x})y(i) = \sum (x(i) - \bar{x})\hat{y}(i) = \sum (x(i) - \bar{x})(\hat{\mu} + \hat{\beta}(x(i) - \bar{x}))$$
$$= \hat{\beta}\sum (x(i) - \bar{x})^2$$

which also give the familiar results for $\hat{\mu}$, $\hat{\alpha}$, and $\hat{\beta}$.

Some algebra for matchers. Matchers come in bundles. Suppose $h = \{h(i)\}$ and $k = \{k(i)\}$ are matchers. Then

$$\sum h(i)y(i) = \sum h(i)\hat{y}(i),$$

and

$$\sum k(i)y(i) = \sum k(i)\hat{y}(i).$$

Adding with various weights gives

Weights 1 and 1: $\sum [h(i) + k(i)]y(i) = \sum [h(i) + k(i)]\hat{y}(i),$

Weights 2 and -3: $\sum [2h(i) - 3k(i)]y(i) = \sum [2h(i) - 3k(i)]\hat{y}(i),$

Weights c_h and c_k: $\sum [c_h h(i) + c_k k(i)]y(i) = \sum [c_h h(i) + c_k k(i)]\hat{y}(i).$

Thus all weighted sums or linear combinations of matchers are themselves matchers.

Our problem is twofold:

◇to get enough matchers so that we have enough independent equations to solve, at least in principle, for all of our unknowns $(\hat{\beta}_1, \hat{\beta}_2, \ldots, \hat{\beta}_k)$;

◇to choose a set of matchers such that solving these equations will not get us into numerical difficulty—and such that the equations will be easy to solve.

Fitting $y = \beta x$. When fitting $y = \beta x$ by least squares, only constant multiples of x are matchers. One is enough. Two would be redundant. Arithmetic is usually minimized by using 1 as the constant multiplier of x. (If x was in half-integers, we might like 2 or 4.)

Fitting y = α + βx. When fitting $y = \alpha + \beta x$ by least squares, all expressions of the form $c + dx$ are matchers. Two will be enough—unless one is a multiple of the other, and thus linearly dependent. Three would be redundant.

But if all x's are small, trying to use

$$1000000 + x \qquad \text{and} \qquad 1000001 + x$$

will lead to a pair of equations so delicately related that, because of rounding errors or numbers of significant figures required, we will soon wish we did not have to solve them. Almost the same is true if $1950 \le x \le 1975$ and we pick 1 and x as our matchers. Instead, choosing $x - \bar{x}$ as a matcher leads, after a little algebra, to

$$\sum [x(i) - \bar{x}]y(i) = \hat{\beta} \sum [x(i) - \bar{x}]^2 \qquad \text{(a pure slope equation)},$$

which is easy to solve and for which the residual form $x(i) - \bar{x}$ has already smoothed the computational difficulty. And using $1 + cx$, where $c = -\sum x(i)/\sum (x(i))^2$, leads to

$$\sum (1 + cx(i))y(i) = \hat{\alpha} \sum (1 + cx(i)) \qquad \text{(a pure intercept equation)},$$

which reduces to

$$n\bar{y} + c\sum x(i)y(i) = \hat{\alpha}(n + cn\bar{x}) \qquad \text{(a pure intercept equation)},$$

again an equation easy to solve for $\hat{\alpha}$.

If we wish to fit $y = \mu + \beta(x - \bar{x})$, matters are even simpler, because taking 1 and $(x - \bar{x})$ as matchers gives

$$\sum y(i) = n\hat{\mu}$$

and

$$\sum [x(i) - \bar{x}]y(i) = \hat{\beta} \sum [x(i) - \bar{x}]^2.$$

If we have k coefficients to fit, we need k equations—k equations that are not linearly dependent. This means that the family of matchers has to be at least k-dimensional. It cannot have larger dimension, either, because $k + 1$ linearly independent equations in k unknowns would be incompatible.

14C. Matchers Tuned to a Single Coefficient—and Catchers

We next develop a special set of matchers that make it easy to solve for the $\hat{\beta}$'s and that catch all the information the data have about $\hat{\beta}$. Suppose we are fitting

$$y = \beta_1 x_1 + \beta_2 x_2 + \cdots + \beta_k x_k$$

by a process that can be described by matchers (all kinds of linear least squares are included). How about a $\hat{\beta}_1$ found by a single matcher? If we can find a matcher $h = \{h(i)\}$ such that

$$0 = \sum h(i)x_2(i) = \sum h(i)x_3(i) = \cdots = \sum h(i)x_k(i), \qquad (*)$$

we are in luck so far as $\hat{\beta}_1$ goes. For if this holds

$$\sum h(i)y(i) = \hat{\beta}_1 \sum h(i)x_1(i), \qquad (\text{**})$$

so that

$$\hat{\beta}_1 = \frac{\sum h(i)y(i)}{\sum h(i)x_1(i)}, \qquad \sum h(i)x_1(i) \neq 0.$$

We may quite properly say that h is *tuned to* $\hat{\beta}_1$, since $\hat{\beta}_2, \hat{\beta}_3, \ldots, \hat{\beta}_k$ do not appear in (**). As "radio stations" they are "tuned out"; they are not "heard." If we can tune out all but one $\hat{\beta}$, we can readily solve for it.

We have already seen simple examples. Thus, if we are fitting $\alpha + \beta x$, then $x - \bar{x}$ is tuned to β—and tunes α out—while, for a very special c, $1 + cx$ is tuned to α—and tunes β out.

More complex examples are available: Suppose that $x_1 \equiv 1$, $x_2 = x$, $x_3 = x^2$, that is, the three variables are the 0th, 1st, and 2nd powers of x, where x takes the integer values 1, 2, 3, 4, 5, 6, 7, 8, 9, 10. Then it turns out that

$$c[22 - 11x + x^2]$$

is tuned to $\hat{\beta}_3$, for any c.

(To verify that

$$0 = \sum_1^{10} (22 - 11x + x^2)1 = \sum_1^{10} (22 - 11x + x^2)x,$$

it may help to recall that

$$\sum_{x=1}^{n} 1 = n, \qquad \sum_1^{n} x = \frac{n(n + 1)}{2}, \qquad \sum_1^{n} x^2 = \frac{n(n + 1)(2n + 1)}{6},$$

$$\sum_1^{n} x^3 = \left[\frac{n(n + 1)}{2}\right]^2.)$$

If we have enough matchers to ensure a unique solution, we will soon show that there is always a matcher tuned to whatever $\hat{\beta}_M$ interests us. For let $h_1, h_2, \ldots, h_k$ be k linearly independent matchers; then all we need is to find a set of d's to satisfy the $k - 1$ equations

$$\sum_i (d_1 h_1(i) + d_2 h_2(i) + \cdots + d_k h_k(i))x_J(i) = 0, \qquad J \neq M \text{ (and } 1 \leq J \leq k).$$

These are equivalent to (*), in that they tune out all but $\hat{\beta}_M$.

While we are at it, let us fix the common multiplying constant that would otherwise be free by asking that

$$\sum_i (d_1 h_1(i) + d_2 h_2(i) + \cdots + d_k h_k(i))x_M(i) = 1.$$

We need the coefficients in this set of k ($= (k - 1) + 1$) equations in the d's to have a nonzero determinant and hence for the equations to have a unique solution. More explicitly, we need the determinant

$$\begin{vmatrix} \sum h_1 x_1 & \sum h_1 x_2 & \cdots & \sum h_1 x_k \\ \sum h_2 x_1 & \sum h_2 x_2 & \cdots & \sum h_2 x_k \\ \cdot & & & \cdot \\ \cdot & & & \cdot \\ \cdot & & & \cdot \\ \sum h_k x_1 & \sum h_k x_2 & \cdots & \sum h_k x_k \end{vmatrix}$$

not to vanish.

Put

$$c_M = d_1 h_1 + d_2 h_2 + \cdots + d_k h_k.$$

The matcher c_M is not only a matcher tuned to $\hat{\beta}_M$, as any constant multiple of c_M would be, it is more; it is a

catcher

for $\hat{\beta}_M$ as we soon explain. When we match y and $\hat{y}$ using this matcher, we get

$$\sum c_M(i) y(i) = \hat{\beta}_M \cdot 1,$$

which is

$$\hat{\beta}_M = \sum c_M(i) y(i).$$

In view of this relation—and the nature of c_M—depending only on the x's and not at all on the y's—we have

$$\text{var } \hat{\beta}_M = \sum \{c_M(i)\}^2 \text{ var } y(i)$$

$$= \sigma^2 \sum \{c_M(i)\}^2.$$

The last form holds provided that all var $y(i) = \sigma^2$.

Accordingly, all that the data are allowed to tell us about $\hat{\beta}_M$, when fitted to satisfy a set of matchers in the presence of the other x's, is to be found from a listing of the pairs of values: $y(i)$ and $c_M(i)$. Thus c_M catches all the information about $\hat{\beta}_M$. Knowing a catcher reduces finding $\hat{\beta}_M$ to a one-parameter regression problem. (We will reduce it to an even more helpful one-parameter regression problem in Section 14E.)

14D. Ordinary Least Squares

We now show that when $x_1, x_2, \ldots, x_k$ are chosen as matchers for the fit, they produce the least-squares fit—the fit that minimizes $\sum (y - \hat{y})^2$.

Suppose now that we are fitting

$$y = \beta_1 x_1 + \beta_2 x_2 + \cdots + \beta_k x_k$$

and that we have—so far more or less arbitrarily—chosen $x_1, x_2, \ldots, x_k$ as matchers for our fit. This means that we have also chosen

$$g_1 x_1 + g_2 x_2 + \cdots + g_k x_k$$

for any $\{g_j\}$ not all zero, as a matcher—because as we have seen, linear combinations of matchers are again matchers. (We use the subscript j here because i is already busy, being used for the ith set of observations.) If $\hat{y}$ and $\hat{\hat{y}}$ are two fits of this form—however chosen—then each is also a matcher for our fit, and so is $\hat{y} - \hat{\hat{y}}$, because $\hat{y}$, $\hat{\hat{y}}$, and $\hat{y} - \hat{\hat{y}}$ are each a linear combination of matchers. Recall that choosing a set of $\hat{\beta}$'s chooses a fit.

We now let $\hat{y}$ be a special fit, the one obtained with our selected matchers $x_1, x_2, \ldots, x_k$. Then,

$$y - \hat{\hat{y}} \equiv (\hat{y} - \hat{\hat{y}}) + (y - \hat{y})$$

whence, as is easy to check,

$$\sum (y - \hat{\hat{y}})^2 \equiv \sum (\hat{y} - \hat{\hat{y}})^2 + 2 \sum (\hat{y} - \hat{\hat{y}})(y - \hat{y}) + \sum (y - \hat{y})^2.$$

The middle term on the right is, setting the factor of 2 aside for a moment,

$$\sum (\hat{y} - \hat{\hat{y}})(y - \hat{y}) = \sum (\hat{y} - \hat{\hat{y}})y - \sum (\hat{y} - \hat{\hat{y}})\hat{y},$$

which vanishes because $\hat{y} - \hat{\hat{y}}$ is a matcher for the fit of $\hat{y}$. ($\hat{y} - \hat{\hat{y}}$ could be used to help determine $\hat{y}$ in terms of y.)

Thus

$$\sum (y - \hat{\hat{y}})^2 = \sum (\hat{y} - \hat{\hat{y}})^2 + \sum (y - \hat{y})^2$$

and, since the first term on the right is positive or zero,

$$\sum (y - \hat{\hat{y}})^2 \geq \sum (y - \hat{y})^2,$$

so that $\hat{y}$ makes the sum

$$\sum (\text{observed} - \text{fitted})^2$$

as small as possible. Hence the term "least squares" for the fits

$$y = \hat{\beta}_1 x_1 + \hat{\beta}_2 x_2 + \cdots + \hat{\beta}_k x_k,$$

in which all

$$g_1 x_1 + g_2 x_2 + \cdots + g_k x_k$$

are matchers.

14E. Tuning for Ordinary Least Squares

In multiple regression when we are good at predicting x_i from the other x's, then the variance of $\hat{\beta}_i$ will be large, because its denominator is the residual sum of squares of $x_i - \hat{x}_i$, that is,

$$\text{var } \hat{\beta}_i = \frac{\sigma^2}{\sum (x_i - \hat{x}_i)^2},$$

where $\hat{x}_i$ is the least-squares estimate of x_i based on the other x's and σ^2 is the common variance of y. We use the method of matchers to prove this in this section.

We also show that all the information about $\hat{\beta}_i$ is contained in the combination of (a) y residuals after taking out of y the costock of x_i and (b) the x_i residuals after taking out of x_i the least-squares fit of the costock of x_i.

Estimating x_i. If we are in the ordinary least-squares case, fitting

$$y = \beta_1 x_1 + \beta_2 x_2 + \cdots + \beta_k x_k$$

with $x_1, x_2, \ldots, x_k$ and all their linear combinations as matchers, we can take $z = x_1$, and fit

$$\hat{z} = \hat{\gamma}_{12} x_2 + \hat{\gamma}_{13} x_3 + \cdots + \hat{\gamma}_{1k} x_k$$

by ordinary least squares. We can take $x_2, x_3, \ldots, x_k$ as the working matchers for this fit, so that the conditions on the fit are

$$\sum x_J(i) z(i) = \sum x_J(i) \hat{z}(i) \qquad \text{for } J = 2, 3, \ldots, k,$$

which can be written

$$\sum x_J(i)(z(i) - \hat{z}(i)) = 0 \qquad \text{for } J = 2, 3, \ldots, k$$

or

$$\sum x_J(i)(x_1(i) - \hat{\gamma}_{12} x_2(i) - \hat{\gamma}_{13} x_3(i) - \cdots - \hat{\gamma}_{1k} x_k(i)) = 0$$

for $J = 2, 3, \ldots, k$.

Now let us look at the residual of x_1 after fitting $x_2, \ldots, x_k$, namely,

$$x_{1 \cdot 23 \cdots k} = x_1 - \hat{\gamma}_{12} x_2 - \hat{\gamma}_{13} x_3 - \cdots - \hat{\gamma}_{1k} x_k = x_1 - \hat{x}_1.$$

(In factor analysis, $x_{1 \cdot 23 \cdots k}$ is sometimes called the anti-image of x_1 in x_1, $x_2, \ldots, x_k$ and the summation of $\hat{\gamma}_{1J} x_J$ is called the image, rather attractive memory aids.) The $k - 1$ conditions we just established are just what is needed to show that

$$x_{1 \cdot 23 \cdots k} = x_1 - \hat{\gamma}_{12} x_2 - \hat{\gamma}_{13} x_3 - \cdots - \hat{\gamma}_{1k} x_k$$

is tuned to $\hat{\beta}_1$ because when we multiply $y = \sum_j \beta_j x_j$ through by $x_{1 \cdot 23 \cdots k}$ and sum both sides over i, the sums $\sum x_j x_{1 \cdot 23 \cdots k}$ (for $j \neq i$) vanish, as we have already arranged. Thus we are left with

$$\sum y x_{1 \cdot 23 \cdots k} = \hat{\beta}_1 \sum x_1 x_{1 \cdot 23 \cdots k}, \qquad \text{the tuned equation for } \hat{\beta}_1.$$

How can $\hat{\beta}_i$ arise? Since

$$\hat{\beta}_1 = \frac{\sum x_{1 \cdot 23 \cdots k} y}{\sum x_{1 \cdot 23 \cdots k} x_1}$$

and since

$$x_1 = x_{1 \cdot 23 \cdots k} + \hat{\gamma}_{12} x_2 + \cdots + \hat{\gamma}_{1k} x_k,$$

we must have

$$\sum x_{1 \cdot 23 \cdots k} x_1 = \sum x_{1 \cdot 23 \cdots k} x_{1 \cdot 23 \cdots k}.$$

Similarly, if we fit y by $\delta_2 x_2 + \cdots + \delta_k x_k$, finding $y_{\cdot 23 \cdots k}$ as the residual, we have

$$y = y_{\cdot 23 \cdots k} + \hat{\delta}_2 x_2 + \cdots + \hat{\delta}_k x_k,$$

so that

$$\sum x_{1 \cdot 23 \cdots k} y = \sum x_{1 \cdot 23 \cdots k} y_{\cdot 23 \cdots k}.$$

So we must have

$$\hat{\beta}_1 = \frac{\sum x_{1 \cdot 23 \cdots k} y_{\cdot 23 \cdots k}}{\sum (x_{1 \cdot 23 \cdots k})^2}.$$

This says that $\hat{\beta}_1$ can be found from the one-parameter regression problem involving the n pairs of values

$$y_{\cdot 23 \cdots k} \qquad \text{and} \qquad x_{1 \cdot 23 \cdots k}.$$

Because $y_{\cdot 23 \cdots k}$ is likely to be much smaller in magnitude than y itself, a plot for this latter regression problem will help us much more than a plot of

$$y \qquad \text{against} \qquad x_{1 \cdot 23 \cdots k}$$

or of

$$y \qquad \text{against} \qquad \text{the catcher for } \hat{\beta}_1.$$

(These two will look exactly alike, since the catcher for $\hat{\beta}_1$ has to be a multiple of $x_{1 \cdot 23 \cdots k}$. Thus $y_{\cdot 23 \cdots k}$ can be equally usefully plotted against either.)

We might think that calculating $y_{\cdot 23 \cdots k}$ and then $y_{\cdot 13 \cdots k}$ (leaving out x_2), then the residual of y after the fit leaving out x_3, and so on to leaving out x_k was more work than we want to face; but it need not be, because the β_1's in

$$\beta_1 x_1 + \beta_2 x_2 + \cdots + \beta_k x_k$$

and in

$$\beta_1 x_{1 \cdot 23 \cdots k} + \beta_2^* x_2 + \cdots \beta_k^* x_k$$

are the same, as we can easily see by substituting in the value of $x_{1 \cdot 23 \cdots k}$. This means that

$$y_{\cdot 123 \cdots k} = y_{\cdot 23 \cdots k} - \hat{\beta}_1 x_{1 \cdot 23 \cdots k}.$$

So the plot of

$$y_{\cdot 23 \cdots k} \qquad \text{against} \qquad x_{1 \cdot 23 \cdots k}$$

is just the same as the plot of

$$y_{\cdot 123 \cdots k} - \hat{\beta}_1 x_{1 \cdot 23 \cdots k} \qquad \text{against} \qquad x_{1 \cdot 23 \cdots k},$$

something that is easy to make, given the final y-residual $y_{.123\cdots k}$ and the x_1-residual, $x_{1.23\cdots k}$. This is the one-parameter regression problem that tells us the most about estimating $\hat{\beta}_1$ when we are fitting $\hat{\beta}_1 x_1 + \hat{\beta}_2 x_2 + \cdots + \hat{\beta}_k x_k$— not only what the estimate is but how hard each data set is working to fix the estimate. (See Larsen and McCleary (1972).)

Residuals get smaller. Since ordinary least squares minimizes

$$\sum (x_1 - \hat{x}_1)^2$$

we have

$$\sum (x_{1.23\cdots k})^2 \le \sum (x_{1.\text{fewer}})^2 \le \sum x_1^2,$$

where "fewer" stands for any proper subset of the variables $2, 3, \ldots, k$.

Unhappiness in many fitting problems comes about when

$$\sum (x_{1.23\cdots k})^2 \ll \sum x_1^2,$$

where "$\ll$" means "much less than". Thus when

$$x_1 = 1, \quad x_2 = x, \quad x_3 = x^2, \quad x_4 = x^3, \quad \text{and} \quad x_5 = x^4, \quad \text{all for}$$
$$x = 1, 2, 3, \ldots, 10,$$

we find

$$\frac{\sum (x_{1.2345})^2}{\sum x_1^2} = 9.16 \times 10^{-3},$$

$$\frac{\sum (x_{2.1345})^2}{\sum x_2^2} = 1.87 \times 10^{-4},$$

$$\frac{\sum (x_{3.1245})^2}{\sum x_3^2} = 2.38 \times 10^{-5},$$

$$\frac{\sum (x_{4.1235})^2}{\sum x_4^2} = 1.70 \times 10^{-5},$$

$$\frac{\sum (x_{5.1234})^2}{\sum x_5^2} = 9.82 \times 10^{-6}.$$

Many, many decimal places that one might have naively thought had been captured have to go down the drain in such situations.

Recall that, since

$$\sum x_{1.23\cdots k}(i)y(i) = \hat{\beta}_1 \sum x_{1.23\cdots k}(i)x_1(i)$$
$$= \hat{\beta}_1 \sum (x_{1.23\cdots k}(i))^2,$$

the variance of $\hat{\beta}_1$ is

$$\text{var}\{\hat{\beta}_1\} = \frac{\sum x_{1.23\cdots k}^2(i)\,\text{var}\{y(i)\}}{(\sum (x_{1.23\cdots k}(i))^2)^2}$$

$$= \frac{\sigma^2}{\sum (x_{1.23\cdots k}(i))^2},$$

the last provided every $\text{var}\{y(i)\} = \sigma^2$.

Step up or down. Notice also that if we write

$$\hat{\beta}_1 x_1 + \hat{\beta}_2 x_2 + \cdots + \hat{\beta}_k x_k$$

as

$$\hat{\beta}_1 x_{1 \cdot 23 \cdots k} + \hat{\beta}_2^* x_2 + \cdots + \hat{\beta}_k^* x_k,$$

the $\hat{\beta}^*$'s are just what we would get by fitting

$$\hat{\beta}_2^* x_2 + \cdots + \hat{\beta}_k^* x_k,$$

leaving x_1 out. This follows because (1) $x_2, x_3, \ldots, x_k$ are matchers for both fits and (2) they tune $x_{1 \cdot 2 \cdots k}$ out.

14F. Weighted Least Squares

In surveying and in astronomy, where least squares originated, investigators long ago recognized that some observations are "better" or "stronger" than others and took appropriate action. This action often assigned differing weights to different observations, either for objective reasons or as a matter of judgment. Thus the history of weighted least squares is almost as extensive as that of ordinary least squares. Today, as we shall see in Section 14H, we have further uses for weighted least squares, further reasons for knowing how it works.

Suppose that we become more arbitrary and pick almost any nonnegative weights $w = \{w(i)\}$ we wish, one for each data point

$$\{x_1(i), x_2(i), \ldots, x_k(i)\},$$

and decide that $wx_1, wx_2, \ldots, wx_k$ are to be our matchers. For example,

$$w(1)x_2(1), \quad w(2)x_2(2), \quad w(3)x_2(3), \quad \ldots, \quad w(n)x_2(n)$$

are then the values of a matcher, namely wx_2.

If, again, $\hat{y}$ and $\hat{\hat{y}}$ are two fits of the form $c_1 x_1 + c_2 x_2 + \cdots + c_k x_k$, however chosen, then $w\hat{y}$, $w\hat{\hat{y}}$, and $w(\hat{y} - \hat{\hat{y}})$ are all matchers, as linear combinations of the wx_i. Let $\hat{y}$ now be the fit obtained with our selected matchers, $wx_1, wx_2, \ldots, wx_k$. We still have

$$(y - \hat{\hat{y}}) \equiv (y - \hat{y}) + (\hat{y} - \hat{\hat{y}}),$$

whence

$$\sum w(y - \hat{\hat{y}})^2 \equiv \sum w(y - \hat{y})^2 + 2 \sum w(\hat{y} - \hat{\hat{y}})(y - \hat{y}) + \sum w(\hat{y} - \hat{\hat{y}})^2.$$

As in Section 14D, because $w(\hat{y} - \hat{\hat{y}})$ is now a matcher, the middle term vanishes when $\hat{y}$ is the fit fixed by our new set of matchers, the wx_i's. By an argument similar to that at the close of the previous section,

$$\sum w(y - \hat{\hat{y}})^2 \geq \sum w(y - \hat{y})^2,$$

so that our new $\hat{y}$ minimizes

$$\sum w(\text{observation} - \text{fit})^2.$$

Unweighting. We have done weighted cases separately from unweighted—meaning "equally weighted"—ones. We did this to get into a frame of mind

◇ where weights are then taken seriously,
◇ where we ask what weights to use.

So far as proofs and mathematics go, however, we can always reduce weighted cases to unweighted ones. Suppose we are fitting

$$y = \beta_1 x_1 + \beta_2 x_2 + \cdots + \beta_k x_k$$

with weight w. The matching equations are

$$\sum w(i)x(i)y(i) = \sum w(i)x(i)\hat{y}(i)$$

for any linear combination $x(i)$ of the $x_J(i)$. We may write them as

$$\sum [\sqrt{w(i)}x(i)][\sqrt{w(i)}y(i)] = \sum [\sqrt{w(i)}x(i)][\sqrt{w(i)}\hat{y}(i)],$$

which corresponds to fitting

$$(\sqrt{w}y) = \beta_1(\sqrt{w}x_1) + \beta_2(\sqrt{w}x_2) + \cdots + \beta_k(\sqrt{w}x_k) \qquad (***)$$

with matchers

$$\sqrt{w}x = \sqrt{w}(\text{any linear combination of } x_1, x_2, \ldots x_k)$$
$$= \text{any linear combination of } \sqrt{w}x_1, \sqrt{w}x_2, \ldots \sqrt{w}x_k),$$

the last form showing that our fit of (***) is unweighted (equally weighted).
Note that, in particular,

$$\sum wyx_{1\cdot23\cdots k} = \hat{\beta}_1 \sum w(x_{1\cdot23\cdots k})^2$$

where $x_{1\cdot23\cdots k}$ is the residual corresponding to the varying weights w. Since this can also be written

$$\sum \sqrt{w}y\sqrt{w}x_{1\cdot23\cdots k} = \hat{\beta}_1 \sum (\sqrt{w}x_{1\cdot23\cdots k})^2,$$

the estimating formula for $\hat{\beta}_1$ can be written, not only as

$$\hat{\beta}_1 = \frac{\sum wyx_{1\cdot23\cdots k}}{\sum w(x_{1\cdot23\cdots k})^2}$$

but also as

$$\hat{\beta}_1 = \frac{\sum (\sqrt{w}y)(\sqrt{w}x_{1\cdot23\cdots k})}{\sum (\sqrt{w}x_{1\cdot23\cdots k})^2}.$$

Accordingly, if we have weights, the plot of

$$\sqrt{w}y \qquad \text{against} \qquad \sqrt{w}x_{1\cdot23\cdots k}$$

or better

$$\sqrt{w}y_{.23\cdots k} \qquad \text{against} \qquad \sqrt{w}x_{1.23\cdots k}$$

has to tell the whole story.

Tuning in weighted least squares. The device of unweighting shows us that if w has been so chosen that

$$w \text{ var} \{y\} = \sigma^2,$$

which implies

$$\text{var} \{\sqrt{w}y\} = \sigma^2,$$

then

$$\text{var } \hat{\beta}_1 = \frac{\sigma^2}{\sum w(x_{1.23\cdots k})^2}$$

for weighted least squares, where $x_{1.23\cdots k}$ is the residual after fitting x_1 with $x_2, x_3, \ldots, x_k$ using weight w.

A Comment

Often we want most weights more or less the same size, interspersed with a few small ones—as if we had a few high-variability observations among a mass of others. When we do have a few high-variability observations, moving the few weights down to the low values they should have corresponds to facing the facts, to taking away "information" that really wasn't there.

In a very extreme case, where three observations out of 100 are extremely variable while the others are much less variable (but equally variable among themselves), keeping all the weights the same will make the answers horribly variable, while making it seem as if we have 100 observations. Weighting the three *zero* will give good results, though we must now admit to only 97 observations. Putting in three small weights will do a little better—maybe we will get the equivalent of 97.13 observations.

The three cases look like this:

Equivalent number of equivariable observations*

Weight for the three bad values	We think we have	We actually have
1	100	Few (maybe 3 or 4 or even 5)
0	97	97
0.041	97.13	97.13

We never really had an equivalent of 100 equally good observations. Any attempt to argue for $99 = 100 - 1$ degrees of freedom for σ^2 is a sham—once

---

* If the third set of weights is optimum.

we know that three observations are extremely variable. Either $96 = 97 - 1$, when we set the three wholly aside, or $96.13 = 97.13 - 1$, when we down-weight them, may be legitimate numbers of degrees of freedom. But if we leave the three in, we will be in much worse shape. Indeed, if they are variable enough, our σ^2 may be worth as little as 3 degrees of freedom, and our other estimates will be very bad. Large weights that should be small can burn us badly—the results both will in fact be bad and will appear better than they actually are (often better than they could possibly be).

★A Generalization (Optional)

Occasionally we may want to be even more general. If we have a two-way array of "weights" $\{w_{ij}\}$, how shall we do the fitting? We might minimize

$$\sum_i \sum_j w_{ij}(y(i) - \hat{y}(i))(y(j) - \hat{y}(j)). \tag{*}$$

We can have symmetry in the weights without loss of generality, because replacing w_{ij} by

$$w'_{ij} = \tfrac{1}{2}(w_{ij} + w_{ji})$$

would provide the symmetry $w'_{ij} = w'_{ji}$, without changing the value of the double sum.

After arranging that $w_{ij} = w_{ji}$, let us take

$$h_1(i) = \sum_j w_{ij}x_1(j),$$

$$h_2(i) = \sum_j w_{ij}x_2(j),$$

$$\cdot$$
$$\cdot$$
$$\cdot$$

$$h_k(i) = \sum_j w_{ij}x_k(j)$$

as the matchers defining $\hat{y}$. Further, let $\hat{\hat{y}}$ be another fit of the same form

$$b_1x_1 + b_2x_2 + \cdots + b_kx_k,$$

that is, another choice of the β's with the same carriers. Then

$$\hat{h}(i) = \sum_j w_{ij}\hat{y}(j)$$

and

$$\hat{\hat{h}}(i) = \sum_j w_{ij}\hat{\hat{y}}(j)$$

are also matchers, and so, of course, is their difference

$$h'(i) = \sum_j w_{ij}[\hat{\hat{y}}(j) - \hat{y}(j)].$$

We plan to show that the $\hat{y}$'s minimize the sum labelled (*). To do this we look at

$$\sum_i \sum_j w_{ij}[y(j) - \hat{y}(j)][y(i) - \hat{y}(i)]. \tag{**}$$

As before, we write

$$y - \hat{y} \equiv (\hat{\hat{y}} - \hat{y}) + (y - \hat{\hat{y}})$$

and apply it in (**) to get:

$$\sum_i \sum_j w_{ij}[\hat{\hat{y}}(j) - \hat{y}(j)][(\hat{\hat{y}}(i) - \hat{y}(i)]$$

$$+ \sum_i \sum_j w_{ij}[\hat{\hat{y}}(j) - \hat{y}(j)][y(i) - \hat{\hat{y}}(i)]$$

$$+ \sum_i \sum_j w_{ij}[y(j) - \hat{\hat{y}}(j)][\hat{\hat{y}}(i) - \hat{y}(i)] \tag{***}$$

$$+ \sum_i \sum_j w_{ij}[y(j) - \hat{\hat{y}}(j)][y(i) - \hat{\hat{y}}(i)].$$

In the second line of (***),

$$\sum w_{ij}[\hat{\hat{y}}(j) - \hat{y}(j)]$$

is a matcher for $\hat{y}$ and so this term vanishes. Because we made $w_{ij} = w_{ji}$, the third line is identical with the second, and therefore it vanishes as well.

We need to introduce a special condition to assure nonnegativeness of the original expression (*). When we deal with sums of squares, the positivity comes on a silver platter, but weighted sums of cross-products can be negative. The condition on the w's we require is that

$$\sum_i \sum_j w_{ij}u_iu_j \geqq 0 \tag{****}$$

for any $\{u_i\}$. (In matrix language, we require the matrix of weights to be positive semidefinite.) Frankly, this condition is not an easy one to check. If the w's happened to be elements of the inverse of a variance–covariance matrix, this condition would be automatically satisfied. At any rate, let us assume that our w's have the nonnegativity property (****).

Then the first line of (***) must be $\geqq 0$. Our situation now is that we have the equality

(**) = first line of (***) + last line of (***).

The nonnegativity assumption ensures that all these parts are $\geqq 0$, but we now emphasize in particular that the first line of (***) is $\geqq 0$. When we replace it by zero we have

(**) $\geqq$ last line of (***),

or, written more fully,

$$\sum_i \sum_j w_{ij}[y(j) - \hat{y}(j)][y(i) - \hat{y}(i)] \geqq \sum_i \sum_j w_{ij}[y(j) - \hat{\hat{y}}(j)][y(i) - \hat{\hat{y}}(i)],$$

which is the desired further generalization of the result of Section 14D:

If the two-way array of weights $\{w_{ij}\}$ yields a nonnegative quadratic form (condition (****)), then the matchers

$$h_m(i) = \sum_j w_{ij} x_m(j), \qquad m = 1, 2, \ldots, k,$$

provide the estimate $\hat{y}$ which is the corresponding doubly-generalized least-squares estimate among all fits to y of the form

$$\beta_1 x_1 + \beta_2 x_2 + \cdots + \beta_k x_k.$$

14G. Influence Curves for Location

To appreciate how a few statistics respond to data, let us see how they behave when one observation in the sample travels smoothly through all possible values. For convenience, let us consider a sample of 11 measurements in all. Let 10 of the measurements have the values

$$10, 7, 3, 3, 3, -2, -5, -5, -6, -8,$$

so that they sum conveniently to zero, and the 11th measurement, x, will move from very negative values to very positive ones. The example, although numerical, should display the general shape, called the *influence curve*, of the effect of the value of one x on each statistic.

A. The Mean, $\bar{x}$.

Because the other 10 measurements sum to zero, the grand total must be x itself. Consequently, the sample mean

$$\bar{x} \equiv \frac{\sum x_i}{n} = \frac{x}{n} = \frac{x}{11}.$$

Exhibit **1** shows in the top panel the straight line with slope $1/11$ that indicates how $\bar{x}$ responds to changes in x in this example.

B. The Median, x.

Because the sample size is odd, the middle observation is the median, here the sixth from either end. The median is

$$
\begin{array}{lll}
-2 & \text{when} & x \leq -2, \\
x & \text{when} & -2 \leq x \leq 3, \\
3 & \text{when} & 3 \leq x.
\end{array}
$$

Consequently, the influence curve as shown in the middle panel of Exhibit 1 is composed of three straight-line segments, two parallel to the horizontal axis, and a third with a slope 1 connecting the other two. The distance between the parallel lines equals the distance between the two middle observations in the basic set of 10.

Exhibit **1** of Chapter 14

Influence curves.

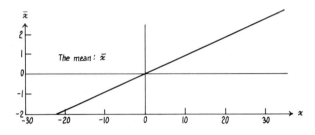

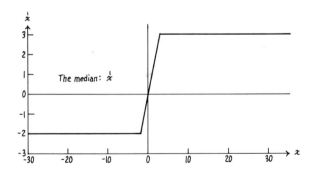

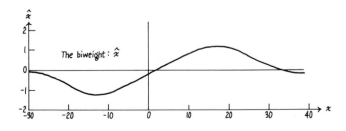

C. The Biweight, $\hat{x}$.

Biweight is an abbreviation for bisquare weight. Let us weight observations according to

$$w(u) = \begin{cases} (1 - u^2)^2 & |u| \leq 1, \\ 0 & \text{elsewhere,} \end{cases}$$

with

$$u_i = \frac{x_i - \hat{x}}{cS},$$

choosing somewhat arbitrarily $c = 6$ and $S = \frac{1}{2} \times$ (interquartile range). For simplicity we have counted in 3 from each end and taken the difference of these observations to get essentially the interquartile range. For distributions near the normal, the interquartile range averages to near $\frac{4}{3}\sigma$, and so $6S$ will average about 4σ; this gives a way of thinking roughly about our choices. Finally the estimate is defined as

$$\hat{x} = \frac{\sum_{i=1}^n w(u_i)x_i}{\sum w(u_i)}.$$

Consider first the value of S. To get the interquartile range, we compute the distance, I, between the observations 3 in from each end:

For:

$-\infty \leq x \leq -6,$	$I = 9,$	$S = 4.5;$
$-6 \leq x \leq -5,$	$I = 3 - x,$	$4.0 \leq S \leq 4.5, \quad S = \dfrac{3 - x}{2};$
$-5 \leq x \leq 3,$	$I = 8,$	$S = 4;$
$3 \leq x \leq 7,$	$I - x + 5,$	$4 \leq S \leq 6, \quad S - \dfrac{x + 5}{2};$
$7 \leq x \leq +\infty,$	$I = 12,$	$S = 6.$

Because $\hat{x}$ depends on the weights and the weights depend on $\hat{x}$, we have to iterate. Exhibit **2** shows illustrative calculations used to find the curve shown in the bottom panel of Exhibit 1.

Note that the bottom panel of Exhibit 1 shows a behavior that we will often like. As the moving observation x walks too far either to the left or to the right, its influence goes to zero, and we get a result based entirely on the other 10 measurements. Approximately, x gets zero influence when $x < -27$ or $x \geq 36$, and, of course, zero effect when $x \approx 0$.

These influence curves show:

◇how the mean walks off toward $+\infty$ or $-\infty$ when one measurement is sufficiently wild;

◇how the median ignores the change in the measurement once it gets outside a narrow range determined by the middle measurements; and

◇how the biweight similarly ignores the changes in the measurement outside a substantial range, and yet responds sensitively to it in the middle portion of this range.

Exhibit **2** of Chapter 14

Illustrative calculations for biweight influence curve $x = -21$, $6S = 27$, $\hat{x}^{(1)} = \bar{x} = -\frac{21}{11} = -1.9091$.

A) First iteration

x_i	$x_i - \hat{x}^{(1)}$	$u_i = \dfrac{x_i - \hat{x}^{(1)}}{6S}$	u_i^2	$1 - u_i^2$	$w_i(u_i) = (1 - u_i^2)^2$ if $\lvert u \rvert \leq 1$	$w_i(u_i)x_i$
10	11.9091	.4411	.1945	.8055	.6488	6.4875
7	8.9091	.3300	.1089	.8911	.7941	5.5587
3	4.9091	.1818	.0331	.9669	.9350	2.8049
3	4.9091	.1818	.0331	.9669	.9350	2.8049
3	4.9091	.1818	.0331	.9669	.9350	2.8049
-2	-.0909	-.0034	.0000	1.0000	1.0000	-2.0000
-5	-3.0909	-.1145	.0131	.9869	.9740	-4.8698
-5	-3.0909	-.1145	.0131	.9869	.9740	-4.8698
-6	-4.0909	-.1515	.0230	.9770	.9546	-5.7277
-8	-6.0909	-.2256	.0509	.9491	.9008	-7.2065
-21	-19.0909	-.7071	.4999	.5001	.2501	-5.2511
Total					9.3014	-9.4640

$$\hat{x}^{(2)} = \frac{\sum w_i x_i}{\sum w_i} = \frac{-9.4640}{9.3014} = -1.0175$$

B) Second iteration: $\hat{x}^{(2)} = -1.0175$

x_i	$x_i - \hat{x}^{(2)}$	u_i	u_i^2	$1 - u_i$	w_i	$w_i x_i$
10	11.0175	.4081	.1665	.8335	.6947	6.9471
7	8.0175	.2969	.0882	.9118	.8314	5.8200
3	4.0175	.1488	.0221	.9779	.9562	2.8686
3	4.0175	.1488	.0221	.9779	.9562	2.8686
3	4.0175	.1488	.0221	.9779	.9562	2.8686
-2	-.9825	-.0364	.0013	.9987	.9974	-1.9947
-5	-3.9825	-.1475	.0218	.9782	.9570	-4.7848
-5	-3.9825	-.1475	.0218	.9782	.9570	-4.7848
-6	-4.9825	-.1845	.0341	.9659	.9331	-5.5983
-8	-6.9825	-.2586	.0669	.9331	.8707	-6.9657
-21	-19.9825	-.7401	.5477	.4523	.2045	-4.2954
Total					9.3144	-7.0509

$$\hat{x}^{(3)} = \frac{-7.0509}{9.3144} = -0.7570$$

➡

Exhibit **2** of Chapter 14 (continued)

C) Third iteration: $\hat{x}^{(3)} = -0.7570$ **D)** Fourth iteration: $\hat{x}^{(4)} = -0.6821$

x_i	w_i	$w_i x_i$
10	.7077	7.0774
7	.8417	5.8921
3	.9617	2.8850
3	.9617	2.8850
3	.9617	2.8850
−2	.9958	−1.9915
−5	.9512	−4.7561
−5	.9512	−4.7561
−6	.9260	−5.5560
−8	.8613	−6.8900
−21	.1917	−4.0267
Total	9.3117	−6.3519

x_i	w_i	$w_i x_i$
10	.7114	7.1145
7	.8446	5.9125
3	.9632	2.8895
3	.9632	2.8895
3	.9632	2.8895
−2	.9952	−1.9905
−5	.9495	−4.7475
−5	.9495	−4.7475
−6	.9239	−5.5435
−8	.8585	−6.8678
−21	.1881	−3.9504
Total	9.3103	−6.1517

$$\hat{x}^{(4)} = \frac{-6.3519}{9.3117} = -0.6821$$

$$\hat{x}^{(5)} = \frac{-6.1517}{9.3103} = -0.6607$$

E) Fifth iteration: $\hat{x}^{(5)} = -0.6607$

x_i	w_i	$w_i x_i$
10	.7125	7.1251
7	.8455	5.9183
3	.9636	2.8907
3	.9636	2.8907
3	.9636	2.8907
−2	.9951	−1.9902
−5	.9490	−4.7450
−5	.9490	−4.7450
−6	.9233	−5.5399
−8	.8577	−6.8614
−21	.1871	−3.9287
Total	9.3100	−6.0947

$$\hat{x}^{(6)} = \frac{-6.0947}{9.3100} = -0.6546$$

To see how such a small change in the measure of scale chosen in the biweight matters, we have also computed the influence curve replacing S in u by the median absolute deviation from the median. We get the result shown in Exhibit **3**. It is extremely close to that shown for the biweight in Exhibit 1.

The reader may wish to refer to an article by F. E. Hampel (1974); see Reference list at end of chapter.

Exhibit **3** of Chapter 14

Influence curves using median absolute deviation from the median as the measure of spread S.

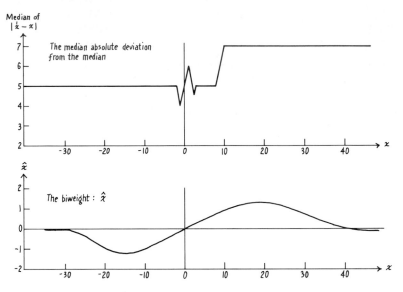

14H. Iteratively Weighted Linear Least Squares

Notation. As before, we use $\hat{y}$ for anything that we could get from ordinary least squares, from weighted least squares, or from the generalization of weighted least squares discussed at the end of Section 14F. We wish to extend this idea to a sequence of fits where

i) each is least squares, and

ii) the weights change.

As long as we are thinking of a step at a time, we continue to use $\hat{y}$, $\hat{\beta}_j$, and so on for least-squares results, reserving the double-hats $\hat{\hat{y}}$, $\hat{\hat{\beta}}_j$, and so on, for fits not involving least squares. For the final result of a sequence of least-squares fits with changing weights, we use y^*, β_j^*, and so on. We also use e_i for a residual following either a fit or a step of a fit.

We remarked, at the beginning of the section on weighted least squares, that weights could be and have been used to reflect the strengths or weaknesses of individual observations. In this section, we learn how weights can be used toward a second end, essentially to ensure the kind of influence curves we want, and thus to ensure the high performance in estimation that goes with such curves.

This second kind of weighting differs from the more historical use in technique and interpretation though not in ultimate purpose—better estimates. We can use both ideas in the same analysis, and when we do, we need to multiply together a weight of one kind and a weight of the other to find the "weight" that enters the final calculation.

Suppose, to begin with, that the weights of the first kind are all the same. We recognize both the advantages and disadvantages of least-squares fitting. It is easy to describe in terms of matchers, is reducible to a single solution of a set of simultaneous equations, and is the subject of much experience and many computer programs (not all trustworthy, but some very safe). Or we may be able to afford to tune the matchers, and so use the solutions to a set of simple equations. All these are very good things. On the other hand, least-squares fitting is relatively inflexible—though *a priori* weights can and do help—and, as our examples show, is quite susceptible to serious perturbation of results when a few wild values are present. How can we have the advantages without the disadvantages?

Suppose we consider:

◇choosing a $w_1 = \{w_1(i)\}$ and finding the corresponding fit $\hat{y}^{(1)}$;

◇then choosing a $w_2 = \{w_2(i)\}$ which may depend on how the first fit came out, and finding the corresponding least-squares fit $\hat{y}^{(2)}$;

◇then choosing a w_3 . . . and so on.

We can either go a prechosen number of steps or stop once the change in fit from one step to the next seems small enough. (We can think about going on to the limit, but we would not, in practice, wish to wear out either pencil or computer, to say nothing of the user.)

The real question is how to choose the successive weights. The first set must be chosen as a matter of initial judgment. For the present, let us assume that the initial weighting is equal weighting for all i.

A very useful way to choose the later weights is to let the ith weight depend on the ratio of the ith residual—in the previous iteration—to some measure of the general size of residuals in that iteration, namely on

$$u_i = \frac{y(i) - \hat{y}^{(k)}(i)}{cS_k},$$

where c is a numerical constant, and S_k is a measure of spread for all the residuals left by the kth fit and $\hat{y}^{(k)}(i)$ is the estimate of $y(i)$ produced by the kth fit.

Biweighting. What dependence of $w(i)$ on u_i should we use? Again there can be a variety of good choices. On a computer, we like to take the bisquare weighting, abbreviated biweighting,

$$w(u) = (1 - u^2)^2 \qquad u^2 < 1,$$
$$= 0 \qquad\qquad u^2 \geq 1.$$

The general idea is that deviations that are small get larger weights and, as the deviations grow, their weights get smaller and finally there is a rather smooth but complete cutoff when $u \geq 1$.

We often use

$$S_k = \text{median of the } |y(i) - \hat{y}^{(k)}(i)|,$$
$$c = 6 \text{ or } 9,$$

but lots of other choices work well, too. Note that S_k is obtained by computing the absolute value of each residual for y in the kth fit and then taking the median of these absolute values. Thus S_k measures variation, essentially around the middle of the data set. Then u_i looks at each deviation in terms of this measure of variation. In particular, c determines how soon the weight becomes zero. The choice of $c = 6$ gives 0 weight to deviations more than $6S_k$ from the latest estimate of location.

Example. Outlier. We can expect to do only the simplest examples by hand, so let us start with fitting a line through the origin. Exhibit **4** shows a plot of five points. Exhibit **5** applies the $w(u) = (1 - u^2)^2$ fitting procedure. Since the fifth point is rather out of line with the others, the first task of the fitting procedure is to modify $\hat{\beta}^{(1)} = 1.13$ until this point has little weight. Here this takes two steps of fitting, after which the successive estimates are $\hat{\beta}^{(4)} = 1.02$, $\beta^{(5)} = 1.00$, and $\hat{\beta}^{(6)} = 0.99$. If the reader will carry the calculation one step farther, he will find 0.99 again. To our 2-decimal precision the calculation has converged.

The number of steps required in this example is more than we would like. Is there any easy way to deal with this problem? In our example, yes—in others, sometimes. We can here look at Exhibit 4 and say to ourselves that the fifth point is out of line, so we will start with weights 1 on the first four, but 0 on the fifth.

Setting the weight initially to zero does not set the point permanently aside. A point can, in principle, regain a nonzero weight in the course of the iteration. Here we find $\hat{\beta}^{(1)} = 29.6/30 = 0.987$, which one further step of biweight fitting moves hardly at all. In general, we cannot expect to know in advance which points should get low weight, since we cannot expect so simple a plot.

Exhibit **4** of Chapter 14

Line through the origin.

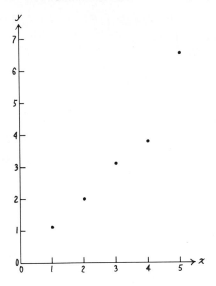

Exhibit **5** of Chapter 14

The arithmetic of biweight fitting a line through the origin.

A) Data and simple sums

x	y	(w)	xy	x^2
1	1.1	(1.0)	1.1	1
2	2.0	(1.0)	4.0	4
3	3.1	(1.0)	9.3	9
4	3.8	(1.0)	15.2	16
5	6.5	(1.0)	32.5	25
			62.1	55

$\hat{\beta}^{(1)} = 62.1/55 = 1.13$

Exhibit **5** of Chapter 14 (continued)

B) Second step—median residual in boldface type; two decimals only

x	y	fit$^{(1)}$ = 1.13x	$\lvert e \rvert$ = $\lvert y - 1.13x \rvert$	$\lvert u \rvert$ = $\lvert e \rvert/6(.29)$	$w^{(1)}$ = $(1 - u^2)^2$ or 0	$w^{(1)}xy$	$w^{(1)}x^2$
1	1.1	1.13	.03	.02	1.00	1.10	1.00
2	2.0	2.26	.26	.15	.96	3.84	3.84
3	3.1	3.39	**.29**	.17	.94	8.74	8.46
4	3.8	4.52	.72	.41	.69	10.49	11.04
5	6.5	5.65	.85	.49	.58	18.85	14.50
						43.02	38.84

$\hat{\beta}^{(2)} = 1.11$

C) Third step—likewise

x	y	fit$^{(2)}$ = 1.11x	$\lvert e \rvert$ = $\lvert y - 1.11x \rvert$	$\lvert u \rvert$ = $\lvert e \rvert/6(.23)$	$w^{(2)}$ = $(1 - u^2)^2$ or 0	$w^{(2)}xy$	$w^{(2)}x^2$
1	1.1	1.11	.01	.01	1.00	1.10	1.00
2	2.0	2.22	.22	.16	.95	3.80	3.80
3	3.1	3.33	**.23**	.17	.94	8.74	8.46
4	3.8	4.44	.64	.46	.62	9.42	9.92
5	6.5	5.55	.95	.69	.27	8.78	6.75
						31.84	29.93

$\hat{\beta}^{(3)} = 1.06$

D) Fourth step—likewise

x	y	fit$^{(3)}$ = 1.06x	$\lvert e \rvert$ = $\lvert y - 1.06x \rvert$	$\lvert u \rvert$ = $\lvert e \rvert/6(.12)$	$w^{(3)}$ = $(1 - u^2)^2$ or 0	$w^{(3)}xy$	$w^{(3)}x^2$
1	1.1	1.06	.04	.06	.99	1.09	.99
2	2.0	2.12	**.12**	.17	.94	3.76	3.76
3	3.1	3.18	.08	.11	.98	9.11	8.82
4	3.8	4.24	.44	.61	.39	5.93	6.24
5	6.5	5.30	1.20	1.67	0	0	0
						20.28	19.81

$\hat{\beta}^{(4)} = 1.02$

➤

Stepweighting. For hand work we might like to save arithmetic by introducing a discrete set of weights. Hand work is thinkable only for the simplest problems. We suggest the weights

$$w(u) = \begin{cases} 4 & |u| \leq 0.2, \\ 3 & 0.2 < |u| \leq 0.4, \\ 2 & 0.4 < |u| \leq 0.6, \\ 1 & 0.6 < |u| \leq 0.8. \\ 0 & 0.8 < |u| \end{cases}$$

Exhibit **6** has the arithmetic for the example of the line through the origin pictured in Exhibit 4.

Exhibit 5 of Chapter 14 (continued)

E) Fifth step—likewise

x	y	fit$^{(4)}$ = 1.02x	$\|e\|$ = $\|y - 1.02x\|$	$\|u\|$ = $\|e\|/6(.08)$	$w^{(4)}$ = $(1 - u^2)^2$ or 0	$w^{(4)}xy$	$w^{(4)}x^2$
1	1.1	1.02	.08	.17	.94	1.03	.94
2	2.0	2.04	.04	.08	.99	3.96	3.96
3	3.1	3.06	.04	.08	.99	9.20	8.91
4	3.8	4.08	.28	.58	.44	6.69	7.04
5	6.5	5.10	1.40	2.92	0	0	0
						20.88	20.85

$\hat{\beta}^{(5)} = 1.00$

F) Sixth step—likewise

x	y	fit$^{(5)}$ = 1.00x	$\|e\|$ = $\|y - 1.00x\|$	$\|u\|$ = $\|e\|/6(.10)$	$w^{(5)}$ = $(1 - u^2)^2$ or 0	$w^{(5)}xy$	$w^{(5)}x^2$
1	1.1	1.0	.10	.17	.94	1.03	.94
2	2.0	2.0	.00	.00	1.00	4.00	4.00
3	3.1	3.0	.10	.17	.94	8.74	8.46
4	3.8	4.0	.20	.33	.79	12.01	12.64
5	6.5	5.0	1.50	2.50	0	0	0
						25.78	26.04

$\hat{\beta}^{(6)} = 0.99$

Exhibit **6** of Chapter 14

Stepweight fitting the example of Exhibit 4 (with $c = 6$).

A) Panel A as in Panel A of Exhibit 5.

B) Second step

x	y	fit$^{(1)}$ = $1.13x$	$\lvert e \rvert$ = $\lvert y - 1.13x \rvert$	$\lvert u \rvert$ = $\lvert e \rvert / 6(.29)$	$w^{(1)}$	$w^{(1)}xy$	$w^{(1)}x^2$
1	1.1	1.13	.03	.02	4	4.4	4.
2	2.0	2.26	.26	.15	4	16.0	16.
3	3.1	3.39	**.29**	.17	4	37.2	36.
4	3.8	4.52	.72	.41	2	30.4	32.
5	6.5	5.65	.85	.49	2	65.0	50.
						153.0	138.

$\hat{\beta}^{(2)} = 1.11$

C) Third step

x	y	fit$^{(2)}$ = $1.11x$	$\lvert e \rvert$ = $\lvert y - 1.11x \rvert$	$\lvert u \rvert$ = $\lvert e \rvert / 6(.23)$	$w^{(2)}$	$w^{(2)}xy$	$w^{(2)}x^2$
1	1.1	1.11	.01	.01	4	4.4	4.
2	2.0	2.22	.22	.16	4	16.0	16.
3	3.1	3.33	**.23**	.17	4	37.2	36.
4	3.8	4.44	.64	.46	2	30.4	32.
5	6.5	5.55	.95	.69	1	32.5	25.
						120.5	113.0

$\hat{\beta}^{(3)} = 1.07$

D) Fourth step

x	y	fit$^{(3)}$ = $1.07x$	$\lvert e \rvert$ = $\lvert y - 1.07x \rvert$	$\lvert u \rvert$ = $\lvert e \rvert / 6(.14)$	$w^{(3)}$	$w^{(3)}xy$	$w^{(3)}x^2$
1	1.1	1.07	.03	.04	4	4.4	4.
2	2.0	2.14	**.14**	.17	4	16.0	16.
3	3.1	3.21	.11	.13	4	37.2	36.
4	3.8	4.28	.48	.57	2	30.4	32.
5	6.5	5.35	1.15	1.37	0	0	0
						88.0	88.

$\hat{\beta}^{(4)} = 1.00$

We have now reduced the arithmetic till most of it is involved in finding the residuals, something we cannot do without.

If we now put $c = 5$ in the stepweight procedure, we can reduce the arithmetic still further, since the rule for weights becomes:

$$w_i = 4 \quad \text{if} \quad |e_i| \leq S_k,$$

$$w_i = 3 \quad \text{if} \quad S_k < |e_i| \leq 2S_k,$$

$$w_i = 2 \quad \text{if} \quad 2S_k < |e_i| \leq 3S_k,$$

$$w_i = 1 \quad \text{if} \quad 3S_k < |e_i| \leq 4S_k,$$

$$w_i = 0 \quad \text{if} \quad 4S_k < |e_i|.$$

The arithmetic reduces now to that of Exhibit 7. At each step, beyond finding residuals we have to find S_k, the median |residual|, and its multiples by 1, 2, 3 and 4; to see whether w_i is 4, 3, 2, 1, or 0 (by comparing $|e_i|$ with these multiples in the (*) column); to form simple integer multiples of xy and x^2; and to find the corresponding sums and $\hat{\beta}^{(k+1)}$ as their ratio. If we can afford to do a basic least-squares fit—which ought always involve getting residuals—by hand, we can almost always afford going on to a stepweighted fit.

Exhibit **7** of Chapter 14

The arithmetic for stepweight fitting with $c = 5$.

A) Panel A in Exhibit 5.

B) Second step

x	y	fit$^{(1)}$ = 1.13x	$\|e\| =$ $\|y - 1.13x\|$	(*) Mults. of S_1	w	$w^{(1)}$	$w^{(1)}xy$	$w^{(1)}x^2$
1	1.1	1.13	.03	(.29)	4	4	4.4	4.
2	2.0	2.26	.26	(.58)	3	4	16.0	16.
3	3.1	3.39	.29	(.87)	2	4	37.2	36.
4	3.8	4.52	.72	(1.16)	1	2	30.4	32.
5	6.5	5.65	.85		0	2	65.0	50.
							153.0	138.

$\hat{\beta}^{(2)} = 1.11$

➔

We will sometimes do better with $c = 9$ and a somewhat different set of steps, so that

$$w_i = 4 \quad \text{if} \quad |e_i| \leq 2.4S_k,$$
$$w_i = 3 \quad \text{if} \quad 2.4S_k < |e_i| \leq 4.2S_k,$$
$$w_i = 2 \quad \text{if} \quad 4.2S_k < |e_i| \leq 5.6S_k,$$
$$w_i = 1 \quad \text{if} \quad 5.6S_k < |e_i| \leq 7.4S_k,$$
$$w_i = 0 \quad \text{if} \quad 7.4S_k < |e_i|.$$

Once these 4 multiples of S_k are calculated, we proceed as before.

Exhibit **7** of Chapter 14 (continued)

C) Third step

x	y	fit$^{(2)} =$ 1.11x	$|e| =$ $|y - 1.11x|$	(*) Mults. of S_2	w	$w^{(2)}$	$w^{(2)}xy$	$w^{(2)}x^2$
1	1.1	1.11	.01	(.23)	4	4	4.4	4.
2	2.0	2.22	.22	(.46)	3	4	16.0	16.
3	3.1	3.33	**.23**	(.69)	2	4	37.2	36.
4	3.8	4.44	.64	(.92)	1	2	30.4	32.
5	6.5	5.55	.95		0	0	0	0
							88.0	88.

$\hat{\beta}^{(3)} = 1.00$

D) Fourth step

x	y	fit$^{(3)} =$ 1.00x	$|e| =$ $|y - 1.00x|$	(*) Mults. of S_3	w	$w^{(3)}$	$w^{(3)}xy$	$w^{(3)}x^2$
1	1.1	1.00	.10	(.10)	4	4	4.4	4.
2	2.0	2.00	.00	(.20)	3	4	16.0	16.
3	3.1	3.00	**.10**	(.30)	2	4	37.2	36.
4	3.8	4.00	.20	(.40)	1	3	45.6	48.
5	6.5	5.00	1.50		0	0	0	0
							103.2	104.

$\hat{\beta}^{(4)} = .99$

Weights on weights. Suppose we had advance knowledge that different data points had different precision, so that we would want to start with different $w(i)$'s. What changes would we need to make?

If we had weighted in inverse proportion to anticipated variance, as we would in the simplest case, we would want to look at suitably weighted residuals, namely the

$$\sqrt{w(i)}(y(i) - \hat{y}(i))\text{'s},$$

the sum of whose squares we had in fact, minimized. Taking, at the kth step

$$e_i^{(k)} = \sqrt{w(i)}[y^{(k)}(i) - \hat{y}^{(k)}(i)],$$
$$S_k = \text{median of the } |e_i^{(k)}|,$$
$$c \quad \text{as before,}$$
$$u_i^{(k)} = e_i^{(k)}/cS_k,$$
$$w(u) \quad \text{as before,}$$

we would want to take as weights for the next step, not just $w(u_i^{(k)})$ but the product

$$w(u_i^{(k)}) \cdot w(i)$$

of the new weights and the initial weights. These are readily computed. Note that $w(i)$ appears in $e_i^{(k)}$ but $w(u_i^{(k-1)})$ does not.

★ 14I. Least Absolute Deviations (Optional)

Sometimes, instead of least-squares fitting, we wish to minimize

$$(\text{constant}) \sum | y(i) - \hat{y}(i)| = \sum \frac{\text{constant}}{|y(i) - \hat{y}(i)|}(y(i) - \hat{y}(i))^2$$

to obtain a

least absolute deviation fit,

which we will refer to as a least-absolutes fit. In one way this is better than least squares: It does not pay excessive attention to large residuals. In another it is worse, since it pays undue attention to very small ones. (We will see how to avoid this difficulty shortly.)

If we try the iterative fitting process of Section 14G with

$$w_{k+1}(i) = \frac{\text{constant}}{y(i) - \hat{y}^{(k)}(i)}$$

and the process converges to $\hat{y}(i)$, we must have minimized

$$\sum w_{\text{Last}}(i)(y(i) - \hat{y}(i))^2 = \sum \frac{\text{constant}}{|y(i) - \hat{y}(i)|}(y(i) - \hat{y}(i))^2.$$

Hence $\hat{y}(i)$ is a least-absolutes fit. Thus we see that iterative weighted least squares can lead to least-absolute deviation fits. We also have had brought to our attention an unfavorable feature of least-absolute deviations: great weight on the observations with smallest residuals, because a small $|y(i) - \hat{y}(i)|$ in the denominator produces a great weight.

Example* 1. *One outlier. Given the five points listed in Exhibit **8**, obtain the least-absolute deviations fit of the form $y = \beta x$.

Discussion. The figure at the bottom of Exhibit 8 suggests one attractive line as $y = x$, or $\hat{\beta} = 1$. If we start our interpretation there, we fit the four points nearest the origin exactly, and this gives us difficulty in taking reciprocals. The interpretation is that infinite weight should be given to four of the points and only a finite weight to the fifth, which thus has zero relative weight. This does not bother us because we like $\hat{\beta} = 1$. But suppose we had put the line through

Exhibit **8** of Chapter 14

Fitting a slope near $\beta = 1$.

x	y	$\hat{\beta}x = 1 \cdot x$	$|y - x|$	$\hat{\beta}x = (1 + d)x$	$|y - (1 + d)x|$
1	1	1	0	$1 + d$	d
2	2	2	0	$2 + 2d$	$2d$
3	3	3	0	$3 + 3d$	$3d$
4	4	4	0	$4 + 4d$	$4d$
5	6	5	1	$5 + 5d$	$1 - 5d$
				Total	$1 + 5d$

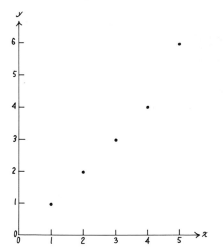

the fifth point instead. Then it would get infinite weight, the other four would not matter, and the iteration would be stuck again, but this time in what we would all agree was the wrong place. Computationally this is unacceptable.

If we wish to find the least-absolutes fit, we can take $\hat{\beta}$ as $1 + d$, where d is a small number (less than 0.2). The residuals are computed for this fit in Exhibit 8 and we see that the sum of their absolute values is $1 + 5d$. Therefore we can minimize the sum of absolute deviations by setting $d = 0$, and the desired minimizing β is 1.

Example 2. Difficulties. We could illustrate the iterative procedure either on this example or on the example in Section 14G. The illustration of Section 14G is carried out in Exhibit **9** for the interested reader. We started with $\hat{\beta}^{(1)} = 1.02$.

The main point is that very heavy weights given to points with small residuals not only are troublesome computationally, but are also unsatisfactory

Exhibit **9** of Chapter 14

Fitting a line through the origin using least absolute deviations.

A) Least-squares estimate: $62.1/55 \approx 1.13$

x	y	$\hat{\beta}^{(1)}x =$ 1.02x	$\|y - 1.02x\|$ = e	Weight .08/e	wxy	wx^2
1	1.1	1.02	.08	1.0	1.10	1.00
2	2.0	2.04	.04	2.0	8.00	8.00
3	3.1	3.06	.04	2.0	18.60	18.00
4	3.8	4.08	.28	.286	4.35	4.58
5	6.5	5.10	1.40	.0571	1.86	1.43
			1.84		33.91	33.01

$$\hat{\beta}^{(2)} = \frac{33.91}{33.01} = 1.0273, \text{ round up to } 1.028$$

y	$\hat{\beta}^{(2)}x =$ 1.028x	$\|y - 1.028x\|$ = e	Weight .072/e	wxy	wx^2
1.1	1.028	.072	1.00	1.10	1.00
2.0	2.056	.056	1.285	5.14	5.14
3.1	3.084	.016	4.5	41.85	40.50
3.8	4.112	.312	.231	3.51	3.69
6.5	5.140	1.360	.0529	1.72	1.32
		1.812		53.32	51.65

$$\hat{\beta}^{(3)} = \frac{53.32}{51.65} = 1.0323$$

from a statistical view. Even if (i) we want to minimize a sum of terms, and (ii) we like absolute deviations as a measure of size for large residuals, we need to use a more reasonable measure of size for small residuals. We want to give the middle measurements substantial weight and then reduce the weights as deviations get large.

Flattened weights. We can get such a measure by using.

$$w_{h+1}(i) = \begin{cases} 1, & \text{if } |y(i) - \hat{y}^{(h)}(i)| \le k, \\ \dfrac{k}{|y(i) - \hat{y}^{(h)}(i)|}, & \text{otherwise} \end{cases}$$

where k is some moderate value, perhaps the median absolute residual, called S_k in Section 14H.

The result is to minimize

$$\sum \Psi(y - \hat{y})$$

where

$$\Psi(u) = \begin{cases} u^2 & \text{where this is } \le k^2 \text{ (for } -k \le u \le k\text{)}, \\ k|u| & \text{elsewhere.} \end{cases}$$

Example 3. Weights. Let us apply the suggestion just given to the data of Example 2.

Discussion. Let us begin with a rounded version of our latest estimate of β, namely 1.03. From the fourth column of Exhibit **10**, we see that the median of the errors is 0.07, and we set this equal to k. This leads to unit weights for the three points nearest the origin, a modest weight for the fourth point, and a small weight for the last point. This last point matters even with this small weight because the difference between its weighted xy and x^2 accounts for a notable share of the difference of the final slope from 1.000.

Exhibit **10** of Chapter 14

Flattened weights related to absolute deviations.

x	y	$\hat{\beta}^{(1)}x =$ 1.03x	$\|y - \hat{\beta}^{(1)}x\|$ $= \|e\|$	.07/|e|	$w^{(2)}$	$\|w^{(2)}xy\|$	$\|w^{(2)}x^2\|$
1	1.1	1.03	.07	1.00	1	1.1	1.0
2	2.0	2.06	.06	1.17	1	4.0	4.0
3	3.1	3.09	.01	7.00	1	9.3	9.0
4	3.8	4.12	.32	.22	0.22	3.34	3.52
5	6.5	5.15	1.35	.05	0.05	1.63	1.25
						19.37	18.77

$$\hat{\beta}^{(2)} = \frac{19.37}{18.77} = 1.032.$$

Since the movement from a slope of 1.03 to 1.032 is so slight, it does not seem worth iterating further.

Other powers. If we want to minimize

$$\sum |y - \hat{y}|^{2-p}$$

for any p, we see that we have only to use

$$w_{h+1}(i) = \frac{\text{constant}}{|y(i) - \hat{y}^{(h)}(i)|^p}.$$

Again it would pay us to flatten off the peak so as not to weight the closely fitted values so heavily.

Weights on weights. Suppose we want to minimize, not just

$$\sum |y - \hat{y}|^{1.5}$$

but rather

$$\sum w(i)\,|y(i) - \hat{y}(i)|^{1.5},$$

where the $w(i)$ are given—and do not depend on the values $y - \hat{y}$. To do this, we need only change, as suggested in Section 14H,

$$w_{h+1}(i) = \frac{\text{constant}}{|y(i) - \hat{y}^{(h)}(i)|^{0.5}}$$

into

$$w_{h+1}(i) = w(i) \cdot \frac{\text{constant}}{|y(i) - \hat{y}^{(h)}(i)|^{0.5}}$$

as the reader can verify.

14J. Analyzing Troubles

When $\sum x_{1\cdot 23\ldots k}^2$ is small, the variance of $\hat{\beta}_1$ is large. There is nothing that can be done to stop this happening. All that ordinary least squares can tell us about β_1 is present in a simple picture, the plot of $y_{\cdot 23\ldots k}$ against $x_{1\cdot 23\ldots k}$. If, as may happen, we find, out of 200 values of i,

$$-0.0002 \leq x_{1\cdot 23\ldots k}(i) \leq 0.0003 \qquad \text{for 197 values of } i,$$
$$1.97 \leq x_{1\cdot 23\ldots k}(i) \leq 2.01 \qquad \text{for 2 values of } i,$$
$$x_{1\cdot 23\ldots k}(i) = 47.34 \qquad \text{for 1 value of } i,$$

we can be fairly sure there have been three large errors—perhaps in measurement, perhaps in copying, perhaps in computation—and for what i's these have occurred. Moreover we can be sure that if we insist on fitting

$$y = \beta_1 x_1 + \beta_2 x_2 + \cdots + \beta_k x_k$$

instead of

$$y = \beta_2 x_2 + \cdots \beta_k x_k,$$

then what we find for $\hat{\beta}_1$—and, regrettably, but also automatically, for $\hat{\beta}_2, \hat{\beta}_3 \cdots \hat{\beta}_k$—will be almost entirely determined by these three identifiable data-sets with the large errors. This usually ought to be quite unacceptable—so we ought to know when it is happening. To find out, we need routinely to look at the distribution of the values of $x_{1 \cdot 23 \ldots k}$ (and at those of its brothers and sisters).

At first blush, we might hope to escape by setting aside these three points. It will be a good thing to set them aside—at least our results will not be controlled by errors. But we still have serious trouble. For our assumed values of $x_{1 \cdot 23 \ldots k}$ we have

$$(x_{1 \cdot 23 \ldots k})^2 \leqq 0.00000009 < 10^{-7},$$

so that

$$\text{var } \hat{\beta}_1 > \frac{\sigma^2}{n \cdot 10^{-7}} = 1{,}000{,}000 \text{ times } \frac{10\sigma^2}{n},$$

which is likely to be devastatingly large.

Moreover, the variances of the other $\hat{\beta}_J$, if found for a fit including a $\hat{\beta}_1$, will be correspondingly large.

If only a very few values of $x_{1 \cdot 23 \ldots k}$ are large enough to be useful, a least-squares fit of

$$\beta_1 x_1 + \beta_2 x_2 + \cdots + \beta_k x_k$$

is not likely to be what it seems.

Occasionally we know that these few unusual values are correct. Then writing the fitted regression as

$$\hat{\beta}_1 x_{1 \cdot 23 \ldots k} + \hat{\beta}_2^* x_2 + \cdots + \hat{\beta}_k^* x_k,$$

where the β^*'s are the coefficients of the fit with x_1 left out (see the last section), makes it easier to warn ourselves what we do and do not know. Although $\hat{\beta}_1$ still depends almost entirely on a few data points, we can, by looking at $x_{2 \cdot 3 \ldots k}$ and its brothers and sisters, find out which β^*'s depend on a lot of data points and which on a few.

After setting some aside. If we believe these few large values come from errors which may be in any or several of the x_i's, then we need to set aside these data points and fit the others. We are not likely to be happy about the smallness of

$$\sum (x_{1 \cdot 23 \ldots k})^2$$

and the corresponding largeness of var $\{\hat{\beta}_1\}$, but that is just a fact of life.

If again we put

$$x_{1 \cdot 23 \ldots k} = x_1 - \hat{\gamma}_{12} x_2 - \cdots - \hat{\gamma}_{1k} x_k,$$

we have

$$\hat{\beta}_j = \hat{\gamma}_{1j}\hat{\beta}_1 + \hat{\beta}_j^*, \qquad j = 2, 3, \ldots, k;$$

and if simultaneously (1) $\hat{\gamma}_{1j}$ is not small and (2) var $\{\hat{\beta}_j^*\}$ is not large, then

$$\text{var}\{\hat{\gamma}_{1j}\hat{\beta}_1\} \gg \text{var}\{\hat{\beta}_j^*\},$$

so that var $\{\hat{\beta}_1\}$ and all other var $\{\hat{\beta}_j\}$ will still be large. It may still pay to look at

$$\hat{\beta}_1 x_{1\cdot23\cdots k} + \hat{\beta}_2^* x_2 + \cdots + \hat{\beta}_k^* x_k$$

instead of

$$\hat{\beta}_1 x_1 + \hat{\beta}_2 x_2 + \cdots + \hat{\beta}_k x_k.$$

The other coefficients, $\hat{\beta}_2^*, \ldots, \hat{\beta}_k^*$ may still be well determined in one form, and if we reveal how small the sum of squares of $x_{1\cdot23\cdots k}$ now is (after setting aside the few unusual points), we will have made clear which coefficients we know little about (anyone that involves more than $\hat{\beta}_2^*, \ldots, \hat{\beta}_k^*$) and why $\sum(x_{1\cdot23\cdots k})^2$ is too small.

General Comments

Rewriting regression expressions so that some coefficients are well determined and others less may, therefore, sometimes be desirable. The new coefficients will have new meanings, and this can be true even if their numerical values happen not to change. The hope is that we will have learned a bit more about what the data tell us and what they don't.

One way to rewrite regression expressions is by dropping out one or more terms. This may be wise or it may be an evasion of responsibility. When we fit a regression without interpreting the coefficients, not even to compare them with coefficients arising from a similar fit to other data, then pulling out a variable whose coefficient is ill determined is likely good practice. If we fit a regression whose coefficients are to be interpreted, and we felt there was reason to include x_1, then taking out x_1, perhaps without comment, may be irresponsible. It may be better to leave x_1 in, in the form $x_{1\cdot23\cdots k}$, and recognize what is not known about its coefficient.

Which variable should be special? Whether we go to a "fully dotted" variable like $x_{1\cdot23\cdots k}$ or leave x_1 out, we are likely to have a choice as to which x is to play the special role. We may be able to choose among

$$\hat{\beta}_1 x_{1\cdot23} + \hat{\beta}_2^* x_2 + \hat{\beta}_3^* x_3$$

and

$$\hat{\beta}_1^{**} x_1 + \hat{\beta}_2 x_{2\cdot13} + \hat{\beta}_3^{**} x_3$$

and

$$\hat{\beta}_1^{***} x_1 + \hat{\beta}_2^{***} x_2 + \hat{\beta}_3 x_{3\cdot12},$$

or among

$$\hat{\beta}_2^* x_2 + \hat{\beta}_3^* x_3$$

and

$$\hat{\beta}_1^{**} x_1 + \hat{\beta}_3^{**} x_3$$

and

$$\hat{\beta}_1^{***} x_1 + \hat{\beta}_2^{***} x_2.$$

How shall we choose? We suggest that you write down the alternatives, see what you would say about each, and then try to see which story is most helpful. Unfortunately, fixing up one variable may not be enough; we may need to handle two or more x's specially.

Weights. As we noticed, we can extend the definition of $x_{1 \cdot 23 \dots k}$ to a case with weights $w(i)$—either weights inherent in what we believe about how var$\{y(i)\}$ depends on σ, or weights chosen iteratively to reach more desirable fits or combinations of both. Here, too, if all but a few values of $x_{1 \cdot 23 \dots k}$ are very small, we are in trouble.

The only way that we can get out of trouble with small values of $\sum (x_{1 \cdot 23 \dots k})^2$ is therefore to get additional information.

An inequality. If we fit, successively,

$$y = \beta_1 x_1 \qquad\qquad \text{(to find } \hat{y}^{(1)}),$$
$$y = \beta_1 x_1 + \beta_2 x_2 \qquad\qquad \text{(to find } \hat{y}^{(2)}),$$
$$y = \beta_1 x_1 + \beta_2 x_2 + \beta_3 x_3 \qquad \text{(and so on),}$$
$$\vdots$$
$$y = \beta_1 x_1 + \beta_2 x_2 + \cdots + \beta_k x_k,$$

then (in addition to the great changes in the meaning of β_1 as we go from one fit to another) we have

$$\sum x_1^2 \geqq \sum (x_{1 \cdot 2})^2 \geqq \sum (x_{1 \cdot 23})^2 \geqq \cdots \geqq \sum (x_{1 \cdot 23 \dots k})^2.$$

We do not know var$\{y\}$. All we can do is to estimate it. We usually do this from the values of $y - \hat{y}$—more precisely from the values of whichever of $y - \hat{y}^{(1)}$, $y - \hat{y}^{(2)}$, $y - \hat{y}^{(3)}, \dots$ is appropriate. Because we are doing least squares,

$$\sum (y - \hat{y}^{(1)})^2 \geqq \sum (y - \hat{y}^{(2)})^2 \geqq \sum (y - \hat{y}^{(3)})^2 \geqq \cdots \geqq \sum (y - \hat{y}^{(k)})^2.$$

Now to estimate var $\{y\}$, we usually divide

$$(y - \hat{y}^{(i)})^2 \quad \text{by} \quad (n - i),$$

so our estimates of var $\{y\}$ can grow slowly as we fit more terms. In fact, however, they are not likely to grow.

If our estimates of var $\{y\}$ were all the same, which is what would happen, on average, if $x_2, x_3, \ldots, x_k$ had only random explanatory power for y, we would have

$$\hat{\text{var}}\{\hat{\beta}_1 \mid \beta_1 x_1\} < \hat{\text{var}}\{\hat{\beta}_1 \mid \beta_1 x_1 + \beta_2 x_2\} < \cdots$$
$$< \hat{\text{var}}\{\hat{\beta}_1 \mid \beta_1 x_1 + \beta_2 x_2 + \cdots + \beta_k x_k\}.$$

Thus, in this situation, the more terms we add to the fit, the more variable the coefficients entered earlier become.

This is not what we are used to thinking about. Often we find adding more terms makes the $y - \hat{y}$ much smaller. So we happily take credit for this with a smaller estimate of σ^2 for the long fit than for the short. Let us think of a longish list of terms, where we can stop fitting after varying numbers of terms. What is going on is that

$$\sum (y - \hat{y})^2$$

involves first, systematic variations (because we haven't yet put in enough variables to take them away), and second, some relatively random variation that doesn't depend very much, if at all, on how much we fit. Suppose, then, that

$$\text{est var}\{y\} = \text{systematics}^2 + \text{variability},$$

where systematics^2, coming from what we have not yet fitted, goes down as we fit longer expressions. The apparent variance of $\hat{\beta}_1$, say, is

$$\frac{\text{est var}\{y\}}{\sum x_{1\cdot\text{some}}^2} = \frac{(\text{systematics after some})^2 + \text{variability}}{\sum x_{1\cdot\text{some}}^2},$$

where the first term in the numerator is decreasing so that the numerator can race the denominator to see which shrinks fastest. Often the numerator will win for a while, but this will come to an end. Eventually the denominator becomes smaller faster, and the fraction starts to rise again.

Thus we can go too far. When we recognize that adding x's always redefines the meanings of the coefficients of the x's we already had, it is clearly easy to go too far without recognizing it, if we need meaningful coefficients.

We can also go too far toward making the average of var $\{\hat{y}\}$ larger instead of smaller, although overshooting happens slowly, so that results are less sensitive to the exact choice. (See the discussion in Section 15A and the references given at the close of Chapter 15 for the relevant techniques.)

14K. Proof of the Statement of Section 13B

We now return to the proof that if

$$y_{\cdot 1x} = y - a - bx$$
$$t_{\cdot 1x} = t - c - dx$$

and the least squares fit to $y_{\cdot 1x}$ is

$$y_{\cdot 1x} \sim e + ft_{\cdot 1x},$$

then

$$a + bx + f(t - c - dx)$$

is the least-square fit of y to 1, x, and t. We complete the proof by building up the result from scratch. Because $y_{\cdot 1x}$ is a residual from a least-squares fit that includes 1 (we fitted $a \cdot 1 + bx$), 1 is a matcher, so that we also know that the sum of the y-residuals vanishes, so that

$$\sum 1(y_{\cdot 1x}) = 0$$

and so, for the same reason, must

$$\sum 1(t_{\cdot 1x}) = 0.$$

Consider the "first" normal equation when we regress $y_{\cdot 1x}$ on $t_{\cdot 1x}$ and 1,

$$\sum 1(y_{\cdot 1x} - (e + ft_{\cdot 1x})) = 0,$$

whence

$$\sum y_{\cdot 1x} - e \sum 1 + f \sum t_{\cdot 1x} = 0;$$

and since the sums of the y-residuals and of the t-residuals vanish, we find that e has to be zero. Our last fit must therefore be

$$y_{\cdot 1x} \sim ft_{\cdot 1x}.$$

What happens if we start putting things together? We first fitted

$$a + bx$$

to y, and last fitted

$$ft_{\cdot 1x}$$

to what was left. We ought to look at the combination

$$a + bx + ft_{\cdot 1x},$$

which can also be written in more extended form as

$$a + bx + f(t - c - dx) = (a - fc) + (b - fd)x + ft.$$

What can we say about it?

We know that

$$\sum 1(y) = \sum 1(a + bx),$$

and we have just seen that

$$\sum 1(t_{\cdot 1x}) = 0.$$

Thus we could write, by "adding zero" in the form $f \sum t_{\cdot 1x}$,

$$\sum 1(y) = \sum 1(a + bx + ft_{\cdot 1x}).$$

This shows that

$$\sum 1(y - a - bx - ft_{\cdot 1x}) = 0. \qquad (*)$$

We know that

$$\sum x(y) = \sum x(a + bx)$$

where "$x(y)$" means "x times y", and

$$\sum x(t_{\cdot 1x}) = 0.$$

Thus again "adding zero" in the form $f \sum xt_{\cdot 1x}$, we get

$$\sum x(y) = \sum x(a + bx + ft_{\cdot 1x}).$$

This shows that

$$\sum x(y - a - bx - ft_{\cdot 1x}) = 0. \qquad (**)$$

We know that

$$\sum 1(t_{\cdot 1x}) = 0,$$
$$\sum x(t_{\cdot 1x}) = 0,$$

so that

$$\sum t_{\cdot 1x}(a + bx) = 0.$$

Hence, because $y_{\cdot 1x} = y - a - bx$,

$$\sum t_{\cdot 1x}(y_{\cdot 1x}) = \sum t_{\cdot 1x}(y - a - bx) = \sum t_{\cdot 1x}y.$$

Next we show that, because the value of f was chosen to make the second of the following expressions equal to zero,

$$\sum t_{\cdot 1x}(y - a - bx - ft_{\cdot 1x}) = \sum t_{\cdot 1x}(y_{\cdot 1x} - ft_{\cdot 1x}) = 0.$$

Because $t_{\cdot 1x} = t - c - dx$, we have $t = c + dx + t_{\cdot 1x}$, and because we have shown that (*) and (**) hold, we have now proved that

$$\sum t(y - a - bx - ft_{\cdot 1x}) = 0. \qquad (***)$$

Thus we have shown the vanishing of all three sums

$$\sum \begin{Bmatrix} 1 \\ x \\ t \end{Bmatrix} (y - a - bx - ft_{\cdot 1x}) = 0, \qquad \begin{matrix} (*) \\ (**) \\ (***) \end{matrix}$$

so that the form we get from adding the step-by-step fits

$$a + bx + ft_{\cdot 1x} = (a - fc) + (b - fd)x + ft$$

has been proved to be the *least-squares multiple regression* of y on 1, x, and t.

How should we interpret the coefficient of t? The construction we have just gone through shows that it is the regression of

$$y\text{-adjusted-linearly-for-}x \equiv y_{\cdot 1x}$$

$$\text{on}$$

$$t\text{-adjusted-linearly-for-}x \equiv t_{\cdot 1x}.$$

A more detailed description for f is that it is

"the regression coefficient obtained when y (linearly adjusted for 1 and x) is regressed on t (linearly adjusted for 1 and x)."

General case. We could have left out 1, doing

$$y \sim Bx,$$
$$t \sim Dx,$$
$$y_{\cdot 1x} \sim Ft_{\cdot 1x},$$

without any change in the argument.

We could have had several x's instead of one, with only the simplest change in the argument (putting in each x_i in turn in place of x alone). Such an approach is part of some computational methods for stepwise regression. The general result is that quoted at the end of Section 13B.

Summary: Mechanisms for Fitting Regression

Many mechanisms of fitting, either as a whole or stage by stage (i) involve adding up across data sets and (ii) can be defined in terms of a family of *matchers*, for each of which the relation

$$\sum (\text{matcher})y = \sum (\text{matcher})\hat{y},$$

involving summation across data sets, is a condition that can help to define the fit.

All forms of least squares, including varied forms of weighting, which may be either fixed or changing from stage to stage of an iteration, can be described in terms of matchers.

When we fit a general straight line, selected from the stock {all $\alpha + \beta x$}, by *ordinary* least squares the matchers are {all $c + dx$}.

For any coefficient β_i, appearing in any expression of a stock $\{\text{all } \sum \beta_i x_i\}$ to be fitted by ordinary least squares, there is a special matcher worthy of being called a *catcher* because it satisfies not only

$$\sum (\text{catcher})y = \sum (\text{catcher})\hat{y}$$

but also

$$\hat{\beta}_i = \sum (\text{catcher})y,$$

where the summations are over the data sets.

When we have $\{\text{all } \sum \beta_i x_i\}$ as the stock, we can take, as the collection of all matchers, all the potential fits in this stock. If we do, the sum of squares of the deviations, $\sum (y - \hat{y})^2$, will be a minimum. Thus it will be the fit by ordinary least squares.

To understand many of the troubles that may be concealed in fitting from $\{\text{all } \sum \gamma_i x_i\}$ it is useful to look hard at the values of

$$x_{i\cdot} = x_i - \hat{x}_i$$

for each and every i, as well as at the y-residuals, $y - \hat{y}$.

The variance of fitting β_i in $\{\text{all } \sum \beta_i x_i\}$ by ordinary least squares is

$$\text{var } \hat{\beta}_i = \frac{\sigma^2}{\sum (x_i - \hat{x}_i)^2}$$

when each y satisfies

$$y = \text{unknown fit} + \text{perturbation},$$

provided that the perturbations are uncorrelated and have a common variance σ^2, and that $\hat{x}_i$ is the ordinary least-squares fit to x_i in terms of the other carriers (the ordinary least-squares fit from the costock of x_i).

Since we have taken a fit from the costock out of x_i, we may as well do the same for y, forming residuals $y_{\cdot\text{all but } i}$, after which an ordinary least-squares fit of a straight line to the $(x_i - \hat{x}_i, y_{\cdot\text{all but } i})$ pairs will give us $\hat{\beta}_i$.

Labor in calculating $y_{\cdot\text{all but } i}$ need not be great, since we may write

$$y_{\cdot\text{all but } i} = y_{\cdot\text{all}} + \hat{\beta}_i(x_i - \hat{x}_i),$$

something that saves effort when we are using this approach to understand (i) how the value of $\hat{\beta}_i$ is related to the details of the data and (ii) how we might alter the mechanism of our fit to better match the idiosyncrasies of our data. Since

$$\sum x_i^2 \geq \sum x_{i\cdot 1}^2 \geq \sum x_{i\cdot 12}^2 \geq \cdots$$

and since these decreases may be by large factors, the precision of fitting $\hat{\beta}_i$ can only decrease as more and more carriers are added to the stock concerned.

This decrease can be covered up, for a certain number of increases, by a decrease in the s^2 we calculate from $\sum (y - \hat{y})^2$ as more flexible fits offer a better chance to match the actual behavior of ave $\{y\}$ as a function of the x's.

If we fit from $\{\text{all } \sum \beta_i x_i\}$ by choosing a weight for each data set and taking $\{\text{all (weight) } (\sum \gamma_i x_i)\}$ as our matchers, the resulting fit will minimize

$$\sum (\text{weight})(y - \hat{y})^2$$

and will thus be the fit by weighted least squares with the chosen weights.

This result means that weighted least-square fitting of y from $\{\text{all } \sum \beta_i x_i\}$ is equivalent to ordinary least-square fitting of $y\sqrt{\text{weight}}$ from $\{\text{all } \sqrt{\text{weight}} \sum \beta_i x_i\}$.

Accordingly, if we choose weights so that

$$(\text{weight}) \times \text{var}\{y\} = \sigma^2,$$

then

$$\text{var }\hat{\beta}_i = \frac{\sigma^2}{\sum (\text{weight}) x_{i\,\text{dot}}^2},$$

where $x_{i\,\text{dot}}$ is the residual after fitting, with the chosen weights, x_i from its costock.

If we introduce a "matrix weight" $\{w_{ij}\}$ and choose as our matchers all the linear combinations

$$\left\{ \text{matcher at } i = \sum_j w_{ij} (\text{potential fit at } j) \right\},$$

one for each potential fit, we will minimize

$$\sum \sum w_{ij} (y_i - \hat{y}_i)(y_j - \hat{y}),$$

thus obtaining a generalized weighted least-square fit.

If we are to consider how expressions calculated from data respond to the values involved, we can do well by looking at the influence curve generated when all but one of the data values are constant and that one varies as it wishes.

The influence curve of the biweight shows just the behavior we would like to have in a harsh, realistic world, namely: (i) near straight-line behavior when the wandering value is near the middle of the other values, (ii) flattening off as it moves further away, and (iii) a final turn over and decrease to the point where the presence or absence of the value matters not, once it has moved sufficiently far away.

Using a sequence of fittings in which each stage is an instance of weighted least squares, but where the weights in each step depend upon the residuals in the previous stage, is a very flexible method of fitting regressions.

Using a biweight procedure, say with $S = $ median absolute deviation from the median (or, alternatively, perhaps half the spread between the hinges—defined in Chapter 10) and $c = $ something like 9 to perhaps 6, offers an effective fitting procedure that is resistant and robust of efficiency.

When we want both (i) to weight to compensate for understood changes in variability of y from data set to data set and (ii) to biweight to provide resistance and robustness of efficiency, then we should:

a) take (weight) × (biweight) as our working weight,

b) take as the numerator of our biweights the product $\sqrt{\text{weight}}$ $(y - \hat{y}$ at last stage).

Even least absolute deviation fits—or modifications (flattened weights) of them to allow more effective use of data sets with small $y - \hat{y}$'s—can be made by iterative least-squares fitting with weights based upon the residuals of the previous stage.

We can:

◇make biweight fits using the computer,

◇approximate biweight fitting by hand, using a stepweighting scheme,

◇include weights compensating for understood changes in variability from data set to data set in such fittings.

References

Hampel, F. E. (1974). "The influence curve and its role in robust estimation." *J. Amer. Statist. Assoc.*, **69**, 383–393.

Larsen, W. A., and S. J. McCleary (1972). "The use of partial residual plots in regression analysis." *Technometrics*, **14**, 781–790.

Chapter 15/Guided Regression

Frequently problems arise where

◇we have several, even many, carriers;

◇we do not expect to use all of them;

◇we are content, after some preliminary modification, to let the eventually chosen stock be defined by a subset, conceivably any subset, of our carriers. Possibly more than one subset could be available at the close of the operation.

We could not be much concerned with the coefficients of our fit, because if we were, we would have to worry about just which stocks made sense. The purpose must then be either regression as description or regression as exclusion.

In this chapter, we describe only techniques based entirely on least squares, either equally weighted or with some fixed set of weights. We include some resistant versions. This does not mean that we are opposed to methods other than least squares.

We do want methods that let the data guide its own analysis, at least as far as selecting the stocks. In practice, this means guiding the computer, because the required volume of arithmetic soon drowns the hand calculator. The combination computer–graphical display–human operator has not yet been followed through far enough to let us judge how much it can do.

15A. How Can We Be Guided in What to Fit?

Because we consider many alternative stocks, we must choose among them. Naturally we want to measure how well each alternative performs so that our choice can be guided. To develop such a measure, let us begin with some "ideal conditions" because we plan to compute as if these were true. As usual, the ith data point is $(x_{1i}, x_{2i}, \ldots, x_{ki}, y_i)$.

Ideal Conditions

1. **The model.** The actual values of y can be expressed as

$$y_i = \eta_i + \text{error}_i.$$

2. **Unbiasedness.** Each error_i has average (expected value) zero.

3. *Homogeneity.* Each error$_i$ has the same variance σ^2.

4. *Lack of correlation.* Each distinct pair of errors, error$_i$, and error$_I$, for $i \neq I$, has zero covariance.

5. *The structure.* The "true" values can be expressed linearly in the x's as

$$\eta_i = \sum_j \beta_j x_{ji}.$$

(Thus the stock is adequate to describe all systematic effects).

Although these ideal conditions may well be far from true, they are often a reasonable starting point. Indeed, working as if they were true often gives us a useful answer. (Making a reasonable fit, whether or not we have these ideal conditions, is ordinarily the first step in looking to see how reasonable they are.)

If these conditions hold for any stock, they also hold for any larger stock containing the first stock. When we want to avoid unnecessarily large stocks, it is appropriate for us to consider the problem of choosing among stocks, for all of which the ideal conditions hold. Note carefully that we are choosing between stocks rather than fits; for this chapter we agreed earlier to use least squares with a fixed set of weights.

In any event, the convenience and simplicity of these ideal conditions, when combined with an ideal minimand, lead us to a classical basis for guiding our fitting. This ideal minimand is the expected sum of squared residuals— squares of fitted value MINUS true value,

$$\text{ave} \sum (\hat{y}_i - \eta_i)^2$$

which measures an ideal, average (not directly measurable) lack of fit. It gives the long-run average sum of squares of the deviations of the estimates $\hat{y}_i$ from the true values η_i for a particular stock satisfying the ideal conditions.

A practical question arises at once: We are not likely to know the η's; indeed finding out something about them is ordinarily why we are fitting. We must turn, then, to estimating this ideal lack of fit—to replacing the ideal minimand with a minimand we can handle.

Estimating the lack of fit. Let us recall that under the ideal conditions $\hat{y}_i$ is an unbiased estimate of η_i, that is,

$$\text{ave } \hat{y}_i = \eta_i.$$

As a result, the mean square error of $\hat{y}_i$ from η_i is the variance of $\hat{y}_i$,

$$\text{ave } (\hat{y}_i - \eta_i)^2 = \text{var } \hat{y}_i.$$

Using this result, the ideal minimand can be written

$$\sum_i \text{var } \hat{y}_i.$$

What we plan to show is, first, that the ideal minimand is:

$$\sigma^2 \times (\# \text{ of carriers})^*$$

and, second, that an estimate of this is

$$s^2 \times (\# \text{ of carriers}),$$

where s^2 is the familiar estimate of the common σ^2,

$$s^2 = \widehat{\sigma^2} = \frac{\Sigma(y_i - \hat{y}_i)^2}{\# \text{ of data points MINUS } \# \text{ of carriers}}.$$

For example, when we fit $y = \beta_0 \cdot 1 + \beta_1 x$, we use two carriers, 1 and x, and so the denominator on the right is $n - 2$, where n is the number of data points. We turn now to the derivation.

Derivation. We are still talking about one stock, and so we can change the x_j without changing the stock, so that

$$\sum_i x_{ji} x_{Ji} = 0, \qquad j \neq J, \tag{1}$$

$$\sum_i x_{ji}^2 = 1.$$

We suppose the change in x's has been made and that (1) now holds. Since $\hat{y}_i$ stays the same function of $y_1, \ldots, y_n$, this does not change $\Sigma \operatorname{var} \hat{y}_i$. Since the x's are fixed, we recall first that

$$\hat{y}_i = \sum_j \hat{\beta}_j x_{ji}$$

and second that

$$\hat{\beta}_j = \sum_i x_{ji} y_i.$$

Using these facts, we have a sequence of further results. First,

$$\operatorname{cov}(\hat{\beta}_j, \hat{\beta}_J) = 0, \qquad j \neq J,$$

and

$$\operatorname{var}\{\hat{\beta}_j\} = \sigma^2.$$

---

* The symbol $\#$ stands for "number."

Furthermore,

$$\text{var } \hat{y}_i = \sum_j x_{ji}^2 \text{ var } \{\hat{\beta}_j\}$$

$$= \sigma^2 \sum_j x_{ji}^2,$$

so that

$$\sum_i \text{var } \hat{y}_i = \sigma^2 \sum_i \sum_j x_{ji}^2 = \sigma^2 \sum_j \sum_i x_{ji}^2 = \sigma^2 \sum_j 1$$

$$= \sigma^2 \times (\# \text{ of carriers}).$$

Moreover, because least squares leads, under our ideal conditions, to

$$\text{cov } (y_i - \hat{y}_i, \hat{y}_i) = 0,$$

we have

$$\text{var } y_i = \text{var } (y_i - \hat{y}_i) + \text{var } \hat{y}_i$$

$$= \text{ave } (y_i - \hat{y}_i)^2 + \text{var } \hat{y}_i,$$

which becomes

$$\sigma^2 = \text{ave } (y_i - \hat{y}_i)^2 + \sigma^2 \sum_j x_{ji}^2.$$

Then, transposing gives

$$\text{ave } (y_i - \hat{y}_i)^2 = (1 - \sum_j x_{ji}^2)\sigma^2,$$

so that

$$\sum_i \text{ave } (y_i - \hat{y}_i)^2 = \sum_i \left(1 - \sum_j x_{ji}^2\right)\sigma^2 = \left(n - \sum_j 1\right)\sigma^2,$$

since $\sum_i x_{ji}^2 = 1$, which gives

$$\sum_i \text{ave } (y_i - \hat{y}_i)^2 = (\# \text{ of data points MINUS } \# \text{ of carriers})\sigma^2.$$

Since we assumed that the y's have a common σ^2, these results lead us to the usual estimate

$$s^2 = \widehat{\sigma^2} = \frac{\Sigma(y_i - \hat{y}_i)^2}{\# \text{ of data points MINUS } \# \text{ of carriers}}$$

and to an estimate of our actual minimand as

$$s^2 \times (\# \text{ of carriers}),$$

which is what we started out to show.

You may want to reconsider the merits of this after we have reached our discussion of C_P below.

If we like either minimand, then we dislike unnecessary carriers because they increase their size.

If we like our actual minimand, either for itself or as an estimate of the ideal one, then we prefer—among all stocks with a given number of carriers—those that make s^2 smallest. Such stocks produce the closest fits to the data. (If the ideal assumptions weren't true, some stocks might have smaller σ^2's than others. Then a stock that made σ^2 smallest would produce the closest fits to the true values. But we cannot know this.) Such a stock *appears* to come closest, among all with the given number of carriers, to meeting the ideal assumption that we can express

$$\eta_i = \text{ave}\,\{y_i\}$$

exactly.

The idea we require may be more familiar for sample and population means than for sums of squares of deviations. If we have many samples from many different populations all with about the same true mean, then the chances are good that the population with the largest observed mean does not have the largest true mean. Yet if we are to pick one we hope has a large mean, we will pick the one with the largest observed mean. In both situations we have a problem of *multiplicity*, more specifically of *competition*. Turning now to the stocks, the corresponding problem is that the stock that actually fits best among all our stocks with k carriers may well not be the one that apparently fits best with our data. When we fit a great many alternative stocks, many fits may have ideal minimands (expected values for the real minimands) that are close together, and chance will then help determine which stock will be *observed* to fit best in the sample. If many stocks are nearly equally close in ideal terms, then sampling variation could make it almost certain that the apparently best fitting one is *not* the ideally best fitting one.

Choice among k-carrier stocks. Lacking other information, however, we are likely to choose—among all our k-carrier stocks, or among all the k-carrier stocks we have tried—the one that yields the smallest s^2.

Some may wish to pick out several of the k-carrier stocks with small s^2, and then go on to other considerations.

Choice of k. How then do we choose k, the number of carriers? A variety of criteria have been suggested, often with immediate words of warning that it would be well not to follow them blindly. We mention three here as reasonable possibilities:

◇ Mallows's C_p,
◇ Anscombe's (Tukey's) $s^2/(n - k)$,
◇ Allen's PRESS.

In the context of a rather predetermined sequence of stocks, each including the previous ones, Anscombe (1967) produced a criterion for the degree of fit which Tukey, in discussing the same paper, simplified. The quantity they want to be small is

$$\frac{\text{residual sum of squares}}{(n-k)^2} = \frac{s^2}{n-k} = \frac{\text{mean square}}{\text{residual df}}.$$

Several investigators have developed an approach to selecting sub-sets of predictor variables, usually labeled PRESS, meaning prediction sum of squares. Anderson, Allen, and Cady (1972) and Allen (1974) are good sources.

Their context is that of forward and backward stepwise regression, but with modification. In the forward stepwise approach, after a new variable is entered into a subset, then each old variable is reconsidered for dropping. Their criterion is most simply understood as

$$\text{PRESS}_{\text{stock}} = \sum_{i=1}^{n} (y_i - \hat{y}_{(i)\text{stock}})^2.$$

This quantity is computed for each stock under consideration. The residual is the observed y_i minus the estimated y_i using the given stock, but with the ith observation omitted from the calculations leading to the estimate. If this formula is used directly, the calculations can be rather heavy. It turns out that alternative ways of writing this expression show $\text{PRESS}_{\text{stock}}$ to be a weighted sum of squares of the ordinary residuals, $y_i - \hat{y}_{(i)\text{stock}}$, and the calculations of the weights reduce the work.

In the context of reviewing methods of subset selection, C. L. Mallows (1973) offers a most instructive set of examples and further references. His index for appraising stocks is

$$C_p = \frac{1}{\hat{\sigma}^2}(\text{residual sum of squares for stock}) - n + 2p,$$

when n is the number of observations, p the number of carriers in the stock, and $\hat{\sigma}^2$ some unbiased estimate of the true variance. We use p here instead of our k because Mallows' C_p has come to have that name. And so $p \equiv k$ in these discussions. Mallows is careful to avoid using C_p to determine stopping or choices; he emphasizes the desirability of often looking at more than a few fits, including two or more using the same number of carriers. In a personal communication he says "that inspection of a C_p-plot may sometimes suggest that a particular k be chosen, but equally it may indicate that any one choice would be foolish." Others who want a rule might wish to use C_p.

He also points out that C_p is derived in a way similar to that of the estimate we gave earlier for the ideal minimand. We could write

$$J = \text{ave}\,(\hat{y}_i - \eta_i)^2 = \text{bias} + \text{variance}, \tag{*}$$

where

$$\text{bias} = \sum(\text{ave}\,\hat{y}_i - \eta_i)^2,$$
$$\text{variance} = (n-p)\sigma^2.$$

Further, we have the average value of the residual sum of squares (RSS) as

$$\text{ave RSS} = \text{bias} + p\sigma^2.$$

By substituting $\text{RSS} - p\sigma^2$ for bias in (*) and s^2 for σ^2, we get, as an estimate of J,

$$\hat{J} = \text{RSS} - (n - 2p)s^2.$$

This can be rewritten as

$$ps^2 + \text{RSS} - (n - p)s^2.$$

The first term is the measure we gave earlier as the ideal minimand and the last two terms form an estimate of the bias from the specific stock used. And this bias may be nonzero when the stock chosen is not adequate. The reader may then prefer C_p to ps^2 for guidance.

The whole area of guided regression is fraught with intellectual, statistical, computational, and subject-matter difficulties.

We cannot expect to know just which fit would prove most useful to us; this often applies to which carriers appear in the fit, as well as just what coefficients they appear with.

Letting the formal procedures produce a number of fits, and then deciding among them on the basis of both apparent quality of fit and other considerations, is often a wise thing to do.

It is now time to see whether we can drop one carrier of our triple with little cost. If so, we do, and then try adding a new third to our new pair. The new triple cannot be worse than the first triple and is sometimes much better. Thus moving both forward (more carriers) and backward (fewer) can help.

15B. Stepwise Techniques

What shall we do if we face 10 or 100 or 1000 possible carriers and a y?

If we must use only one carrier, we have little choice. We try each carrier separately, compare the s^2's they leave, and take the carrier that leaves the smallest one. Equivalently, we would look at the reduction in sum of squares due to fitting each carrier alone,

$$\text{RSS}_j = \frac{\left(\sum_i x_{ji} y_i \right)^2}{\sum_i x_{ji}^2}, \qquad j = 1, 2, \ldots, k.$$

and pick the carrier with the largest RSS. (Either choice picks the carrier having the highest squared correlation with y.)

If we plan to use exactly two carriers, we might try all pairs. For $k = 10$, with 45 pairs, we might well do just this. For $k = 1000$, where there are 499,500 pairs, we cannot expect to look at them all.

One approach would be to try all 1000 alone, see which does best, try that one in combination with all 999 others, and see which of these 999 pairs does best. There is no guarantee that any of these 999 pairs is even close to best among all 499,500 pairs, but we may not do too badly. How can we try to do a little better?

Having found our first pair, we can try all 998 ways of expanding it to a triple, and then try all pairs contained in the triple. This may do better—it cannot do worse.

Standard stepwise procedures. Standard programs for stepwise regression usually operate in the following way:

◇*Forward step*: Try every carrier not now in the regression, select the one with the greatest reduction in residual sum of squares, then test; if test is favorable, add this carrier to the regression (else halt); go to either another forward step or a backward step (depending on program).

◇*Backward step*: Try removing from the regression every carrier now in, select the one yielding the least increase in residual sum of squares, then test; if test is favorable, delete this carrier from regression; go to forward step.

Here the usual tests calculate "F-statistics",

$$F = \frac{\text{change in residual sum of squares}}{s^2 \text{ for larger stock}},$$

and compare each such value with value(s) arbitrarily selected by the user (if backward steps are taken, the test F for them can usually be chosen separately). If possible, we recommend a procedure that halts only after such a test has been "failed" for several—say at least three—successive forward steps. In any case, the output should show all useful results for the regression found after each step, forward or backward.

The statistic used is analogous to formal tests of significance for the presence of a carrier with a nonzero coefficient in a nonstepwise problem. Naturally in the sequential situation of stepwise analysis, the statistic remains a useful measure, but its nominal level of significance loses its exact frequency interpretation. Although the classical levels of significance such as 5% and 1% are often used in stepwise fitting, it may be better to move up to, say, 10% or 20%, to avoid missing good candidates when the effectiveness of many of the variables is small.

The user's responsibility. With such output, the user is prepared to use one or more of the criteria mentioned in Section 15A to help decide when to stop. Mechanizing this step seems dangerous. The user needs some contact with what is going on.

Desirable supplements. What more might a reasonable user ask of a stepwise program?

The user should ask for information about the residuals after each selected fit. A minimum would be (a) information about their distribution and (b) a listing of the dozen or so largest (in absolute value). If only a few substantially exceed the others, the investigator will probably want (a) to return the data with these few data-sets omitted and (b) to study these unusual data-sets carefully. He or she might find some mistakes—or some really new phenomenon.

Displaying what the coefficients of each carrier are (or would be if the carrier were to be added to the stock) at each step costs little and may save some grief. (The necessary calculation is a natural step toward finding the reduction in residual-sum-of-squares, already needed for guidance.) Sudden changes—or even slower drifts—warn us about near dependencies, multicollinearities, among the variables.

Example. Macdonald and Ward (1963) studied the internal structure of the magnetic-character figure, C_i, one of several measures of disturbance of the earth's magnetic field. These disturbances are clearly related to solar activity, almost certainly through the arrival of energetic particles from the sun. As might be expected, results for one day considerably resemble those of another. Days roughly 27 days apart—27 days being roughly the time for one rotation of the sun's atmosphere—also resemble each other.

With a view both to predicting and to producing relatively structureless residuals, Macdonald and Ward studied the regression of

$$y = C_i \quad \text{for day } t + k$$

on any subset of

$$x_1 = C_i \quad \text{for day } t,$$
$$x_2 = C_i \quad \text{for day } t - 1,$$
$$x_3 = C_i \quad \text{for day } t - 2,$$
$$\cdot \quad \cdot$$
$$\cdot \quad \cdot$$
$$\cdot \quad \cdot$$
$$x_{98} = C_i \quad \text{for day } t - 97.$$

Using a stepwise program with a particular stopping rule, they applied the technique to separate five-year blocks of data, each involving either 1826 or 1827 daily values of t.

For $k = 1$, predicting tomorrow, they found

$$y = 0.3746 + 0.570x_1 - 0.174x_2 + 0.113x_{27}$$
$$+ 0.071x_{54} - 0.063x_{86} - 0.057x_6$$

for the period 1915–1919. Here x_1 and x_2 are C_i for the previous two days, their coefficients accounting for some persistence in both level and slope. The next two terms, involving x_{27} and x_{54}, refer to times one and two solar revolutions earlier. Finally the x_{86} and x_6 terms do not have a clear meaning.

The results for four five-year periods are compared in Exhibit **1**. The exhibit shows considerable consistency in the regressions selected for the four periods. To summarize: we should use (a) nearby days, (b) days about 27 days back, (c) possibly a day or days about 54 days back.

That we can give fairly convincing reasons why particular terms appear makes this an unusual application of regression, and especially of stepwise regression. (But we did not try to interpret the sizes of the coefficients. In such unusual examples, where the stocks are fairly definitely fixed, we might try this.)

The weighted case. Most stepwise programs are written to cover only equiweighted fits, but there is no difficulty—either in theory or in practice—in

Exhibit **1** of Chapter 15

The results of stepwise autoregression of C_i (prediction 1 day ahead)

A) The TERMS from NEARBY DAYS

Epoch	Terms
1915–19	$.570x_1 - .174x_2 - .057x_6$
1920–24	$.513x_1 - .114x_2$
1925–29	$.586x_1 - .137x_2$
1930–34	$.527x_1 - .123x_2$

B) The TERMS for LAGS near 27 DAYS (one solar rotation)

1915–19	$+ .113x_{27}$
1920–24	$+ .110x_{26} + .140x_{27}$
1925–29	$+ .102x_{26}$
1930–34	$+ .181x_{26} + .109x_{27} + .070x_{28}$

C) The TERMS for LAGS near 54 DAYS (two solar rotations)

1915–19	$+ .071x_{54}$
1920–24	nil
1925–29	nil
1930–34	$+ .056x_{54}$

D) OTHER TERMS

1915–19	$- .063x_{86}$
1920–24	nil
1925–29	nil
1930–34	nil

extending them to cover arbitrary weights. Simple sums of squares—and their reductions—are replaced by weighted sums of squares—and their reductions. (The interpretation of the F values changes somewhat, but ordinarily not enough for us to change our choice of critical values.)

Working with variance-dependent weights makes no essential change, then, except that point (4) of the ideal conditions becomes:

4. **Variance proportionality.** For each i, the product of the weight W_i and the variance of error$_i$ is the same—σ^2.

Resistant stepwise fitting. We know little about resistant stepwise fitting, but we can suggest techniques from which we can anticipate a satisfactory performance. Since the simplest resistant fits known involve repeated changes of weight, we can revise our stepwise calculation to allow for this. Our suggestion, chosen from many without much regard for computer time, should be relatively reliable.

The suggested procedure is:

1. Choose weights, if desired.
2. Run a conventional stepwise regression using these as fixed weights, and choose a stock.
3. Refit this stock iteratively and resistantly, ending up with (a) a fit, (b) residuals, and (c) second weights, which are products of initial weights and weights emerging from the resistant technology. Examine the residuals and resistant weights carefully. If they seem reasonable, then:
4. Repeat step (2) using second weights;
5. Refit as in step (3), reaching third weights;
6. Repeat step (2) using third weights;
7. Refit as in step (4), reaching fourth weights.

How many repetitions are likely to be needed will have to be learned by experience for each type of problem.

15C. All-subset Techniques

If we have exactly three carriers, we can fit the $8 = 2^3$ possible regressions (including the empty one) where we "take out"

$$\text{Nothing}$$
$$x_1$$
$$x_1 \text{ and } x_2$$
$$x_2$$
$$x_2 \text{ and } x_3$$
$$x_3$$
$$x_3 \text{ and } x_1$$
$$x_3 \text{ and } x_1 \text{ and } x_2$$

Thus we consider all subsets of the variables in the original state. In this unusual ordering, moving from one to the next always takes either a standard forward step (moving one carrier in) or a standard backward step (moving one carrier out). Thus, if we had access to the pieces of a stepwise regression program, we could instruct those pieces to get us all eight regressions with no waste motion.

Such arrangements can be constructed for more than three carriers. Daniel and Wood (1971) report the existence and availability of a program to do all $4096 = 2^{12}$ regressions where the carriers come from a chosen set of 12. Because we do not ordinarily want all 4096 sets of answers, Daniel and Wood's program reports a selected few, using Mallows' C_p to guide the selection.

When we can afford this approach, we need not worry about the quality of our guidance—we used none. (We could miss something valuable only in the unlikely event that C_p did not work well for us.) For examples of using this technique, see the Daniel and Wood book (1971).

An algorithm by Furnival (1971) allows the extension of this approach to 18 or 20 carriers. Even more important is the development, by Furnival and Wilson (1974), that allows the streamlined calculation of the m best subsets, each of k carriers, for various values of k. This seems applicable to sets of as many as 35 carriers or more.

The weighted case. Again we have no difficulty in inserting any fixed set of weights, specifically variance-dependent weights, into the calculations. Equiweighted sums of squares and cross-products are replaced by generally-weighted sums of squares and cross-products. If possible, a display of residuals against weights (probably in the form "log |residual| against log weight") should be made and examined for satisfactory behavior. We would like something like the plot of "2 log |residual| + log weight" against "log weight" not very tilted. This would correspond roughly to having successfully weighted inversely as the variance. We also look for outliers, special patterns, and other troublesome or helpful features. (If desired, this could be applied for the fit by all carriers concerned.)

Resistant versions. It is not so easy to convert all-subsets fits from fixed least squares to iteratively reweighted least squares. Until we know more about various alternatives, however, a reasonably safe program, when resistant fits are in order and there are not too many carriers, would seem to be:

1. Fit all the carriers together resistantly and record the final weights from the resistant technology.

2. Examine the final weights from the resistant technology and residuals for reasonableness.

3. If the examination uncovers no serious abnormalities, run a weighted

all-subset procedure—or a Furnival and Wilson procedure—using the final weights of the resistant technology from (1) throughout.

If variance-dependent weights are desired, the modifications are not hard to add.

15D. Combined Techniques

What plan should we use when we have many carriers available?

◇one regression with all carriers in?

◇all regressions on all subsets?

◇a stepwise calculation, using a path of selected subsets?

If no one approach looks promising, we often combine approaches.

Many times when we look hard at our carriers, we can profitably sort them something like this:

◇*Key carriers*: a few (none to six) that we want to include in any regression;

◇*Promising carriers*: a second set, perhaps up to twelve, or more if we use Furnival and Wilson, that deserve somewhat special attention;

◇*The haystack*: a motley collection of other carriers that deserve limited attention.

If this sorting seems appropriate, we can arrange the procedure in a few large steps:

1. *Remove the key carriers.* Take the key carriers and regress them out of both y and all other carriers by doing this resistantly. Plan that all further work deals with residuals from these regressions. Examine, if possible, the distribution of each kind of residual and decide whether it is variable enough that we can use it in further work and/or whether we have few enough large residuals to call for their special examination and, possibly, a reanalysis with the corresponding data-sets put aside.

2. *Choose the promising set and select a promising subset.* Take the residuals from the first step, and pick out 0 to 12 or more carriers (other than those used in the first step) that seem to deserve special attention. Apply an all-subsets procedure, perhaps Furnival and Wilson. Select a promising subset as a basis for further work. (When in doubt, include one or two more carriers.) Regress this subset, also, out of (the residuals of) y and (the residuals of) the remaining carriers. Examine and act upon the new residuals as in step 1.

3. *Stepwise search in the haystack.* Run a stepwise regression of the y-residuals after step 2 on the x-residuals after step 2. Use this search to select a few more carriers to be added to the collection. Examine the residuals of y very carefully, and act on them as in step 1 when required.

4. *All-subsets check of nonkey carriers.* Decide whether to stop or check once more. To check, take up to 12 or more of

    ⋄the carriers selected at step two,

    ⋄the carriers selected at step three,

    ⋄the carriers considered, but not selected, at step two,

using this priority order, and run an all-subsets analysis (on residuals from step one).

    Nothing can be guaranteed, but this offers a fairly thorough program that includes the contributions of the subject matter and exploits the freedom of exploration that stepwise regression offers.

The weighted case. All this can be equally well done with any fixed weights. (We have discussed the simplicity of the required changes in the all-subset and stepwise procedures above.)

The resistant case. Here the first step would naturally be a resistant fit of the key and promising carriers together, so as to fix a set of weights from the resistant technology to be used through the fitting and refitting of these carriers.

    The stepwise calculation would then proceed as described at the end of Section 15B, using all key, selected promising, and so-far-stepwise-included carriers in each resistant refit.

    Another alternative would be to confine the resistant calculations to steps 1 and 4. If this is done, we will need to at least try adding each unincluded carrier singly and resistantly.

15E. Rearranging Carriers—Judgment Components

To use many carriers burdens us with two costs:

    ⋄looking at more carriers takes more computing;

    ⋄fitting more coefficients leaves the $\hat{y}$'s more variable (if the fit to η was adequate with fewer).

How can we reduce these costs?

    If we can "boil down" our carriers so that we use a smaller number, we can save on both costs.

    If we can modify our carriers, while preserving their number (and often their overall stock), so that fewer of them appear in our final regression, we will not only save on computing, but we will also save on the variances of the $\hat{y}$'s. Let us see how.

Modifying carriers. Often we have a variety of carriers that measure much the same thing. Thus we might, when studying wide receivers for professional football teams, be likely to include a variety of carriers that measure size, including, perhaps:

1. standing height
2. height to extended right arm
3. height to extended left arm
4. length of legs
5. length of shorter arm

6. length of longer arm
7. length of right hand
8. length of left hand
9. length of fingers of right hand
10. length of fingers of left hand

In addition, we might include carriers for speed, experience, weight, jumping ability, pupillary reflex, and the like.

If we thought we were trying to settle which of the size variables mattered, we probably should stop before we begin. The various size measures will be closely correlated (because professionals tend to be large, the closeness will be reduced because of cutoff or "attenuation"; with high-school players the correlation might be higher). But if we want to do nearly as well as we can in predicting wide-receiver performance from physique, we can go ahead.

What about the ten size carriers we listed? Should we use them as they stand? Surely not. Indeed, most of the simple correlations among them will be high. We might take logarithms of the measurements and combine the logs. Let then x_j be the log of the jth measure in our list.

Eight of the ten come in pairs. We are likely to gain by going to sums and differences, using

$$x_{23} = \tfrac{1}{2}(x_2 + x_3) \qquad \text{and} \qquad x_2 - x_3 = x_{32}$$

$$x_{56} = \tfrac{1}{2}(x_5 + x_6) \qquad \text{and} \qquad x_5 - x_6 = x_{65}$$

$$x_{78} = \tfrac{1}{2}(x_7 + x_8) \qquad \text{and} \qquad x_7 - x_8 = x_{87}$$

$$c_{90} = \tfrac{1}{2}(x_9 + x_{10}) \qquad \text{and} \qquad x_9 - x_{10} = x_{09}$$

Next, we have now six measures of overall size: $x_1, x_{23}, x_4, x_{56}, x_{78}, x_{90}$. Let us combine these into three measures, one of overall size, one of arm length, one of hand size, as follows:

$$x_{1234567890} = \tfrac{1}{6}(x_1 + x_{23} + x_4 + x_{56} + x_{78} + x_{90}),$$

$$x_{567890} = \tfrac{1}{3}(x_{56} + x_{78} + x_{90}),$$

$$x_{7890} = \tfrac{1}{2}(x_{78} + x_{90}).$$

Then we can pick up the pieces with what is left of seven of the other carriers after regression on these three.

An alternative, possibly more reasonable, would be to use

in place of

$$x_{56}$$

$$x_{567890},$$

on the ground that we suspect that, where arm length matters, we lose, not gain, by including hand and finger size. We only suspect, we do not know.

In any event, we have a reasonable hope—based on our general insight about how people vary and how the sizes of wide receivers affect their performance—that if we regress performance on our new set of size carriers we will need no more carriers than before—perhaps we can get along with fewer—to come close to the best regression.

If, for example, we came out with

$$1.74x_{567890} + 0.12x_{1234567890}$$

as a satisfactory regression, we would be paying—in terms of var$\{\hat{y}(i)\}$—for fitting only 2 coefficients. The fact that this regression can be written, first, as

$$\frac{1.74}{3}(x_{56} + x_{78} + x_{90}) + \frac{0.12}{6}(x_1 + x_{23} + x_4 + x_{56} + x_{78} + x_{90})$$

$$= 0.02x_1 + 0.02x_{23} + 0.02x_4 + 0.60x_{56} + 0.60x_{78} + 0.60x_{90}$$

and then as

$$0.02x_1 + 0.01x_2 + 0.01x_3 + 0.02x_4 + 0.30x_5 + 0.30x_6$$
$$+ 0.30x_7 + 0.30x_8 + 0.30x_9 + 0.30x_{10}$$

is irrelevant to what we have to pay for fitting coefficients. We fitted only 2, not 10. Of course, if one did a lot of preliminary fitting, plotting, and stepwise regressions looking for small variances before choosing such a plan, it might be possible to choose a remarkably good subset on a basis that sounds like the present analysis but that was *actually guided by the data* rather than judgment. If that approach was used, one should be charged for using all the variables; there is no *reduction* in the variance of the coefficients. The variance will be larger, and *we will not know what it is.*

Judgment components. We need a name for purely judgment-constructed combinations of the carriers with which we started. By analogy with the "principal components" we will meet in the next section, we will call them

judgment components.

When volunteered by persons of sound insight, they may be better than those that mechanical processes construct on the basis of the data being analyzed. The good judge may succeed sometimes, but not always, because he bases his judgment on many other similar sets of data carefully analyzed in the past.

Thus the judge may not always be using only intuition: sometimes his judgment will be broadly based upon data, even hard data.

Nonlinear combinations. As we have observed, sometimes a number of carriers tend to move together—a number of measures of general economic conditions is perhaps the most plausible case. If

◇ we can scale these carriers so their values are very nearly alike, and

◇ we fear that unusual phenomena, perhaps like oil embargoes, are likely to make a few of them much less effective for our purposes,

we may want to make up a judgment component in a way that will be insensitive to the localized distortions that we fear. To do this we might adjust the values of the different carriers (mainly by scaling) to come out almost the same. Then, treating these values like a sample, we replace them by a single value, using either median or bisquare—rather than mean—to form our first judgment component. (One thing to say for the mean—both good and bad—is that it combines different values while paying attention to them all and, sometimes to our sorrow, neglecting none.)

15F. Principal Components

How can we do something with sets of carriers whose mutual relations we must assess from the numbers in the data before us, that is, when we lack the insights such as we had about the sizes of wide receivers? Given a stock described by specific carriers, how can we find linear combinations of the given carriers that will be likely, or unlikely, to pick up regression from some as yet unidentified y? Can we find a way that pays for fewer fitted coefficients?

As before, we cannot offer a guaranteed answer. A malicious person who knew our x's and our plan for them could always invent a y to make our choices look horrible. But we don't believe nature works that way—more nearly that nature is, as Einstein put it (in German), "tricky, but not downright mean." And so we offer a technique that frequently helps.

One approach that seems likely to work is called

principal components.

If we begin by scaling our x's in a natural way (details later), various linear combinations

$$\sum c_j x_j$$

will have various variances. If we look, instead, at the ratio of such a variance to a measure of scale of the c_j's, more specifically at

$$\frac{\operatorname{var}(c_1 x_1 + c_2 x_2 + \ldots + c_k x_k)}{c_1^2 + c_2^2 + \ldots + c_k^2},$$

or, more precisely, at our estimate

$$\frac{\hat{\text{var}}\,(c_1 x_1 + c_2 x_2 + \ldots + c_k x_k)}{c_1^2 + c_2^2 + \ldots + c_k^2}$$

of this ratio, we might believe that linear combinations with higher variances will be likely to attract more regression than those with low variances, especially if those with low variances are really almost entirely made up of "errors and fluctuations". Those with high variances may be responding to some common cause, one that may also help to drive the y's up and down.

The components. What we propose to do is to compare two quadratic forms in the c_j's, namely

$$c_1^2 + c_2^2 + \ldots + c_k^2,$$

which is the variance of the weighted sum when the x's are uncorrelated and have a common scale, and

$$\hat{\text{var}}\,\{c_1 x_1 + c_2 x_2 + \ldots + c_k x_k\} = c_1^2\,\hat{\text{var}}\,x_1 + c_2^2\,\hat{\text{var}}\,x_2 + \ldots + c_k^2\,\hat{\text{var}}\,x_k$$
$$+ 2c_1 c_2\,\hat{\text{cov}}\,\{x_1, x_2\} + 2c_1 c_3\,\hat{\text{cov}}\,\{x_1, x_3\}$$
$$+ \ldots + 2c_{k-1} c_k\,\hat{\text{cov}}\,\{x_{k-1}, x_k\},$$

the natural estimate of the variance when we do not assume zero correlations.

It is a well-known result that such a comparison leads naturally to a sequence

$$(c_{11}, c_{21}, \ldots, c_{k1}),\qquad (c_{12}, c_{22}, \ldots, c_{k2}),\qquad \ldots,\qquad (c_{1k}, c_{2k}, \ldots, c_{kk})$$

of sets of coefficients in which the relative size

$$\frac{\text{one quadratic form}}{\text{other quadratic form}}$$

grades steadily from one end to the other.

If A and B are the symmetric matrices representing the quadratic forms in some coordinate system, the desired components are determined by the solutions of

$$|A - \lambda B| = 0.$$

In our case, we can take

$$A = \begin{pmatrix} \hat{\text{var}}\,x_1 & \hat{\text{cov}}\,(x_1, x_2) & \ldots & \hat{\text{cov}}\,(x_1, x_k) \\ & \hat{\text{var}}\,x_2 & \ldots & \ldots \\ & & \ldots & \ldots \\ & & & \hat{\text{var}}\,x_k \end{pmatrix}$$

and

$$B = \begin{pmatrix} 1 & 0 & 0 & - & - & 0 \\ & 1 & 0 & - & - & 0 \\ & & 1 & - & - & 0 \\ & & & - & - & \\ & & & & - & - & 1 \end{pmatrix}$$

where all below the main diagonal is to be filled in to mirror what is above that diagonal.

Solving such determinantal equations can be tiresome, but computer programs fortunately are now widely available for getting these sets of coefficients.

Choice of scale. If we know nothing about the x's, it is often especially convenient to scale them so that

$$\hat{\text{var}}\, x_j = 1 \qquad \text{for all } j.$$

This choice puts 1's on the main diagonals of A and correlation coefficients elsewhere.

If we can estimate the measurement errors of the x_j's, we may do considerably better if we scale the x's so that

$$\text{Judged measurement variance of } x_j = 1.$$

Having done this, we can look at all components with too small a variance, say,

$$\hat{\text{var}}\, \{c_1 x_1 + c_2 x_2 + \ldots + c_k x_k\} \le k \qquad (\text{or perhaps } k/2 \text{ or perhaps } 2k)$$

with a jaundiced eye. A linear combination that is no more variable than would be accounted for by independent, judged measurement errors is not likely to be telling us much unless its excess size is concentrated on a few data sets.

If an individual number is mostly measurement error, its value is not likely to help us very much. If the average squared deviation of all values is mainly made of measurement error, the only time we are likely to be told anything helpful by the array of values is when a few of them are much bigger than the general size of all the rest. Only for such unusual values can we be relatively sure that only a small part of their value is measurement error. Only these will be telling us something helpful.

Measurement covariances. Tacitly, this argument assumes that the

$$\text{MEASUREMENT covariances among the } x_j \approx 0. \qquad (*)$$

(Otherwise we ought to put the measurement correlations in the off-diagonal elements of B.) In problems where such correlations are modest, as they frequently are, we often get a reasonable approximation; but it can be unsatisfactory.

Assumption (*) above differs substantially from

$$\text{OBSERVED covariances among the } x_j \approx 0. \qquad (**)$$

If we do a good job of separately measuring things that are naturally correlated, we can easily have assumption (*) hold when assumption (**) does not.
Thus (*) has a better chance of being nearly true.

Small components. When assumption (*) holds, we can be relatively sure that
omitting the small components from further analysis loses us little. Even if we
did not have any reasonable basis to judge measurement variances, some of the
small components often look small enough that we are willing to set them
aside. In either case, we are using principal components as a boiling-down
process—as a way to replace more carriers by fewer.

An application. If we had a friend who was about to start stepwise regression
in a situation involving 20 variables concerned with each of home environment, school environment, and community environment, we would feel bad to
have him start with those 60 (= 20 + 20 + 20) variables as his initial carriers.
If all 60 seemed reasonable variables, well measured, we would much rather
see him start with 1, 2, or 3 principal components for each bunch of 20—this is
perhaps as well as we know how to help him in general.

If, after taking out what he could with these selected combinations, he
insisted on trying again combining the selected components and all the individual variables, we would tend to discourage him—but we would admit, if
pressed, that this would be better than starting with only individual variables.

If only some of each set of 20 looked good, we would certainly urge our
friend to start with principal components for the subbunches of good variables,
certainly doing these first. (Once they had been tried, a trial with a few largest
components from each good subbunch and a very few from each bad subbunch
might be reasonable.)

Factor analysis. Psychometricians and economists have developed a variety of
techniques, other than principal components, that can be used for boiling-down
stocks that require few to many carriers to describe. Some were designed to be
easy by hand calculation; some take advantage of resistant techniques. Most
have been created as the first step in a "factor-analysis" procedure, whose final
aim is to identify some linear combinations thought to be more legitimate than
others in serving as a coordinate system for the boiled-down stock.

If we follow Section 15D. If we follow the general approach of Section 15D,
and want to boil down our carriers by applying principal components to either
all our carriers together or to each of several groups of carriers, the pressures
on us are relaxed, because we can draw from each use of principal components
a fourfold partition of these components:

◇a few (may be none), to be "must" carriers;

◇a few more, to be "all subsets" carriers;

◇a few to several more, to be "stepwise" carriers;

◇the remainder, to be set aside.

With four parts, of which only the last is set aside, we may do a fair job of partitioning.

Reshaping. Separating big components from little components can be effective—although middle components are not likely to be sharply separated from either. The distinctive names of

⋄the largest component,

⋄the second-largest component, and

⋄the third-largest component,

may be merely pseudonyms. Even when we analyze a large amount of data, the largest component often fluctuates considerably *in direction.* and so do the others.

As a result, reshaping the large components by

⋄ choosing new linear combinations of the few largest principal components so as to make them more interpretable,

⋄dropping carriers that appear with relatively small coefficients,

⋄ adjusting coefficients to well-selected standard values (see References for recent work of B. F. Green),

should not make our components much less useful and may make them considerably more interpretable. Thus, we can profitably reshape in these ways.

15G. How Much Are We Likely to Learn?

We started this chapter with a limited aim: to do a reasonable job of regression as description or regression as exclusion. In the regression chapters, we have warned repeatedly that we usually dare not take the coefficients seriously. We warn again. Nevertheless, let us sketch some circumstances where we might pay attention to some of the coefficients.

Stable coefficients. Earlier chapters showed that we cannot count on the coefficient b_J of x_J in a fit involving many x's to resemble, either in size or sign, the coefficient b_J of x_J when x_J is fitted either alone or accompanied by a different set of carriers. If, nevertheless, we carry out a stepwise regression and find, when we watch b_J, that b_J changes relatively little all the way from beginning to end, we are appropriately tempted to think that some useful meaning *may* be given to its value.

If this constancy starts only after a certain set of other carriers is fitted, we will usually benefit from thinking hard about the interpretation of that coefficient when fitted in the presence of those other carriers. If the interpretation thus reached is likely to be relevant to our problems, we can afford to pay heed to this coefficient.

Judgment or principal components

The nice situations just described are more likely when we have used judgment components or principal components. They are also likely to be flagged by coefficients relatively large for the first component (or first few components) and dropping rapidly for the others. Such coefficient patterns often signal both a well selected set of components and at least one possibly meaningful coefficient.

Warning. Without such conditions, interpreting coefficients can be most misleading.

15H. Several y's or Several Studies

Given several responses in a single study—or several studies with the same response—or both—we have a choice of trying to do well with each separately, or to do well with all together.

To treat things together, we allow ourselves separate regression coefficients on any x_J in the regression—for each y, for each study, for both—but we add up (after the most reasonable scaling we know how to do) the sums of squares of residuals, or the reductions in such sums of squares. Except for this, and the need to have parallel pictures to examine separately the possible needs to re-express different y's in our study or studies, all is as before. (We should not expect to use different expressions for the same response in different studies using the same carriers. Should the data suggest this, we ought to examine it very carefully indeed.)

15I. Regression Starting Where?

To suppose that everything has to be discovered anew in each new data set breeds waste and loses efficacy of analysis. Given the idea of a single analysis, which is laid down in concrete in advance no matter how inappropriate the data may show it to be, we would have some excuse for forgetting the past. Legal constraints may occasionally produce such straitjackets, but those will be rare exceptions. In most applications of regression we did not control the values of the x's—and no one analysis is forced onto the data. Thus we ought to expect to look both to the past and to parallel bodies of data in search of information, hints, and intuitions.

At the earliest stage, such sources often equip us with a list of x's that ought to be considered, come what may. Often, too, we can have a list of likely candidates—and a list of those not so likely.

Next, the past can often give us guidance whether the first-aid expressions of our variables are the best we can do—at least before questioning carefully the data before us.

Next, the past may well be able to tell us what values certain coefficients in our regression are expected to take—before we look at the present data.

If we are taking a flexible approach, where we let the data guide us as to which carriers stay in the regression (and which are allowed to fall by the wayside), we can use our advance information—or appropriate theory, trustworthy or dubious—to set initial values for certain coefficients. Thus if we expected

$$y \sim \text{constant} + 17x_1 + ?x_2 + 5x_3 - ?x_4 + 3x_5,$$

we would do well to analyze

$$y - (17x_1 + 5x_3 + 3x_5),$$

rather than analyzing y.

Suppose that we find this composite variable to be fitted by

$$(1.2 \pm 1.5)x_1 + (4 \pm 1)x_2 + (-3 \pm 2)x_3 + (7 \pm 3)x_4 + (1 \pm 0.1)x_5,$$

where the numbers after the $\pm$ are estimated standard deviations of the estimated coefficients before the same $\pm$. We might well reduce this fit to

$$4x_2 + 7x_4 + x_5,$$

or, more precisely, to

$$(4 \pm 1)x_2 + (7 \pm 3)x_4 + (1 \pm 0.1)x_5.$$

If we did, then the fit to y is reasonably reported as

$$y \sim 17x_1 + 4x_2 + 5x_3 + 7x_4 + 4x_5,$$

where two terms come from our original subtraction, two from the fit, and the fifth, more clearly written as

$$4x_5 = x_5 + 3x_5,$$

from both.

When it comes time, as it should, to quote uncertainties for these coefficients, it seems reasonable to consider that when

$$b \pm s_b$$

is replaced by zero, we ought perhaps to consider a mean-square error of

$$\begin{array}{lll} b^2 & \text{when} & |b| \geq |s_b|, \\ s_b^2 & \text{when} & |b| \leq |s_b|, \end{array}$$

so that the result would be

$$y \sim (17 \pm 1.5)x_1 + (4 \pm 1)x_2 + (5 \pm 3)x_3 + (7 \pm 3)x_4 + (4 \pm 0.1)x_5,$$

where, for x_3 the ± 3 is $\pm b$, though ± 1.5, ± 1, ± 3, and ± 0.1 are $\pm s_b$. Note that at this stage we attend to the b and s_b of the fitted part rather than the sum of the "advance" coefficient and the fitted coefficient. If we had not reduced x_1 and x_3 out of the fit, the result would have been

$$y \sim constant + (18.2 \pm 1.5)x_1 + (4 \pm 1)x_2$$
$$+ (2 \pm 2)x_3 + (7 \pm 3)x_4 + (4 \pm 0.1)x_5.$$

(Throughout this discussion we have acted as if the correlations between the estimated coefficients can be neglected. This is, of course, not always the case—some would even say "not commonly the case".)

(An early form of this approach appeared in a presentation by E. C. Harrington at a November 1956 symposium at North Carolina State College.)

The broad philosophy is clear—science builds on previous work. Estimation of coefficients often needs to do the same.

15J. Arbitrary Adjustment

Sometimes we are concerned with regression as removal and we find that the precision of the important coefficients, when estimated from the data before us, is not as high as we need. At such a time, we should ask ourselves whether external coefficients intuited from past knowledge does not prescribe some or all of the relevant coefficients with greater precision than do the data before us. If they do, we ought to take seriously the possibility of adjusting our y's according to such external coefficients. Sometimes, indeed, either theory or semiquantitative understanding gives us coefficients that are better than even the consolidation of past knowledge. When either is so, we ought not to use the coefficient or coefficients estimated from our particular piece of data—unless the estimates are demonstrably significantly different from those provided by the past, by theory, or by semiquantitative understanding. (For one example see Cox (1957).)

Again, not everything must be reproved in every study.

Summary: Guided Regression

One approach to the question of how many carriers to include in our stock leads to choosing to minimize:

$$\frac{\Sigma(y - \hat{y})^2}{\# \text{ of data sets MINUS } \# \text{ of carriers}}.$$

Better considered approaches lead to the minimization of such expressions as Mallows' C_p, Allen's PRESS, or the additionally divided ratio (mean square)/(degrees of freedom).

One way to avoid the penalties incurred by fitting coefficients of many carriers is to recombine our carriers into linear combinations, either by pure judgment or by the determinantal equations of the method of principal components.

We can apply stepwise techniques of fitting in which carriers are added to—or removed from—the fit one by one, using any of a variety of standard computer programs.

Some possible approaches are:

◇ a plausible pattern of calculation when we want a resistant stepwise fit.

◇ Daniel and Wood's all-subsets procedure to up to 12 carriers.

◇ Furnival and Wilson's "best m subsets for each k" procedure to up to 35 carriers.

◇ modifications of these procedures to accommodate variance-compensating weights.

◇ a plausible pattern of calculation when we want resistant analogs of the "all-subsets" or "best m for every k" procedures.

◇ a combined approach to data with many potential carriers, for which the first step is to divide the carriers into (i) key carriers, (ii) promising carriers, (iii) the haystack, and in which the successive steps make careful use of this separation.

◇ replacing bundles of significantly intercorrelated carriers by linear combinations selected by judgment, of which there may be either fewer or an equal number.

◇ replacing bundles of significantly intercorrelated carriers by linear combinations selected by applying principal components to these carriers (*not* including the response).

◇ combining either of these latter approaches with the key–promising–haystack approach.

◇ observing how a "given" coefficient evolves through the steps of a stepwise procedure.

◇ starting our regression analyses, not from all coefficients zero, but rather from each coefficient at its best value, before looking at the data at hand.

References

Allen, D. M. (1971). "Mean square error of prediction as a criterion for selecting variables." *Technometrics*, **13,** 469–475.

Allen, D. M. (1974). "The relationship between variable selection and data augmentation and a method of prediction." *Technometrics*, **16,** 125–127.

Anderson, R. L., D. M. Allen, and F. B. Cady (1972). "Selection of predictor variables in linear multiple regression," in Bancroft, T. A. (Ed.), *Statistical Papers in Honor of George W. Snedecor*, Ames, Iowa: Iowa University Press.

Anscombe, F. J. (1967). "Topics in the investigation of linear relations fitted by the method of least squares." *J. Roy. Statist. Soc.*, Series B, **29,** 1–29.

Cox, D. R. (1957). "The use of a concomitant variable in selecting an experimental design." *Biometrika*, **44,** 150–158.

Daniel, C., and F. S. Wood (1971). *Fitting Equations to Data.* New York: Wiley and Sons.

Furnival, G. M. (1971). "All possible regressions with less computation." *Technometrics,* **13,** 403–408.

Furnival, G. M., and R. W. Wilson, Jr. (1974). "Regression by leaps and bounds." *Technometrics,* **16,** 499–511.

Green, B. F., Jr. "Parameter sensitivity in multivariate methods." Submitted for publication in *Psychometrika.*

Macdonald, N. J., and F. Ward (1963). "The prediction of geometric disturbances indices: 1. The elimination of internally predictable variations." *J. Geophys. Res.,* **68,** 3351–3373.

Mallows, C. L. (1973). "Some comments on C_p." *Technometrics,* **15,** 661–675.

Tukey, J. W. (1967). "Discussion of Anscombe's paper." *J. Roy. Statist. Soc.,* Series B, **29,** 47–48.

Chapter 16/Examining Regression Residuals

Chapter index on next page

In Section 3J we treated economical plots of residuals against any chosen x. Such plots are intended for diagnosis, not for definitive results.

We need diagnosis after fitting a regression of prechosen form to data. We ask "Are we through? How can we look ahead and guess which possible additions to our regression may be important?" These are good questions, but we may not be able to address them directly.

We ask first about y, $\hat{y}$, $y - \hat{y}$, and then about other variables. Because the y and $\hat{y}$ problem is more general than regression, we discuss it first.

16A. Examining $\hat{y}$

Suppose that we had just y and x, and had fitted bx, finding

$$\text{ave } (y - bx)^2 = s^2.$$

To find out what is left, we plot the residuals $y - bx$ against x or bx. We do not plot them against y, because this is likely to be misleading, as we now explain.

Suppose y is highly variable, as illustrated in the leftmost panel of Exhibit 1, but bx varies only modestly, staying close to the constant value c; then

$$Y \equiv (y - bx) + bx \approx (y - bx) + c,$$

and all the points $(y, y - bx)$ are close to the line containing the points

Exhibit **1** of Chapter 16

The rightmost panel illustrates the misleading character of the $y - \hat{y}$ versus y plot when the relation between y and x is slight.

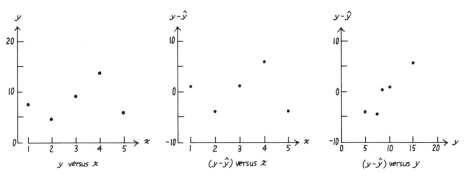

407

$(t + c, t)$. This is illustrated in the rightmost plot of Exhibit 1. The tightness and strong slope of this line would seem to indicate that it is possible to improve the original linear fit. However, here the original linear fit, though poor, is the best linear fit possible, so the indication from the plot of the residuals against y is fallacious. When we plot $y - bx$ against x, as in the middle panel of Exhibit 1, we have a vertically-wide blur and no indication of tilt, thus giving us correct insight into both the adequacy of the choice of b and the substantial size of the residuals.

Suppose, instead, that we thought x would do in place of bx. Then we would plot $y - x$ against something. The slope of $y - x$ against y would show, by a slope close to 1, when the fit is poor, but not whether the slope is right or wrong. The slope of $y - x$ against x would show, by its slope of $b - 1$, when we need to use some bx instead of x. Therefore we plot residuals $y - \hat{y}$ against x, not y.

The next question we ask, once we have begun with a fit $\hat{y}$, is "Are we getting full value from $\hat{y}$? More specifically, will a function of $\hat{y}$ fit y better?" If the answer is *yes*—that a modification will fit better—then we usually have two choices:

◇ fit the original y with a function of $\hat{y}$,

◇ re-express y so that it can be fitted more easily by something new drawn from the stock that produced $\hat{y}$.

Let us illustrate with an example.

Example. Scattered bank deposits. Exhibit **2** gives one amount of loans outstanding for cach of 16 quarters for the years 1923 through 1927. Each amount is for all banks in one Federal Reserve district, the districts varying haphazardly among Boston, New York, Philadelphia, and Richmond. A fit, of the form

$$\hat{y} = \text{linear in time PLUS district effect}$$

is shown at the bottom of Exhibit 2, together with the residuals in the rightmost column.

Exhibit **3** shows the results of ordering on $\hat{y}$ and then smoothing both $\hat{y}$ and $y - \hat{y}$. After this has been done, these two can be added together again, and the result, $h(\hat{y})$, smoothed, giving a possible function of $\hat{y}$.

Exhibit **4** shows the smoothed $y - \hat{y}$ against $\hat{y}$, which is clearly concave-upward—we call ∪ or ◡ or ⌞ concave-upward and ⌢ or ⌐ or ⌐ convex-upward—suggesting either:

◇ using a concave-upward function of $\hat{y}$ with the present y, or

◇ using a convex-upward function of y and fitting it with a new $\hat{y}$.

One natural choice for the convex-upward function of y would be log y. We explore this shortly. This corresponds to using a concave-upward function of $\hat{y}$ of the form

$$ae^{b\hat{y}}.$$

Exhibit **5** shows our smoothed $h(\hat{y})$ in comparison with

$$1500e^{0.00019\hat{y}} = \hat{\hat{y}}.$$

Exhibit **2** of Chapter 16

16 scattered values of loans outstanding (all banks in a Federal Reserve District at a call date near the close of the quarter indicated; in millions of dollars)

District	Quarter name and #	$y =$ Loans	$*\hat{y}$	$\lvert y - \hat{y} \rvert$
Boston	4Q23(−8)	3146	2740	406
New York	1Q24(−7)	8229	9251	−1022
Phila.	2Q24(−6)	1940	1624	316
Richmond	3Q24(−5)	1751	1269	482
New York	4Q24(−4)	9119	9578	−459
Boston	1Q25(−3)	3487	3285	202
Richmond	2Q25(−2)	1804	1596	208
Phila.	3Q25(−1)	2294	2169	125
Phila.	4Q25 (0)	2368	2278	90
Richmond	1Q26 (1)	1873	1923	−50
Boston	2Q26 (2)	3796	3830	−34
	3Q26 (3)			
New York	4Q26 (4)	10976	10450	526
Richmond	1Q27 (5)	1829	2359	−530
Phila.	2Q27 (6)	2509	2932	−423
New York	3Q27 (7)	11731	10777	954
Boston	4Q27 (8)	4031	4484	−453

* Fit used:

$$\hat{y} = 109 \times (\text{Quarter \#}) + \begin{cases} 3612 & \text{(Boston)} \\ 10014 & \text{(New York)} \\ 2278 & \text{(Phila.)} \\ 1814 & \text{(Richmond)} \end{cases}$$

S) SOURCE
Several annual reports of the Federal Reserve Board.

Our interest here is directed toward the residuals, rather than toward the fitting procedure that produced the particular values of the coefficients a and b. Thus we do not present the fitting process here, only the final results. Clearly, while the fit is not perfect, it is encouraging.

Exhibit **6** gives the numerical values of $1500e^{0.00019\hat{y}}$ and the corresponding residuals $y - \hat{y}$. Values of log y are given in Exhibit 6, as well as the fitted values and residuals obtained from the particular fit displayed at the bottom of the exhibit. Exhibit **7** shows the smoothing of both sets of residuals, after sorting on the fit. Exhibit **8** plots the residuals of the form

$$y - 1500e^{0.00019\hat{y}} = y - \hat{y}.$$

They show no noticeable structure.

Exhibit **3** of Chapter 16

Smoothing $y - \hat{y}$ against $\hat{y}$.

$\hat{y}$ ordered	$\hat{y}$ smoothed*	$y - \hat{y}$	$y - \hat{y}$ smoothed		$h(\hat{y})$**	$h(\hat{y})$ smoothed
1269	T	482			1751	
1596		208	316		1912	1832
1624	h	316	208		1832	1912
1923		−50	125		2048	
2169	e	125	90		2259	
2278		90	90		2368	
2359		−530	90		2449	
2740		406	−423	90	2830	
2932		−423	202	−34	2898	
3285	s	202	−34		3251	
3830		−34			3796	
4484	a	−453			4031	
9251		−1022	−459		8792	
9578	m	−459			9119	
10450		526			10976	
10777	e	954			11731	

*Smoothing is by running medians of three until no more changes occur.
** $h(\hat{y})$ = smoothed $\hat{y}$ + smoothed $(y - \hat{y})$.

Exhibit **4** of Chapter 16

Smoothed $(y - \hat{y})$ plotted against $\hat{y}$ for the values from Exhibit 3.

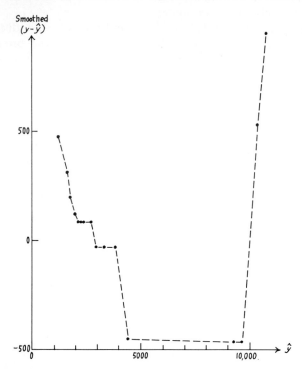

Exhibit **5** of Chapter 16

Smoothed $h(\hat{y})$ compared with $1500e^{0.00019\hat{y}}$, both plotted against y. Dots are $(\hat{y}, h(\hat{y}))$.

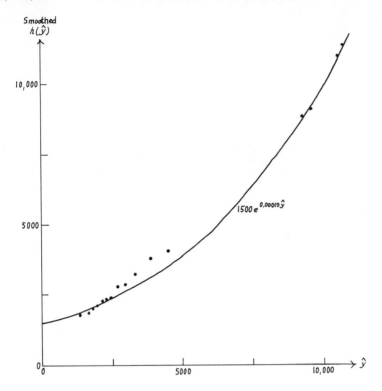

Exhibit 6 of Chapter 16

The exponential fit of $\hat{y}$ to y and the linear fit of $\log_{10}y$ to $\log_{10}y$

District	Quarter #	Based on raw y				Based on $\log_{10}y$		
		y	$\hat{y}$	$1500e^{0.00019\hat{y}}$	Resid. $=$ $y - ae^{b\hat{y}}$	$\log_{10}y$	(*) $\log_{10}\hat{y}$	$\log_{10}y - \log_{10}\hat{y}$
Boston	-8	3146	2740	2525	621	3.498	3.495	.003
New York	-7	8229	9251	8699	-470	3.915	3.946	-.031
Phila.	-6	1940	1624	2040	-100	3.288	3.312	-.024
Richmond	-5	1751	1269	1909	-158	3.243	3.220	.023
New York	-4	9119	9578	9256	-137	3.960	3.970	-.010
Boston	-3	3487	3285	2800	687	3.542	3.534	.008
Richmond	-2	1804	1596	2025	-221	3.256	3.243	.013
Phila.	-1	2294	2169	2265	29	3.361	3.350	.011
Phila.	0	2368	2278	2312	56	3.374	3.358	.016
Richmond	1	1873	1923	2162	-289	3.273	3.267	.006
Boston	2	3796	3830	3105	691	3.579	3.573	.006
New York	4	10976	10450	10924	52	4.040	4.032	.008
Richmond	5	1829	2359	2348	-519	3.262	3.298	-.036
Phila.	6	2509	2932	2618	-109	3.400	3.405	-.005
New York	7	11731	10777	11624	107	4.069	4.056	.013
Boston	8	4031	4484	3516	515	3.605	3.621	-.016

*Fit used: $\log_{10}\hat{y} = 0.00786 \times$ (Quarter #) PLUS $\begin{cases} 3.558 & \text{(Boston)} \\ 4.001 & \text{(New York)} \\ 3.358 & \text{(Phila.)} \\ 3.259 & \text{(Richmond)} \end{cases}$

Exhibit **7** of Chapter 16

The smoothed values of the two sets of residuals from Exhibit 6.

Based on raw y

Quarter #	$\hat{y}$	Resid.	Smoothed resid.
-5	1269	-158	
-2	1596	-221	-158
-6	1624	-100	-221 -158
1	1923	-289	-100
-1	2169	29	
0	2278	56	29
5	2359	-519	56 29
-8	2740	621	-109 56
6	2932	-109	621
-3	3285	687	
2	3830	691	687
8	4484	515	
-7	9251	-470	-137
-4	9578	-137	
4	10450	52	
7	10777	107	

Based on $\log_{10} y$

Quarter #	$\log_{10} y$	Resid.	Smoothed resid.	$h(\log_{10} y)$	$h(\log_{10} y)$ smoothed
-5	3.220	.023		3.243	t
-2	3.243	.013		3.256	
1	3.267	.006		3.273	h
5	3.298	-.036	-.024	3.274	
-6	3.312	-.024		3.287	
-1	3.350	.011		3.361	e
0	3.358	.016	.011	3.369	
6	3.405	-.005	.003	3.408	
-8	3.495	.003		3.498	s
-3	3.534	.008	.006	3.540	
2	3.573	.006		3.579	a
8	3.621	-.016		3.605	
-7	3.946	-.031	-.016	3.930	m
-4	3.970	-.010		3.961	
4	4.032	.008		4.040	e
7	4.056	.013		4.069	

$h(\log_{10} y)$ = smoothed $\log_{10} y$ + smoothed $(\log_{10} y - \log_{10} y)$.

Exhibit **9** plots the residuals of the form

$$\log_{10} y - \widehat{\log_{10} y}.$$

We see that their median is 0.003, a constant we might add to $\widehat{\log_{10} y}$; but beyond this, as with the residuals from the exponential fit, we see very little new. Having noticed the unstructured appearance of the residuals when plotted against $\widehat{\log_{10} y}$ in Exhibit 9, we plot the smoothed $h(\widehat{\log_{10} y})$ against $\widehat{\log_{10} y}$ in Exhibit **10** and recognize that we again have a plot close to a straight line, with a hint of a spur at the lower left.

The conclusion of this example is that we can fit raw values of loans outstanding with the form

$$ae^{b\hat{y}},$$

where $\hat{y}$ is of the form

(straight line in date) PLUS (district effect),

or that we can use this form directly to fit log (loans outstanding).

Exhibit **8** of Chapter 16

The smoothed residuals $y - 1500e^{0.00019\hat{y}}$ plotted against $\hat{y}$.

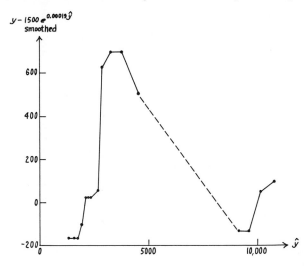

Exhibit **9** of Chapter 16

The smoothed residuals $\log_{10} y - \widehat{\log_{10} y}$ plotted against $\widehat{\log_{10} y}$.

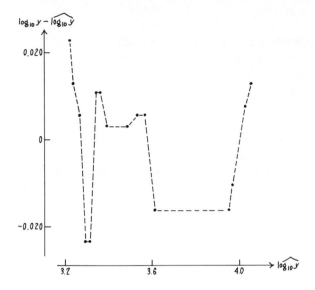

Exhibit **10** of Chapter 16

The smoothed function $h(\widehat{\log_{10} y})$ plotted against $\widehat{\log_{10} y}$.

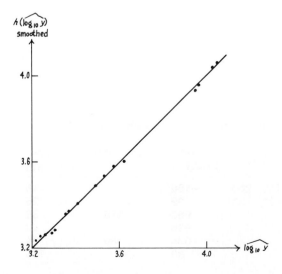

More points? The example we have just seen, involving 16 data points, was small enough for us to work point by point. Even 40 points would be enough to make an individual-point plot of

$$(\hat{y}, y - \hat{y})$$

less than satisfactory for answering our question about full value from $\hat{y}$.

In Section 3J, we looked deeper at such questions of large bodies of data by grouping the $y - \hat{y}$ values according to the values of $\hat{y}$, summarizing the groups, smoothing the summaries, and plotting the result. Exhibit **11** shows the corresponding numbers for our last example. We try both (1) to keep the sizes of the groups of residuals, the $y - \hat{y}$ groups, somewhat the same size, and (2) to keep the lengths (and separations) of the $\hat{y}$ intervals also about the same. As the example illustrates, we often cannot do both perfectly. The resulting plot is displayed in Exhibit **12**. From this picture we can perceive much of the basic structure shown in the slightly more time-consuming analysis of Exhibit 7 pictured in Exhibit 8. The final rise at the end has, however, disappeared. (It would have required very small groups to preserve it.)

What then are the possibilities for examining $y - \hat{y}$ with respect to $\hat{y}$? Arranged in rough order of increasing effort, we can:

1. Make a plot of extremes, 10 high and 10 low points.

2. Smooth the data and plot the smoothed result. Of course, if we have a lot of points, we can use the device of Section 3J.

3. Combine (1) and (2).

4. Plot all the points.

5. Combine (2) and (4).

6. Do (2); then plot all deviations from this smooth.

Exhibit **11** of Chapter 16

Grouping and medianing the residuals of the raw y (from Exhibit 7).

$\hat{y}$ (in hundreds)	Residual = z $= y - 1500e^{0.00019\hat{y}}$	Median resid. = z'	Smoothed z'	Median $\hat{y}$ (in hundreds)
12–19	−158, −221, −100, −289	−190		16
21–29	29, 56, −519, 621, −109	29		25
32–38	687, 691	689	515	35
44	515	515		44
92–95	−470, −137	−304	80	94
104–107	52, 107	80		106

Once there are more than a few points, we prefer any of the others to (4), and all but (4) to (5). Since (4) and (5) are relatively more work than most, we don't expect to do them often. If we want to work hard, (6) is often the thing to do.

16B. Variables—and Other Carriers

We are used to re-expressing variables. For example, we may replace t ($t > 0$) by log t, or replace s by s^3 or $-1/s$. What are the essentials of re-expression? For us they are:

⋄ability to calculate the values of either expression from the other;

⋄(usually) a monotone relationship between the expressions, so that when one goes up so does the other, and when one goes down so does the other.

What happens when a function of a variable is not a re-expression?

Most often, probably, it involves some kind of folding. Thus t^2 is folded once at zero, with $-t$ and t being put together, since $(-t)^2 = t^2$. If θ, for $-\pi \leq \theta \leq +\pi$, is our variable, then cos θ is folded once (at zero) while sin θ is folded twice (at $\pi/2$ and $-\pi/2$). If we consider t for all real values, then cos t is folded at every integer multiple (positive, zero, or negative) of π, while sin t is folded at every t of the form "($\pi/2$) + an integer multiple of π". In more general cases, of course, the folding involves stretching or shrinking of one side of the fold as compared to the other.

Exhibit **12** of Chapter 16

Smoothed medians z' of residuals grouped according to $\hat{y}$ plotted against median $\hat{y}$ (in hundreds) for values from Exhibit 11.

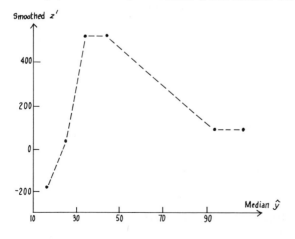

Besides re-expressing single variables, we could concern ourselves with re-expressing pairs of variables. Now things are less simple. In particular, there seems to be no handy package of re-expressions corresponding to the one we use for single variables: started powers (and their logarithmic and exponential limiting forms). We are not yet ready to treat the re-expression of pairs of variables generally and simply.

Carriers and variables. When we fit a quadratic, for example, we have choices as to how to write it. We can write

$$13 + 3t - 2t^2,$$

or, equivalently,

$$18 + 0(t - 12) - (2t^2 - 3t + 5).$$

In the first expression, we say the carriers are 1, t, and t^2, and in the second, 1, $(t - 12)$, and $(2t^2 - 3t + 5)$. More generally, we could speak of the set of all quadratics in t with carriers 1, t, and t^2 as

$$\{\text{all } a + bt + ct^2\}$$

or, equivalently, of

$$\{\text{all } d + e(t - 12) + f(2t^2 - 3t + 5)\}$$

with carriers 1, $(t - 12)$, and $(2t^2 - 3t + 5)$.

We now want to relate stocks or fits to variables. Let us begin by relating, to certain variables, a stock—or fit—as in this example:

$$\{\text{all } a + bt_1 + ct_1^2 + dt_2 + et_2^3\}.$$

Here the carriers are 1, t_1, t_1^2, t_2, t_2^3, each of which is fixed either by a value of t_1 or by a value of t_2 (or, in the case of 1, by a value of either). We can take the variables involved as (t_1, t_2) or (t_1^3, t_2^3) or (t_1, t_2^3), all simple re-expressions of one another. (We cannot take t_1 and t_2^2 as the variables if negative t_2 can occur, since the value of t_2^3 is not determined by the value of t_2^2.)

Simplicity would ordinarily lead us to say merely that the variables are t_1 and t_2, a choice that is not forced but is likely to be helpful.

If we change how the stock—or the fit—is carried, it can be natural to change what we think of as the variables. In

$$\{\text{all } a + bt_1 + ct_2\}$$

or

$$19 - 5t_1 + 25t_2,$$

it is natural to think of t_1 and t_2 as the variables. In the expressions equivalent to those just given,

$$\{\text{all } a + d(t_1 + t_2) + e(t_2 - t_1)\}$$

or

$$19 + 10(t_1 + t_2) + 15(t_2 - t_1),$$

it is natural, but not necessary, to think of $t_1 + t_2$ and $t_2 - t_1$ as the variables. Are these different?

In one sense, they are the same. After all, $(t_1 + t_2, t_2 - t_1)$ is a very simple re-expression of (t_1, t_2). In another sense, they are very different. In analogy to Lincoln and Barnum's famous saying, perhaps, most of us always think one variable at a time—and all of us do this in most of our thoughts. So a re-expression that mixes up variables is not trivial in its impact on our thinking. Why should we make such re-expressions?

Basically, we re-express to make matters simpler. To write down, look at, or think about

$$1000(t + s) - 0.004(t - s)$$

is simpler than to write down, look at, or think about

$$999.996t + 1000.004s.$$

Similarly

$$\{\text{all } a + b(t + s) + c(t + s)^2 + d(t + s)^3 + e(t - s)\}$$

is simpler than

$$\{\text{all } a + b(t + s) + c(t^2 + 2st + s^2) + d(t^3 + 3t^2s + 3ts^2 + s^3) + e(t - s)\},$$

especially if the first is written

$$\{\text{all } a + bu + cu^2 + du^3 + ev\},$$

where $u = t + s$ and $v = t - s$.

Habits of mind. In every field where data are collected and analyzed, we have habits of mind about which are "the" variables. These habits change with time, and data have their impact on these changes. For the last few decades "Gross National Product" has been "a variable" for economists, businessmen, and newspaper readers. Yet not too long ago, long explanations would have been necessary both about how values that were separately observed were put together and about what the result might mean. For another example, turn back to a century and a half ago when Benjamin Rumford, the German count from New England, studied the behavior of "heat" and separated the ideas of "temperature" and "quantity of heat," thus introducing variables that, on the one hand, made the behavior of the data simpler and, on the other, laid the foundation for modern thermodynamics.

A reasonable rule for variables is: Use the variables currently popular, unless some new variables make the behavior of the data noticeably simpler; never stick with popular variables if others make the data much simpler.

After all, different people are entitled to have different ways of thinking about the same problem—sometimes one way may be much better than the others, though perhaps only time will tell.

Naturalness of variables. In practice, our understanding of the world often prescribes how natural we will judge certain variables to be. The combination

Age of mother   PLUS   # of living children

is not likely to be accepted willingly as a natural variable. On the other hand, the pair

Age of husband ± Age of wife

seems, if anything, more acceptable than the separate variables

Age of husband,

Age of wife,

especially since the difference in ages is much commented on in daily life. The commoner pair of carriers are

Age of husband,

Age of husband − Age of wife,

or, perhaps,

Age of wife,

Age of wife − Age of husband.

This is not just a matter of algebraic structure. While we would often accept

# of boys   PLUS   # of girls

as a natural family variable, we are much less likely to accept

# of boys   MINUS   # of girls.

Nonlinear stocks. In the linear cases discussed so far, we note that the carriers in an expression can be obtained by taking the partial derivatives with respect to the coefficients. Thus the carriers in

$$f(t) = a + bt$$

are

$$\frac{\partial f}{\partial a} = 1, \qquad \frac{\partial f}{\partial b} = t.$$

If we use this approach, what happens when we have nonlinear stocks or fits? If, for example, our stock is

$$\{\text{all } ae^{-bt}\},$$

then if the carriers are the derivatives with respect to the parameters, we get

$$e^{-bt} \qquad \qquad \text{(derivative w.r.t. } a)$$

$$-ate^{-bt} \qquad \qquad \text{(derivative w.r.t. } b)$$

(equivalently, e^{-bt} and te^{-bt}). In general, then, in the nonlinear case the carriers depend upon where we are in the stock; here, on the value of b.

Similarly, if the fit is

$$13e^{-7t},$$

we take the carriers as

$$e^{-7t} \quad \text{and} \quad 13te^{-7t}$$

or, equivalently, as

$$e^{-7t} \quad \text{and} \quad te^{-7t}.$$

16C. The Next Step: Looking with Regard to an Old Variable, t_{old}

We can go on with the diagnosis of residuals from regression more easily if we have a standard setup. So let us suppose that we have fitted

$$\hat{y} = b_0 + b_1 x_1 + \cdots + b_k x_k,^*$$

where each x is a function of exactly one of the variables, any of which we can call t_{old},

$$t_1, t_2, \ldots, t_h \quad (\text{for} \quad h \leq k).$$

There are likely to be other variables around as well, so let any one such be

$$t_{\text{new}}.$$

Some examples (of everything but t_{new}) may help. One, with $k = 3$ and $h = 1$, is

$$\hat{y} - 13 - 2t + 3t^2 - 4t^3,$$

where the apparent x's are $1, t, t^2, t^3$, and we need only the one variable t. Another, with $k = 5$ and $h = 3$, is

$$\hat{y} = 8 - 3t + 4t^2 - 3s + 7s^2 - st,$$

where the apparent x's are $1, t, t^2, s, s^2, st$, and we need three variables, most naturally, t, s, and st. (We need st because we are sticking to "each carrier as a function of one variable".) Still another example, with $k = 5$ and $h = 5$, is

$$\hat{y} = 4 + 3x_1 + 2x_2 + x_3 - 7x_4 - 6x_5,$$

where the apparent x's are $1, x_1, x_2, x_3, x_4, x_5$, and the natural t's are x_1, x_2, x_3, x_4, x_5.

Looking at a t_{old}. How do we find out whether we are getting full value from one of the variables already represented in the fit? Whichever of $t_1, t_2, \ldots, t_h$ this variable may be, let us call it t_{old}. What it is reasonable to do depends in

---

* It is not important whether b_0 is included; we have put it in because it appears more often than not in practice.

part on how frequently t_{old} is represented—or nearly represented—among the carriers of our fit.

If t_{old} is represented only once, or perhaps twice, we may be able to proceed very much as we did with $\hat{y}$. Ordering the $y - \hat{y}$ according to t_{old} and then smoothing the $y - \hat{y}$ values may be revealing. If we have many points, we will want to take medians of groups before smoothing.

To do this, we do not need to care much about how t_{old} is initially expressed. If we change expression, the sorting order and the smoothing of $y - \hat{y}$ will be unaffected. All that can happen to our plot is a tendency to "crowd up" or spread out somewhere. If our plot does crowd up, we re-express t_{old} and re-plot without any need for recomputing the smoothed $y - \hat{y}$ values. (We need to re-plot to avoid crowding.)

When will this simple and direct approach be satisfactory? Basically, when we can get enough well-determined points to plot, either initially or by taking medians of groups. If we have three or four times as many points as times when the systematic behavior appears to change sign, and if the perturbations to which each point appears to be subject are small compared to the systematic behavior, we will likely see what is going on. If we have only one point for each interval (now unrevealed) of constant sign of the systematic behavior, we will not see enough. Equally, if there are three or four points in each interval of constant sign but the apparent perturbations are large compared to the systematic behavior, we will not see much either. We are forced to do something better than "smooth and plot" when the data are not generous with information about what to do next.

Next carrier. If we need to squeeze the data for information about unaccounted-for dependence on t_{old}, which often happens when we already have several representations of t_{old} in our fit, we will have to work with a specific carrier—or two specific carriers. Our first task is to pick this carrier, or these carriers, by picking specific functions of t_{old}.

Suppose, for example, that $t_{old} = t_1$, and that

$$\hat{y} = a + bt_1 + ct_1^2 + dt_2 + et_3.$$

It might now seem natural to add t_1^3 as the next carrier involving t. Taken properly this is quite correct, but we dare not take it too literally. Plotting $y - \hat{y}$ against t_1^3, for example, is not going to show us anything much, even if there is a lot there. For t_1^3 is just a re-expression of t_1, and we held out little hope for good results from plotting against t_1.

Suppose the fit were

$$\hat{y} = 1 + 23t_1 - 5t_1^2$$

(where we drop the t_2 and t_3 terms for a moment for simplicity) and we choose t_1^3 as a carrier. The new fit would be something of the form

$$\hat{y} = 1 + 23t_1 - 5t_1^2 + h(t_1^3 - j_2t_1^2 - j_1t_1 - j_0),$$

where the expression in parentheses

$$x_{dot} = t_1^3 - j_2 t_1^2 - j_1 t_1 - j_0,$$

is the result of "eliminating" the earlier carriers from t_1^3. ("Eliminating" here means fitting

$$t_1^3 \quad \text{by} \quad j_2 t_1^2 + j_1 t_1 + j_0$$

using the same sort of fitting procedure—possibly least squares—and forming the difference. Such elimination is sometimes called *orthogonalization*.) If the change in the fit is by a multiple of x_{dot}, then we should use x_{dot}, and not t_1^3, in examining the behavior of the $y - \hat{y}$.

If we go back to the original, slightly more complicated example, with the t_2 and t_3 terms included, we will, of course, have to include these terms in x_{dot}, making it

$$x_{dot} = t_1^3 - \widehat{t_1^3},$$

where

$$\widehat{t_1^3} = j_2 t_1^2 + j_1 t_1 + j_0 + k_2 t_2 + k_3 t_3.$$

Thus our routine procedure, if t_{old} appears more than once or perhaps twice among the carriers, begins with these steps:

⬦ pick a next carrier (or two next carriers—problems do occur, for example, where adding t_1^3 has no effect but adding t_1^4 helps).

⬦ find the corresponding x_{dot} (or the two corresponding x_{dot}'s) as the result of orthogonalizing the next carrier(s) to all the carriers we have already used.

⬦ sort the data points on x_{dot} and smooth the sorted $y - \hat{y}$'s (or do this first for one x_{dot} and then for the other).

It is possible that finding just one x_{dot} and plotting the 10 high and 10 low extremes against x_{dot} will be revealing; but if it is not, we go on to do the appropriate smoothing and afterward, as seems necessary and effective, to:

2. plot the smoothed result;

3. plot the extremes and the smooth, or

6. do (2); then plot all deviations from this smooth.

The extra work of (6) often does pay off.

Note that we have not considered plotting the deviations of the residuals from the smooth against the selected carrier. Plotting such deviations would not tell us about such matters as how the spread of the apparent fluctuations seems to depend on t_{old}. To ask such questions we would plot against t_{old} and not against x_{dot}.

Example. Fitting an unperturbed exponential. The left section of Exhibit **13** shows 10 values of an exponential function, at equally spaced values of t_{old}, the least-squares quadratic fit, and the corresponding residuals. These residuals are plotted against t_{old} in Exhibit **14**, and will be plotted against x_{dot} in Exhibit **15**. Since we are working with unperturbed values whose natural behavior is very smooth, we have no difficulty in seeing the nature of the behavior in Exhibit 14, as well as in Exhibit 15. (In both exhibits we have shown a dotted trace through the results of repeated median smoothing of adjacent sets of three points, called 3R smoothing. For Exhibit 15, the residuals from the quadratic fit are ordered on x_{dot} and then smoothed.)

What additional fit do these exhibits suggest? In Exhibit 14 we appear to have zeros at t_{old} values of 0.8, 4.6, and 8.3, suggesting using a multiple of

$$P_3 = (t_{old} - 0.8)(t_{old} - 4.6)(t_{old} - 8.3).$$

Exhibit **13** of Chapter 16

Residuals when an exponential, with and without perturbations, is fitted by a quadratic, and x_{dot} based on $(t_{old})^3$.

t_{old}	Raw	Quadratic fit[2]	Residuals	Raw	Quadratic fit[4]	Residuals	Exp. PLUS perturbation; residuals	x_{dot} [5]
		Exponential[1]			Perturbations[3]			
0	1.000000	1.006394	−.006394	−.0035	−.0006	−.0029	−.0093	−42
1	1.105171	1.103277	.001894	.0012	−.0007	.0019	.0038	14
2	1.221403	1.216113	.005290	.0050	−.0007	.0057	.0110	35
3	1.349859	1.344901	.004958	−.0069	−.0007	−.0062	−.0012	31
4	1.491825	1.489642	.002183	.0051	−.0007	.0058	.0079	12
5	1.648721	1.650336	−.001615	−.0056	−.0006	−.0050	−.0067	−12
6	1.822119	1.826982	−.004863	−.0040	−.0004	−.0036	−.0085	−31
7	2.013753	2.019581	−.005828	.0057	−.0002	.0059	.0000	−35
8	2.225541	2.228132	−.002591	−.0020	.0000	−.0021	−.0046	−14
9	2.459603	2.452636	.006967	.0008	.0003	.0005	.0074	42

Notes

(1) Exponential is e to the power $t_{old}/10$.
(2) $\hat{y} = 0.0079763x^2 + 0.0889069x + 1.0063940$, and for the normal equations, $\sum x = 45$, $\sum x^2 = 285$, $\sum x^3 = 2025$, $\sum x^4 = 15333$, $\sum xy = 86.778199$, $\sum x^2 y = 589.15937$, $\sum y = 16.337995$.
(3) Perturbations are from *A Million Random Digits with 100,000 Normal Deviates*, by the RAND Corporation; The Free Press, N.Y., 1955, p. 47, last column, upper two blocks.
(4) $\hat{y} = -0.00002x^2 - 0.0001x - 0.0006$.
(5) x_{dot} is here taken as a convenient multiple ($\frac{5}{3}$; see Fisher, R. A., and F. Yates, (1953). *Statistical Tables for Biological, Agricultural and Medical Research*, 4th edition. London: Oliver and Boyd. Table XXIII, p. 80) of the result of orthogonalizing t_{old}^3 to 1, t_{old} and t_{old}^2.

At the extremes and the local maximum and local minimum, this takes the following values, which we compare with those of the residuals by taking ratios:

t_{old}:	0	2.5	7.0	9.0
$y - \hat{y}$:	−0.0064	0.0057	−0.0059	0.0070
P_3:	−30.544	20.706	−19.344	25.256
Ratio:	0.00021	0.00028	0.00030	0.00028

Since these ratios indicate what multiple of P_3 might be a good fit, this simple approach would lead us to think (a) that a multiple of P_3 might be a good thing to add to $\hat{y}$, and (b) that 0.00027 would be a good first guess for the multiplier. Turning now to Exhibit 15, we see a very clear dependence of $y - \hat{y}$ on x_{dot}. If we take the endpoints, for which

$$\frac{(0.0070) - (-0.0064)}{(42) - (-42)} = 0.00016,$$

we are led to think of

$$0.00016 x_{dot}$$

as our choice to supplement $\hat{y}$.

Exhibit **14** of Chapter 16

Plot of $y - \hat{y}$ against t_{old} for the unperturbed exponential of Exhibit 13. Smoothed points (changes shown by small x's where different) are connected by dotted line.

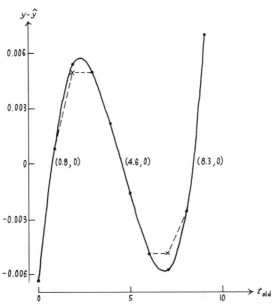

Clearly, using either Exhibit 14—plotting against t_{old}—or Exhibit 15—plotting against x_{dot}—has worked well in this example, although the latter has given an even stronger regression.

Example. Fitting a perturbed exponential. The middle section of Exhibit 13 gives some possible perturbations, drawn from a table of random numbers, for the 10 values of t_{old}, fits a quadratic in t_{old}, and finds the residuals. Since we are, in these two examples, fitting by ordinary least squares, we have

fit to (exponential   PLUS   perturbations)

EXACTLY EQUALS

(fit to exponential)   PLUS   (fit to perturbations),

and hence

residuals for (exponential   PLUS   perturbations)

EXACTLY EQUALS

(residuals for exponential)   PLUS   (residual for perturbations).

The residuals for the perturbed exponential are shown on the right side of Exhibit 13. They are plotted against t_{old} in Exhibit **16**—and will be shown

Exhibit **15** of Chapter 16

Plot of $y - \hat{y}$ against x_{dot} for the unperturbed exponential. Smoothed points (shown by small x's where different) are connected by dotted line.

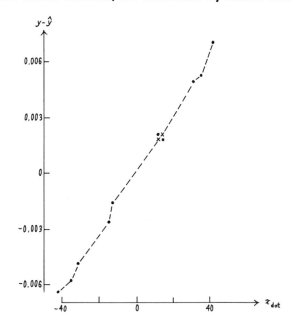

plotted against x_{dot} in Exhibit **17**. When we look at the points alone in Exhibit 16, we can hardly detect any systematic behavior. Using the (3R) smoothed values would leave us with a picture suggesting sign-changes near 0.4, 5.2, and 7.8, not far from those found for the unperturbed exponential.

In Exhibit 17, we have only to look at the points to see a clear dependence of $y - \hat{y}$ on x_{dot}. (If we look at the smooth, the appearance is strengthened.)

Exhibit **16** of Chapter 16

Plot of $y - \hat{y}$ against t_{old} for the perturbed exponential of Exhibit 13. Smoothed points are connected by dotted line.

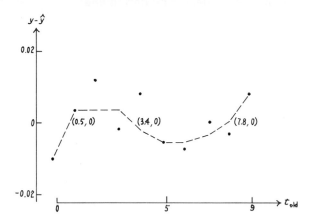

Exhibit **17** of Chapter 16

Plot of $y - \hat{y}$ against x_{dot} for the perturbed exponential of Exhibit 13. Smoothed points are connected by dotted lines.

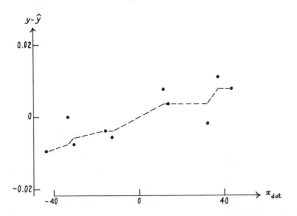

Example. Doubly perturbed exponential. To continue the process of perturbation to larger disturbances, Exhibits **18** and **19** plot the residuals for

$$(\text{exponential}) + 2(\text{perturbation})$$

against t_{old} and x_{dot}, respectively. We can see nothing in the plot against t_{old}, but the plot against x_{dot} still gives a reasonable suggestion of dependence—one

Exhibit **18** of Chapter 16

Plot of $y - \hat{y}$ against t_{old} for the exponential (of Exhibit 13) PLUS 2 × perturbation. Smoothed points are connected by dotted line.

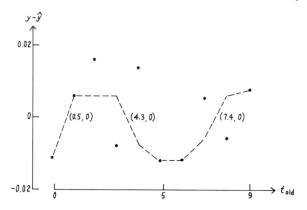

Exhibit **19** of Chapter 16

Plot of $y - \hat{y}$ against x_{dot} for the exponential (of Exhibit 13) PLUS 2 × perturbation. Smoothed points are connected by dotted line.

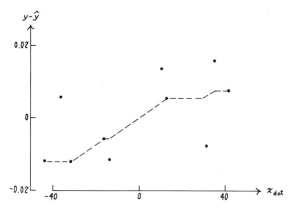

somewhat weaker than Exhibit 17—whether we look at only the points or at their 3R smooth.

Conclusions from the three examples. These examples suggest solving some problems by plotting $y - \hat{y}$ against t_{old}, but also suggest that hard problems can be dealt with only by plotting against a suitable x_{dot}. Smoothing has made the plots more effective, and allowed problems of intermediate difficulty to be handled by plotting against t_{old}, where otherwise, we would have had to plot against x_{dot}.

In addition to better detectability, the plot against x_{dot} has one further advantage: We can find the coefficient (of x_{dot} or of t_{old}; they are the same) directly—as the apparent slope of $y - \hat{y}$ against x_{dot}. Note that even when fitting a polynomial, as in the analysis based on Exhibit 14, getting at the coefficient of t_{old} from a plot of $y - \hat{y}$ against t_{old} is not straightforward.

If we have kept the values of the coefficients of the other carriers in x_{dot}, and found the coefficient c of x_{dot}, to get the new fit we have only to form

$$\hat{y} + cx_{dot},$$

expanding x_{dot} in terms of the original carriers if we wish.

16D. Looking with Regard to a New Variable, t_{new}

What should we do to ask whether a t_{new} will help us? We did two different things when responding to this question about a t_{old}:

◇ direct plotting against t_{old},

◇ selection of a "next carrier", conversion to an x_{dot}, then plotting against x_{dot}.

We also found that the second approach may well work when the first does not.

We ought to expect to have the same two possibilities when asking about t_{new}, although we ought not to expect the question "Will direct plotting work?" to necessarily have the same sort of answer. Let us turn first to an example, and then go on to discussion.

Example. Railroad returns.   Exhibit **20** sets out 20 years of values for (y) dollars of freight revenue to U.S. railroads per ton of freight hauled and (t_{new}) average miles of haul. A cubic fit in time (t) is shown for each, as are the residuals. Exhibit **21** plots $y - \hat{y}$ against t_{new}, and shows at most a very weak tendency to dependence. Exhibit **22** plots $y - \hat{y}$ against x_{dot}, here the residuals of t_{new} after fitting a cubic in t. The dependence here is now quite clear. There is more that we ought to take out with a further regression. Obviously we have learned much more from the plot against x_{dot}.

Exhibit **20** of Chapter 16

Dollars per ton hauled and average length of haul for all U.S. (class I, II, and III) railroads combined.

Year t	Dollars per ton			Miles hauled		
	Actual y	Cubic fit	Residuals $y - \hat{y}$	Actual t_{new}	Cubic fit	Residuals x_{dot}
1957	6.26	6.11	.15	429.20	441.60	−12.40
1956	5.97	6.18	−.21	428.08	429.60	−1.52
1955	5.95	6.18	−.23	430.67	421.65	9.02
1954	6.19	6.14	.05	431.65	417.18	14.47
1953	6.27	6.05	.22	420.66	415.57	5.09
1952	6.16	5.92	.24	426.93	416.25	10.68
1951	5.66	5.75	−.09	419.99	418.60	1.39
1950	5.58	5.56	.02	416.32	422.04	−5.72
1949	5.57	5.34	.23	412.02	425.98	−13.96
1948	5.12	5.12	.00	405.64	429.81	−24.17
1947	4.43	4.88	−.45	407.82	432.95	−25.13
1946	4.10	4.64	−.54	415.48	434.79	−19.31
1945	4.43	4.41	.02	458.14	434.75	23.39
1944	4.53	4.19	.34	473.28	432.23	41.05
1943	4.41	3.99	.42	469.07	426.64	42.43
1942	4.02	3.81	.21	427.76	417.37	10.39
1941	3.48	3.67	−.19	368.54	403.84	−35.30
1940	3.35	3.56	−.21	351.13	385.45	−34.32
1939	3.45	3.49	−.04	351.21	361.61	−10.40
1938	3.54	3.48	.06	356.05	331.72	24.33

S) SOURCE

y is series Q85, t_{new} is series Q83, both from page 431 of the *Historical Statistics of the United States, Colonial Times to 1957*. U.S. Bureau of the Census.

Exhibit **21** of Chapter 16

Plot of $y - \hat{y}$ against t_{new} for freight revenues (3R smoothed values shown by dotted line).

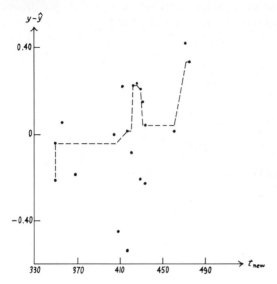

Exhibit **22** of Chapter 16

Plot of $y - \hat{y}$ against x_{dot} for freight revenues (3R smoothed values indicated by dotted line).

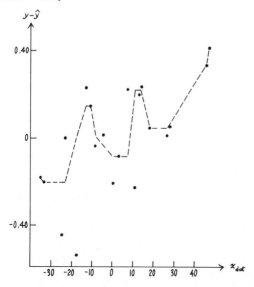

Discussion and Comments

Why has x_{dot} been more useful? Exhibit **23** shows the time histories of $y - \hat{y}$, t_{new}, and x_{dot}. Here we can see more. Fitting the cubic in time has taken out whatever slow motion there was in y; what remains in $y - \hat{y}$ is oscillatory, waving rapidly up and down around zero. When we look at t_{new}, we see a combination of things: a bulge during World War II (1942–1945), combined with a general upward trend. When we look at x_{dot}, we see that the removal of a cubic in t has also made x_{dot} oscillatory.

Exhibit **23** of Chapter 16

Plots of $y - \hat{y}$, t_{new}, and x_{dot} against date.

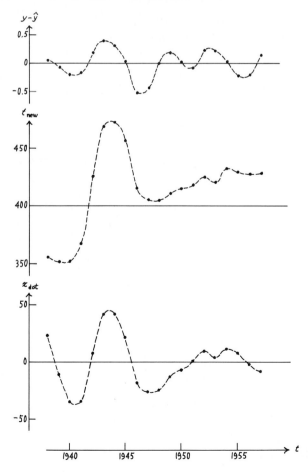

The existence of a trend in t_{new} makes it not directly useful in fitting $y - \hat{y}$, because $\hat{y}$ is a trend. Only after we have taken trend out of t_{new} by fitting a cubic in t, to obtain x_{dot} as the residual, do we have something that can directly fit $y - \hat{y}$.

Another way to say the same thing is: We cannot gain much by adding a multiple of t_{new} to $\hat{y}$ unless we are free to readjust the coefficients in $\hat{y}$ appropriately.

In this example, the similarity between t_{new} and the stock leading to $\hat{y}$ was more easily visible because it consisted of slowly changing variation, while the similarity between t_{new} and $y - \hat{y}$, which is the same (except for a multiplying constant) as the similarity between x_{dot} and $y - \hat{y}$, is rapidly changing. To have this happen makes the example easier to understand, which is good. Beware, however, of thinking that such similarities always are slowly changing and rapidly changing; they often differ in other ways.

The central points remain:

1. A similarity between t_{new} and the stock already fitted stands in the way of adding $f \cdot t_{new}$ to $\hat{y}$ unless we can readjust the constants in $\hat{y}$.

2. The easiest way to allow for such readjustment is to think of adding $f \cdot x_{dot}$ to $\hat{y}$. Since the similarity between t_{new} and $\hat{y}$ has been eliminated in forming x_{dot}, we can do this freely and easily. Thus, the plot of $y - \hat{y}$ against x_{dot} is directly useful.

Earlier, when we discussed improving a fit of the form

$$\hat{y} = a + bt_1 + ct_1^2 + dt_2 + et_3$$

by including further dependence on t_1, we were careful to insist that x_{dot} was to have "eliminated" from it

⋄not only 1, t_1, t_1^2, for the reasons discussed in Section 16C,

⋄but also t_2 and t_3.

We now see the reasons that "elimination" of t_2 and t_3 may be important; they are those we have just been discussing.

16E. Looking for Additional Product Terms

Multiplicative adjustments. So far we have looked for things that could be added to our fit. Sometimes we look for things that might be multiplied. We first consider a method that is not likely to help. If the fit is already reasonably close, then a change from

$$\hat{y} \qquad \text{to} \qquad \hat{y}(1 + u),$$

where u is, for instance, of the form

$$c_0 + c_1 x_1 + c_2 x_2 + \cdots + c_k x_k$$

and relatively small, corresponds to a change on the log scale from

$$\log \hat{y} \qquad \text{to} \qquad \log (\hat{y}(1 + u)) = \log \hat{y} + \log (1 + u)$$
$$= \log \hat{y} + (\log e) \log_e (1 + u)$$
$$\doteq \log \hat{y} + (\log e)u,$$

where $e = 2.71828 \ldots$ (When we write the formula this way, the logarithms can be taken to any desired base. When natural logs are used, in the approximation the second term on the righthand side becomes u.) Thus we can look for a gain from such a fit by relating

$$\log y - \log \hat{y}$$

either to old t's or a t_{new}.

Unfortunately, the resemblance between $y - \hat{y}$ and $\log y - \log \hat{y}$ will usually be high enough for such an approach to be mainly looking again for what we already sought.

A different possibility. The possibility of gaining by modifying $\hat{y}$ to

$$\hat{y}_{\text{med}} + (\hat{y} - \hat{y}_{\text{med}})(1 + u),$$

where $\hat{y}_{\text{med}}$ is the median of the values of $\hat{y}$, is different in character. It leads to

$$y - (\hat{y}_{\text{med}} + (\hat{y} - \hat{y}_{\text{med}})(1 + u)) = y - \hat{y} + (\hat{y}_{\text{med}} - \hat{y})u,$$

and to making inquiry into the possible regression of

$$y - \hat{y} \qquad \text{on} \qquad (\hat{y} - \hat{y}_{\text{med}})u.$$

This is a fresh inquiry, since $\hat{y} - \hat{y}_{\text{med}}$ takes both positive and negative signs, about equally often.

We may well set aside data points for which $\hat{y} - \hat{y}_{\text{med}}$ is small, because the product $(\hat{y} - \hat{y}_{\text{med}})u$ will be even smaller for such points. We shall locate the hinges (or quartiles) of $\hat{y}$ and set aside data points for which $\hat{y}$ falls between them. Let us consider two approaches, one through replacing $y - \hat{y}$ by $Q_{\hat{y}}$ and one through replacing $y - \hat{y}$ by $q_{\hat{y}}$, where the new quantities are defined as

$$Q_{\hat{y}} = \begin{cases} (y - \hat{y})/(\hat{y} - \hat{y}_{\text{med}}) & \text{(top and bottom quarters of } \hat{y}\text{),} \\ \text{(nothing)} & \text{(middle half of } \hat{y}\text{),} \end{cases}$$

and as

$$q_{\hat{y}} = \begin{cases} +(y - \hat{y}) & \text{(top quarter of } \hat{y}\text{),} \\ \text{(nothing)} & \text{(middle half of } \hat{y}\text{),} \\ -(y - \hat{y}) & \text{(bottom quarter of } \hat{y}\text{).} \end{cases}$$

The latter uses only the sign of $\hat{y} - \hat{y}_{\text{med}}$ and not its magnitude; thus it is a little easier to construct. It often suffices.

It now makes good sense for us to examine the relation of either $Q_{\hat{y}}$ or $q_{\hat{y}}$ to either old or new t's.

Example. Honolulu high tides. Our next example deals with the heights and times of high tides predicted for Honolulu in 1969. To make the data more manageable, we will confine our analysis to high tides falling on the 2nd, 4th, 6th, 8th, 10th, 12th or 14th of the months of January, February, March, April, May, and June. This yields 77 tides. We take time of high tide as x, predicted height of high tide as y, and date–hour as t.

To do anything like a thorough job with Honolulu tides would require more space—and more of your time—than this example deserves in this chapter of this book. In particular, it would be desirable to take at least all the 700-odd high tides in one year (more years would be still better). It would be desirable, too, to give reasonable attention to the physics of the problem—as always, subject-matter knowledge should be considered. But we are trying to illustrate a couple of special techniques and so we do not want to go more deeply into this example than it is worth to the reader. We will try only to show how q's or Q's can help us cut into what starts as apparently a mass of messy data, and how they can lead to quite specific phenomena that deserve further attention.

Exhibit **24** gives, for these tides, the following:

◇the predicted height, y, in tenths of a foot (above mean lower low water),

◇time of day, x, both in tenths of an hour and in degrees (where $360° = 24$ hours),

◇the epoch, t, as an integer for the month number plus a fraction for the day and hour converted into thousandths of a 31-day month (all multiplied by 1000).

Here a tide at 3:48 A.M. (written as 038, since 48 minutes is 0.8 hour) on 2 January would occur

$$(2 - 1) + \frac{38}{240} = 1.158 \text{ days}$$

after the beginning of January, since January begins on the midnight preceding the first day of the month. This corresponds to

$$\frac{1.158}{31} = 0.037 \quad \text{of a 31-day month,}$$

so the entry is

$$1000(1 + 0.037) = 1037.$$

This sets midnight before New Year's Day at 1000 (rather than at 0000).

We were lazy in using fractions of a thirty-one day month. This means that 3.2 days, for example, contributes an equal fraction in any month, no matter what its length. (It also means that 0.96 of the way through February, for example, never occurs. We thought the simplification worth such complexities for our present purposes. In a more careful analysis we would have to admit that months were of different lengths, something that would probably force us to work in days or in fractions of a year.)

As a first step, we divide the heights according to the angle of the day. Exhibit **25** gives the results of using 20° intervals. The smoothed medians of

Exhibit **24** of Chapter 16

Predicted high tides, Honolulu, 1969 (Days 2, 4, 6, 8, 10, 12, 14 of Jan., Feb., Mar., Apr., May, June).

y Ht. (*)	x Time (**)	x Time (***)	t Epoch (****)	y Ht. (*)	x Time (**)	x Time (***)	t Epoch (****)	y Ht. (*)	x Time (**)	x Time (***)	t Epoch (****)
22	038	57°	1037	20	034	51°	3037	8	037	56°	5037
5	153	230°	1053	9	156	234°	3053	22	168	252°	5055
23	048	72°	1103	18	043	64°	3103	5	052	78°	5104
5	166	249°	1119	12	167	250°	3119	23	184	276°	5122
21	059	88°	1169	14	052	78°	3168	3	075	112°	5171
6	181	271°	1186	15	181	272°	3186	21	204	306°	5189
18	069	104°	1235	9	062	93°	3234	6	111	166°	5241
9	201	302°	1253	17	200	300°	3253	18	225	338°	5256
13	080	120°	1301	18	226	339°	3321	11	132	198°	5308
13	226	339°	1321	4	122	183°	3371	12	011	16°	5356
8	098	147°	1368	21	017	26°	3422	16	144	216°	5374
22	016	24°	1422	7	142	213°	3438	9	024	36°	5423
5	128	192°	1436	14	037	56°	4037	20	156	234°	5440
22	044	66°	2038	17	164	246°	4054	4	052	78°	6039
7	164	246°	2054	10	047	70°	4103	24	182	273°	6057
20	153	79°	2104	20	178	267°	4121	4	077	116°	6107
9	177	266°	2121	6	060	90°	4169	21	199	298°	6124
15	062	93°	2170	20	197	296°	4187	8	107	160°	6176
12	194	291°	2187	3	085	128°	4237	16	216	324°	6190
10	072	108°	2235	19	221	332°	4256	14	127	190°	6243
15	218	327°	2255	6	125	188°	4307	11	234	351°	6257
22	013	020°	2357	18	011	16°	4356	9	002	3°	6291
4	131	196°	2372	11	141	212°	4374	19	141	212°	6309
24	029	43°	2423	15	025	38°	4423	6	018	27°	6357
6	150	225°	2440	15	153	230°	4440	21	153	230°	6375
								5	032	48°	6424
								22	164	246°	6441

(*) **Height (above mean lower low water) in tenths of a foot.**
(**) **Time (24-hour clock) in tenths of an hour.**
(***) **Time (360° per day).**
(****) **Date in thousandths of a month (see text).**

Exhibit **25** of Chapter 16

Heights of tide (from Exhibit 24), displayed according to angle of the day.

(*) Tag for angle	Mean angle	Height	Median height	Smoothed median	(**)	Resid₁ = smoothed median − (**)	Smoothed resid₁	(***)	Resid₂ = smoothed resid₁ − (***)
0	10°	18, 12, 9	12		11.4	.6	1.9	3.0	-1.1
2	30°	22, 22, 21, 15, 9, 6	18	17	14.0	3.0	.6	1.0	-.4
4	50°	22, 24, 20, 14, 8, 5	17		16.6	.4		-1.0	1.4
6	70°	23, 22, 20, 18, 14, 10, 5, 4	16		17.9	-1.9		-3.0	1.1
8	90°	21, 15, 9, 6	12		17.5	-5.5		-4.6	-.9
10	110°	18, 10, 3, 4	6	8	15.4	-7.4	-5.5	-5.6	.1
12	130°	13, 3	8		12.6	-4.6		-6.0	1.4
14	150°	8	8		10.5	-2.5		-5.6	2.5
16	170°	6, 8	7		10.1	-3.1	-3.1	-4.6	1.5
18	190°	5, 4, 4, 6, 11, 14	6	7	11.4	-4.4	-3.1	-3.0	.1
20	210°	7, 11, 16, 19	14	12	14.0	-2.0	-2.6	-1.0	-1.6
22	230°	5, 6, 9, 15, 20, 21	12	14	16.6	-2.6		1.0	-3.6
24	250°	5, 7, 12, 17, 22, 22	14		17.9	-3.9	-2.6	3.0	-5.6
26	270°	6, 9, 15, 20, 23, 24	18		17.5	.5		4.6	-4.1
28	290°	12, 20, 21	20	18	15.4	3.4		5.6	-2.2
30	310°	9, 17, 21	17		12.6	4.4		6.0	-1.6
32	330°	13, 15, 18, 19, 18, 16	17		10.5	6.5	4.4	5.6	-1.2
34	350°	11	11	12	10.1	1.9		4.6	-2.7
			(12)			(.6)			
			(18)			(3.0)			
			(17)			(.4)			

(*) Tag: 0 represents 0° to 19°; 2 represents 20° to 39°; and so on.
(**) = 14 + 4 cos 2(θ − 75°).
(***) = 6 cos (θ − 310°).

Note: Mean angles would be reduced by 0.5° if carried to an additional decimal place.

the corresponding groups of heights are plotted using small circles in Exhibit **26**. The prominent features of this plot are:

◇peaks near 60° and 270° (= 180° + 90°) .

◇valleys near 170° (= 90° + 80°) and 360° (= 270° + 90°).

Let us remove a rough fit involving sines or cosines of 2θ, because these will repeat every 180°. If we use a multiple of $\cos 2(\theta + \text{phase constant})$, we can fit any linear combination of $\cos 2\theta$ and $\sin 2\theta$ exactly. This function has a peak when $2(\theta + \text{phase constant})$ is 0°. The data do not peak at 0°, and therefore, we need a nonzero phase constant. Let us choose 75°, so that the carrier of our fit is $\cos 2(\theta - 75°)$. The peaks of the smoothed medians rise to 17 and 18, an average of 17.5 or about 18, and go as low as 12 and 7, averaging 9.5, or about 10. So an overall level of 14, which is close to the median 13.0 of the smoothed medians, offers a start for a first fit. Thus we are fitting $14 + a \cos 2(\theta - 75°)$. The distance between the peaks and the valleys is about $18 - 10 = 8$, and so a reasonable value for a is 4. The final choice is

$$14 + 4 \cos 2(\theta - 75°).$$

Exhibit **26** of Chapter 16

Smoothed medians from Exhibit 25 plotted against θ and a first fit against θ (14 + 4 cos 2(θ − 75°), shown dotted).

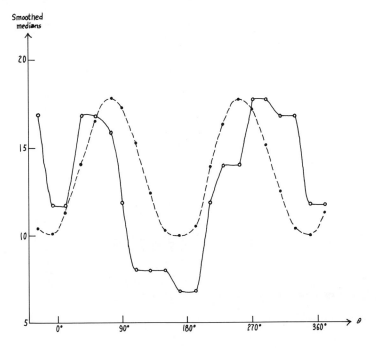

This fit is shown in column (**) of Exhibit 25, and subtracted from the medians of 20° slices. The new residuals are plotted in Exhibit **27**. Most of what we now see is a single valley and a single peak, so we fit a cosine or sine of θ itself. The same rough-and-ready technology we used before suggests $6 \cos(\theta - 310°)$ as the second fit. The slightly high fitted points around $\theta = 310°$ originate in the use of the coefficient 6, rounded from 5.5.

This second fit is also shown and subtracted (still from the medians of 20° slices) in Exhibit 25. We could improve our fit, now of the form

$$14 + 6 \cos(\theta - 310°) + 4 \cos 2(\theta - 75°)$$

somewhat further, but we turn instead to the Q and q analyses.

Analysis by $q_{\hat{y}}$. Exhibit **28** extends Exhibit 25, where the tidal heights of Exhibit 24 were classified by mean angles of the earth's rotation, by giving the more precise values of $\hat{y}$ evaluated at the specific angles.

The values of $\hat{y}$ at the specific angles are recorded in Exhibit **29** in chronological order (of increasing $t =$ epoch) and no longer in a classification ordered by angles. Here $y - \hat{y}$ is shown, as are the values of $q_{\hat{y}}$ and $Q_{\hat{y}}$, which are defined only for the outer quarters of $\hat{y}$, where $\hat{y} < 9.8$ or $\hat{y} > 17.6$. For facilitating the computation of $Q_{\hat{y}}$, the relevant values of $\hat{y} - \hat{y}_{med}$ are noted. The median $\hat{y}_{med}$ of $\hat{y}$ is 15.2.

Exhibit **27** of Chapter 16

The first residuals plotted against θ and a second fit against θ (6 cos(θ − 310°), shown dotted)

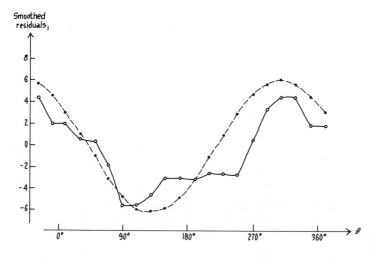

Exhibit **28** of Chapter 16

Extension of Exhibit 25 to ŷ for the specific angles given in Exhibit 24

ŷ at mean angle*	ŷ at tag**	Specific angles	t = epoch	ŷ at specific angles
14.4	14.6	16°, 16°, 3°	4356, 5356, 6291	14.7, 14.7, 14.6
15.0	14.7	24°, 20°, 26°, 38°, 36°, 27°	1422, 2357, 3422, 4423, 5423, 6357	14.8, 14.7, 14.9, 15.2, 15.2, 14.9
15.6	15.3	57°, 43°, 51°, 56°, 56°, 48°	1037, 2423, 3037, 4037, 5037, 6424	15.2, 15.3, 15.2, 15.2, 15.2, 15.3
14.9	15.2	72°, 66°, 79°, 64°, 78°, 70°, 78°, 78°	1103, 2038, 2104, 3103, 3168, 4103,	14.7, 14.3, 15.1, 14.2, 15.1, 14.6,
	13.9		5104. 6039	15.1, 15.1
12.9	11.4	88°, 93°, 93°, 90°	1169, 2170, 3234, 4169	12.9, 12.3, 12.3, 12.6
9.8	8.2	104°, 108°, 112°, 116°	1235, 2235, 5171, 6107	10.8, 10.1, 9.5, 8.8
6.6	5.8	120°, 128°	1301, 4237	8.2, 7.2
4.9	5.2	147°	1368	5.6
5.5	7.0	166°, 160°	5241, 6176	5.7, 5.2
8.4	10.7	192°, 196°, 183°, 188°, 198°, 190°,	1436, 2372, 3371, 4307, 5308, 6243	9.2, 10.0, 7.6, 8.5, 10.3, 8.8
13.0	15.3	213°, 212°, 216°, 212°	3438, 4374, 5374, 6309	13.7, 13.5, 14.4, 13.5
17.6	19.2	230°, 225°, 234°, 230°, 234°, 230°	1053, 2440, 3053, 4440, 5440, 6375	17.2, 16.3, 18.0, 17.2, 18.0, 17.2
20.9	22.5	249°, 246°, 250°, 246°, 252°, 246°	1119, 2054, 3119, 4054, 5055, 6441	20.7, 20.2, 20.8, 20.2, 21.2, 20.2
22.1	21.6	271°, 266°, 272°, 267°, 276°, 273°	1186, 2121, 3186, 4121, 5122, 6057	22.0, 22.2, 22.0, 22.2, 21.8, 21.9
21.0	19.8	291°, 296°, 298°	2187, 4187, 6124	20.6, 20.2, 20.0
18.6	17.4	302°, 300°, 306°	1253, 3253, 5189	19.6, 19.8, 19.1
16.1	15.4	339°, 327°, 339°, 332°, 338°, 324°	1321, 2255, 3321, 4256, 5256, 6190	15.5, 16.7, 15.5, 16.2, 15.6, 16.0
14.7	(14.6)	351°	6257	15.3
(14.4)				

* = (**) PLUS (***) from Exhibit 25
  = $14 + 6(\theta - 310°) + 4 \cos 2(\theta - 75°)$ evaluated at θ = mean angle
** = mean between adjacent values in column *, corresponding to ŷ evaluated at $\theta = 0°, 20°, \ldots$

Exhibit **29** of Chapter 16

Values of y (from Exhibit 25, the tidal example), ŷ, and y − ŷ, at specific epochs (from Exhibit 28); values of $q_{\hat{y}}$ and $Q_{\hat{y}}$.

t epoch	y Ht	$\hat{y}$	$y - \hat{y}$	$q_{\hat{y}}$	$\hat{y} - \hat{y}_{med}$	$Q_{\hat{y}}$
1037	22	15.2	6.8			
1053	5	17.2	−12.2			
1103	23	14.7	8.3			
1119	5	20.7	−15.7	−15.7	5.5	−2.8
1169	21	12.9	8.1			
1186	6	22.0	−16.0	−16.0	6.8	−2.4
1235	18	10.8	7.2			
1253	9	19.6	−10.6	−10.6	4.4	−2.4
1301	13	8.2	4.8	−4.8	−7.0	−.7
1321	13	15.5	−2.5			
1368	8	5.6	2.4	−2.4	−9.6	−.2
1422	22	14.8	−14.8			
1436	5	9.2	−4.2	4.2	−6.0	.7
2038	22	14.3	7.7			
2054	7	20.2	−13.2	−13.2	5.0	−2.6
2104	20	15.1	4.9			
2121	9	22.2	−13.2	−13.2	7.0	−1.9
2170	15	12.3	2.7			
2187	12	20.6	−8.6	−8.6	5.4	−1.4
2235	10	10.1	−.1			
2255	15	16.7	−1.7			
2357	22	14.7	7.3			
2372	4	10.0	−6.0			
2423	24	15.3	8.7			
2440	6	16.3	−10.3			
3037	20	15.2	4.8			
3053	9	18.0	−9.0	−9.0	2.8	−3.2
3103	18	14.2	3.8			
3119	12	20.8	−8.8	−8.8	5.6	−1.6
3168	14	15.1	−1.1			
3186	15	22.0	−7.0	−7.0	6.8	−1.0
3234	9	12.3	−3.3			
3253	17	19.8	−2.8	−2.8	4.6	−.6
3321	18	15.5	2.5			
3371	4	7.6	−3.6	3.6	−7.6	.5
3422	21	14.9	6.1			
3438	7	13.7	−6.7			
4037	14	15.2	−1.2			
4054	17	20.2	−3.2	−3.2	5.0	−.6
4103	10	14.6	−4.6			
4121	20	22.2	−2.2	−2.2	7.0	−.3

→

Exhibit **30** plots the values of $q_{\hat{y}}$ against t. The six months appear as six groupings, as they must. Even a casual look shows:

◇a general upward trend, and

◇wider spreads toward the left, except for February.

Exhibit **29** of Chapter 16 (continued)

4169	6	12.6	−6.6			
4187	20	20.2	−.2	−.2	5.0	0
4237	3	7.2	−4.2	4.2	−8.0	.5
4256	19	16.2	2.8			
4307	6	8.5	−2.5	2.5	−6.7	.4
4356	18	14.7	3.3			
4374	11	13.5	−2.5			
4423	15	15.2	−.2			
4440	15	17.2	−2.2			
5037	8	15.2	−7.2			
5055	22	21.2	.8	.8	6.0	.1
5104	5	15.1	−10.1			
5122	23	21.8	1.2	1.2	6.6	.2
5171	3	9.5	−6.5	6.5	−5.7	1.1
5189	21	19.1	1.9	1.9	3.9	.5
5241	6	5.7	.3	−.3	−9.5	0
5256	18	15.6	2.4			
5308	11	10.3	.7			
5356	12	14.7	−2.7			
5374	16	14.4	1.6			
5423	9	15.2	−6.2			
5440	20	18.0	2.0	2.0	2.8	.7
6039	4	15.1	11.1			
6057	24	21.9	2.1	2.1	6.7	.3
6107	4	8.8	4.8	−4.8	−6.4	−.8
6124	21	20.0	1.0	1.0	4.8	.2
6176	8	5.2	3.2	−3.2	−10.0	−3.2
6190	16	16.0	0			
6243	14	8.8	5.2	−5.2	−6.4	.8
6257	11	15.3	−4.3			
6291	9	14.6	−5.6			
6309	19	13.5	5.5			
6357	6	14.9	−8.9			
6375	21	17.2	3.8			
6424	5	15.3	−10.3			
6441	22	20.2	1.8	1.8	5.0	.4

Furthermore, early months (Jan., Feb., Mar., Apr.) show a very steady trend within months.

The purpose of any routine process of exploration, like the use of $q_{\hat{y}}$—or $Q_{\hat{y}}$ or q_t or Q_t—is not to check if a particular simple addition to the fit will help (we can check this by just going and fitting). It is rather to find out whether a general direction offers opportunities. Here the general direction, before the $q_{\hat{y}}$ analysis, was toward:

⬦not too complicated combinations of $\hat{y}$ and t—or θ and t.

We have been told, quite correctly, by the $q_{\hat{y}}$ plot, that this general direction is promising. It is now up to us to look harder and identify more specific directions we can explore.

The plot of $q_{\hat{y}}$ against t tells us that we can now improve our fit by combining

θ, which describes time of day, and

t, which describes time of year (within 1969).

Exhibit **30** of Chapter 16

$q_{\hat{y}}$ from Exhibit 29 against date–hour combination t. The dotted lines connect results for alternate days of a month.

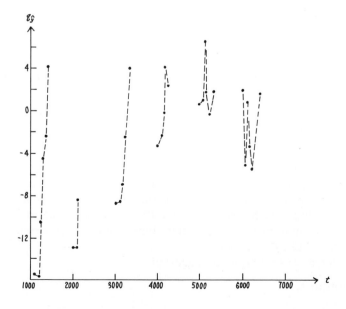

This is rather striking since up to this point y had depended on θ alone. The most useful combination is unlikely here to be of the simple form

$$d(\hat{y} - e)(t - t_{med})$$

corresponding to a single straight line in Exhibit 30, though an additional term of this form would be of some help. (To see roughly how much, plot $Q_{\hat{y}}$ against y.)

Parting comment. We have used these data to illustrate various techniques for breaking into a nonobvious structure. We have not tried to take the analysis of these data nearly as far as we could have.

Other products. To assess products of carriers, one of which is already a carrier in the fit, we choose an appropriate x_j, call it x, find its hinges (or quarters), set aside data points for which x_j falls between its quartiles, and form, for the other half of the data sets, either or both of

$$Q_x = (y - \hat{y})/(x - x_{med}),$$
$$q_x = (y - \hat{y}) \, \text{sgn} \, (x - x_{med}),$$

where

$$\text{sgn} \, z = \begin{cases} +1 & \text{when} \quad z > 0, \\ -1 & \text{when} \quad z < 0, \\ 0 & \text{when} \quad z = 0. \end{cases}$$

We can then explore the apparent behavior of Q_x or q_x just as we could that of $y - \hat{y}$, of $Q_{\hat{y}}$ or of $q_{\hat{y}}$.

If we wish to consider products involving carriers representing two t_{new}'s, say, x_{new} and x_{newer}, we have only to take $x = x_{new}$ and relate $Q_{x_{new}}$ or $q_{x_{new}}$ to x_{newer}.

16F. In What Order?

We have provided several ways to explore possibilities for further reducing the size of residuals by increasing the complexity of the fit. We cannot expect to try all of them on every fit we make. It would be good to have a (1) general indication of which to try first, which second, . . . and (2) a general indication of where to stop. The first of these needs we can meet fairly well, but experience does not yet seem to tell about the second.

The following order of examination is suggested:

1. $y - \hat{y}$ in relation to $\hat{y}$.

2. $y - \hat{y}$ in relation to t_{old}'s, sometimes in relation to the appropriate x_{dot}'s, that seem important.

3. $y - \hat{y}$ in relation to t_{new}'s, orthogonalized to become x_{dot}'s, that we strongly suspect of being important.

4A.* $y - \hat{y}$ in relation to t_{old}'s (still often as x_{dot}'s) not thought to be important.

4B.* $Q_{\hat{y}}$ or $q_{\hat{y}}$ in relation to t_{old}'s (still often as x_{dot}'s) that seem important.

4C.* Q_t or q_t for t an important t_{old}, and other important t_{old}'s (still often as x_{dot}'s).

5A. $y - \hat{y}$ in relation to t_{new}'s (as x_{dot}'s) not thought to be important.

5B. $Q_{\hat{y}}$ or $q_{\hat{y}}$ in relation to t_{new}'s (as x_{dot}'s) not thought to be important.

While guidance as to where to stop the overall examination is not at hand, we can offer guidance about how to behave at various stages, namely:

◇If we find an appreciable apparent dependence at stage (1), we either re-express y or fit further terms to absorb this appearance into the fit. (We can then start again.)

◇If we find a large apparent dependence in one of individual comparisons of stages (2), (3), (4A), (4B), or (5A), we proceed immediately to refitting and starting again.

◇If we find a moderate or small dependence in an individual comparison of one of these stages, we complete all the individual comparisons of that stage before choosing which one or few carriers to add to the new fit, refitting and restarting.

◇If we find a high or moderate dependence in stages (4C) or (5B), we proceed immediately to refitting and starting again.

◇If we find a small dependence in either of these stages, we think hard about what to do next.

Note that:

◇The x_{dot} for a particular t, whether a t_{old} or a t_{new}, is the same for $y - \hat{y}$ or any Q or any q (as is the need/no need decision about going to an x_{dot}).

◇$q_{\hat{y}}$ and the q_t's are easy to find (set aside half the data sets, and change the sign of half the $y - \hat{y}$'s that remain).

Accordingly, having done stage (2), it is easy—and often worthwhile—to do stage (4B) for $q_{\hat{y}}$ and stage (4C) for at least some q_t's.

---

* The order of choice among 4A, 4B and 4C (or between 5A and 5B) depends on the situation in ways about which we cannot yet give general indications.

Summary: Examining Regression Residuals

Displaying $y - \hat{y}$ against y is misleading.

Displaying $y - \hat{y}$ against $\hat{y}$ is often helpful.

We may have a variety of choices as to which are "the variables" once the stock, or even the carriers, are fixed.

The carriers in a nonlinear fitting situation are defined by the partial derivatives of the fit with respect to the coefficients.

A useful sequence of steps is:

⬦to display $y - \hat{y}$ against $\hat{y}$.

⬦to smooth $y - \tilde{y}$ against $\hat{y}$ and to add it to smoothed $\hat{y}$, in order to suggest an $h(\hat{y})$ that might be a better fit than $\hat{y}$.

⬦to reflect our understanding of such an $h(\hat{y})$ by either (i) re-expressing $\hat{y}$ or (ii) re-expressing x.

⬦to work with groups of points falling in (narrow ranges of) y when we have enough points to make this useful.

We may have to use one or two selected "next carriers" when examining the need to use further a variable already represented by, say, two or more carriers, and to regress present residuals against such next carriers.

We may use either display against t_{new} or display against x_{dot} (here the residuals of t_{new} after fitting with the carriers in the present fit) to enquire into the apparent importance of adding some function of t_{new} as an additional carrier. (Here we expect x_{dot} to be more powerful, especially if t_{new} was suitably re-expressed before x_{dot} was formed).

We hunt for additional product terms by displaying $y - \hat{y}$ against either $Q_{\hat{y}}$ or $q_{\hat{y}}$ (see early Section 16E for formulas).

We hunt for other additional product terms by displaying against Q_x or q_x, where x is a well-chosen carrier already in the fit (see late Section 16E for formulas).

Appendix/Details about the Need to Re-express

A. The General Case

The question before us is:

"Do we need to change from a less satisfactory expression of x (described below as 'the raw carrier') to a more satisfactory expression (described below as 'the straightened carrier')?"

Our assumption is that we know the straightened carrier to be better, that the question is only "Is it *enough* better to be worth the trouble—where the trouble may be a matter of calculation, but is more likely to be a matter of exposition and description?" We often want to answer this question on the basis of what we have learned by fitting a regression on the raw carrier—to answer it without going to the trouble of fitting the (admittedly somewhat better) regression on the straightened (more satisfactory) carrier.

Then our basic data have to be:

◇How much are we willing to worsen the fit by using the raw carrier?

◇How alike are the raw and straightened carriers?

◇How good was the fit using the raw carrier?

◇How many data points were there?

We measure the first three of these as follows:

◇worsening of the fit by δ, where the mean-square failure of fit (a modification of the mean-square error of fit, to be explained in Section E) is $(1 + \delta)$ times as large when the raw carrier is used as when the straightened carrier is used.

◇similarity between raw and straightened carriers, either by $r_{carriers}$, the correlation coefficient between them, or by ϵ, where

$$r_{carriers}^2 = \text{correlation}^2(x_{raw}, x_{straightened})$$

$$= \frac{1}{1 + \epsilon^2}.$$

(ϵ is often an easier basis for tabulation or discussion. Its interpretation is discussed in Section E.)

449

◇the quality of fit, using the raw carrier, by the corresponding correlation coefficient r_{rawfit}, where

$$r_{\text{rawfit}}^2 = \text{correlation}^2 (y, x_{\text{raw}}),$$

$$1 - r_{\text{rawfit}}^2 = \frac{\text{variance of residuals from fit of } x_{\text{raw}}}{\text{variance of } y \text{ before regression}}.$$

Three questions now arise:

◇How do we judge what δ is acceptable?

◇How are we to appraise r_{carriers}^2 (or ϵ) in practice?

◇How do we assess the actual δ, using, say, n, r_{rawfit}^2, and ϵ?

As mentioned above, we answer the second question for the most frequently-arising re-expressions by relating r_{carriers}^2 to:

◇(mainly) the ratio "largest x/smallest x" for the raw carrier.

◇(small amount of modulation) whether the x-values are, compared to the normal, stretch-tailed, neutral, or squeezed-tailed (as they, for example, will be if they are uniformly spaced).

The most important relation is of r_{carriers}^2 to the ratio of largest to smallest x, but we should allow a small amount of modulation from the shape of the x-values. We return to the discussion of the ratio in Section C and of the spacing in Section D.

We can express δ as follows in terms of ϵ^2, n, and r_{rawfit}^2:

$$\delta = \frac{\epsilon^2}{1 + \epsilon^2} \left(\frac{(1 + (n - 1)\epsilon^2)r_{\text{rawfit}}^2}{1 - (1 + \epsilon^2)r_{\text{rawfit}}^2} - 1 \right).$$

We plan to work in terms of three values of δ, namely,

$\delta = 1$,   where we care only moderately about fit quality and do not mind doubling the mean-square failure;

$\delta = 0.1$,   where we care quite a lot about fit quality but are willing to let the mean-square failure rise by 10%;

$\delta = 0.01$, where we care extremely strongly about fit quality, and are only willing to let the mean-square failure rise by 1%.

We think of the two end values, $\delta = 1$ and $\delta = 0.01$, as rather extreme, and the middle one $\delta = 0.1$ as a moderate value.

Sometimes we will not want to calculate δ, preferring a quick look at a table of r_{rawfit}^2 for the three particular values of δ. Exhibit **1** gives these values.

Some examples. Suppose we have $n = 100$ and $r_{\text{carriers}}^2 = 0.9412$, so that $\epsilon = 0.25$. (Both $1 - r_{\text{carriers}}^2$ and ϵ^2 show that about 6% of the variance of the raw carrier would be left over if we fitted the raw carrier with the straightened carrier.) Suppose, further, that in the population, r^2_{rawfit} is about 0.50. What can we find from this exhibit?

If we look at Exhibit 1, for $n = 100$ and $\epsilon = 0.25$, we find that: (A) it would take an r^2_{rawfit} of 0.68 to make $\delta = 1$ and $1 + \delta = 2$; (B) an r^2_{rawfit} of 0.27 to make $\delta = 0.1$ and $1 + \delta = 1.1$. If we were willing to allow an increase in mean-square failure of up to $\delta = 1$, we need not re-express; but if we do not wish to allow an increase beyond $\delta = 0.1$, we must re-express.

Exhibit **1** of Appendix

Values of (true) r^2_{rawfit} between response and raw carrier such that mean-square failure in finding slope is (A) doubled, (B) increased by 10%, (C) increased by 1%.

	$\epsilon=.6$	$\epsilon=.5$	$\epsilon=.4$	$\epsilon=.3$	$\epsilon=.25$	$\epsilon=.2$	$\epsilon=.15$	$\epsilon=.1$	$\epsilon=.06$	$\epsilon=.03$	$\epsilon=.01$
A) Mean-square failure—times 2 ($= 1 + \delta$)											
$n = 5$	N	N	N	(.84)	(.88)	.923	.956	.980	.9928	.9982	.999800
10	N	(.56)	(.69)	.81	.87	.917	.954	.980	.9928	.9982	.999800
30	N	(.38)	.54	.73	.82	.893	.945	.978	.9925	.9982	.999800
100	N	(.18)	.31	.54	.68	.82	.916	.971	.9916	.9981	.999799
300	N	N	.14	.31	.46	.66	.841	.953	.9891	.9980	.999797
1000	N	N	N	.12	.22	.39	.654	.895	.9804	.9974	.999790
B) Mean-square failure—times 1.1 ($= 1 + \delta$)											
$n = 5$	N	N	N	N	N	N	(.82)	(.906)	.963	.9902	.99890
10	N	N	N	N	(.61)	(.71)	.81	.902	.962	.9902	.99890
30	N	N	N	(.37)	.48	.61	.76	.888	.960	.9901	.99890
100	N	N	N	(.18)	.27	.41	.62	.84	.952	.9895	.99889
300	N	N	N	N	(.12)	.22	.41	.73	.930	.9879	.99887
1000	N	N	N	N	N	.08	.19	.50	.860	.9825	.99880
C) Mean-square failure—times 1.01 ($= 1 + \delta$)											
$n = 5$	N	N	N	N	N	N	N	N	N	.923	.9901
10	N	N	N	N	N	N	N	(.64)	.78	.922	.9901
30	N	N	N	N	N	(.36)	.46	.61	.77	.921	.9901
100	N	N	N	N	N	(.20)	.31	.50	.73	.917	.9900
300	N	N	N	N	N	N	.16	.33	.64	.904	.9898
1000	N	N	N	N	N	N	(.06)	.15	.45	.864	.9891

Entries of N almost always, and entries in parentheses usually, indicate that the raw fit is never really appropriate (either because it is too weak (nonsignificant) or because the fit with the straightened carrier is better). (See "Relation to significance/nonsignificance" near the end of this section, for further explanation.)

If we choose to calculate δ, we have

$$\delta = \frac{0.0625}{1.0625} \left(\frac{(1 + (100 - 1)(0.0625))(0.50)}{1 - (1 + 0.0625)(0.5)} - 1 \right) = 0.39.$$

For even a moderate quality of fit (one that practitioners from some fields might think good, while those from others would think horribly poor), we cannot accept $\epsilon = 0.25$ if we want to do a reasonably careful job of fitting.

What if ϵ were 0.1 and the observed r^2_{rawfit} were still 0.50? For $n = 100$, we find from Exhibit 1 that true r^2_{rawfit}'s of 0.971, 0.84, and 0.50, respectively, correspond to $\delta = 1, 0.1$, and 0.01. Thus our supposed observed $r^2_{\text{rawfit}} = 0.50$ falls right at $\delta = 0.01$. If we do not wish to worsen our fit by more than 1%, use of the raw carrier is just acceptable. (If our original fit was somewhat better, $r^2_{\text{rawfit}} = 0.6$ or 0.7, we would fall in between two of these values, worsening by more than 1% and less than 10%, by using the raw carrier; if $r^2_{\text{rawfit}} = 0.84$, we would worsen by 10%.)

The closer the fit to the raw carrier, the better carriers we need to make the fit reasonably effective.

Relation to Significance/Nonsignificance

We understand that raising the quality of the fit, as measured by r^2_{rawfit}, all else being the same, raises the value of δ. For given r^2_{carriers} or ϵ, and given n, then, making r^2_{rawfit} large enough will make the use of the raw carrier unacceptable. Other things happen as we start from $r^2_{\text{rawfit}} = 0$ and make r^2_{rawfit} larger and larger. In particular, the fit—originally nonsignificantly different from zero—first becomes significant at 5% and then significant at 1%.

If we do not take fits seriously until they reach or nearly reach significance, the sequence as fits become closer can be the following:

◇ first, fits not taken seriously, because they are not beyond or close to significance.

◇ then, fits taken seriously but not worth straightening

◇ finally, fits worth straightening.

These three cases will arise if significance comes for a lower r^2_{rawfit} than does the profitability of straightening.

If, on the other hand, the need for straightening manifests itself at a smaller r^2_{rawfit} than does significance, the middle case does not arise, and our choice is only between "don't take it seriously" and "straighten it at once".

Thus it matters quite a lot which happens for a lower r^2_{rawfit}: significance or a need for straightening. If the need for straightening comes first, every fit that we take at all seriously should be redone using the straightened carrier.

Let us review our main ideas. Let us fix, for a moment, (i) sample size and (ii) ϵ^2 or r^2_{carriers} and consider what happens as r^2_{rawfit} changes. If r^2_{rawfit} is small enough, the fit with the raw carrier will not be significant, suggesting that either we forget this kind of fit or try the straightened carrier to see whether it might be significant. If, on the other hand, r^2_{rawfit} is large (close to 1), the value of δ will also be large and there will be a value of r^2_{rawfit} so large that the corresponding value of δ is intolerable. Thus, the range of values of r^2_{rawfit} for which it is reasonable to use the fit by the raw carrier is limited at both ends—at one by nonsignificance, at the other by intolerable δ. As we stiffen our standards of tolerability, decreasing the δ we are able to accept, this usable range of values shrinks, and eventually disappears.

Cases where no useful values of r^2_{rawfit} are possible, using significance at 5% to limit the range of useful values, are shown in Exhibit 1 by the entry of an "N"—to stand for "Nonsignificant or intolerable". We set ( )'s around additional values for which the interval of useful values vanishes when significance at 1% is taken to set its lower limit.

In practice, of course, we have only a sample value of r^2_{rawfit}, and we use this to enter, for example, Exhibit 1, whose entries are population (true) values of r^2_{rawfit}. It has seemed accurate enough to do this, especially since our choice of δ is rarely really precise. We recommend ignoring this further complication in using Exhibit 1.

We can extract some rough guidance in simple form from the values in Exhibit 1 and the appearances of N's and ( )'s:

1. If doubling of the mean-square failure by using the raw carrier is allowable,

   ◇ and if detection of a marginal dependence between response and raw carrier is sought, one barely significant at 5%, we may get away with ϵ almost as large as 0.6 (which means an r^2_{carriers} of about 0.74) (this takes n's like 10 to 100);

   ◇ but if we want to use a much closer raw fit, a larger r^2_{rawfit}, we cannot use nearly so large an ϵ, and must have r^2_{carriers} much closer to 1;

   ◇ and if the fit we seek is of the quality regarded as "not so good" for many experiments in a freshman physics laboratory, corresponding, say, to residuals about 1% as large as $(y - \bar{y})$'s, so that $1 - r^2_{\text{rawfit}} \approx 0.0001$, we will need to hold ϵ down to something like 0.007 (an r^2_{carriers} of 0.99995).

2. If we want a fit of reasonable quality ($\delta = 0.1$),

   ◇ when the closeness between response and raw carrier is marginally significant at 5%—just enough to detect the fit—we may be able to go as far as ϵ a little smaller than 0.3, to an r^2_{carriers} of 0.92 (this takes n's like 30 to 100);

   ◇ where the closeness is much higher, we will need a smaller ϵ, perhaps 0.1 (an r^2_{carriers} of 0.99 between raw and straightened);

⋄when the closeness is that of a nice experiment in freshman physics, corresponding, say, to residuals about 0.1% as large as $(y - \bar{y})$'s, we will need to hold ϵ down to 0.0003 or less (to an r^2_{carriers} of 0.999999 or more between raw and straightened).

3. If we want a very careful fit ($\delta = 0.01$),

⋄where the closeness is marginal, we may be able to go as far as something like $\epsilon = 0.15$, $r^2_{\text{carriers}} = 0.98$ (again for n's like 30 to 100);

⋄when the closeness is much higher, we may need to stay below $\epsilon = 0.05$, and above $r^2_{\text{carriers}} = 0.997$.

The better we try to do—that is, the smaller δ we demand—and the closer the fit (the higher the r^2_{rawfit}) we start with, the more careful we must be. We then need to keep ϵ smaller or keep $(1 - r^2_{\text{carriers}})$ smaller for the relation between raw and straightened carriers, if we are to decide reasonably that the raw carrier will do.

B. The Case of a Very Good Fit

If we deal with very good fits (large r^2_{rawfits}'s), and hence with small ϵ, the ratio

$$\frac{1 - r^2_{\text{rawfit}}}{1 - r^2_{\text{carriers}}}$$

is nearly constant, given δ.

Exhibit 2 shows the values of this ratio for small ϵ and a variety of n's, still for $\delta = 1$, 0.1, and 0.01. To a very adequate approximation:

⋄to do no worse than doubling by using the raw carrier, we have to make sure that $(1 - r^2_{\text{carriers}})$ is no more than one-half of $(1 - r^2_{\text{rawfit}})$;

⋄to do no worse than 10% excess, we have to hold $(1 - r^2_{\text{carriers}})$ to no more than one-tenth of $1 - r^2_{\text{rawfit}}$ (really about one-eleventh);

⋄to do no worse than 1% excess, we have to hold $1 - r^2_{\text{carriers}}$ to no more than about one-hundredth of $(1 - r^2_{\text{rawfit}})$.

In the range of close fits, matters are easily described. We see that we often need

$$1 - r^2_{\text{carriers}}$$

that is substantially or very considerably smaller than

$$1 - r^2_{\text{rawfit}}$$

if we are not to be compelled to change from x_{raw} to $x_{\text{straightened}}$.

C. Do We Need to Take Logs?

So far we have answers in terms of what r^2_{carriers} we can stand. We need answers in even simpler terms, such as these: If we believe $\log x$ is right, will x do? If we believe $\sqrt{x}$ is right, will x do? If we believe $-1/x$ is right, will x do?

Relatively simple answers to these questions rest on the ratio

$$\frac{\text{largest } x}{\text{smallest } x},$$

at least for well-behaved sets of x's.

To be sure of this, we have only to calculate the r^2_{carriers} between a set of $n\,x$'s and the n corresponding values of $\log x$. This needs to be done over a range of values of n—and a variety of shapes of batch. We have chosen to work with 3 basic shapes of batch:

◇ equispaced (squeezed-in tails);

◇ roughly normal (spacing between the ith and $(i + 1)$st ordered value proportional to $1/i(n - i)$);

◇ stretched-out tails (values of the tangent of a suitable equispaced batch).

Each of these can be used to describe either the shape of x or the shape of $\log x$, thus somewhat exploiting the effect of skewness.

Exhibit **2** of Appendix

Ratios of $(1 - r^2_{\text{rawfit}})$ to $(1 - r^2_{\text{carriers}})$ for small ϵ

	$\epsilon = .01$	$\epsilon = .003$	$\epsilon = .001$
A) for doubling, $\delta = 1$			
$n = 5$	2.00	2.00	
1000	2.10	2.01	2.00
3000	2.23	2.02	
B) for 1.1 times, $\delta = 0.1$			
$n = 5$	10.98	11.00	
1000	11.97	11.09	11.00
3000	13.97	11.27	
C) for 1.01 times, $\delta = 0.01$			
$n = 5$	99.0	100.8	
1000	108.7	101.8	101.0
3000	128.0	103.6	101.2

Exhibit **3** displays the batches of sizes 10 and 20. The diversity of shape involved is quite clear. The extremes are probably more extreme than most examples that will arise in practice.

Exhibit **4** shows the apparent dependence of r^2_{carriers} on the ratio

$$\frac{\text{largest } x}{\text{smallest } x}$$

for three groups of situations, only the outermost curves being shown for each—namely, from NW to SE:

Exhibit **3** of Appendix

The three shapes of batch for batches of 10 and 20. (Relative scales not important.)

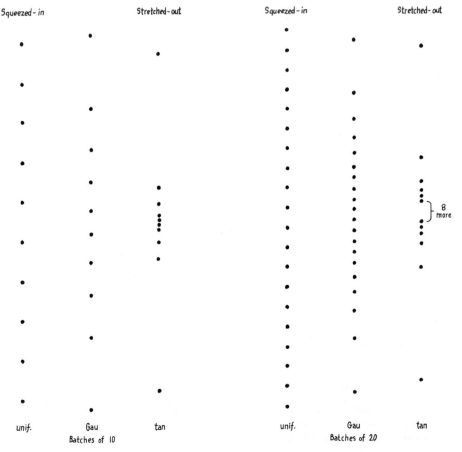

See continuation on next page. ➡

◇stretched-out tails for log x symmetric and $n = 10$, and for x symmetric and $n = 10, 20, 40$ (dashed boundaries)

◇neutral ($\sim$normal) tails for log x symmetric and $n = 10$, and for x symmetric and $n = 5, 10, 20, 40$ (solid boundaries)

◇squeezed-in tails for log x symmetric and $n = 10$, and for x symmetric and $n = 10, 20, 40$ (dotted boundaries).

Exhibit 4 covers ratios of about 15 down to 1.05 and corresponding $r^2_{carriers}$ values from about 0.85 to 0.9999. The values of $r^2_{carriers}$ are sufficiently similar for different n—and their change from $n = 20$ to $n = 40$ is so small—that we feel relatively happy using Exhibit 4 for any value of n that comes along.

Exhibit **3** of Appendix (continued)

Column one: x's are equally spaced.

Column two: x's are spaced proportionally to:

Differences	1/9,	1/16,	1/21,	1/24,	1/25,	1/24,	1/21,	1/16,	1/9
Values	−.2829	−.1718	−.1093	−.0617	−.0200	.0200	.0617	.1093	.1718	.2829

Column three: x's are proportional to:

Values	tan 85°,	tan 66.11°,	tan 44.22°,	tan 28.33°,	tan 9.44°
Values	11.4301	2.2578	1.0807	.5392	.1663

and their negatives

Column four: x's are equally spaced.

Column five: x's are spaced proportionally to:

Diff.	1/100	1/99	1/96,	1/91,	1/84,	1/75,	1/64,	1/51,	1/36,	1/19
Values	±.00500	±.01510	±.02552	±.03651	±.04841	±.06174	±.07737	±.09698	±.12476	±.17739

Column six: x's are spaced like:

Values $\tan 85°$, $\tan \left(\dfrac{17}{19}\right) 85°, \ldots$

Values   ±11.4301, ±4.0265, ±2.3679, ±1.6102, ±1.1589, ±.8470
Values of points missing in the figure (±.6084, ±.4115, ±.2386, ±.0782)

Note that overall scale is neither fixed nor relevant.

For the particular illustrations of worsening (dependent on δ) and of closeness of fit (measured by r^2_{rawfit}) considered in Section A, we find, if log x is the good carrier to use:

1. and if doubling is permitted,

   ◇and marginal fits are to be considered, then (since we saw that r^2_{carriers} might go down to 0.74) we are off the top of Exhibit 4 and might get along with (largest x)/(smallest x) ratios of somewhat more than 20 and not need to re-express;

Exhibit **4** of Appendix

Dependence of r^2_{carriers} on $\dfrac{\text{largest } x}{\text{smallest } x}$ when log x is the straightened carrier. (Solid lines = neutral tails; dashed lines = stretched tails; dotted lines = squeezed tails; see text for details.)

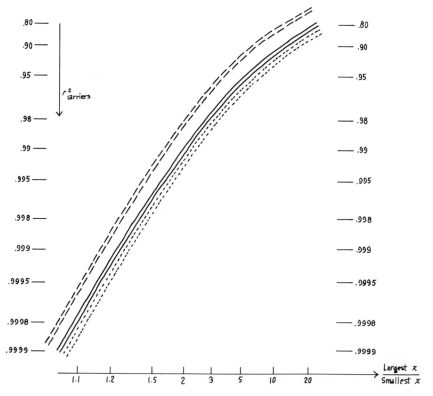

◇but we want to use our raw carrier for closer fits, we will have to have (see Section A) an $r^2_{carriers}$ of perhaps 0.8 or 0.9, and will be in the upper part of Exhibit 4 where one can accept (largest x)/(smallest x) of perhaps 5;

◇and we have quality of fit that would be "not so good" for many freshman physics experiments, calling for $r^2_{carriers} = 0.99995$ (see Section A), then we are off the bottom of Exhibit 4 and have to be uncomfortable at (largest x)/(smallest x) values of 1.05 or 1.1.

2. and if reasonable efficiency ($\delta = 0.1$) is desired,

◇for marginal detectable fits, we are in the upper portion of Exhibit 4 ($r^2_{carriers} = 0.92$ or 0.93) and can live with (largest x)/(smallest x) ratios of perhaps 8 (neutral), 10 (squeezed-in to equispacing), or 4 (highly stretched);

◇for much closer fits, we are nearer the middle of Exhibit 4 ($r^2_{carriers} = 0.99$) and can live with (largest x)/(smallest x) ratios of perhaps 2 (neutral), 2.5 (squeezed-in to equispacing), or 1.8 (highly stretched);

◇when the closeness is that of a nice experiment in freshman physics, so that (see Section A) we need an $r^2_{carriers}$ of perhaps 0.999999, we are far off Exhibit 4 to the left and can hardly expect to use x_{raw}, even if (largest x)/(smallest x) is as small as 1.02.

3. and if we want a very careful fit ($\delta = 0.01$),

◇where the closeness is marginal, we are a little below the top on Exhibit 4 ($r^2_{carriers} - 0.98$) and can live with (largest x)/(smallest x) ratios of perhaps 2.5 (neutral), 3 (squeezed-in to equispacing), or 2 (highly stretched);

◇with closeness much better than marginal, we are below the middle of Exhibit 4 ($r^2_{carriers} = 0.997$) and can live with (largest x)/(smallest x) ratios of perhaps 1.4 (neutral), 1.5 (squeezed-in to equispacing), or 1.3 (highly stretched).

Clearly the ratio of largest x/smallest x we can live with depends very much upon how good a fit we have—and, to an only somewhat lesser extent, on how careful we wish to be, that is, on how small we want to keep δ.

D. And What About $\sqrt{x}$ and $-1/x$?

We have answered in some detail when $\log x$ "should" be used. What if $\sqrt{x}$ "should" be used? Or $-1/x$?

Exhibit 5 shows a set of curves for all three cases. For simplicity, only the squeezed-in and stretched-out curves are shown. In view of our finding for log x, that batch size does not matter very much, only the values for $n = 10$ are shown. To make the relation to Exhibit 1 and its accompanying discussion as close as possible, the vertical scale is given in terms of ϵ (rather than $r^2_{\text{carriers}} = 1/(1 + \epsilon^2)$). This exhibit, when combined with Exhibit 1 and its accompanying discussion, tells us how important it may be to re-express an x, once we know what re-expression would be sound.

Exhibit **5** of Appendix

Relation between ϵ and (largest x)/(smallest x) when $-1/x$, log x, or $\sqrt{x}$ is the correct expression. (Drawn for $n = 10$; useful for most reasonable n.)

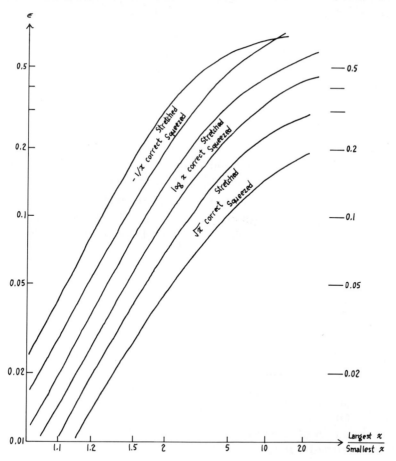

Thus, for one example, if $\epsilon = 0.1$ is the quality we require, we can probably stand omitting re-expression for ratios of (largest x)/(smallest x) no larger than:

◇from 1.3 (stretched) to 1.4 (squeezed) if $-1/x$ is sound;

◇from 1.7 (stretched) to 2.0 (squeezed) if $\log x$ is sound;

◇from 3.5 (stretched) to 5 (squeezed) if $\sqrt{x}$ is sound.

Turning back to Exhibit 1 we see that $\epsilon = 0.1$ corresponds to:

◇doubling the mean-square failure by using the raw carrier above $r_{\text{rawfit}}^2 = 0.98$ ($n \approx 5$ or 10), 0.97 ($n \approx 100$), or 0.95 ($n \approx 300$).

◇ adding 10% to the mean-square failure above $r_{\text{rawfit}}^2 = 0.90$ ($n \approx 10$), 0.84 ($n \approx 100$), or 0.73 ($n \approx 300$).

◇adding 1% to the mean-square failure above $r_{\text{rawfit}}^2 = 0.60$ ($n \approx 30$), 0.50 ($n \approx 100$), or 0.33 ($n \approx 300$).

Proceeding similarly, we can get reasonable guidance in almost any situation. In particular:

◇although the method was derived for single-carrier regressions, we have no qualms about using this guidance for multiple regression.

◇if we need an $\epsilon < 0.01$, we would re-express without stopping to look at Exhibit 5.

◇while some extreme circumstances may appear to let us use ϵ's of more than 0.5, we do not expect to try to take advantage of them—if we don't have $\epsilon < 0.5$, we will re-express anyway.

◇when looking at Exhibit 5 still leaves us in doubt, we go ahead and re-express.

E. Derivation

We return now to our prototype situation, and develop the formulas. Let us suppose that:

1. x and z have zero means, equal variances, say 1, and zero covariance;

2. x should be the carrier used, but we are planning to use $x + \epsilon z$ instead, where ϵ measures the amount of irrelevance introduced. Thus, $x = x_{\text{straightened}}$, $x + \epsilon z = x_{\text{raw}}$, in our previous language. Then

3. how much do we lose in quality of estimation of a regression coefficient b by using $x + \epsilon z$ instead of x?

We lose no generality by assuming that the mean, 0, and variance, 1, is common to both x and z, and, further, that the response we are regressing is x itself, perturbed by independent errors σe_i of mean 0 and variance σ^2, so that the desired value of the regression coefficient, b, is 1.00. That is, the response is

$$y_i = x_i + \sigma e_i, \qquad i = 1, 2, \ldots, n.$$

We shall judge the quality of estimate of b, not by the mean square error, which is

$$(\text{bias of } b)^2 + (\text{actual variance of } b),$$

but rather by a quantity we call the *mean-square failure*

$$(\text{bias of } b)^2 + \text{ave}(\text{apparent variance of } b),$$

because we lose when we think the variance of b is greater than it actually is. By ave( ), we mean the average (expected value) of the quantity in parenthesis.

We use the average of the apparent variance because any systematic dependence of y on x that cannot be accounted for by a linear dependence has to contribute to the residuals, $y - \hat{y}$, and hence, to the sum of their squares. Thus, the s^2 upon which our conventional estimate of var $\{b\}$ is based will have

$$\text{ave } \{s^2\} = \sigma^2 + \frac{1}{n-1} \sum (\text{systematic deviations})^2,$$

and we will thus ordinarily have

$$\text{ave } \{s^2\} > \sigma^2,$$

where σ^2 is the average variance of y (around the sum of straight-line fit and systematic deviation).

Since we do not know, in practice, how the average of s^2 is divided between σ^2 and the systematic part, we have to suffer with an apparent variance for b whose average value is greater than the actual variance of b. The mean-square failure takes some account of this suffering (while the mean-square error does not).

We now derive an expression for the mean-square failure under our hypotheses. By definition, we have

$$b = \frac{\sum (x_i + \sigma e_i)(x_i + \epsilon z_i)}{\sum (x_i + \epsilon z_i)^2}.$$

We have a fixed set of x's and z's, so that our assumptions about means, variances, and covariances imply that

$$\sum x_i = \sum z_i = 0, \qquad \frac{\sum x_i^2}{n-1} = \frac{\sum z_i^2}{n-1} = 1, \qquad \sum x_i z_i = 0.$$

To get the expected value of b, we merely multiply out and sum up terms in numerator and denominator, and then take expectations in the numerator.

Thus

$$\text{ave } b = \text{ave} \left[\frac{\sum x_i^2 + \sigma \sum x_i e_i + \epsilon \sum x_i z_i + \sigma \epsilon \sum e_i z_i}{\sum x_i^2 + 2\epsilon \sum x_i z_i + \epsilon^2 \sum z_i^2} \right]$$

$$= \frac{\sum x_i^2}{\sum x_i^2 + \epsilon^2 \sum z_i^2} = \frac{1}{1 + \epsilon^2}.$$

Because we have selected the true coefficient as 1,

$$\text{bias in } b = 1 - \text{ave } b = \frac{\epsilon^2}{1 + \epsilon^2}.$$

As we know from Chapter 14, the variance of the regression coefficient is the residual variance divided by the sum of squares of the carrier, and so

$$\text{var } b = \frac{\sigma^2}{\sum (x_i + \epsilon z_i)^2} = \frac{\sigma^2}{(n - 1)(1 + \epsilon^2)}.$$

The residuals. The residuals become

$$x_i + \sigma e_i - b(x_i + \epsilon z_i) = (1 - b)x_i + \sigma e_i - b\epsilon z_i.$$

We need to square, take expected values, and sum over i. Squaring gives

$$(1 - b)^2 x_i^2 + \sigma^2 e_i^2 + b^2 \epsilon^2 z_i^2 + 2\sigma(1 - b)x_i e_i - 2(1 - b)b\epsilon x_i z_i - 2\sigma b\epsilon e_i z_i. \tag{1*}$$

Taking expected values and summing the first three terms of (1*) gives:

$$\sum x_i^2 \text{ ave } (1 - b)^2 + \sigma^2 \text{ ave } \sum e_i^2 + \epsilon^2 \sum z_i^2 \text{ ave } b^2. \tag{2*}$$

Among the final three terms of (1*), the middle term sums to zero because $\sum x_i z_i = 0$, according to our covariance assumption. The remaining terms can be regrouped as:

$$2\sigma x_i \epsilon_i - 2\sigma(x_i + \epsilon z_i)b e_i.$$

The term $2\sigma x_i e_i$ has expected value zero because ave $e_i = 0$ and x_i is fixed. We need, then, deal only with

$$-2\sigma(x_i + \epsilon z_i)b e_i.$$

We can write the expected value of the sum as:

$$-2\sigma \sum (x_i + \epsilon z_i) \text{ cov } (e_i, b), \tag{3*}$$

where the term for the product of the means of e_i and b vanishes because ave$(e_i) = 0$.

Assembling our results, we get the expected sum of squares as:

$$\sum x_i^2 \text{ ave } (1 - b)^2 + \sigma^2 \text{ ave } \sum e_i^2$$
$$- 2\sigma \sum (x_i + \epsilon z_i) \text{ cov } (e_i, b) + \epsilon^2 \left(\sum z_i^2 \right) \text{ ave } b^2. \tag{4*}$$

Let us take up the terms one by one. In the first term, $\sum x_i^2 = n - 1$, and we already know the expected value of b and the variance of b.

$$\text{ave } (1 - b)^2 = \{\text{ave } (1 - b)\}^2 + \text{var } b$$
$$= \left(\frac{\epsilon^2}{1 + \epsilon^2}\right)^2 + \text{var } b$$

And so the first term of (4∗) becomes

$$(n - 1)\left[\left(\frac{\epsilon^2}{1 + \epsilon^2}\right)^2 + \frac{\sigma^2}{(n - 1)(1 + \epsilon^2)}\right]. \tag{1}$$

In the second term of (4∗), ave $e_i^2 = 1$, and so the term has value

$$n\sigma^2. \tag{2}$$

To evaluate the third term of (4∗), we need an expression for cov (e_i, b). When we expand the sum in the numerator of b, we have the sum of terms like

$$x_j^2 + \sigma e_j x_j + \epsilon x_j z_j + \sigma \epsilon e_j z_j, \qquad j = 1, 2, \ldots, n.$$

These must be multiplied by e_i, and the expected value taken. The only value of j that makes a contribution is $j = i$, and only the second and fourth terms contribute. They give, since ave $e_i^2 = 1$,

$$\sigma(x_i + \epsilon z_i),$$

and so the required covariance is

$$\text{cov } (e_i, b) = \frac{(x_i + \epsilon z_i)\sigma}{\sum (x_i + \epsilon z_i)^2}.$$

We use this to reduce the third term of (4∗), and we get:

$$-2\sigma^2 \tag{3}$$

The fourth term of (4∗) gives

$$(n - 1)\epsilon^2\left(\frac{\sigma^2}{(n - 1)(1 + \epsilon^2)} + \left(\frac{1}{1 + \epsilon^2}\right)^2\right). \tag{4}$$

Adding up (1), (2), (3), and (4) and simplifying gives, as our expected sum of squares of residuals:

$$(n - 1)\left(\frac{\epsilon^2}{1 + \epsilon^2} + \sigma^2\right).$$

If s^2 is the sample variance of the residuals, then its expected value is

$$\text{ave } s^2 = \sigma^2 + \frac{\epsilon^2}{1 + \epsilon^2}.$$

The expected value of the estimated variance of b is

$$\text{ave} \frac{s^2}{\sum (x_i + \epsilon z_i)^2} = \frac{\sigma^2 + \dfrac{\epsilon^2}{1 + \epsilon^2}}{(n - 1)(1 + \epsilon^2)}.$$

And so, at last, the

$$\text{mean-square failure} = (\text{bias of } b)^2 + \text{ave (estimated variance of } b)$$

$$= \left(\frac{\epsilon^2}{1 + \epsilon^2}\right)^2 + \frac{\sigma^2 + \dfrac{\epsilon^2}{1 + \epsilon^2}}{(n - 1)(1 + \epsilon^2)}. \tag{MSF}$$

Size of the irrelevant contribution. One way to think about the mean-square failure is as a proportional change from what we would have had under perfection, namely, when $\epsilon = 0$. Then the mean-square failure is $\sigma^2/(n - 1)$. We can write the mean-square failure as $(1 + \delta)$ times this quantity, and then δ measures the proportional increase.

Writing

$$(1 + \delta) \frac{\sigma^2}{n - 1} = \left(\frac{\epsilon^2}{1 + \epsilon^2}\right)^2 + \frac{\sigma^2 + \dfrac{\epsilon^2}{1 + \epsilon^2}}{(n - 1)(1 + \epsilon^2)},$$

we can solve for σ^2 and simplify, to get:

$$\sigma^2 = \frac{n - 1 + \dfrac{1}{\epsilon^2}}{\delta \left(\dfrac{1 + \epsilon^2}{\epsilon^2}\right)^2 + \dfrac{1 + \epsilon^2}{\epsilon^2}}. \tag{5*}$$

Correlations. We want to find out the proportion of the variance of y explained by regression given a value for ϵ^2. To make the relation, first recall that, without error, the variance of x is 1 and that, with error, it is $(1 + \sigma^2)$. And so we write

$$\text{mean-square residual} = (1 - r_{\text{rawfit}}^2)(\text{variance of } y)$$

or

$$\sigma^2 + \frac{\epsilon^2}{1 + \epsilon^2} = (1 - r_{\text{rawfit}}^2)(1 + \sigma^2).$$

Solving this for r_{rawfit}^2 gives:

$$r_{\text{rawfit}}^2 = \frac{1}{(1 + \epsilon^2)(1 + \sigma^2)}. \tag{6*}$$

where σ^2 is a function of n, δ, and ϵ, as shown in (5*).

Expression for δ. Solving (5*) for δ gives

$$\delta = \left(\frac{(n-1) + \dfrac{1}{\epsilon^2}}{\sigma^2} - \frac{1 + \epsilon^2}{\epsilon^2}\right)\left(\frac{\epsilon^2}{1 + \epsilon^2}\right)^2$$

$$= \frac{\epsilon^2}{1 + \epsilon^2}\left(\frac{(n-1)\epsilon^2 + 1}{(1 + \epsilon^2)\sigma^2} - 1\right).$$

Solving (6*) for σ^2 gives

$$\sigma^2 = \frac{1 - r_{\text{rawfit}}^2 - \dfrac{\epsilon^2}{1 + \epsilon^2}}{r_{\text{rawfit}}^2}.$$

So

$$(1 + \epsilon^2)\sigma^2 = (1 + \epsilon^2)\frac{1 - r_{\text{rawfit}}^2 - \dfrac{\epsilon^2}{1 + \epsilon^2}}{r_{\text{rawfit}}^2}$$

$$= \frac{1 - (1 + \epsilon^2)r_{\text{rawfit}}^2}{r_{\text{rawfit}}^2}.$$

Therefore we can write

$$\delta = \frac{\epsilon^2}{1 + \epsilon^2}\left(\frac{(1 + (n-1)\epsilon^2)r_{\text{rawfit}}^2}{1 - (1 + \epsilon^2)r_{\text{rawfit}}^2} - 1\right), \qquad (7*)$$

which can be converted to

$$\delta = (1 - r_{\text{carriers}}^2)\left(\frac{1 + (n-1)(1 - r_{\text{carriers}}^2)}{r_{\text{carriers}}^2 - r_{\text{rawfit}}^2}\right)r_{\text{rawfit}}^2.$$

Summary: When is Re-expression Profitable?

We can assess whether to bother to go from x to $\log x$ (or to $\sqrt{x}$, or to $-1/x$) when we know that re-expressing the carrier is a good thing. The questions that have to be answered are: How good a thing? Is the good worth the effort, often mostly the effort of explanation?

We approach the first question in terms of two correlations: r_{carriers}, the correlation between x and $\log x$ (or the other preferred carrier); and r_{rawfit}, the correlation between x (or the several x's) and y.

Over much of the range, the answer is: "It will pay considerably to change, unless $1 - r_{\text{carriers}}^2$ is only a small fraction of $1 - r_{\text{rawfit}}^2$."

Exhibit 1 gives more detail, and may help us decide when this rule is not good enough.

We assess the size of r_{carriers}^2 well enough by noticing the value of (largest x)/(smallest x) and turning to Exhibit 4 (for $\log x$ only) or Exhibit 5.

Problems

Problems are numbered by chapter and section, and by problem number within the section. Thus 7C4 indicates Problem 4 of Section 7C.

A superscript C as in 2B4^C means that if this problem is assigned it might be helpful for the group to discuss its solution.

A starred problem * means that the solution may require the student to have more than the minimal preparation for the course.

A P as in 16P4 means the problem is a project.

A Q as in 14Q1 is a problem directly tied to another chapter.

Exhibits for problems—for example, Exhibit 1 for Problem 1E3—often appear near the problems. We also have several sets of data in the section entitled Data Exhibits for Problems that appears at the end of the problems.

Chapter 1

1A1) State four stages in organizing data analysis, and suggest some pros and cons of traversing through these stages in opposite directions.

1A2) a) What staircase of steps did Student shortcut, and
b) What did he use in place of data?

1A3) Discuss whether, and when, Student's shortcut was a good idea.

1A4) Give three different instances, putting each in the framework of a specific application, of Student's t. Discuss the possible "contemplated values" in each case. (Make clear what a contemplated value is.)

1A5*) Distinguish "nonparametric" and "distribution-free" procedures.

1B1) For about how many degrees of freedom does Student's t give a 95% confidence limit 2.29 times as long as a $\frac{2}{3}$ confidence limit?

1B2) Find the smallest value of

$$\frac{\text{Length of 95\% confidence limit}}{\text{Length of 2/3 confidence limit}}$$

offered by Student's t. What sort of degrees of freedom make this ratio largest? What are its three largest values (for integer numbers of degrees of freedom)?

1B3) Find the standard confidence points, based on Student's t, for (a) 15 degrees of freedom, (b) 60 degrees of freedom (see Note 2 of Exhibit 1 of Chapter 1).

1B4) In what did the great value of Student's work lie? In notable changes in critical values?

1B5) Discuss the drawbacks—then, now, and in the future—of Student's step ahead.

1C1) Is a Gaussian distribution "normal"? Why and why not?

1C2) Do all normal distributions have the same "shape"?

1C3) Sketch roughly the probability density function of the normal distribution with $\mu = 1$, $\sigma = 2$. Be sure to label the axes. (Exhibit 3 of Chapter 1 should help.)

1C4) Name a family of distributions that contains distributions with widely differing shapes.

1C5) When are the tails of a distribution said to be "more stretched out"? When "more straggly"? What implicit standard of comparison underlies such verbal expressions?

1C6) Can we rely on either the center or the tails of a distribution to tell us about the other? Why/why not? When yes/when not?

1D1) For the normal distribution what are Q_1 and Q_3? For a uniform distribution running from 0 to 1?

1D2) Precisely what are skewness and kurtosis?

1D3) In large samples, about what fraction of observations lie within $0.25s$ of $\bar{x}$? Beyond $3.1s$?

1D4) *Class project.* Find a large sample of measurements and try out some techniques used by Wilson and Hilferty on Peirce's data. (A source is: Michelson, A. A., F. H. Paese, and F. Pearson (1935). "Measurement of the velocity of light in a partial vacuum." *Astrophysical Journal*, Vol. 82, 26–61. See Table VI, 51–54.)

1E1) Do you think the irregularities in Exhibit 6 of Chapter 1 are important? Why/why not? When/when not?

1E2) What happens at the ends of the curves in Exhibit 7 of Chapter 1? Answer the same question for Exhibit 8 of Chapter 1.

1E3) Why may deviations from the normal shape be important in the tails?

1E4) What do we prefer to safety in statistical analyses?

1E5) Name and explain at least two kinds of robustness.

1E6[c]) What is a "wild shot"?

1E7) How is behavior not in the tails related to behavior in the tails? Is this important? Why/why not?

1E8^C) How much difference do you think 1% in efficiency makes?

1E9) A neurotic throws away half the values in his/her sample at random. How efficient is the mean of the half that is left relative to the mean of those thrown away? Relative to the mean of all the observations?

1F1) List and discuss three vague statistical concepts, at least one of which is not mentioned in Section 1F or 1G.

1G1^C) List the pros and cons of the sample median as a numerical summary.

1G2) Do 1G1 for the sample mean instead of the sample median.

1G3) List three instances of indications, two of which are not discussed in section 1G.

1H1) Discuss the value of data analysis taken only as far as indication.

1H2) Can chi-square, used as an indication of poor fit, be converted into an estimator?

1H3) Distinguish between indication, determination, and inference.

Chapter 2

2A1) Why is the recipe of the second paragraph of Section 2A "a complete flight from reality"?

2A2) List with explanations three instances of indication not covered, directly or implicitly, in Section 2A.

2B1) Find three selected instances of indication in your daily paper. Discuss the importance or lack of importance of not stopping with indication in each case.

2B2) Find, and discuss, two instances of problems of multiplicity that are of particular interest to you.

2B3) A careful investigator tested each of 137 chemicals (in mice) for their possible effect in making the occurrence of one kind of cancer less frequent. Each chemical was compared separately with an inert material, so that $2 \times 137 = 274$ groups of mice were involved. The results were 15 chemicals at least significant at 5%, 2 at least at 1%, 1 at least at 0.1%. Discuss the investigator's results. What do they seem to indicate?

2B4^C) An agronomist tested out 11,273 new strains of a plant, obtained by hybridization. When arranged in order, the individual significance of certain strains were: best 0.03%, 5th best 0.17%, 25th best 0.72%, 125th best 2.8%, 625th best 10.3%. Another agronomist tested 1492 strains of another plant with substantially the same results for the significance of the best, 5th best, ... If you had charge of deciding (a) whether either agronomist were to retest some selected strains of his plant and (b) how many were to be retested, what would you do?

2B5) Given a single body of data to be used for both exploration and confirmation, would you choose one-third for exploration and hint-searching, two-thirds for confirmation as a proper balance? Why/why not?

2B6) Ten investigators study the same problem; no one finds an individually significant result, but 9 of the 10 lean in the same direction. What would you conclude?

2C1) Find at least three instances of concealed "informal" inference in print. Copy out the relevant sentences and discuss why you do or do not find the inference "obvious".

2D1) Is the mean of a population a parameter? In which sense?

2D2) Can the family of Gaussian (normal) distributions be parametrized by μe^{σ^2} and $\log \mu$? By μe^{σ^2} and $\sqrt{\mu^2 + \sigma^2}$? Why/why not?

2D3) If you knew just what question you wanted to ask, and that each of two estimators would provide estimates responsive to this question, what else would you like to know to guide your choice between these two estimates? Would the relative amounts of computing effort required influence your decision? Why/why not? When/when not?

2D4) Which of these functions of χ^2, used as a measure of goodness of fit (for a total of n counts) to a model of equal numbers in each of $k = \nu + 1$ categories, has an estimand:

$$\chi^2, \quad \chi^2/k, \quad \sqrt{n}(\chi^2 - \nu), \quad (\chi^2 - \nu)/\sqrt{n}, \quad (\chi^2 - \nu)/n\nu$$

Why/why not/which estimand?

2D5) Suppose an estimator: (1) in samples of two always gives either the answer "0" or the answer "∞" while (2) in largish samples (say, ≥100) it gives answers close to μe^{σ^2}. What is its estimand in samples of two? In samples of 2000? Explain your answers.

2E1) Suppose the midmean (a trimmed mean with 25% trimmed from each end) is used instead of the 10% trimmed mean discussed in Section 2E. What Gaussian efficiency would we expect? What is the ratio of efficiency for the midmean to efficiency for a 10% trimmed mean in the Gaussian situation?

2E2) (Continues 2E1) If the situation deviates from Gaussian shape in the direction of more straggling tails, what do you think will happen to the distribution of efficiencies?

2E3^C) Find a table of (uniform) random numbers and use Exhibits 1 and 2 for Problem 2E3 to generate samples of 20 from both the extremely-stretched-tail distribution of Exhibit 1 for Problem 2E3 and the stretched-tail distribution of Exhibit 2 for Problem 2E3. (Each individual participating should prepare an agreed-upon number of samples, at least 3, and should use a different part of the random-number table.) For each sample calculate (a) its median, (b) its midmean (see Problem 2E1), (c) its 10% trimmed mean. Collect the class's results for each of the six combinations of two tail stretchings with three estimators. Discuss the results and decide what comparative performance of the three estimates is indicated for each degree of tail stretching. (SAVE BOTH SAMPLES AND ESTIMATES, tagged by individual responsible, FOR LATER USE.)

Exhibit **1** for Problem 2E3

Table for constructing samples from an extremely stretched-tailed distribution from random strings of digits

   A) MODE OF USE

Draw digits from a table of uniform random numbers until the string is neither all nines nor all zeros. Example (not very random): 1, 7, 2, 4, 90, 001, 4, 6, 94, 98, 07, 0000001, 999992. Here the table below is entered first with 1 giving −220, then with 7 giving .24, then with 2, then with 4, then with 90 giving 2.2×10^2, then 001 giving L, and so on.

   B) TABLE—001 to 998 (values marked "L" are $\leq -10^{100}$, those marked "H" are $\geq +10^{100}$).

Enter		Enter		ENTER		Enter		Enter	
000?	L		←		←	90	2.20×10^2	990	2.69×10^{41}
001	L	01	-2.69×10^{41}	1	−220.	91	6.69×10^2	991	1.80×10^{44}
002	L	02	-5.18×10^{19}	2	−1.45	92	2.68×10^3	992	1.94×10^{52}
003	L	03	-3.00×10^{12}	3	−.24	93	1.60×10^4	993	1.10×10^{60}
004	L	04	-7.20×10^8	4	−.07	94	1.73×10^5	994	2.41×10^{70}
005	-7.33×10^{84}	05	-4.85×10^7	5	.00	95	4.85×10^7	995	7.33×10^{84}
006	-2.41×10^{70}	06	-1.73×10^5	6	.07	96	7.20×10^8	996	H
007	-1.10×10^{60}	07	-1.60×10^4	7	.24	97	3.00×10^{12}	997	H
008	-1.94×10^{52}	08	-2.68×10^3	8	1.45	98	5.18×10^{19}	998	H
009	-1.80×10^{44}	09	-6.69×10^2	→		→		999?	H

   C) EXTREME VALUES—rarely needed

If numerical values are needed for cases marked "L" or "H" in panel B, convert strings of digits to a fraction u, and use

$$\text{Value} \doteq -e^{1/u}/100 = -10^{(.434/u)-2} \qquad \text{for} \quad u \leq .004,$$

$$\text{Value} \doteq e^{1/(1-u)}/100 = 10^{(.434/(1-u))-2} \qquad \text{for} \quad u \geq .996,$$

where exponents can be taken to the nearest integer with adequate accuracy. Thus:

String	Fraction u	.434/[u or $1-u$]	Rounded MINUS 2
001	.001	434.0	432
0000001	.0000001	434000.0	433998
99999	.99999	4340.0	4338
003	.003	144.7	143

   D) UNDERLYING FORMULA—using fraction "u"

$$\text{Value} = \frac{1}{100} \left(e^{1/(1-u)} - e^{1/u} \right)$$

2E4^C) Make up a further table, like that of Exhibits 1 and 2 of Problem 2E3, but less extreme.

2E5^C) (Continues 2E4) Use the table constructed in 2E4 in the same way that Exhibits 1 and 2 were used in 2E3. (SAVE RESULTS.)

2F1) A sports writer claims that, to predict the difference in the number of runs between two teams in a particular game, you should take the mean difference for all the other games between these two teams played in the same season. Given a season's detailed record, can we cross-validate this procedure? If so, what exactly would we do? If not, why not?

2F2) Another sports writer proposes to use the same procedure for the difference in points between two football teams. Same questions.

2F3) What differences would it make in 2F1—or in 2F2—if only games already played were used for prediction?

Exhibit **2** for Problem 2E3

Table for constructing samples from a stretched-tailed distribution from random strings of digits

A) MODE OF USE—see Panel A of Exhibit 1 for Problem 2E3.

B) TABLE—0001 to 9998 (for extreme values see panel C).

Enter		Enter		Enter		ENTER		Enter		Enter		Enter	
0000?	L		←		←		←	90	.98	990	100	9990	1.00×10^4
0001	-1.00×10^6	001	−10000	01	−100.	1	−.98	91	1.22	991	123	9991	1.23×10^4
0002	-2.50×10^5	002	−2500	02	−25.0	2	−.23	92	1.55	992	156	9992	1.56×10^4
0003	-1.11×10^5	003	−1111	03	−11.1	3	−.09	93	2.03	993	204	9993	2.04×10^4
0004	-6.25×10^4	004	−625	04	−6.25	4	−.03	94	2.77	994	278	9994	2.78×10^4
0005	-4.00×10^4	005	−400	05	−3.99	5	.00	95	3.99	995	400	9995	4.00×10^4
0006	-2.78×10^4	006	−278	06	−2.77	6	.03	96	6.24	996	625	9996	6.25×10^4
0007	-2.04×10^4	007	−204	07	−2.03	7	.09	97	11.1	997	1111	9997	1.11×10^5
0008	-1.56×10^4	008	−156	08	−1.55	8	.23	98	25.0	998	2500	9998	2.50×10^5
0009	-1.23×10^4	009	−123	09	−1.22	→		→		→		9999?	H

C) EXTREME VALUES—RARELY NEEDED

Strings 0000 . . . : Multiply values in lefthand column of panel B by one factor of 100 for each additional 0 beyond three.
Strings 9999 . . . : Multiply values in righthand column of panel B by one factor of 100 for each additional 9 beyond three.

D) FORMULA

With u, the corresponding fraction between 0 and 1, we have

$$\text{Value} = \frac{1}{100}\left(\frac{1}{(1-u)^2} - \frac{1}{u^2}\right)$$

2F4^c) What would double cross-validation mean in the examples of 2F1 and 2F2?

2F5) Collect 10 (x, y) pairs of numbers that interest you (not perfectly linearly related). Fit $y = a + bx$ by the best method you know. Then divide the 10 pairs, randomly, into halves, fitting $y = a + bx$ to each half and cross-validating it on the other. How much has $\sum(y_{observed} - y_{fitted})^2$ changed, in each half, from one situation to the other? How would you explain this?

2F6) (Continues 2F5 but with more extensive calculations) Fit $y = a + bx$ to each set of 9 out of 10 of the pairs of 2E5. Cross-validate each on the pair left out. Compare results with those of 2F5.

2F7) Do the same as in 2F5, where (x, y) now equals $(i, (i - 5.5)^2)$ for $i = 1, 2, \ldots, 10$.

2F8) (Continues 2F7, but with heavier calculations) Do the same as in 2F6 where (x, y) now equals $(i, (i - 5.5)^2)$ for $i = 1, 2, \ldots, 10$.

2F9^c) Suppose we have eight values $a, b, c, \ldots, h$; in how many different ways, each division symmetrically related to each of the others, can you divide them 4 and 4 (into halves)?

2F10) What advantages are shared by all objectively formalized methods of fitting—and by no methods that fail to be "objective"? Discuss.

Chapter 3

3A1) In

```
2 |01
3 |
4 |67
```

what are the stems? What are the leaves? What set of values is represented?

3A2, 3A3, 3A4) Make up the stem-and-leaf display corresponding to the population of the 50 U.S. states plus the District of Columbia in (3A2) 1960, (3A3) 1965, and (3A4) 1970 (Data Exhibit 3 for Problems).

3A5) (Continues 3A2–3A4) Compare the stem-and-leaf displays of 3A2, 3A3, and 3A4.

3A6) Construct the stem-and-leaf displays corresponding to births and to deaths in the 50 U.S. states from 1960–1970 (Data Exhibit 3 for Problems).

3A7) Compare the stem-and-leaf displays of 3A6 with each other and with those of 3A2–3A4, if you did them, expressing similarities and differences in words.

3A8) Exhibit **1** for Problem 3A8 shows the sizes of some of the cities in the U.S. with an estimated population of more than 250,000 as of July 1, 1973. Make up the corresponding stem-and-leaf display with unsplit stems, but changes in stem depth (as in Exhibit 2A of Chapter 3).

3B1) What is a median?

3B2) What is the median value of 24 values? of 25 values?

3B3) Find the median values of the batches in Panel B of Exhibit 1 and Panel B of Exhibit 2 of Chapter 3.

3B4) Find the hinges and eighths of the batches in Problems 3A2 and 3A4.

3B5) Complete a skeleton letter-value display (like Exhibit 3 of Chapter 3) for the data of Exhibit 1 of Chapter 3.

3B6, 3B7, 3B8) (continues 3A4) Complete a skeleton letter-value display for the data (3B6) of 3A2, (3B7) of 3A3, and (3B8) of 3A4.

Exhibit **1** for Problem 3A8

Population of U.S. cities above 250,000 in 1973.

Metropolitan area	Population July 1, 1973 (in thousands)	Metropolitan area	Population July 1, 1973 (in thousands)
Akron, Ohio	677	Hartford, Connecticut	733
Albany, New York	800	Honolulu, Hawaii	686
Albuquerque, N. Mex.	376	Houston, Texas	2168
Atlanta, Georgia	1748	Indianapolis, Ind.	1137
Austin, Texas	375	Jackson, Mississippi	275
Baltimore, Maryland	2128	Jacksonville, Florida	661
Birmingham, Alabama	787	Jersey City, N.J.	598
Boston, Mass.	2898	Kansas City, Mo.–Kans.	1299
Bridgeport, Connecticut	397	Lancaster, Pa.	335
Buffalo, New York	1345	Las Vegas, Nevada	308
Charleston, S.C.	352	Lexington, Kentucky	282
Charleston, W. Va.	256	Los Angeles, California	6924
Chicago, Illinois	7002	Louisville, Ky.–Ind.	886
Cincinnati, Ohio	1383	Madison, Wis.	301
Cleveland, Ohio	2006	Miami, Florida	1370
Columbia, S.C.	349	Milwaukee, Wis.	1417
Columbus. Ohio	1057	Minneapolis–St. Paul, Minn.	2000
Dayton, Ohio	848	Nashville, Tenn.	732
Denver, Colorado	1377	Nassau–Suffolk, N.Y.	2630
Detroit, Michigan	4446	New Haven, Connecticut	415
El Paso, Texas	390	New Orleans, La.	1083
Erie, Pennsylvania	273	New York, N.Y.–N.J.	9739
Fort Lauderdale, Florida	756	Newark, N.J.	2053
Fresno, California	435	Philadelphia, Pa.–N.J.	4806
Harrisburg, Pa.	425	Phoenix, Arizona	1127
		Pittsburgh, Pa.	2365

3C1) Give a letter-value display, including spread, for the U.S. population in 1965 and the births and deaths during 1960–1970 (see batches in 3A2 and 3A4).

Re 3D. Data Exhibit 1 for Problems (at the end of Problems) gives various useful kinds of data for 152 cities of 25,000 or more (in 1960) in three census regions—West North Central, West South Central, and Mountain—of the United States. Each solver's next task will be to select his or her own subsamples of (about) 50, of (about) 20, and of (about) 5 from this collection. Before doing this, however, a few questions are best answered.

3D1[c]) If you have access to a table of random numbers (random decimal digits), how can you use such a table to draw out the subsamples just mentioned?

3D2[c]) If all you have are index cards and implicit confidence (probably unwarranted) in your own ability to shuffle cards until they are random, how could you prepare to get subsamples as nearly random (still of sizes close to 50, 20, and 5) as your shuffling will permit?

3D3) Two analysts, who lacked both random numbers and confidence in their shuffling, went about things in the following way: First, keeping the order of the cities in Data Exhibit 1 for Problems, they divided the 152 into 51, 50, and 51. Then they divided each 51, still in the same order, into three 17's, and each 50 into 17, 16, and 17. Then they thirded these into either 6, 5, and 6, or 6, 5, and 5.

One of them claimed that taking the third 51, the fifth 16, and the 7th 6 was a satisfactory way to pick out subsamples. The other argued that there were obviously too many Texas cities in the subsample of 49, and that everyone knew that Texas was different. A better approach to getting a subsample of 50 or 51, she claimed, was to begin with the 1st, 2nd, or 3rd city in the list, choosing among these "at random" and then to take this city and every third city thereafter.

You are now to give judgments stating not only which you prefer, but what pros and cons you can see for each—and whether you would prefer still another approach.

3D4[c]) Given two different coins, say a penny and a nickel, how can we flip them and pick out one of three alternatives with substantially equal probability? (*Optional:* Could three different coins be used to do this more safely?)

3D5[c]) Would it have helped the two analysts mentioned in Problem 3D3 to have had a copy of Moses and Oakford's (reference after Chapter 10) table of random permutations?

3D6) Draw your own subsamples of about 50, about 20, and about 5 from the 152 cities of Data Exhibit 1 for Problems, using the best technique available to you. SAVE these SAMPLES for LATER USE—for example, by clearly marking your copy of Data Exhibit 1 for Problems.

3D7[c]) A third analyst said, "I want to save duplication, so I want my sample of about 5 to be part of my sample of about 20, and my sample of about 20 to be part of my sample of about 50." Discuss the pros and cons of accepting such a "box-within-box" requirement.

3E1^C) (A variable, not median family income, to be assigned to each solver.) For the 152 cities in Data Exhibit 1 for Problems, $k = \sqrt{152} = 12^+$. Focus your attention on (a) the 12 cities with the highest values of the variable assigned to you, (b) the 12 cities with the lowest values of the variable assigned to you. Thin down to 10 each, by omitting the 5th and 9th most extreme in each set. For the two sets of 10 remaining, make plots like those in Exhibit 5 of Chapter 3. Plot your assigned variable against two others, one being median family income. (Class) Collect all the plots; discuss what they seem to show.

3E2) (Continues 3E1, with same assignment of variables to solvers.) Take the sample of about 50 cities you selected in 3D6. Plot your assigned variable against median family income for these "50." How well did the corresponding 10-and-10 plot of 3E1 allow you to forecast the fifty? And vice versa?

3E3) Do 3E2 again, but using your sample of about 20. Answer the same question.

3F1) (Requires assignment of variable, possibly the same as in 3E1, not median family income, to each solver.) Using your sample of 50 cities, make a stem-and-leaf of the 50 values of median family income, and then use this to help you write out the 50 cities in order of median family income. If any cities are tied, bracket them. Then turn to your assigned variable and write down its values (in the same order, replacing values for bracketed cities by their median) as y. You now have either 50 (if there were no ties) or somewhat fewer than 50 (if there were ties) values of y. Now smooth this sequence by running medians of 3 according to the pattern of Exhibit 6 of Chapter 3. Plot the result. Discuss what the result means to you.

3F2) Do the same as 3F1 with a different variable as y.

3F3) Do the same as 3F1, choosing a variable given in Data Exhibit 1 for Problems (at the end of Problems) *other than* either median family income or the variable assigned to you (or the variable chosen in 3F2) as the basis for ordering your 50 cities, but taking the same y's. Discuss the similarities and differences, both as to appearance and as to meaning, between this result and that of 3F1.

3F4) Combine the changes in 3F2 and 3F3. Same instructions as 3F3.

3F5) "Age heaping." It is a well known demographic phenomenon that people tend to report their ages rounded to the nearest multiple of 5 years. The reported female age distribution for Mexico in 1960 (*Source:* Mexico. Direction General de Estadistica, VIII censo general de poblacion, Cuadro 7, 1962) for ages 0 to 75 is given in Exhibit 1 on page 477. Analyze this data by means of running medians. Does it make any difference whether running medians of length 3, 5, or 7 are used? What patterns show up in the residuals? Do they support the age-heaping hypothesis?

3F6) Repeat Problem 3F5 working with the square roots of the numbers reporting each age.

3G1) (Continues 3F1.) Eye-fit a straight line

$$y \sim a + b \times \text{(median family income)}$$

to the smoothed results of 3F1. Adjust each of the 50 individual y's by subtracting this straight line. Smooth the resulting residuals by repeated medians of 3. Plot the result. Do you believe the straight line captured all the nonrandom relation of y to median family income? Why/why not? Add back the same straight line to the smoothed residuals, and plot on the same graph with the results of 3F1. How do the two sets of results compare? Why do you think this happens?

3G2) (Continues 3F2.) Do the same for the data of 3F2 (instead of 3F1).

Exhibit **1** for Problem 3F5

Reported female age distribution for Mexico, 1960 (in thousands)

Age	# at Age	Age	# at Age	Age	# at Age
0	558				
1	513	26	243	51	42
2	582	27	220	52	86
3	604	28	283	53	61
4	584	29	182	54	66
5	566	30	412	55	148
6	562	31	113	56	69
7	524	32	208	57	46
8	529	33	163	58	83
9	430	34	146	59	48
10	497	35	310	60	245
11	369	36	168	61	20
12	455	37	130	62	43
13	398	38	224	63	32
14	404	39	129	64	32
15	382	40	331	65	103
16	366	41	51	66	29
17	346	42	136	67	23
18	403	43	91	68	40
19	300	44	77	69	16
20	409	45	231	70	109
21	226	46	90	71	9
22	325	47	77	72	25
23	294	48	148	73	15
24	289	49	78	74	14
25	380	50	281	75	48

3G3) Exhibit 16 of Chapter 3 (in Section 3H) omits (in comparison with Exhibit 15 of Chapter 3) the 60 points corresponding to even numbers divisible by 6. Write the Goldbach counts for these 60 numbers down (in order of the even numbers) and smooth them by repeated running medians of 3. Subtract the solid "curve" from Exhibit 16 and plot the differences.

3G4) (Continues 3G3.) Take the differences just formed and smooth them once again by repeated running means of 3. Plot the result. Can you find a simple approximation?

3H1) (3H1 to 3H3 can also be usefully shared across the class.) Make a plot like Exhibit 15 of Chapter 3, using the Goldbach counts from 252 to 430 (instead of from 2 to 180).

3H2) Do the same from 9502 to 9680.

3H3) Do the same from 9752 to 9930.

3H4 to 3H6) (Continues corresponding problem above.) Do the same for Exhibit 16 of Chapter 3, including a broken line for the result of smoothing (as in 3F or 3G) the Goldbach counts for even numbers divisible by three (as in 3G3).

3I1) (Class exercise.) The table from which Exhibit 19 of Chapter 3 was taken, which also appears at page 248 of the 1973 *World Almanac*, shows, for the same 30-year period: (a) normal minimum January temperatures; (b) normal July maximum temperatures; (c) normal July minimum temperatures; (d) highest temperature recorded; (e) lowest temperature recorded; (f) normal annual precipitation. These seven variables should be shared out, about one-seventh of the class receiving each. Then plots like Exhibit 7 of Chapter 3 should be made.

3I2) (Continues 3I1.) Eye-fit a slope to the plot just made and plot the residuals against longitude as in Exhibit 18 of Chapter 3. (Those who like the west on the left may run their horizontal scale from right to left.) Discuss your result.

3I3) (Continues 3I2.) Setting aside any cities for which the residuals of 3I2 seemed clearly unusual, seek out the best—most useful, most regression-producing—variable you can find against which to plot the remaining residuals. Make the plot and discuss your results.

3J1) For the about 50 cities in your own sample, relate median family income to percent housing in single-unit structures (a) in terms of 10–plus–10, and (b) in terms of the smoothed median of the medians of successive blocks of (about) 3 cities (16 blocks). Plot the analog of Exhibit 21 of Chapter 3 and discuss your results (i) by themselves and (ii) in comparison with that exhibit.

3J2) Do the same (except for (ii)) relating median family income to the variable assigned to you.

Chapter 4

4A1) Where would $z = t^{1/3}$ fit in the ladder of powers? Or would it?

4A2) Where would $z = \sqrt{t + 4}$ fit in the ladder of powers? Or would it?

4A3) Which rungs in the ladder of powers are also usefully defined for $t < 0$?

4B1) Consider the curve defined (in part, but well enough for our purposes) by the following table:

x:	.73	1.27	1.73	2.31	2.50	2.72	2.91
y:	.05	.26	.65	1.54	1.95	2.52	3.08

What re-expression of y would straighten out the curve? (Explore as in part 1 of Section 4B.)

4B2) For the same curve of Problem 4B1, what re-expression of x would straighten out the curve? (Explore as in part 2 of Section 4B.)

4B3) For the data in Problem 4B1, try log y and "starting" as in part 3 of Section 4B.

4B4) Re-express y to straighten out the curve implied by the following table:

x:	.73	1.27	1.73	2.31	2.50	2.72	2.91
y:	1.11	1.33	1.48	1.63	1.67	1.72	1.76

4B5) Re-express x to straighten out the curve of the table of Problem 4B4.

4B6) Try log y and "starting" to straighten out the curve of the table of Problem 4B4.

4B7) Re-express x to straighten the curve implied by the following table:

x:	.73	1.27	1.73	2.31	2.50	2.72	2.91
y:	2.22	2.34	2.44	2.55	2.59	2.63	2.67

4B8^C) Find the best re-expression you can for the table of 4B7.

4C1) Exhibit **1** shows various curves. Tell which way in y and, alternatively, which way in x, we should move along the ladder of powers to straighten (at least partially) the solid curve in the NW panel (see next page).

4C2) Same for the dashed curve in the NW panel.

4C3) Same for the dotted curve in the NW panel.

4C4/5/6) Same for each of the three curves in the NE/SW/SE panel.

4C7) Given below are the "q_x" values for the United States in 1970. These are the probabilities that a person, aged x, will die in the next five years.

What re-expression of x, q_x, or both seems to straighten this curve best? (x = age, q_x = probability of dying)

x	q_x	x	q_x
5	.0021	45	.029
10	.0020	50	.043
15	.0056	55	.066
20	.0074	60	.096
25	.0072	65	.138
30	.0097	70	.198
35	.012	75	.290
40	.019	80	.407
		85	.555

Exhibit **1** for Problem 4C1

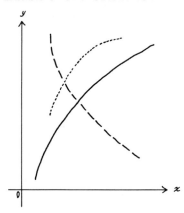

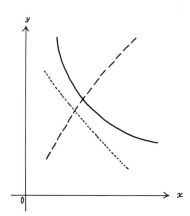

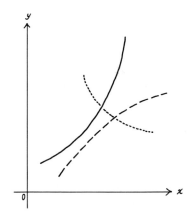

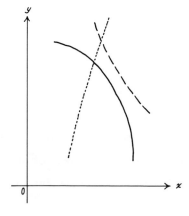

4D1) The following table gives the number of primes less than n for selected values of n:

n	4	16	64	256	1024	4096	16384
#	2	6	18	54	172	564	1900

Plot log # against log n. What does the plot suggest for straightening out the relation between # and n?

4D2) The following table gives values of cot θ for selected values of θ:

θ:	0.1°	0.2°	0.5°	1°	2°	5°	10°	20°	50°
cot θ:	1145.9	286.3	114.59	57.29	28.63	11.43	5.67	2.75	.84

Plot log cot θ against log θ. What does the plot suggest for straightening out the relation?

4D3) Using the table of Problem 4D1 as a base, plot #/n against n. What does this suggest for straightening out the relation? Then choose between plotting log (#/n) against n and plotting #/n against log n. What does your chosen plot suggest?

4D4) Using the table of Problem 4D2 as a base, plot θ cot θ against θ. What does this suggest? Choose re-expressions of θ against which to plot θ cot θ. Which gives the most helpful result? What would you plot next?

4E1[C]) (One eighth of the class to be assigned each variable other than median family income.) Go back to Exhibits 4 and 5 of Chapter 3, and add the ten urban unincorporated places (given in Exhibit 1 for Problem 4E1) with median family incomes at and near the medians of all 88 places.

→

Exhibit **1** for Problem 4E1

10 urban unicorporated places near median for median income, 1959 (see Exhibit 4 of Chapter 3 for further detail)

Place	Median family income	|212|	|229|	|244|	|246|	|248|	|249|	|256|	|264|
Braintree, Mass.	7474	31.0	60.7	52.2	7.7	5.8	88.8	19.4	6.7
Ross, Pa.	7475	31.1	52.6	4.0	12.9	5.7	86.1	28.6	6.5
Elmont, N.Y.	7494	31.1	68.2	44.2	33.2	5.6	88.5	16.6	19.0
Framingham, Mass.	7495	29.1	44.5	53.0	5.9	5.6	79.8	32.7	21.0
Arlington, Mass.	7538	34.8	61.8	60.4	27.7	5.8	56.8	22.3	7.8
Natick, Mass.	7550	28.7	57.8	55.7	9.1	5.9	81.6	23.7	5.4
Ewing, N.J.	7597	31.1	56.0	48.7	6.8	5.6	91.5	23.0	24.0
Middletown, Pa.	7656	22.6	19.2	56.3	7.9	6.1	99.2	35.8	24.3
Catonsville, Md.	7662	32.0	50.5	64.5	14.9	6.0	82.3	25.9	14.8
Hamden, Conn.	7741	35.5	59.4	55.3	13.7	5.7	84.5	21.4	10.6

Replace each set of ten places by the point (median of variable assigned to you, median of median family income). Seek a re-expression that will put these three points nearly along a single line. Discuss. (Bring all results for each of the 8 variables together in class, and discuss.)

Chapter 5

5A1[C]) Label each of the following according to whether they are amounts, counts, balances, counted fractions, ranks, or grades: the concentration of salt in sea water, the concentration of carbon dioxide in the atmosphere, the fraction of persons of Hispanic descent in Manhattan, the number of tons of salt in the sea, the number of molecules of carbon dioxide in the atmosphere, the average number of years of life (the expectation of life) for females born in Germany in a given year, the income for tax purposes of a large corporation, the income for tax purposes of an individual, the class of a baseball league, the "points" by which a football game is won, the position of a football team in the standings (how many ahead and how many tied), the letter grades given in a mathematics course, how far "east of Boston" is a town in Vermont, how far above water is a submerged volcano, or a volcanic island, how many fingers an injured man has lost, how many tires on an old car are badly worn.

5A2) If you wanted to re-express each number in 5A1, what re-expression would you try first?

5B1) Explain in detail how to use Exhibit 1 of Chapter 5 to obtain a two-decimal log of 321.98. Of 0.000098321.

5B2) Two analysts were debating whether to use $\log (y + 0.01)$ or $\log (y + 0.02)$. For which y's would their decision really matter?

5B3) Another pair were debating between $\log (y + 1)$ and $\log (y + 3)$. For which y's would this matter? How could $\log (y + 1)$ or $\log (y + 3)$ be easily found?

5B4) Evaluate $\log 17\frac{1}{6}$; $\log 23\frac{1}{6}$.

5C1) Explain how to use Exhibit 3 of Chapter 5 to obtain a two- or three-digit square root for 321.98, and for 0.000098321.

5C2[C]) Describe the pros and cons of working with each of $1000/y$, $1/y$, $-1/y$, $-1000/y$.

5C3) Yet another pair of analysts were arguing about choosing $\sqrt{y + .05}$ or $\sqrt{y + .1}$. For what y's would this choice matter?

5C4) A fourth pair of analysts were torn between choosing $\sqrt{y + 1}$ and $\sqrt{y + 2}$. For what y's would this choice matter?

5C5) Describe how to use Exhibit 4 of Chapter 5 to obtain a two- or three-digit negative reciprocal for 321.98, and for 0.000098321.

5C6) What are the values of $\sqrt{17\frac{1}{6}}$, $\sqrt{23\frac{1}{6}}$, $\sqrt{47\frac{1}{6}}$, and $\sqrt{289\frac{1}{6}}$?

5C7) What are the values of $-1000/17\frac{1}{6}$, $-1000/23\frac{1}{6}$, $-1000/47\frac{1}{6}$, and $-1000/289\frac{1}{6}$?

5D1) What are the matched re-expressions of 86.5%? Of 13.5%? Of 98.9%? Of 1.1%?

5D2) Describe in detail how it is easy to calculate

$$\log\text{ (one count)} - \log\text{ (the other count)}$$

when the counts are both started by $\frac{1}{6}$ and the raw counts were $17 + 133 = 150$.

5D3) Do the same as in Problem 5D2 for

$$\sqrt{\text{one count}} - \sqrt{\text{the other count}}$$

for the same start and raw values.

5D4^C) Discuss the pros and cons for having various re-expressions of fractions matched near 50%.

5E1) Why do we have to pick a value at which to match powers and logs when we didn't seem to do anything like this when matching re-expressions of fractions?

5E2) If we know that a matched re-expression of y does not differ from y by more than ±1 for all y between 280 and 320, what matched powers could be providing such a re-expression?

5E3) Same question as Problem 5E2 with agreement to ±10 for $2800 \le y \le 3200$.

Warning: Problems in section F require substantial calculations.

5F1) Hartman (quoted by K. A. Mather, 1949, *Statistical Analysis in Biology*, page 196) gave counts of the number of men and the number of women who could not taste phenylthiocarbamide in various strengths. The counts were

Men: 15 A 35 B 46 C 31 D 23 E 13 F 9 G 7 H 10 I 13 J 25 K 63 Total 290

Women: 42 A 52 B 38 C 30 D 19 E 17 F 6 G 5 H 10 I 19 J 33 K 43 Total 314

Here A is the strongest solution, which 15 of the men could not taste, while 35 more could taste A but not B, and so on to the 63 men who could taste the weakest solution. Guided by Exhibit 10 of Chapter 5, use Exhibit 11 of that chapter to obtain two re-expressions for A, B, ..., K, one for men and one for women.

5F2) Davis and Richards (*Journal of Ecology*, **21**: 350–384 and **22**: 106–155) give counts of trees of seven species (one of the seven is two species hard to distinguish) by six size classes in a mixed rain forest (jungle) at Moraballi Creek, British Guiana. The size classes were measured, but you are not being told what diameters cut off the classes. The counts, from smallest class to largest class are as follows: *Eschweilera (decolorans* or *pallida)* (128, 19, 14, 7, 1, 1); *Eschweilera sagotiana* (48, 5, 6, 14, 13, 1); *Licania laxiflora* (24, 18, 13, 5, 6, 0); *Licania heteromorpha* (176, 29, 12, 0, 0, 0); *Licania venosa* (128, 26, 16, 9, 10, 0); *Ocotea rodiaei* (16, 8, 2, 5, 10, 3); *Pentaclethra macroloba* (80, 31, 20, 16, 6, 0). Find a re-expression for each size class (omitting empty size classes) using, in turn, each species. Do the seven re-expressions seem to agree fairly well?

5F3) (Continues 5F2) For each size class in 5F2, find the mean re-expression for the three smallest-size classes (chosen as occupied for all species) and subtract the result for all values of that re-expression. For each size class you will now have three to seven such adjusted re-expressions. Take their median as a first approximation to an overall re-expression for that size class. Plot each original re-expression against this approximate overall re-expression. Do the plots look reasonably like straight lines?

5F4) (Continues 5F3) Eye-fit straight lines to each of the seven plots with which Problem 5F3 closed. Calculate the values of

$$\text{Second adjustment} = \frac{\text{Residual from line}}{\text{Slope of line}}$$

for each combination of size class and species. Take medians over species. Do these seem meaningfully large? Form an improved overall re-expression for each size class as

approximate overall PLUS median second adjustment.

How nearly equally spaced do the size classes seem to be?

5G1) Re-expression of ranks is sometimes useful when the data given is numerical. The 1960 population (POP) and median family income (MFI) of the counties of Arizona, of the counties of each of the (1964) Congressional districts of Arkansas, and of each of three "stripes" along California are given in Exhibit 1 for Problem 5G1, ordered by 1960 population. For the state or portion assigned to you, find the re-expression $\log(i - \frac{1}{3})$ MINUS $\log(n + 1 - i - \frac{1}{3})$ for both the POP's and MFI's. Plot both raw MFI against raw POP and re-expressed MFI against re-expressed POP. Which plot seems more useful?

5G2) Take 3 points from each of the eight county groupings given in Exhibit 1 for Problem 5G1—the second (in smaller groupings) or the third (in larger groupings) from each end and the middle point are convenient choices— and explore to see what power (or log) of the 1960 population appears to behave in a reasonably straight-line way with the $\log(i - \frac{1}{3})$ MINUS $\log(n + 1 - i - \frac{1}{3})$ re-expression (used in 5G1).

5G3) (Continues 5G2) For the county groupings assigned to you, plot your selected power re-expression of the 1960 population against its rank-based re-expression.

Exhibit 1 for Problem 5G1

Populations and median family incomes (MFI) of counties in selected areas

Arizona 1960			First CD Arkansas 1960		
County	POP	MFI	County	POP	MFI
Maricopa	663,510	5896	Mississippi	70,174	2725
Pima	265,660	5690	Crittenden	47,564	2506
Pinal	62,673	4412	Craighead	47,303	3408
Cochise	55,039	5107	Phillips	43,997	2360
Yuma	46,235	5360	St. Francis	33,303	1973
Coconino	41,857	5398	Poinsett	30,834	2591
Navajo	37,994	4237	Greene	25,198	2654
Apache	30,438	2832	Clay	21,258	2633
Yavapai	28,912	5197	Lee	21,001	1710
Gila	25,245	5087	Cross	19,551	2480
Graham	14,045	4593			
Greenlee	11,059	5168			
Santa Cruz	10,808	4620			
Mohave	7,736	5111			

Second CD Arkansas 1960			Third CD Arkansas 1960		
County	POP	MFI	County	POP	MFI
Pulaski	242,980	4935	Sebastian	66,685	3089
White	32,795	2893	Washington	55,797	3683
Lonoke	24,551	2708	Benton	36,272	3180
Faulkner	24,303	2968	Crawford	21,318	3122
Arkansas	23,355	3348	Pope	21,177	3046
Jackson	22,843	2995	Boone	16,116	2837
Independence	20,048	2502	Logan	15,957	2376
Monroe	17,327	2162	Johnson	12,421	2484
Lawrence	17,267	2255	Yell	11,940	2600
Conway	15,430	2751	Carroll	11,284	2555
Woodruff	13,954	1902	Franklin	10,213	2611
Randolph	12,520	2497	Baxter	9,943	2800
Prairie	10,515	2853	Madison	9,068	1928
Cleburne	9,059	2137	Searcy	8,124	2066
Izard	6,766	2699	Scott	7,297	2168
Fulton	6,657	1886	Van Buren	7,228	1968
Sharp	6,319	1902	Marion	6,041	2210
Stone	6,294	1740	Newton	5,963	1666
Perry	4,927	2217			

➡

5G4) (For those who wish to explore further, and who have access to a 1962 *County and City Data Book*.) Other columns that might be interesting to examine, separately or together, are (4) population density, (5) population % growth, (13) median age, and the ratio of (107) number of food stores to (86) number of manufacturing establishments.

5H1) How much does section 5H tell you that the earlier sections have not told you?

5H2) You have a large number of triangles, each of which has one side 10 units long and one side 5 units long. Under first-aid rules, how would you re-express the length of the third side?

Exhibit 1 for Problem 5G1 (continued)

Fourth CD Arkansas 1960			Tidw* California 1960		
County	POP	MFI	County	POP	MFI
Jefferson	81,373	3671	Los Angeles	6,038,771	7046
Union	49,518	4361	San Diego	1,033,011	6545
Garland	46,697	3511	Alameda	908,209	6786
Miller	31,686	3372	San Francisco	740,316	6717
Ouachita	31,641	3686	Santa Clara	642,315	7417
Saline	28,956	4483	Sacramento	502,775	7100
Columbia	26,400	3438	San Mateo	444,387	8103
Ashley	24,220	3432	Contra Costa	409,030	7327
Hot Springs	21,893	3881	San Joaquin	249,919	5889
Clark	20,950	3127	Ventura	199,138	6466
Desha	20,770	2430	Monterey	198,351	5770
Hempstead	19,661	2676	Santa Barbara	168,962	6833
Chicot	18,990	2013	Solano	134,597	6190
Drew	15,213	2614	Humboldt	104,892	6282
Lincoln	14,447	1911	Santa Cruz	84,219	5325
Bradley	14,029	3069	Napa	65,890	6524
Polk	11,981	2694	Mendocino	51,058	5803
Lafayette	11,030	2245	Del Norte	17,771	6277
Howard	10,878	3033			
Nevada	10,700	2538			
Dallas	10,522	2809			
Sevier	10,156	3089			
Little River	9,211	2725			
Grant	8,294	2985			
Pike	7,864	2614			
Cleveland	6,944	2363			
Calhoun	5,991	2394			
Montgomery	5,370	2572			

➡

5H3[C]) A maker of quality resistance units selects from all units produced those with resistance values between 99% and 101% of nominal. How would you re-express the measured resistances of nominal 500-ohm units? Explain your answer.

5H4) Standard patterns of mortality. William Brass ("On the Scale of Mortality", pp. 69–110 in *Biological Aspects of Demography* (W. Brass, ed.), London: Taylor and Francis, Ltd., 1971) has proposed that there is a "standard" pattern of mortality which, when modified by only two parameters, adequately describes most human mortality experiences. Given in Exhibit 1

Exhibit 1 for Problem 5G1 (continued)

Adj** California 1960			Remaining California 1960		
County	POP	MFI	County	POP	MFI
Orange	703,925	7219	Tulare	166,403	4815
San Bernardino	503,591	5998	Butte	82,030	5408
Fresno	365,945	6603	Shasta	59,468	5989
Riverside	306,191	5693	Madera	40,466	4596
Kern	291,981	5933	Yuba	33,859	5031
Stanislaus	157,294	5260	Nevada	20,911	5419
Sonoma	147,325	5725	Tuolumne	14,404	5602
Marin	146,820	8110	Lassen	13,597	5861
Merced	90,448	4806	Colusa	12,075	5604
San Luis Obispo	81,011	5659	Inyo	11,689	5837
Imperial	72,105	5507	Plumas	11,260	5834
Yolo	65,727	6240	Modoc	8,308	5709
Placer	56,468	5989	Mariposa	5,064	4704
Kings	49,954	4957	Sierra	2,247	5863
Sutter	33,380	5670	Mono	2,213	6321
Siskiyou	32,885	5558	Alpine	397	——
El Dorado	29,390	6603			
Tehama	25,305	5589			
Glenn	17,245	5290			
San Benito	15,396	5538			
Lake	13,786	4438			
Calaveras	10,289	5824			
Amador	9,900	5636			
Trinity	9,706	6210			

* Tidw = tidewater = all counties with largest city or town on tidal water
** Adj = adjacent = all counties bounded by tidewater counties

S) SOURCE
1962 *County and City Data Book.*

are three modern "life tables", the expected proportion of people in each population who live until the indicated age, and Brass's proposed standard. Brass has noted that plots of the logits of each life table against the logits of the standard are nearly linear. Dealing with each country separately, is the logit the best re-expression? Which others could be tried? Considering them jointly, how well does the logit hold up? Do you believe there is a "standard" pattern of mortality, based on these data?

5I1) Two analysts had data giving the number of traffic accidents hour by hour for three nearby large cities for the 168 hours of a particular week. One wanted to analyze the logs of the counts; the other objected because there were so many zeros. What two quite different proposals could the first analyst make that would avoid all difficulty with logs of zero?

5I2) An investigator, studying slopes of curves made by a recording device, thinks logs would make sense—in part because running the paper faster would divide all slopes by the ratio of the paper speeds. You are asked how to avoid trouble with both horizontals and verticals—with both zero slopes and infinite slopes. What do you say?

5I3) In 5I2, would your answer depend on whether the investigator had 2 zeros and 3 infinities in 1000, or 97 zeros and 43 infinities in 1000? Why/why not?

Exhibit 1 for Problem 5H4

Proportion surviving to indicated age (Brass)

Age	Brass's Standard	Sweden Females, 1959	Italy Females, 1901–11	Japan Males, 1959
1	.850	.987	.848	.964
5	.769	.984	.739	.952
10	.750	.982	.721	.947
20	.713	.979	.699	.938
30	.652	.974	.641	.917
40	.590	.965	.592	.892
50	.511	.943	.541	.846
60	.396	.891	.466	.738
70	.238	.757	.317	.524
80	.076	.447	.107	.214

(Data reprinted by permission of the author and of the publisher.)

Chapter 6

6–1) If we would gain somewhat from re-expression, is it always worth the bother? Why/why not?

6–2) If you had observations 1, 1.5, 3, 5, . . . , 63, 89, 121, and believed that taking logs would give a better expression, is it likely to be worthwhile to do this? Why/why not? When/when not?

6–3) Answer Problem 6–2 if the values range from 23 and 29 to 365 and 513.

6–4) Answer Problem 6–2 if the values range from 51 and 54 to 473 and 551.

6–5) Answer Problem 6–2 if the values range from 1.04 and 1.09 to 1.73 and 1.81.

Exhibit 1 for Problem 6–6

Proportions of Congressional seats and popular vote (Democratic), 1900–1972

Year	Democrat % Votes	Democrat % Seats	Year	Democrat % Votes	Democrat % Seats
1900	46.60	43.59	1936	58.48	78.91
1902	48.68	46.23	1938	50.82	60.79
1904	43.66	35.23	1940	52.97	62.24
1906	46.55	42.49	1942	47.66	51.51
1908	48.11	43.99	1944	51.71	56.12
1910	50.50	58.46	1946	45.27	43.32
1912	57.11	69.54	1948	53.24	60.60
1914	50.34	54.48	1950	50.04	54.04
1916	48.88	49.30	1952	49.94	49.08
1918	45.10	44.63	1954	52.54	53.33
1920	37.67	30.56	1956	50.97	53.79
1922	46.40	47.92	1958	56.10	64.91
1924	42.09	42.56	1960	54.97	59.95
1926	41.57	45.14	1962	52.42	59.45
1928	42.84	37.91	1964	57.50	67.82
1930	45.87	49.77	1966	51.33	57.01
1932	56.87	72.79	1968	50.92	55.86
1934	56.18	75.76	1970	54.32	58.62

S) SOURCE

See E. R. Tufte, "The relationship between seats and votes in two-party systems." *American Political Science Review*, 68, June 1974, 540–554. (Data reprinted by permission of the author and of the journal.)

6–6) *Seats and Votes* (Project) In a two-party parliament or congress, one does not expect that, if a party gets x percent of the aggregate vote, it will get x percent of the seats. In fact, various models have been proposed to describe this relationship. The simplest model is known as the "Cube Law", i.e.,

$$\frac{S}{1 - S} = \left(\frac{V}{1 - V}\right)^3,$$

where S is the proportion of seats, and V is the proportion of votes. A more complicated "logit" model is:

$$\log\left(\frac{S}{1 - S}\right) = \beta_0 + \beta_1 \log\left(\frac{V}{1 - V}\right).$$

E. R. Tufte proposes a simpler model:

$$S - .5 = \beta_0 + \beta_1(V - .5),$$

where β_0 has a simple interpretation as the "bias" of a system, and β_1 is described as the "swing ratio", i.e., the percent of seats gained for one percent more votes. What theoretical objections are there to the first and third model? What is the relationship between the Cube Law and the logit model? The proportions of seats and votes for 36 U.S. Congressional elections in the 20th century are given in Exhibit **1**. Based on these data, and bearing in mind the theoretical aspects, which model seems most appropriate? In light of its simple interpretation, is the linear model "accurate enough" for most purposes? Can you propose a better model?

Chapter 7

7A1) Go back to Exhibit 5 of Chapter 5 and collect, for days 2 to 24, the apparent variance of a daily mean (1) as based on differences between the daily means and (2) as the mean of the interval estimates, $(s_{\bar{x}})^2$, of the variances of daily means. If

$$\text{True variance of daily mean} = \frac{s^2}{n_{\text{eff}}},$$

then

$$\frac{n}{n_{\text{eff}}} = \frac{\text{Variance based on differences (external)}}{\text{Variance based on } (s_{\bar{x}})^2 \text{ (internal)}}$$

Find an estimate of n/n_{eff}. In view of $n \sim 500$, about how big is n_{eff}? What does this mean?

7A2^C) Why is it impossible that a valid measure of uncertainty be routinely found by looking up a formula?

7B1) F. A. Palumbo and E. S. Strugala (*Industry Quality Control*, November 1945, 6–8) give data on the fraction defective (in vulcanization) of battery adapters used in Handie-Talkies on 32 consecutive lots. Each data pair gives (sample size, number defective). $(140, 77)$, $(140, 19)$, $(140, 24)$, $(140, 20)$, $(140, 27)$, $(155, 0)$, $(155, 0)$, $(210, 0)$, $(155, 0)$, $(210, 0)$, $(50, 50)$, $(50, 4)$, $(50, 17)$, $(90, 0)$, $(105, 0)$, $(105, 4)$, $(155, 8)$, $(155, 2)$, $(155, 0)$, $(210, 4)$, $(155, 5)$, $(155, 7)$, $(105, 3)$, $(210, 12)$, $(190, 9)$, $(125, 7)$, $(125, 5)$, $(125, 2)$, $(75, 0)$, $(75, 4)$, $(125, 1)$, $(125, 2)$. There were 9 lots with sample sizes of 50 to 105, 10 with sample sizes of 125 or 140, 13 with sample sizes of 155 or more. For each of these three groups of lots, calculate an internal estimate (average of pq/n) for the variance of a lot p, and an external estimate (the usual s^2 based on the p values for those lots). Compare and discuss the comparison.

7C1) In a study of National Football league wide receivers, various physical dimensions are to be collected for later comparison with pass-catching performance. Incidentally, we want to study differences in wide-receiver heights among the four conferences into which the NFL is divided. Limitations on effort restrict us to two teams per conference. For some reason or other it has been decided to study wide receivers in both the *most* successful team in each conference and the *least* successful team in each conference. What is the most nearly appropriate error term to use in comparing conferences? Why? If this error term is biased, in which direction is it likely to be off?

7C2) In a study of 5 New England regional cookbooks, one aspect to be studied was a score for ease of preparation. Two proposals were made for the sampling of recipes to be kitchen-tested. In one, 20 recipes were to be selected from each book at random; in the other the recipes were to be divided into 5 to 10 categories: meat, fish, vegetables, sauces, and gravies, etc., and a fixed number of recipes were to be selected at random from each category. What would you consider the correct error term if each of these plans were adopted?

7C3) (Continues 7C2) What if it were necessary to use several combinations of cooks and kitchens, which could, however, be balanced across cookbooks?

7C4) A comparison of 6 supermarkets on prices for food not on special sale involved dividing the food sold into 20 categories, and selecting 10 relatively standard items in each category. These were then priced twice a week (on rotating days) for a year, so that 124,800 prices were collected. What would you recommend as the proper basis for an error term? How would your answer depend on whether the results were supposed to be for the year of observation, or to be used to guide further purchases?

7D1) Two analysts provided alternative plans for a study. These plans allocated a fixed total amount of money differently. One collected more data in fewer independent groups—a total of 1000 observations in five groups. The other collected less data in more groups—a total of 700 observations in ten groups. If the number of groups used affects only the number of degrees of freedom available—not the size of group-to-group variation, not the appropriateness of group-to-group variation—which of these plans will, on average, produce the shorter 95% interval?

7D2) (Continues 7D1) And what of a plan that collects 900 observations in 9 groups?

7E1) F. Proschan (*Technometrics*, **5**: 376 (1963)) gives time intervals between failures in the air-conditioning system of thirteen type 720 jet aircraft as follows (we give airplane # before a ";", followed by successive times between failures, stopping at any major overhaul): (7907; 194, 15, 41, 29, 33, 181), (7908; 413, 14, 58, 37, 100, 65, 9, 169, 447, 184, 36, 201, 118), (7909; 90, 10, 60, 186, 61, 49, 14, 24, 56, 20, 79, 84, 44, 59, 29, 118, 25, 156, 310, 76, 26, 44, 23, 62), (7910; 74, 57, 48, 29, 502, 12, 70, 21, 29, 386, 59, 27), (7911; 55, 320, 56, 104, 220, 239, 47, 246, 176, 182, 33), (7912; 23, 261, 87, 7, 120, 14, 62, 47, 225, 71, 246, 21, 42, 20, 5, 12, 120, 11, 3, 14, 71, 11, 14, 11, 16, 90, 1, 16, 52, 95), (7913; 97, 51, 11, 4, 141, 18, 142, 68, 77, 80, 1, 16, 106, 206, 82, 54, 31, 216, 46, 111, 39, 63, 18, 191, 18, 163, 24), (7914; 50, 44, 102, 72, 22, 39, 3, 15, 197, 188, 79, 88, 46, 5, 5, 36, 22, 139, 210, 97, 30, 23, 13, 14), (7915; 359, 9, 12, 270, 603, 3, 104, 2, 438), (7916; 50, 254, 5, 283, 35, 12), (8044; 487, 18, 100, 7, 98, 5, 85, 91, 43, 230, 3, 130), (8045; 102, 209, 14, 57, 54, 32, 67, 59, 134, 152, 27, 14, 230, 66, 61, 34). For each airplane calculate (a) median time between failures, (b) fraction of failure intervals less than 20, (c) median length of intervals less than 20, (d) median length of intervals equal to or larger than 20, and apply Student's t to set limits on the averages of each of these four over time and airplanes.

7E2) R. L. Anderson (*Journal of the American Statistical Association*, **42**: 612–634 (1947)) studied the logs of the ratios of hog prices at Cincinnati and Louisville for 2 weight classes by 12 months, by 5 days of the week, by 5 years (1937–41). Among his summary figures are these: Months (January to December; 147.4, 194.3, 190.5, 208.9, 215.6, 200.9, 192.1, 185.9, 146.6, 137.6, 144.9, 162.8). Days (Monday to Friday; 175.2, 174.0, 180.0, 173.6, 183.8). Years (1937 to 1941; 169.5, 195.3, 220.4, 168.7, 132.6). Treating each of months, days, and years in turn as a basis for direct error assessment, use Student's t to set limits on the overall mean. On abstract principles, which ought to be the most appropriate choice as an error term? Explain your answer. After seeing the data, what changes would you make in your answer? Why/why not?

7F1) (Teaser = means for solution not given in text.) Anderson's hog-price data (see 7E2 above) include the following averages for day of the week and year combinations, given here (rounded) in the form (year; Monday, Tuesday, Wednesday, Thursday, Friday) for one size class: (1937; 156, 168, 173, 158, 173), (1938; 202, 192, 200, 195, 201), (1939; 207, 210, 211, 214, 219), (1940; 144, 118, 132, 131, 151), (1941; 125, 130, 131, 123, 132). Pretend, if necessary, that both years and days of the week should contribute to the variance assigned to the grand mean. Show the details of a calculation in which an estimated variance allows properly for both of these specific sources of variation, as well as for residual variation. Compare the corresponding 95% confidence intervals with those found when either one of the sources of variation is allowed for alone.

7G1) A certain chemical analysis is known, from long historical studies, to give answers that vary from day to day within a week ($\sigma^2 = 2 \times 10^{-5}$), from week to week within a month ($\sigma^2 = 1 \times 10^{-5}$), and from month to month within a year ($\sigma^2 = 3 \times 10^{-5}$). Sixteen determinations were made on one day and gave $s_x^2 = 0.0000173$. What variance should be assigned the mean of these sixteen as reflecting (a) a mean over a month or (b) a mean over a year?

7G2) (Continues 7G1) Another sixteen determinations were made in a day and gave s^2 (not s_g^2) $= 0.000279$. What variance should be assigned the grand mean of the two sets of 16 (under both (a) and (b) above) if (i) the two days were the same, (ii) the two days were different, but in the same week, (iii) the two days were not in the same week, but in the same month, (iv) the two days were not in the same month, but in the same year?

7G3) (Continues 7 G2) Should your answer to 7G2 depend on the target to which the estimate is regarded as aimed—for example, a weekly mean, a monthly mean, a yearly mean, a long-run mean? Illustrate your answer numerically for case (ii) of 7G2.

Chapter 8

8A1) Name two important uses of the jackknife. Why is the method called the jackknife?

8A2) (a) What is y_{all}? (b) What is $y_{(j)}$? (c) What is y_{*j}? (d) What is y_*?

8A3) Examine equation (1) of Section 8A and explain how a pseudo-value can be regarded as analogous to a value of y.

8A4) Let the expected value of a particular statistic based on a sample of size n be $\mu + (a/n)$. Thus the statistic is a biased estimate of μ. When $n = k$, find the expected value of y_{*j} in equation 1. (Note that $E(y_{(j)}) = \mu + [a/(k-1)]$.) Explain how this result supports the second remark of the first paragraph of Section 8A.

8A5) When the statistic is the sample mean, derive y_*. (That is, when $y_{all} = \sum y_j/k$, and
$$y_{(j)} = \frac{(\sum y_j) - y_j}{k-1}.)$$

8A6) We are to estimate the variance of a normal distribution with unknown mean. Show that the bias in the estimator $y_{all} = \sum(x - \bar{x})^2/n$ is of order $1/n$. Is the bias in y_j bigger than that in y_{all}? Show that y_* is unbiased.

8B1) The geometric mean of k measurements is defined as $g_k = \sqrt[k]{y_1 y_2 \cdots y_k}$. Four measurements have been drawn: 1, 2, 2, 4. Use the jackknife to estimate the population geometric mean and to set $\frac{2}{3}$ confidence limits on it.

8B2) Use logs in Problem 8B1 to get confidence limits on log g_k, and thence for g_k.

8B3) For fitting a line through the origin, an investigator has 6 points (x_i, y_i), $i = 1, 2, \ldots, 6$. They are

$$(2, 1) \quad (2, 2) \quad (4, 2)$$
$$(2, 2) \quad (3, 2) \quad (5, 4)$$

To estimate the slope, he uses the estimate

$$m = \frac{\sum y_i}{\sum x_i}.$$

Use the jackknife to set $\frac{2}{3}$ confidence limits on the true slope.

8B4) (Continues 8B3) As an alternative method of estimating the slope, the investigator considers

$$m_1 = \frac{1}{k}\left(\frac{y_1}{x_1} + \frac{y_2}{x_2} + \cdots + \frac{y_k}{x_k}\right).$$

Jackknife this estimate for the data in Problem 8B3 and compare the s_*^2 for the two methods. Of the two estimators, which seems preferable?

8C1) In Exhibit 7 of Chapter 8, check the $i = 3$ column.

8C2) In Exhibit 7 of Chapter 8, why is the line "1000 (rounded $z_{*i} - 0.100$)" introduced?

8C3) Use the individual groups given in Exhibit 3 of Chapter 8 as a basis for jackknifing the 20% trimmed mean, and set confidence limits on its estimand. Recall that this mean is obtained by deleting the largest 20% and the smallest 20% of the observations and taking the average of the remaining observations.

8C4) What does this trimmed mean estimate?

8C5) If we doubled the number of observations per individual but kept the number of individuals the same in Problem 8C3, would the expected value of y_* be changed? Why/why not?

8D1) What is the purpose of the discriminant function in the Federalist example of Section 8D?

8D2) How does the discriminant function here relate to an ordinary regression equation? Especially explain what y is.

8D3) In the discriminant function, why doesn't it matter what the values of A and B are $(A \neq 0)$? What would happen if A were doubled?

8D4) Usually there is a cutoff point in discriminant problems. What is a natural one to use here?

8D5) Sometimes the distributions associated with the two groups of observations in a discriminant problem look like this:

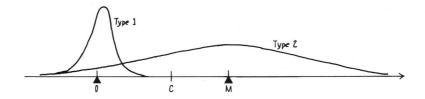

Might you consider cutoffs other than midway between the means? What might you gain?

8D6) If the means for groups 1 and 2 were 0 and M and the standard deviations σ_1 and σ_2, and the distribution of the discriminants is approximately normal, find a formula for the cutoff point, C, that makes the number of errors in classification equal for the two groups.

8D7) Usually in jackknifing we leave one observation out at a time, but in the Federalist problem a Hamilton paper was paired with a Madison paper rather arbitrarily and the pair was omitted. Why was this done? Was it necessary? Was it desirable? Was something lost?

8D8) Use the rates of the word "of" (Exhibit 9 of Chapter 8) to discriminate between Hamilton and Madison's authorship. Essentially set up $y = Ab_3X_3 + B$ to estimate 0 and 1 for Madison and Hamilton, respectively.

8D9) (Continues 8D8) Jackknife your discrimination method, continuing through the equivalent of Exhibit 11 of Chapter 8.

8D10) Use the words "of" and "and" to set up a discriminant function for the Hamilton and Madison problem. Essentially, carry out Exhibit 10 of Chapter 8 for these two words.

8D11) (Continues 8D10) Set up Exhibit 11 of Chapter 8 for "of" and "and".

8E1) (Continues 8D9) Now cross-validate your estimates.

8E2) (Continues 8E1) Compare the performance of "of" alone with that of the 5 words in the text.

8E3) (Continues 8D11) Carry out Exhibit 12 of Chapter 8 for the "of" and "and" approach.

8E4) (Continues 8E3) Compare the performance of "of" and "and" with the 5 words and with that of "of" alone.

8E5) From Exhibit 12 of Chapter 8 get the estimated values of the misclassification probability for Hamilton and that for Madison papers.

8F1) (Project) (Continues 8D8, 8D9, 8E1, 8E2) Carry out the two simultaneous uses of "leave-out-one" given in Section 8F for the discriminant, using "of" in the Federalist example.

Chapter 9

Warning. These problems require heavy calculations and careful checking. It is wise to choose a problem (1, 2, 5, 6, 9, or 10) and plan to carry it throughout the chapter. Problems 9A1 and 9A2 have advantages.

9A1) U. Behn (*Annalen der Physik*, (4) **1**: 257–269 (1900)) gives the mean "atomic heat" for a number of elements, over three temperature ranges. We now give (element; for −180° to −79°, for −79° to +18°, for +18° to +100°) in order of decreasing atomic weight: (Pb; 6.0, 6.2, 6.4), (Pt; 5.4, 6.1, 6.3), (Sb; 5.5, 5.8, 6.0), (Sn; 5.8, 6.1, 6.5), (Cd; 5.6, 6.0, 6.3), (Ag; 5.4, 5.9, 6.0), (Pd; 5.2, 6.0, 6.3), (Zn; 5.2, 5.8, 6.1), (Cu; 4.5, 5.6, 6.0), (Ni; 4.3, 5.8, 6.4), (Fe; 4.0, 5.6, 6.3), (Al; 4.2, 5.3, 6.0), (Mg; 4.6, 5.7, 6.1). Analyze the resulting 3×13 table along the lines suggested in Section 9A. Present the results in standard form. What can you see?

9A2[c]) P. B. Stam, R. F. Kratz, and H. J. White (*Textile Research Journal*, **22**: 448–465 (1952)) give the fractional diameter increase of dry human hairs as the relative humidity is first increased and then decreased. We give now (sample #; at 10%, at 40%, at 60%, at 90%, at 100%, at 90%, at 60%, at 40%, at 10%, at 0%), where "at *x*%" indicates the increase in diameter (in thousandths of the original diameter) after stabilization at *x*% relative humidity. (#9; 25, 57, 63, 89, 142, 83, 62, 50, 20, 9), (#9A; 25, 43, 57, 84, 115, 87, 61, —, 21, 6), (#11; 54, 72, 110, 133, 171, 120, 79, 69, 42, 0), (#12; 10, 29, 44, 80, 116, 100, 54, 38, 16, 1), (#14; 30, 75, 101, 142, 164, 141, 109, 84, 40, 3). Analyze the resulting 5×10 table as suggested by Section 9A. Present your results in standard form. What can you see?

9A3) Repeat 9A2, using the square roots of the values given there.

9A4) Repeat 9A2, using the logs of 1 PLUS the values given there. (Thus, for example, 25 becomes $\log 26 = 1.41$.)

9A5 and 9A6) H. M. Brown and J. S. Graham (*Textile Research Journal*, **20**: 418–425) present a table of the fineness (μg/in) of cotton as a function of variety and % mature fibers. Selected from this comes the following, in the form (variety; at 65%, at 70%, at 75%, at 80%): (Half & Half; 5.10, 5.40, 6.00, 6.81), (Deltalpine 11; 3.95, 4.25, 4.60, 5.07), (Stoneville 5; 4.35, 4.55, 4.68, 4.78), (Stoneville 2B (8275); 3.40, 3.82, 4.22, 4.65), (Hibred; 4.33, 4.73, 5.14, 5.60), (Qualla; 4.63, 4.88, 5.10, 5.28), (Startex; 4.45, 4.82, 5.30, 6.05), (Mexican Big Bowl; 4.10, 4.32, 4.59, 4.88), (Farm Relief; 4.68, 4.95, 5.15, 5.32), (Cleveland (Wannamaker); 4.90, 5.16, 5.30, 5.36), (Arkansas; 3.85, 4.08, 4.34, 4.67), (Cook 912; 4.25, 4.52, 4.81, 5.10), (Rogers Acala; 3.72, 3.95, 4.13, 4.30), (Triumph (759); 4.56, 4.85, 5.15, 5.48), (Triumph (44); 4.47, 4.83, 5.23, 5.70). Analyze the 5×4 table from the first five varieties as suggested by Section 9A. Present your results in standard form. What can you see?

9A6) Analyze the full 15 × 4 table of the data of Problem 9A5 and tell what you see.

9A7) Repeat 9A6 using logs of values rather than raw values.

9A8) (Continues 9A6 and 9A7) How would you compare and relate the analysis of 9A6 and 9A7?

9A9) W. J. Youden (*Analytical Chemistry*, **19**: 946–950) reports the final crucial digits in a precision measurement of the ratio of iodine to silver (one basis for atomic-weight calculations), where 2 iodine preparations were used in most, but not all combinations, with 5 silver preparations. We give medians of comparisons in the form (Iodine sample used; for silver A, for silver B, for silver C, for silver D, for silver E) in units of the 7th significant digit: (Iodine I; 24*, 41*, 29**, 50*, 55), (Iodine II; 18**, 18*, —, 61, —) where * = median of 2 observations, ** = median of 3 observations. Make the best two-way analysis you can of this 2 × 5 table with 2 holes. Present in standard form. Comment.

9A10) The *Report of the Royal Society, IGY Antarctic Expedition to Halley Bay*. etc., (Sir David Brunt, ed. 1962) gives (in table 14, page 191) the mean monthly temperatures at various heights (more precisely at the heights where the air pressure had fallen to specified pressures) for balloon ascents made at noon Greenwich meridian time. Exhibit **1** for Problem 9A10 extracts a 14 × 12 table of these temperatures. Analyze this table as suggested in Section 9A, and present the results in standard form. How successful do you think this analysis is in itself? As a basis for further analysis?

Exhibit **1** for Problem 9A10

Mean monthly temperatures (all negative) in tenths of a degree C (all in 1958)

Pressure level (mb)	J	F	M	A	M	J	J	A	S	O	N	D
30	343	423	529	687	787	870	917	885	827	651	394	326
40	354	425	530	665	779	849	901	891	852	704	443	348
50	368	425	517	640	778	837	888	886	850	720	464	350
60	375	428	509	636	761	822	870	877	839	728	483	371
80	385	431	498	610	719	789	843	851	829	731	520	392
100	396	435	496	585	696	761	811	831	820	740	597	403
150	412	435	472	557	644	717	768	792	789	740	601	446
200	423	435	468	551	630	701	750	762	759	720	626	461
250	453	456	492	584	652	666	701	719	708	677	628	540
300	502	496	548	571	614	616	639	654	644	616	586	556
400	430	427	463	472	509	500	521	525	512	499	465	459
500	334	329	361	373	411	396	417	417	399	385	354	355
700	179	186	212	227	270	254	257	272	248	225	204	209
850	93	123	139	177	208	206	207	218	211	154	128	128

9B1) (Continues 9A1) Make a plot like Exhibit 2 of Chapter 9 for the results of analysis in Problem 9A1.

9Bi) (Continues 9Ai, $i = 2, \ldots, 10$) Do the same as Problem 9B1 for a 9A problem you have been assigned.

9B(10 + i)) (Continues 9Bi, $i = 1, 2, \ldots, 10$) Make a plot like Exhibit 3 of Chapter 9 and like Exhibit 4 of Chapter 9, using the residuals previously found for your assigned problem. What can you see?

9Ci) ($i = 1, 2, \ldots, 10$) Rotate the plot of Problem 9Bi through 45° and add "water-level" indications, as in Exhibit 7 of Chapter 9, for the problem 9Bi you were assigned.

9C(10 + i)) ($i = 1, 2, \ldots, 10$) (Continues 9Ai) Make a plot like that of Exhibit 8 of Chapter 9 based on the residuals of the analysis of Problem 9Ai, whichever you were assigned. What do you see?

9Di) (Continues 9Ai, $i = 1, 2, \ldots, 10$) Polish the analysis of Problem 9Ai as indicated in Section 9D. Use medians. How much has this analysis been improved?

9D(10 + i)) (Continues 9Ai, $i = 1, 2, \ldots, 10$) As in 9Di, but use something other than medians.

9D21) Infant mortality rates, the number of deaths per 1000 live births, are given in Exhibit **1** for Problem 9D21 for some regions of the United States, broken down by education of the father. Analyze these data by means of an additive fit, and analyze the residuals.

9Ei) (Continues 9Ai or 9Di, $i = 1, 2, \ldots, 10$) Produce a plot like Exhibit 19 of Chapter 9 based on the analysis of Problem 9Di or 9Ai. What do you conclude?

Exhibit **1** for Problem 9D21

Infant mortality rates, U.S., 1964–66

Region	Education of father in years				
	≤8	9–11	12	13–15	≥16
Northeast	25.3	25.3	18.2	18.3	16.3
North Central	32.1	29.0	18.8	24.3	19.0
South	38.8	31.0	19.3	15.7	16.8
West	25.4	21.0	20.3	24.0	17.5

S) SOURCE

"Infant Mortality Rates: Socioeconomic Factors, United States," U.S. Department of HEW, NCHS, Vital and Health Statistics, Series 22, Number 14.

9E(10 + i)) (Continues 9Ei, i = 1, 2, ..., 10) If there appeared to be any slope in the plot of Problem 9Ei, calculate a further analysis as in Exhibit 20 of Chapter 9.

9F1) Carry Problem 9A3 through the chapter if you haven't already done it.

9P1) *Parity Progression Ratios* (Project). Parity progression ratios are defined as the proportion of women at a given parity level (number of previous live births) who eventually go on to have another birth. Bean and Wood give estimates of these ratios for three ethnic groups in the Southwestern United States in 1960 and 1970. These ratios are shown in Exhibit 1 for Problem 9P1.

These data can be treated in two ways. First, treat the six rows as three different populations and do a two-way PLUS analysis of the entire table. Alternatively, we can do separate analyses, one on the 1960 data and one on the 1970 data. How do the column (or parity) effects compare in three analyses? What about the row (or population) effects? Does the second method give sufficiently cleaner residuals to warrant the extra trouble?

What do the residuals and diagnostic plots indicate about re-expression? Do they point to the same thing in each of the subtables as in the entire table?

9P2) *Infant Mortality Rates* (Project) Exhibit 1 for Problem 9P2 gives infant mortality rates (deaths in first year/1000 live births) broken down by age of mother and the birth order (number of older siblings + 1) of the children. Which re-expressions seem helpful in fitting this data? Is there any structure remaining in the residuals?

9P3) (Teaser—perhaps a class project) P. B. Maek, J. A. Balog, and M. N. Jordon (*Textile Research Journal*, 22: 30–42) reported on the percentage changes

Exhibit 1 for Problem 9P1

Parity progression ratios

Parity	|0|	|1|	|2|	|3|	|4|	|5|
1960						
Anglos	.879	.797	.558	.484	.480	.531
Mexicans	.939	.920	.818	.796	.789	.774
Blacks	.748	.778	.787	.789	.774	.759
1970						
Anglos	.929	.882	.646	.531	.476	.496
Mexicans	.961	.943	.839	.803	.748	.713
Blacks	.866	.877	.794	.761	.723	.725

S) SOURCE

F. D. Bean and C. H. Wood, "Ethnic variations in the relationship between income and fertility," *Demography*, November 1974, 629–640. (Data reprinted by permission of the authors and of the journal.)

in breaking strengths of 6 fabrics, 2 types of washing, and 5 numbers of launderings. Their results for wet-strength changes can be given as follows, in tenths of a %, in the form: (fabric, type of washing; warp after 1, warp after 5, warp after 10, warp after 15, warp after 20; filling after 1, filling after 5, filling after 10, filling after 15, filling after 20): (Salyna, sonic; −40, −1, −104, −100, −108; −54, −34, −75, −109, −128), (Salyna, hand; −71, −99, −129, −109, −129; −143, −110, −177, −51, −38), (Gabardine I, sonic; 158, 106, −49, −24, 69; −36, −23, 13, 95, 10), (Gabardine I, hand; −213, −291, −286, −420, −379; −5, −50, −69, −104, −114), (Poplin, sonic; 17, −34, −72, −78, −115; −24, −123, −11, 55, −2), (Poplin, hand; −202, −345, −438, −523, −523; 65, 173, 32, 162, 144), (Gabardine II, sonic; 185, 212, 224, 166, 310; 200, 200, 276, 187, 250), (Gabardine II, hand; −130, −78, −146, −138, −246; 169, 127, 122, 344, 74), (Taffeta, sonic; 0, −58, 10, −48, −58; 0, 260, 150, 30, 120), (Taffeta, hand; −11, −112, −135, −124, −180; −32, −24, 71, −48, 16), (Satin, sonic; 115, 135, 62, 94, 83; 78, 117, 109, 86, 78), (Satin, hand; −197, −193, −160, −181, −354; −190, −95, −48, −114, 152). Notice that the changes extend from −523 (a loss of 52.3% of the initial strength) to 344 (a gain of 34.4% above the initial strength). Go as far as you can with an analysis of this $6 \times 2 \times 5 \times 2$ array of data, display the results of your analysis as well as you can, and discuss their appearances.

Exhibit **1** for Problem 9P2

(Deaths in first year)/(1000 live births)

| | | | Birth order | | | |
Age	1	2	3	4	5	>6
15–19	26.1	42.7	54.7	63.4	96.9	140.0
20–24	17.2	21.8	27.3	35.1	45.2	58.7
25–29	17.5	17.3	18.9	22.4	28.5	39.7
30–34	24.1	19.3	18.2	20.2	23.6	33.5
35–39	27.7	22.8	21.0	20.9	22.0	32.0
40–44	33.4	31.2	26.8	24.1	25.0	35.3

S) SOURCE

"A study of infant mortality from linked records by age of mother, total birth order, and other variables." United States, 1960 Live Birth Cohort, National Center for Health Statistics, Vital and Health Statistics, Series 20, Number 14, 1973.

Chapter 10

10A1) Calculate the mean of the days of survival for the control group, with and without the outlier, and for the experimental group, for the data in Exhibit 1 for Problem 10A1.

10A2) Calculate two types of resistant summary, the median and the biweight (for $c = 6$), for both the control and the experimental groups in the guinea-pig study of Vitamin C.

Exhibit **1** for Problem 10A1 (also used for Problems 10A2, 10D1, 10E1, 10E6)

Survival times (beyond day 10) of infant guinea pigs born of dams in experimental and control groups

Norkus and Russo used guinea pigs to see if high Vitamin C intake by expectant mothers caused ascorbic-acid dependency, and eventually scurvy, in the offspring. The expectant guinea pigs are divided into two groups. The experimental group (4 animals) is maintained on an ascorbic-acid level equivalent to 1500 mg for a 70-kg man, and the control group (5 animals) on one-tenth the amount. The 8 offspring in the experimental group and the 14 in the control are given the same dosage as their mothers for ten days. On the eleventh day all offspring are restricted to essentially no Vitamin C.

The experimenters compare the number of days of survival after day 10 for the two groups—by means of a *separate* standard error for each group. The days of survival, in stem-and-leaf format, are:

Control (Low level of Vitamin C)		Experimental (High level of Vitamin C)	
1		1	4466
2	1345	2	0446888
3	133	3	012
4		4	
5	4	5	

They conclude that, when this essential nutrient is removed from the diet, pups from dams receiving high levels of vitamin C die sooner than those on the control diet.

S) SOURCE

Norkus, E. P., and P. Russo (1975). "Changes in ascorbic acid metabolism of the offspring following high maternal intake of this vitamin in the pregnant guinea pig," *Annals of the New York Academy of Sciences*, 258, Second Conference on Vitamin C, 401–409.

10A3) Clarke ("The data of geochemistry", *Bulletin 3304,* The U.S. Geological Survey, 1908, page 608) gives analyses (in %) of 8 samples of native platinum, as given below, after rounding, for five of the more important constituents. For the constituent assigned to you, find (a) the mean, (b) the median, (c) the midmean, and (d) the biweight. Compare the results.

Constituent	Sample							
	A	B	C	D	E	F	G	H
Platinum	86.2	85.5	82.8	76.4	49.0	68.2	78.4	73.0
Palladium	.5	.6	3.1	1.4	.2	.3	.1	21.8
Rhodium	1.4	1.0	.3	.3	3.3	3.1	1.7	?
Copper	.6	1.4	.4	4.1	1.6	3.1	3.9	0
Iron	7.8	6.8	11.0	11.7	18.9	7.9	9.8	0

10A4) Clarke also (page 523) gives analyses of seven serpentinous rocks. Again rounded, they are, for the more important constituents:

Constituent	A	B	C	D	E	F	G
SiO_2	40.4	39.1	41.9	40.5	31.0	44.9	13.1
Al_2O_3	1.9	2.1	.7	.8	1.0	5.5	1.6
Fe_2O_3	2.8	4.3	0	4.0	4.9	1.8	1.2
FeO	4.3	2.0	4.2	2.0	2.0	3.5	.2
MgO	36.0	39.8	38.6	37.4	38.4	25.6	58.4
$H_2O(l)$	10.7	12.7	14.2	13.8	20.8	5.8	24.8

Imitate 10A3 for the constituent assigned to you.

10A5) Clarke also (page 377) gives analyses of seven leucite rocks. Again rounded, they are, for the most important constituents:

Constituent	A	B	C	D	E	F	G
SiO_2	51.9	44.4	50.2	46.5	46.0	42.6	46.1
Al_2O_3	20.3	20.0	11.2	11.9	17.1	9.1	16.0
Fe_2O_3	3.6	5.2	3.3	7.6	4.2	5.1	3.2
FeO	1.2	2.8	1.8	4.4	5.4	1.1	5.6
MgO	.2	1.8	7.1	4.7	5.3	10.9	14.7
CaO	1.6	8.5	8.0	7.4	10.5	12.4	10.6
Na_2O	8.5	6.5	1.4	2.4	2.2	.9	1.3
K_2O	9.8	8.1	9.8	8.7	9.0	8.0	5.1
$H_2O(l)$	1.1	1.4	2.6	3.6	.4	4.2	1.4

Imitate 10A3 for the constituent assigned to you.

10A6) Clarke also gives (page 317) analyses of 6 samples of augite, a mineral of varying composition. Again rounded, they are, for the most important constituents:

Constituent	A	B	C	D	E	F
SiO_2	45.2	49.1	47.5	48.7	54.9	47.1
TiO_2	4.3	0	3.0	0	0	1.8
Al_2O_3	7.7	8.0	4.1	9.3	6.3	7.8
Fe_2O_3	3.0	0	5.6	3.8	2.9	1.3
FeO	4.1	8.3	6.4	6.4	4.6	8.2
MgO	12.2	12.4	10.0	14.7	14.5	13.5
CaO	23.4	22.6	21.6	16.8	15.9	19.3

Imitate 10A3 for the constituent assigned to you.

10A7) Landolt-Bornstein's *Physikalisch-Chemische Tabellen* of 1923 (Volume 2, page 254) collects the results of 9 investigators for the specific heat of water at various temperatures. The results at temperatures from 5°C to 30°C are given (at 5° intervals) in Exhibit 1 for Problem 10A7. Most experimenters based their measurements on the same kind of standard thermometer, but three used one of two other standards, as indicated in the last column. Imitate 10A3 for the temperature assigned to you. Discuss the results in view of the additional information above about the thermometric standard used.

10A8) The Landolt–Bornstein Tabellen also collect (Volume 2, pages 801–802) 14 values for the Planck's constant h. As multiples of 10^{29} erg-seconds, rounded, these are: 667, 662, 656, 655, 652, 658, 654, 650, 653, 657, 656, 653, 656, 654. Imitate 10A3 for these.

Exhibit 1 for Problem 10A7

Specific heat of water at various temperatures according to nine investigators

Investigator	5°C	10°C	15°C	20°C	25°C	30°C	Different thermometer
Regnoult	.9994	.9997	1.0000	1.0004	1.0008	1.0012	(A)
Liidin	1.0027	1.0010	1.0000	.9994	.9993	.9996	
Dieterici	1.0050	1.0021	1.0000	.9987	.9983	.9984	
Bonsfreld	1.0039	1.0016	1.0000	.9991	.9989	.9990	
Callendor	1.0042	1.0019	1.0000	.9988	.9980	.9975	(T)
Ronland	1.0054	1.0019	1.0000	.9979	.9972	.9969	
Bartollis	1.0041	1.0017	1.0000	.9994	1.0000	1.0016	
Janke	1.0040	1.0016	1.0000	.9991	.9987	.9988	
Jaeger	1.0030	1.0013	1.0000	.9990	.9983	.9979	(T)

10A9) Calculate means, medians, and biweights ($c = 6$) for subjects F and G at Lausanne (Exhibit **1** for Problem 10A9). Find the three differences. Comment on their similarity or dissimilarity.

10A10) Do the same for subject F at Lausanne and Zuoz (Exhibit 1 for Problem 10A9).

10A11) Repeat the biweights in 10A3 using $c = 9$. Comment on size of effect.

10A12) Repeat the biweights in 10A4 using $c = 4$. Comment on size of effect.

10A13) Calculate means, medians, and biweights ($c = 6$) for the 1939 uterus weights at .20 and $.28\gamma$/rat of stilbestrol in Exhibit **1** for Problem 10A13. Find the three differences, comment on their similarity or dissimilarity.

10A14) Do the same for the 1939 and 1940 weights at $.28\gamma$/rat.

10A15) Do the same for the moduli of elasticity for trees at sites 5 and 7 in Exhibit **1** for Problem 10A15.

Exhibit **1** for Problem 10A9 (also used for 10A10, 10D2, 10E3, 10E4)

Patellar reflexes at two elevations

Linder** reports the results of measurements, on 8 successive days, of the patellar reflex of several subjects at Lausanne (550-meter elevation) and then, after a day for adjustment, for 8 days at Zuoz (1750-meter altitude). Part of the data, expressed in logarithmic units (values are 20 times the excess of the common logarithms over 2) is given below.

Subject F		Subject G	
Lausanne	Zuoz	Lausanne	Zuoz
77	58	79	59
82	63	80	59
78	81	65	56
75	63	78	56
76	76	78	56
75	71	83	65
74	95	84	66
73	57	79	84

** Linder, A. (1950), "Statistical analysis of some physiological experiments," *Sankhyā*, 10, 1–12. (Used by C. I. Bliss and D. W. Calhoun 1954, *An Outline of Biometry*, Yale Cooperative Corporation, page 100.) Data reprinted by permission of the author and of the editors of Sankhyā.

Exhibit **1** for Problem 10A13 (and also used for 10A14, 10D3, 10D4, 10E3, 10E4, 10E8, 10E9)

Uterus weights of immature rats after stilbestrol feedings (in 1000(log wt. −1.4))

Lee, Robbins and Chen[***] gave varied doses of stilbestrol to immature rats and then weighed their uterus with, among others, the following results (expressed as 1000 (log of wt. − 1.4)):

December 1939		April 1940
.20γ/rat	.28γ/rat	.28γ/rat
−38	83	−2
100	138	122
122	173	130
139	197	191
144	197	321
149	232	
214	251	

[***] Lee, H. M., Robbins, E. B., and Chen, K. K. (1942). "The potency of stilbestrol in the immature female rat," *Endocrinology* 30: 469–473. (Quoted by C. I. Bliss and D. W. Calhoun 1954. *An Outline of Biometry*, Yale Cooperative Corporation, New Haven, at page 103.) Data reprinted with permission of journal and author.

Exhibit **1** for Problem 10A15 (and also used for 10D5, 10D6, 10E5, 10E10)

Red pine elasticity in trees from Connecticut plantation

Kraemer[**] measured the modulus of elasticity of the outside pieces of red pine trees from different sites in a 25–30-year-old plantation in Connecticut. Some of his results were as follows (expressed as 1/10 the modulus):

Site 1	Site 5	Site 7
136	68	95
138	114	117
140	120	118
143	132	124
146	147	126
149	150	132
150	159	133
157	163	146
159	164	150
	178	152
	197	161

[**] Kraemer, J. H., Dissertation, Yale University, 1943. (Quoted by C. I. Bliss and D. W. Calhoun, 1954. *An Outline of Biometry*, Yale Cooperative Corporation, New Haven, at page 59.)

10A16[C]) Martin* measured the length of survival of mice infected with tubercle bacilli. Bliss and Calhoun* combined Martin's experiments into two groups, A and B, with the results shown in Exhibit 1 of Problem 10A16. Treating 14^+ as 14.5 days, etc., calculate as many as you can of mean, median, biweight, midmean, and trimean for each group. Make as many comparisons as you can. Discuss your results.

10A17) Would you be able to calculate more of the measures of location specified in 10A16 if you worked with 1/(time of survival) instead of (time of survival)?

10A18) Make the computations of 10A16 for −100/(days of survival).

10B1) Define robustness of efficiency. (Refer to Section 1E as well.)

10B2[C]) One estimate is 92% as efficient as another. Suppose both have estimates of their variances that we can afford to calculate and that the sample size is so large that we can neglect the variability of these estimates—can treat them as being without error. What will the ratio of their estimated variances be? What will the ratio of the lengths of the corresponding confidence intervals be? Is this enough to care about? To worry about? Why/why not?

10B3/4/5) Same as 10B2 for 50%/80%/10% relative efficiency.

10C1) For small samples from approximately normal distributions without stretched tails, how would you choose among the arithmetic mean, the median, and the biweight?

10C2) To keep the amount of calculation down in large samples, choose among the 3 location estimates of Exhibit 1 of Chapter 10.

Exhibit 1 for Problem 10A16

Mouse survival (in days after inoculation)

Group	14^+	15^+	16^+	17^+	18^+	19^+	20^+	21^+	22^+
A	7	6	12	23	23	43	37	23	17
B	1	—	—	3	2	7	12	14	16

Group	23^+	24^+	25^+	26^+	27^+	28^+	29^+	30^+	≥ 31
A	6	7	8	5	1	2	1	—	3
B	16	24	12	4	0	0	4	4	19

* Martin, A. R. (1946). "The use of mice in the examination of drugs for chemotherapeutic activity against *mycobacterium tuberculosis*," *J. Pathol. Bacteriol.*, 58: 500–585. (Used by C. I. Bliss and D. W. Calhoun, *An Outline of Biometry*, Yale Cooperative Corporation, New Haven, at page 62.)

10C3) For good defense against stretched tails in large samples, what is the location estimate of choice?

10C4^c) A city hires policemen subject to lower and upper limits on their height. A study is being made to compare the heights of traffic-control officers and detectives. Which measure of location would you recommend using?

10C5^c A golf club conducts a hole-in-one tournament each year, participated in by a large number of experts and a smaller number of optimistic dubs. The resting place of every ball has been carefully mapped each year. It is desired to summarize each year's experience (both near-to-far and right-to-left) so that year-to-year variations in the apparent center of impact can be related to year-to-year variations in the placement of the hole on the green. Which measure of location would you recommend using?

10C6) As part of a campaign to strengthen an after-11 P.M. noise ordinance in a middle-sized city, measurements of average noise power for 10-minute intervals have been made in 27 sites in and near the downtown area. These measurements have been made from 11 P.M. to 5 A.M. on each of 53 consecutive days. How would you recommend summarizing the 53 measurements at any one ten-minute interval and place?

10C7) (Continues 10C6.) Would your answer to 10C6 be altered if (a) you were advising the anti-noise committee, (b) you were advising those opposed to stronger noise controls, (c) you knew that a particular site was on a route heavily used by ambulances and police cars? How/why?

10D1) Using the median as the resistant estimate of location for the robust estimate of scale, MAD, complete the following chart for the Vitamin C data given in Exhibit 1 for Problem 10A1; s_{bi}^2 here indicates the modification of the Lax estimate of scale at the close of Section 10D.

	Control	Experimental
Biweight		
Median		
MAD		
s_{bi}^2		

10D2) Fill out a chart analogous to that of 10D1 for subjects F and G at Lausanne (patellar reflex data—see Problem 10A9).

10D3) Do the same for .20 and .28γ/rat in December 1939 (uterus weight data—see Problem 10A13).

10D4) Do the same for .28γ/rat in December 1939 and April 1940 (same).

10D5) Do the same for sites 5 and 7 (pine elasticity modulus data—see Problem 10A15).

10D6)   Do the same for sites 1 and 5 (same).

10D7^C)   Compute as many of hinge-spread ($\sim$ interquartile difference), s^2, and s_{bi}^2 for the two groups of Problem 10A10 as you can.

10D8)   Would you be able to calculate more of the measures of spread specified in Problem 10D7 if you worked with 1/(time of survival) instead of (time of survival)?

10D9)   Make the computations of Problem 10D7 for -100/(days of survival).

10E1)   (Continues 10D1.) Use the results of 10D1 to compare, resistantly, the biweight values for the Vitamin C data from Exhibit 1 for Problem 10A1. Give a 95% confidence interval for the difference.

10E2/3/4/5) (Continues 10D2/3/4/5.) As 10E1, for 10D2/3/4/5.

10E6)   (Continues 10E1.) Make a (nonresistant) comparison of the means in the Vitamin C data using Student's t to set a 95% confidence interval. Compare with the results of 10E1 and discuss.

10E7/8/9/10) Continue Problem 10E2/3/4/5, as in Problem 10E6.

10E11)   For the same constituent as in 10A3, use Student's t both (a) based on $\bar{y}$ and s^2 and (b) based on biweight and s^2, to set 95% confidence limits on a typical value for that constituent. Discuss the difference in the results.

10E12/13/14) Imitate 10E11 for the constituent assigned to you with the data of 10A4/5/6.

10E15/16) Imitate 10E11, for the temperature assigned to you, with the data of 10A7/8.

Re 10G) For the problems on this section see Problems 14Q1 to 14Q4.

10P1)   *National Assessment of Educational Progress* (Project) Sometimes empirical studies can tell us what location estimates are preferable in a given situation. Such opportunities arise especially for large sets of data with repeated use of the same analysis. An example is the study of means, medians, and biweights for data by the National Assessment of Educational Progress, whose researchers set up several different populations based on their empirical findings. They drew 70 observations from a population and computed the three measures of location for that sample. They repeated this operation 400 times for each population. We provide in Exhibit **1** for Problem 10P1 the stem-and-leafs for the biweight, the median, and the mean. The population here was that of the *changes in percent right* on 70 multiple-choice questions for 13-year-olds for the nation, between the first and fourth year examination in the subject Science. Use the stem-and-leaf plots in the exhibit to report on the properties of the distributions of these three measures of location and to help decide which measure of location seems preferable for this particular population.

Exhibit 1 for Problem 10P1

Stem-and-leaf plots

A) For mean

```
 15 | 4
 14 |
 13 | 4
 12 |
 11 | 369
 10 | 002456
  9 | 122248
  8 | 0125677889
  7 | 01124447
  6 | 133567777
  5 | 001122344445678
  4 | 000112444556667899
  3 | 011123444455666778999
  2 | 000001111234455555677778888899
  1 | 0012223333346677777788999
  0 | 0122223333444555666667777778888999
 -0 | 888388777766554444333322222211
 -1 | 99998777665554443333222222111111000
 -2 | 9999998877766555444444322222111000
 -3 | 99999876666655544443333332211100
 -4 | 88777665544333221100
 -5 | 99888776665554333211100
 -6 | 88764332100000
 -7 | 954321110
 -8 | 85221000
 -9 | 9863
-10 | 9641
-11 | 72
-12 | 82
-13 |
-14 |
-15 | 1
```

➡

Exhibit 1 for Problem 10P1 (continued)

B) For median

```
21:6
20:379
19:2278
18:
17:
16:4
15:
14:
13:000
12:9999
11:11111111666666677778
10:5555
 9:000
 8:333333338999
 7:
 6:000000000000000000000000001111111111111111111111111
 5:
 4:77777777777777777777
 3:001111111111144444444444444
 2:
 1:777778888888
 0:000000000000000001111111111111117
-0:666111111111111
-1:
-2:44444444444444444333
-3:977777666
-4:666666666666666554
-5:
-6:97777222222200000000000000
-7:77777777755555555555533333333333333333
-8:888888888888855543333333333333111111
-9:4443333222221111100000000000
-10:9831
-11:911111
-12:88
-13:0
-14:954
-15:000
-16:1
-17:
-18:31
```

➤

Exhibit 1 for Problem 10P1 (continued)

C) For biweight

```
 16 | 2
 15 | 9
 14 |
 13 |
 12 | 12
 11 | 013499
 10 | 002444
  9 | 011367
  8 | 0223455568
  7 | 001123466
  6 | 023333444449
  5 | 00112223334455567778
  4 | 00011223456678999
  3 | 012233333556677889
  2 | 12222334467889
  1 | 0000112334444555556666778888999
  0 | 00001233334444555556667888888999
 -0 | 999888777776655544443222211111
 -1 | 9997777766666666555543332222221111110000
 -2 | 98877777766555533332222
 -3 | 99998888887666555554433222100
 -4 | 99988777666554433332111 0
 -5 | 99888654444332211000
 -6 | 99998766555333311111000
 -7 | 876555310
 -8 | 87520
 -9 | 86200
-10 | 520
-11 | 54
-12 | 80
-13 |
-14 | 4
```

S) SOURCE

Robert Larson and Don Searls, National Assessment of Educational Progress, personal communication.

Chapter 11

11-1) When should standardizing for comparison be used?

11-2) How do randomized trials provide controls?

11A1) What are the distinctions between "crude" and "adjusted" rates?

11A2) What is the standard population (in numbers) for a 50–50 mixture of Easy and Hard in Example 1 of Section 11A?

11A3) Show the calculations for the percentage difference (II − I) with a 50–50 mixture of Easy and Hard.

11A4) What is the standard population (in numbers) for 45% Easy and 55% Hard in Example 1 of Section 11A?

11A5) Let the probabilities of success for Treatment I be: p_{11} in the Easy stratum and p_{21} in the Hard stratum. What would be the crude success rate when Treatment I is applied to a population that is composed of a mixture with proportions t of Easy and $1 - t$ of Hard?

11A6) Let the probabilities of success p_{ij} be as follows

	Treatments	
Strata	I	II
Easy	p_{11}	p_{12}
Hard	p_{21}	p_{22}

with $p_{11} > p_{12}$ and $p_{21} < p_{22}$ What mixtures of Easy and Hard in the population will lead Treatment I to be preferred to Treatment II?

11A7) Using Data Exhibit 2 for Problems (at the end of the problems) on smoking and health, compute the crude death rates for the nonsmokers and the three smoking groups.

11A8) (Continues 11A7) Consider the nonsmoking population as standard and compute standardized death rates R_{non} for nonsmokers, R_{cp} for cigar and pipe smokers only, R_c for cigarette smokers only.

11A9) Compare the results of 11A7 and 11A8.

11A10) Consider the cigarette smokers and nonsmokers in Data Exhibit 2 for Problems. Use ages 0–59 and ages 60^+ as two strata. Calculate the difference, smokers minus nonsmokers, for a 50–50 mixture of the two strata.

11A11) What is direct standardization?

11A12) Assuming no correlation between results for Easy and Hard strata, find Var $(p_{std,I} - p_{std,II})$ when the standard population is as in Question 2 in the text.

11B1) In Section 11A, what are the W's for the directly standardized rates in Questions 1–3 in the text?

11B2) Explain why the greater age of the population in Maine caused its crude death rate to be higher, even though its specific rates in each age class are lower than in South Carolina in Exhibit 2 of Chapter 11.

11B3) In Exhibit 1 of Chapter 11, the crude difference in success rate (Treatment I − Treatment II = 60% − 44%) is different from what it would be if the comparability of the groups were taken into account. In Exhibit 2 of Chapter 11, the crude rates also produce misleading results. Explain how the difficulties are overcome in each example.

11B4) Use the following numbers in each age group for a standard million in the U.S. in 1975 to adjust the death rates in Exhibit 3 of Chapter 11 for Maine and South Carolina by the direct method.

Age (in years)	Standard million for U.S. in 1975	Age (in years)	Standard million for U.S. in 1975
0–4	74,400	25–34	144,600
5–9	81,000	35–44	107,200
10–14	95,500	45–54	111,400
15–19	98,700	55–64	92,200
20–24	90,300	65–74	64,600
		75+	40,200

11B5) (Continues 11B4.) Compare the results of 11B4 for a 1975 standard million to those for a 1940 standard million in Exhibit 3 of Chapter 11.

11B6) In Exhibit 1 for Problem 11B6, Panel B gives the distribution of word lengths for texts of 15 to 18 thousand words from Hamilton and Madison. Panel A gives the frequency (occurrences per 1,000 total words) for one particular word of each length. Combining these, we can find, for each word length, the chance that either author will use the particular word of that length. Panel C gives a standard distribution of word lengths based on Melville's *Moby Dick*. Use it to calculate the chance that any of the three authors will use one of the particular words, standardized for the distribution of word length.

(Exhibit **1** for Problem 11B6 appears on the next page.)

Exhibit **1** for Problem 11B6

Word lengths and word frequency distributions

A) Frequency distributions of words of 1–12 letters

	Rates per 1000 words*	
Word of length k	Hamilton	Madison
$k =$ 1 (a)	22.85	20.22
2 (of)	64.85	57.80
3 (our)	2.27	1.11
4 (what)	1.38	1.15
5 (among)	.39	.84
6 (second)	.18	.37
7 (whether)	.49	.97
8 (language)	.04	.21
9 (direction)	.22	.03
10 (throughout)	.04	.17
11 (destruction)	.13	.01
12 (consequently)	.03	.48

B) Distribution of word lengths

Word length k	Hamilton	Madison
1	423.2	396.1
2	3531.6	3834.7
3	2925.0	3644.7
4	2042.4	2204.9
5	1580.0	1720.2
6	1116.1	1396.4
7	1026.7	1298.7
8	824.5	1027.4
9	805.7	888.1
10	617.6	743.4
11	396.6	450.4
≥12	385.6	483.0
	15,675.0	18,088.0

C) Distribution of word lengths

k	Standard thousand from *Moby Dick***
1	45.0
2	162.0
3	228.1
4	196.5
5	122.3
6	79.5
7	66.0
8	45.3
9	27.9
10	15.1
11	8.1
≥12	4.3

* Source: Mosteller, F., and David L. Wallace (1964). *Inference and Disputed Authorship: The Federalist*. Addison-Wesley: Reading, MA, pp. 244–248, 258, 260. Reprinted by permission of publisher and author.
** Nowlin, A. G. (1973). "Statistical analysis of linguistic word-frequency distributions and word-length sequences," Ph.D. thesis, Princeton University. Reprinted by permission of author.

11B7) In Exhibit **1** for Problem 11B7 the number of deaths after operations from two areas in the country are displayed by age and by sex. For each of the two areas compare the directly standardized death rates for males and females, taking the average male–female population as standard.

Exhibit **1** for Problem 11B7

Surgical deaths

The following counts come from two areas in the United States and refer to a 5-year period. The population is all patients who underwent surgery, and the deaths include all patients who died in the hospital following surgery.

	Area I			
	Population		Deaths	
Age	Males	Females	Males	Females
0–4	2104	1952	34	22
5–14	4272	3911	9	11
15–24	2835	2989	23	5
25–34	2785	2606	19	8
35–44	1930	1886	16	15
45–54	1497	1524	59	40
55–64	960	1013	101	52
65–75	652	855	185	118
76–83	186	287	97	108
84+	69	125	68	103

	Area II			
	Population		Deaths	
Age	Males	Females	Males	Females
0–4	703	689	12	3
5–14	1739	1758	5	2
15–24	1233	1244	14	1
25–34	989	1004	8	3
35–44	897	922	9	13
45–54	921	961	28	15
55–64	686	739	68	37
65–75	611	784	159	73
76–83	189	290	86	88
84+	52	124	70	119

11B8) For Exhibit 1 for Problem 11B7, compare the directly standardized death rates in Areas I and II with a 50–50 mixture of the two populations taken as standard (males and females pooled together in each area).

11C1) Using the standardized binomial, calculate the variance for the standardized difference in word rates from 11B6.

11C2) Use the stratified binomial to improve the calculation of the variance from 11C1.

11C3) Compute, using the improved stratified binomial, the standard error of the differences: $R_{non} - R_{cp}$, $R_{non} - R_c$ for the smoking and health data of 11A8. Based on these standard errors, what can be said about the different rates?

11C4) Are the crude or standardized or stratified binomials appropriate for calculating the standard error?

11D1) In view of Exhibit 4 of Chapter 11, explain why it is important to review any standardizing calculation. What are the potential difficulties?

11D2) A computer program calculates standardized rates. What displays would you like the program to provide to warn you of a wild result?

11D3) In Exhibit 1 for Problem 11B7 on surgical deaths, the estimated death rate for females ages 5–24 is almost zero. What sort of problems does this impose?

11E1) How does indirect standardization differ from direct standardization?

11E2) In Example 5 of Section 11E, why are the reference percent success rates not to be compared with each other but, rather, with their crude success rates?

11E3) Do Treatments I and II perform differently in Example 2 of Section 11A for indirect standardization?

11E4) Given the weighting of treatment totals of Example 6 in Section 11E, are Treatments I and II identical?

11E5) In the calculation of the rough approximation to the variance, when would the standardized number n_{std} not be considered known?

11E6) Based on the data presented in Exhibit 2 of Chapter 11, compute the indirectly standardized death rates for Maine and South Carolina.

11E7) For Data Exhibit 2 for Problems, compute the indirectly standardized lung-cancer ratios for smokers and nonsmokers.

11E8) Describe two awkwardnesses with the indirect approach.

11F1) Why is adjustment necessary?

11F2) In general, what standard difficulties might be selected to adjust for comparison of the Easy and of the Hard groups?

11F3) Why is the logistic used as the distribution of standard difficulty?

11F4) With the standard population as 55% Easy and 45% Hard, what is the difference (Treatment II − Treatment I) for the adjusted groups?

11F5) Does the adjustment work for both directly and indirectly standardized populations?

11F6) Consider the cigarette smokers and nonsmokers in Data Exhibit 2 for Problems. Calculate the centers of gravity for each for the two strata ages 0–59 and ages 60^+.

11F7) Calculate the average difference, smokers minus nonsmokers, for 11F6.

11F8) Compare the results of 11F7 with 11A10.

11F9) Using a mixture of 45% ages 0–59 and 55% ages 60^+ as the standard, find the adjusted difference (smokers minus nonsmokers).

11F10) What advantage is there to calculating percentages of success at the cutting point?

11P1) *Standardized Fertility Indices* (Project) Age distribution of women in the childbearing ages and age-specific fertility rates for 3 modern European countries are given in Exhibit 1 for Problem 11P1. Pooling the data to get distributions and rates, obtain the directly and indirectly standardized fertility rates. Demographers frequently standardize fertility rates indirectly on the fertility of the Hutterites, a religious sect in the western United States and Canada known for their exceptionally high fertility. Compare the three countries based on these rates, given below. By inspection, it seems clear that fertility is highest in France and lowest in the U.K. Do the three methods of standardization bear this out? Based on each method, separately and jointly, would you say that Norway is closer to France or the U.K. in its fertility experience?

Exhibit 1 for Problem 11P1

Age distributions and fertility rates

Age	France (1968) Women (in 1000's)	Rate	Norway (1973) Women (in 1000's)	Rate	U.K. (1973) Women (in 1000's)	Rate	Hutterites Rate
15–19	2070.	.0255	150.8	.0443	1696.	.0432	.300
20–24	1851.	.1580	145.7	.1500	1707.	.1309	.550
25–29	1382.	.1636	150.9	.1373	1799.	.1354	.502
30–34	1514.	.0985	111.8	.0732	1443.	.0635	.447
35–39	1646.	.0479	96.5	.0309	1387.	.0246	.406
40–44	1657.	.0146	101.2	.0073	1424.	.0061	.222
45–49	1561.	.0013	113.0	.0004	1497.	.0004	.061

S) SOURCE

U.N. Demographic Yearbooks, 1972 and 1974; for the Hutterite data, A. J. Coale, "Factors associated with the development of low fertility" in *U.N. World Population Conference,* 1965, Vol. II, U.N.: New York, 1967.

Why do demographers like to use a certain fixed standard with which to compare all populations? Why should they choose the Hutterites, who have the highest reliably recorded fertility known?

Chapter 12

12–1) Compare and distinguish the concepts of asociation, independence, and causation.

12–2) What are 3 sorts of ideas necessary to support the notion of cause?

12–3) Contrast the mathematical and statistical meanings of "dependence".

12–4) Exhibit 1 for Problem 12–4 gives the mean annual level of Lake Victoria Nyanza (x) relative to a fixed standard, and the number of sunspots (y) for the years 1902–1921. Plot y vs. x, and do a simple regression of y on x. Suggest a mechanism that might explain the association of x and y.

12A1) What is the first meaning of regression?

12A2) Describe some situations where one would prefer summaries other than the mean.

Exhibit 1 for Problem 12–4

Mean annual level of Lake Victoria Nyanza and number of sunspots, 1902–1921.

Year	x	y	Year	x	y
1902	−10	5	1912	−11	4
1903	13	24	1913	−3	1
1904	18	42	1914	−2	10
1905	15	63	1915	4	47
1906	29	54	1916	15	57
1907	21	62	1917	35	104
1908	10	49	1918	27	81
1909	8	44	1919	8	64
1910	1	19	1920	3	38
1911	−7	6	1921	−5	25

S) SOURCE

Sir Napier Shaw, *Manual of Meteorology*, Vol. 1: Meteorology in history; Cambridge University Press, London 1942, p. 284. Data reprinted by permission of publisher.

12A3) What is the second meaning of regression?

12A4) What are some advantages and disadvantages of each type of regression?

12A5) Plot length of stay, y, on physical status, x, for the herniorrhaphy data (Data Exhibit 8 for Problems at the end of the book).

  a) Compute the mean of y at each physical status, and plot these points on the graph. In what sense is this a regression? How good a summary do you think it provides?

  b) Repeat part (a) for the median of y at each physical status, and compare. How might the addition of other percentage points improve the regression as a summary?

12A6) Draw a straight line through the points in the graph from the previous problem. What is an equation for your line? Interpret the equation. What advantages and disadvantages does this kind of summary have compared with those of 12A5?

12A7) Exhibit **1** for Problem 12A7 gives the number of icebergs sighted south of Newfoundland (x) and south of the Grand Banks (y) for each month of 1928.

  a) Plot y on x, and draw a line through the points. What is an equation for your line?

  b) Compute the residuals and plot them on x. What can you say about the errors you expect in predictions of y as a function of x?

  c) How do the ideas of association, dependence, and causation apply to this problem?

12B1) Describe five different uses for regression.

12B2) Why don't we care about regression coefficients when fitting for exclusion? What do we care about?

Exhibit **1** for Problem 12A7

Numbers of icebergs sighted monthly south of Newfoundland, x, and south of the Grand Banks, y, for 1920.

Month	J	F	M	A	M	J	J	A	S	O	N	D
x	3	10	36	83	130	68	25	13	9	4	3	2
y	0	1	4	9	18	13	3	2	1	0	0	0

S) SOURCE
Sir Napier Shaw, *Manual of Meteorology*, Vol. 2, p. 407, Cambridge University Press, London, 1942. Data reprinted by permission of publisher.

12B3) Why is it important to remember that many different sets of variables often predict about equally well?

12B4) A regression in the sense of local averages appears in Exhibit **1** for Problem 12B4 for eventual number of children anticipated by women interviewed in national samples taken in 1955, 1960, and 1965. Do women's expectations appear to change from 1950 to 1965? Regress, for the 14 points, the anticipated number of children on year of interview (coded: −1, 0, 1; no constant term) and compute anticipated numbers of children with this trend subtracted.

12B5) Plot length of stay (y) on age in years (x) for the herniorrhaphy data (Data Exhibit 8 for Problems).

a) Draw a line through the points and interpret its equation. Compute lengths of stay adjusted for age.

b) Do you think it wise or unwise to use a single regression formula to adjust for age when the range is from 2 to 80 years, as here?

12B6) Plot appropriations for research to the Air Force (y) against year (x) for Armed Forces Research data (Data Exhibit 10 for Problems).

a) Draw a line through the points and use its equation to get appropriations "adjusted for linear trend."

b) Plot "adjusted y" by year. Have you removed differences over time? What kind of function might work better?

12C1) What does $x_{2;1}$ mean?

Exhibit **1** for Problem 12B4

Anticipated eventual number of children

Age at interview	Year of interview		
	1955	1960	1965
20–24	3.20	3.07	3.26
25–29	3.32	3.46	3.53
30–34	3.32	3.49	3.58
35–39	3.16	3.35	3.59
40–44	—	3.46	3.54

S) SOURCE

N. B. Ryder and C. F. Westoff, *Reproduction in the United States 1965*, Princeton University Press, Princeton, N.J., 1971, p. 42.

12C2) What, in words and formula, is $y_{;1}$?

12C3) What is $x_{2;1}$?

12C4) Exactly what are $y_{;12}$ and $x_{1;25}$?

12C5) Fitting x_2 to $y_{;1}$ is algebraically equivalent to fitting $x_{2;1}$ to $y_{;1}$. What is the advantage of the latter method?

12C6) We learned in Section 12B that very different sets of variables can often produce nearly the same fit. Explain why we often find that very different sets of coefficients for the same variables can produce nearly the same fit as well.

12C7) Use the graphical method of this section to fit age (first) (x_1) and physical status (second) (x_2) to length of stay (y) for the herniorrhaphy data (Data Exhibit 8 for Problems). (If you have done Problem 12B5, the first step has already been done.) Compare your equations

$$y = b_0 + b_1 x_1$$

and

$$y = c_0 + c_1 x_1 + c_2 x_2.$$

12C8) Regress savings on income and interest rate (Moody's Aaa long-term bonds), using the Economics data set (Data Exhibit 4 for Problems). Save your work for later problems.

12C9) a) Which single variable estimates pupil's verbal performance best in the Coleman data (Data Exhibit 7 for Problems)?

b) What second variable seems to do the best job together with your choice in (a)? Does it seem worth having?

12D1) If we wish to adjust y for z and have several measures of z, why may it not help to use all of them in a regression? (More on this topic in Sections 12F and 13G and Chapter 15 including the possible use of composites with coefficients selected by personal judgment.)

12D2) How might two clearly disparate carriers come to be collinear? (See, for example, Problem 12–4.) Thus, collinearity often cannot be predicted from the nature of the variables.

12D3) A "least-squares" fit of y on x_1, x_2, x_3, x_4, x_5 for the Coleman data (Data Exhibit 7 for Problems) yields the formula

$$y = 19.9 - 1.79x_1 + .0432x_2 + .556x_3 + 1.11x_4 - 1.79x_5.$$

a) What is apparently surprising about some of the coefficients? Why is this not unexpected for this stock?

b) Which variables seem to be collinear? Why/why not? Replace the set of collinear variables with one carrier, and refit, using the reduced stock. Discuss how your residuals compare with residuals from the fit with the complete stock.

12D4) For the Economics data (Data Exhibit 4 for Problems), let

$$y = \text{savings,}$$
$$x_1 = \text{income,}$$
$$x_2 = \text{long-term interest rate (Moody's Aaa),}$$
$$x_3 = \text{long-term interest rate (Moody's Bbb),}$$
$$x_4 = \text{short-term interest rate.}$$

Compare the regressions:

a) $y = b_0 + b_1x_1 + b_2x_2$;
b) $y = b_0 + b_1x_1 + b_2x_2 + b_3x_3$;
c) $y = b_0 + b_1x_1 + b_2x_2 + b_3x_3 + b_4x_4$.

12D5) Economists have known for a long time about the empirical relation which we approximate as:

$$(\text{Inflation}) = a + b(\text{unemployment rate}).$$

This is known as the Phillips Curve. For the Economics data (Data Exhibit 4 for Problems), let

$$y = \text{inflation} \left(\text{if } P_t = \text{consumer price index in year } t, \text{ let } y = \frac{(P_t - P_{t-1})}{P_{t-1}} \right),$$

$x_1 = \text{overall unemployment,}$
$x_2 = \text{unemployment for men over 20 years old,}$
$x_3 = \text{unemployment for women over 20 years old.}$

a) For the carrier x_i and pair of carriers (x_i, x_j) assigned to you, fit the regressions

$$y = b_0 + b_i x_i$$

and

$$y = c_0 + c_i x_i + c_j x_j.$$

b) Compare the coefficients and the fits.

12D6) Compare your results in 12D5 with a fit of the form

$$y = d_0 + d_1x_1 + d_2x_2 + d_3x_3.$$

12D7) Drops in barometric pressure are associated with bad weather. In Data Exhibit 6 for Problems, let

$$y = (\text{station pressure}) - (\text{yesterday's station pressure}),$$
$x_1 = \text{relative humidity,}$
$x_2 = \text{fog,}$
$x_3 = \text{precipitation.}$

a) For the carrier x_i and pair of carriers (x_i, x_j) assigned to you, fit the regressions

$$y = b_0 + b_i x_i$$

and

$$y = c_0 + c_i x_i + c_j x_j.$$

b) Compare the coefficients and the fits.

12D8) Compare your results in Problem 12D7 with a fit of the form

$$y = d_0 + d_1 x_1 + d_2 x_2 + d_3 x_3.$$

12D9) Which of the variables in the ancient warfare data (Data Exhibit 12 for Problems) seem to be collinear? Outline the steps you might follow to construct a regression model to estimate the number of months at war.

12E1) Why can two polynomial equations with very different coefficients often fit the same data well? What does this imply about extrapolating beyond the range of the data?

12E2) What happens when a multiple regression is fit to data with collinear carriers? What needs to be done?

12E3) Consider women's anticipated parity from Problem 12B4. Using the parities adjusted for interview year, what can be said about (a) whether women in different births cohorts (1916–20, 1921–25, ..., 1941–45) have different expectations, and (b) whether there is a trend in expectations by age after date of birth is removed?

12E4*) Compute the correlation between X and X^2 if X is uniformly distributed between 0 and A.

12E5) Why may it be advantageous to fit

$$a^* + b^* x + c(x - x_0)^2$$

instead of

$$a + bx + cx^2,$$

even though these are mathematically equivalent?

12E6) Is Problem 13C4, predicting temperatures from date and yesterday's temperature, an example of complete dependence?

12F1) What do we need in order to control for the effect of a variable that is not directly measurable?

12F2) What assumptions do we need to make to do this?

12F3) What is an instrumental variable?

12F4) How does the variability of $y - (b_{yu}/b_{xu})x$ compare with that of $y - b_{yx}x$? Why is this reasonable?

12F5) (Continues 12C8) Suppose we now wish to control for the effect of inflation. The Phillips Curve (inflation = $b_0 + b_1$(unemployment rate)) means that the unemployment rate is a useful measure of inflation. For a more direct measure, we can use the consumer price index (CPI): if the CPI in year t is P_t, let our measure of inflation be $I = (P_t - P_{t-1})/P_{t-1}$. Use I as an instrumental variable to correct the regression of Problem 12C8 for "true" inflation. How do the coefficients for income and interest rates change?

12F6) (Continues 12F5) Which of the assumptions of this method are likely to be wrong?

12F7) (Continues 12F5) Regress the residual savings from Problem 12C8 on I adjusted for income and interest rate. How do your results compare with those for Problem 12F5?

12F8) (Continues 12D4). A common model used in economics is

Savings = $b_0 + b_1$(income) + b_2 (interest rate) + b_3 (last year's inflation).

To adjust the regression in Problem 12D4 for last year's inflation, we can use, as an instrumental variable, the inflation in the consumer price index; the inflation I_t for year t is

$$I_t = (CPI_t - CPI_{t-1})/CPI_{t-1}.$$

The Phillips Curve

$$\text{Inflation} = c_0 + c_1(\text{unemployment rate})$$

suggests that unemployment be used as a second measure of inflation.
a) Adjust the regression for last year's "true" inflation. Discuss the validity of the assumptions in this method.
b) Adjust the regression for I_{t-1} only. How do your results compare?

12F9) In what ways do the data for the United States 1955–74 support or contradict the economic models outlined in Problem 12F8?

12G1) When x and y may be rescaled in different ways, all producing a more or less linear fit, and when reality doesn't care which we choose, how can we decide upon a scale?

12G2) Why do we use these rules when the physical constraints of the problem don't force a choice?

12G3 Which rule takes precedence and why?

12G4) Noyes and co-workers (*Journ. chim. phys.* **6**: 505 (1908) and *Zeits. phys. Chem.* **70**: 350 (1910)) gave equivalent conductivities at various temperatures for solutions of several substances of differing strength, as at the top of page 525.

Concentration	KCl at 18°C	KCl at 100°C	KCl at 306°C	NaCl at 18°C	NaCl at 100°C	NaCl at 306°C	HCl at 18°C	HCl at 100°C	HNO₃ at 306°C
0.0005	128.1		1044	107.5	355	1003	375	835	374.0
.002	126.3	393	1008	105.4	349	955	373.6	826	371.2
.01	122.4	377	910	102.0	335.5	860	368.1	807	365.0
.08	113.5	341.5	720	93.5	301	680	353.0	762	353.7
.1	112.0	336		92.0	290		350.6	754	

a) For the substance and temperature assigned to you, fit both

$$\text{Equivalent conductivity} = a^* + b^*(\text{concentration})$$

and

$$\text{Equivalent conductivity} = a^{**} + b^{**}\sqrt{\text{concentration}}.$$

Which gives the better fit?
b) What would you estimate the equivalent conductivity to be for a very small concentration?
c) What limits can Student's t give you for that answer?

12G5) Klemenc and Remi (*Mon. Chem.* **44**: 307 (1924)) gave viscosities for various mixtures of (a) hydrogen and propane and (b) hydrogen and NO. Find a reasonable fit, in some form, for the case assigned to you. For (a) the (% propane, viscosity) pairs are, rounded: (0, 86), (3.1, 89), (7.8, 94), (8.9, 95), (15, 97), (22.2, 96), (32.7, 92), (51.8, 87), (69.8, 81), (80.4, 77), (100, 75), For (b) the (% NO, viscosity) pairs are (0, 85), (19.8, 142), (23.0, 145), (28.4, 147), (45.1, 160), (70.4, 172), (85.0, 175), (100, 180).

12G6) Exhibit **1** for Problem 12G6 gives the length x_1, width of square cross section x_2, and volume y of 5 bricks. The relation $y = x_1 x_2^2$ holds exactly. How well does the regression

$$y = b_0 + b_1 x_1 + b_2 x_2$$

fit? What properties of this data set make a good fit possible?

Exhibit **1** for Problem 12G6

Dimensions of bricks

x_1	x_2	y
10	5	250
10	6	360
10	4	160
11	5	275
9	5	225

12H1) What are the advantages for beginning analysis of a large body of data using subsamples?

12H2) Which of these advantages remains even when we have a high speed computer to do all the arithmetic?

Chapter 13

13A1) What does the identity

$$117 - 3x + 2x^2 = 109 + 5x + 2(x - 2)^2$$

tell us about interpreting the regression coefficient of x?

13A2) Suppose we fit the regression

$$(\text{Inflation}) = b_0 + b_1(\text{unemployment rate})$$
$$+ b_2(\text{last year's inflation}) + b_3(\text{interest rate}).$$

What is wrong with the statement "the effect of interest rates on inflation is b_3"? *How should b_3 be described?*

13A3) What is a stock?

13A4) What are the major factors influencing b_1 in the fit $y = b_0 + b_1 x_1 + b_2 x_2 + b_3 x_3$?

13A5) Suppose that a chemist wishes to compare the effects of chlorine, bromine, and iodine (halogens) on the boiling points of some alkyl halides. Any alkyl group (e.g., C_4H_9) can combine with any halogen (here Cl, Br, or I) to make an alkyl halide (C_4H_9Cl, C_4H_9Br, or C_4H_9I). We want to study how boiling points vary. The boiling points (in °C) appear in Exhibit **1** for Problem 13A5, with the molecular weights of the alkyl groups. Let

$$y = \text{boiling point}; \qquad x_1 = \text{molecular weight of alkyl group};$$

$$x_2 = \begin{cases} 1 & \text{chlorine halogen} \\ 0 & \text{no chlorine halogen}; \end{cases}$$

and let x_3 and x_4 be similarly defined for bromine and iodine.

a) Graphically or otherwise, fit the regression

$$y = b_1 x_1 + b_2 x_2 + b_3 x_3 + b_4 x_4.$$

(A single plot of y on x_1 will yield good enough estimates for all parameters.)

b) What are "the effects of the halogens"?

13A6) A second chemist may prefer to adjust for the molecular weight of the entire alkyl halide. So set $x_5 =$ molecular weights of alkyl halide. Since the atomic weights of Cl, Br, and I are, respectively, 35.5, 80, 127, we have

$$x_5 = x_1 + 35.5 x_2 + 80 x_3 + 127 x_4.$$

a) Rewrite your fit as

$$y = c_5 x_5 + c_2 x_2 + c_3 x_3 + c_4 x_4.$$

b) What are "the effects of the halogens"?

13A7) This question uses the Municipal Bond Data (Data Exhibit 9 for Problems).

a) Fit the regression $y = b_0 + b_2 x_2$.

b) Fit the regression $y = c_0 + c_1 x_1 + c_2 x_2$, and compare your coefficients with those in (a).

c) Fit the regression $y = d_0 + d_1 x_1$. Do the results agree with your intuition based on the results from (a) and (b)?

13A8) In Data Exhibit 5 for Problems, let

$$y = E \text{ (educational expenditure per public-school pupil)},$$

$$x_1 = SBG \text{ (size of Massachusetts state block grant)},$$

$$x_2 = W \text{ (taxable property value per public-school pupil)}.$$

a) Fit the regressions $y = b_0 + b_1 x_1$ and $y = c_0 + c_2 x_2$.

b) Fit $y = d_0 + d_1 x_1 + d_2 x_2$ and compare with the fits of (a).

13B1) What do we mean by $x_{1;25}$ and by $y_{;1}$ and by $y_{.12}$ and by $x_{2.1}$?

13B2) Suppose we want to regress y on x_1 and x_2 by stages. As a first step, x_1 is fitted to y, leaving $y_{;1}$. Which of the following approaches can be persuaded to produce a 2-carrier fit of x_1 and x_2 to y? Of the approaches that will, which is preferable? Why?

a) Fit x_1 to x_2, leaving $x_{2;1}$. Fit $x_{2;1}$ to $y_{;1}$.

b) Fit x_2 to y.

c) Fit x_1 to x_2, leaving $x_{2;1}$. Fit $x_{2;1}$ to y.

d) Fit x_2 to $y_{;1}$.

13B3) a) How should the coefficients in Problem 12C7 be understood?

b) How should the coefficients in Problem 12C8 be understood?

Exhibit **1** for Problem 13A5

Molecular weights of alkyl groups and boiling points of halogens

Alkyl group	Molecular weight	Halogen		
		Cl	Br	I
C_2H_5	29	12.5	38	72
$n-C_3H_7$	43	47	71	102
$n-C_4H_9$	57	78.5	102	130
$n-C_5H_{11}$	71	108	130	157
$n-C_6H_{13}$	85	134	156	180
$n-C_7H_{15}$	99	160	180	204
$n-C_8H_{17}$	113	185	202	225.5

13B4) In the herniorrhaphy data (Data Exhibit 8 for Problems), let

$$y = \text{length of stay,}$$
$$x_1 = \text{age,}$$
$$x_2 = \text{physical status.}$$

a) Using least-squares at each stage, follow the graphical method to fit $y = b_0 + b_1 x_1 + b_2 x_2$, fitting x_1 before x_2.
b) Compare the least-squares estimates of b_0, b_1, b_2 with those found graphically in Problem 12C7. Compare the fits.

13C1) We have seen the effect of mild disturbances (through rounding) on regression fits, where every point has a chance to be disturbed about the same amount. What is likely to happen if an important variable is left out? If the model is wrong?

13C2) In Example 5 of Chapter 13, we found that there is a "best" least-squares fit, but that the regression coefficients are indeterminate. How can this be? What does this imply about using fits versus interpreting regression coefficients?

13C3) Note that there is no constant term in the regressions of Problem 13A5. If $x_0 \equiv 1$ were included for a constant term, what exact collinearity among x_0, x_1, x_2, x_3, and x_4 has been introduced?

13C4) Suppose we want to predict temperatures for April 2–30 in the weather data (Data Exhibit **6** for Problems). One way to start is to take out a trend by date, since winter is ending, and to use yesterday's temperature as a carrier, since warm spells and cold spells tend to last several days.
a) Use a stepwise regression to do this, taking out the date effect first. What has happened?
b) Does the problem disappear if we try to predict temperatures only on rainy days? Why/why not?

13C5) Discuss how the exact dependencies found in Problems 13C2 and 13C3 affect interpretation of the effects we were trying to measure.

13C6) Exhibit **1** for Problem 13C6 gives values of $y = -1 + x + 0.5x^3$, rounded to 2 decimals, for the x's in Exhibit 1 of Section 13C. Fit a regression $y = b_0 + b_1 x + b_2 x^2$ and discuss the results.

13C7) Repeat Problem 13C6 after rounding y to 1 decimal place.

13C8) Fit a regression $y = b_0 + b_1 x + b_2 x^2 + b_3 x^3$ to the data in Problem 13C6 and discuss the results.

13C9) Repeat Problem 13C8 after rounding y to 1 decimal place.

13C10) a) Find the least-squares estimates for the regression

$$y = b_0 + b_1 x_1 + b_2 x_2$$

using the data of Exhibit 3 of Chapter 12.

b) Add "rounding errors" of $\pm.2$ to x_1, ±500 to x_2, and ±100 to y, using a random-number table, and compare the coefficients of

$$y^* = b_0^* + b_1^* x_1^* + b_2^* x_2^*.$$

c) How has the fit changed?

13C11) Repeat Problem 13C10 using "rounding errors" of ±1 for x_1, ±2000 for x_2, and ±300 for y. How much error do you think these variables have, either due to measurement error or due to measuring something different than what we'd like?

13D1) Suppose that we fit a k-carrier regression

$$y = b_0 + b_1 x_1 + \cdots + b_k x_k$$

and suppose we re-express the k-carriers by linear combinations to z_1, $z_2, \ldots, z_k$ without altering the stock of possible fits. If we now fit a new regression

$$y = c_0 + c_1 z_1 + \cdots + c_k z_k$$

have the coefficients changed? Has the fit changed?

13E1) What do we mean by the phrase "x_1 is a proxy for x_2"?

13E2) Suppose that an investigator finds that scores y on a general knowledge test are correlated strongly with number of years of education, E. Eager to find what kinds of courses were most helpful, the investigator defines new variables $x_1, \ldots, x_k$ of the flavor "number of humanities courses", "number of statistics courses", etc. But the new regression

$$y = b_0 + b_1 x_1 + \cdots + b_k x_k$$

has *no* coefficients very far from 0. What has happened?

Exhibit **1** for Problem 13C6

Rounded values of $y = -1 + x + .5x^3$

x	y
.9	.26
1.0	.50
1.1	.77
1.2	1.06
1.3	1.40
1.4	1.77
1.5	2.19

13E3) Suppose that we regress "number of tackles in a season", T, on height H (in inches) and weight W (in pounds) of defensive tackles in professional football, finding $T = b_0 + .50W - .10H$. Does this mean it helps to be shorter? What can we say about the relative importance of height and weight?

13E4) Let y = municipal bond yield, x_1 = block offer size, and x_5 = college students/population in the municipal bond data (Data Exhibit 9 for Problems).
a) Fit the regressions $y = a_0 + a_1x_1$, and $y = b_0 + b_5x_5$, and $y = c_0 + c_1x_1 + c_5x_5$.
b) Do college students lobby for higher yields on their municipal bonds? What might explain your finding?

13E5) Discuss which variables in the Economics data (Data Exhibit **4** for Problems) might be proxies (a) for unincluded or unmeasurable other variables, or (b) for other variables in the data set.

13E6) Do Problem 13E5 for the Coleman data (Data Exhibit **7** for Problems).

13E7) Do Problem 13E5 for the herniorrhaphy data (Data Exhibit **8** for Problems).

13F1) Can you distinguish "the effect of x_i on y when all other x_j are held constant" from "the effect of x_i, adjusted for all other x's, on y, adjusted for all other x's"? Why/why not?

13F2 If a policy change shifts x_j, when is it often inadequate to substitute the new x_j into the old regression? Why is it often inadequate to substitute all the new $x_1, \ldots, x_k$ into the old regression?

13G1) How are physical science applications of regression often different from applications in other fields?

13G2) Why is regression analysis unable to guarantee control of background variables in an observational setting? Give an example of how this might happen in some field of application you are familiar with.

13G3) Explain why multiple regression with correlated carriers may produce poorly determined individual coefficients even when the fit to the data is close.

13G4) Compute and compare the residuals $(v - \hat{v})$ for the egg data from Exhibit 8 of Chapter 13 when:
a) $\hat{v} = x_2 + 2x_3$;
b) $\hat{v} = .320 + .728x_2 + 1.812x_3$.

13G5) If we have several variables all measuring aspects of the same thing (like "atmosphere at home"), why is it unwise to put them individually into the same regression? How can we use such multiple information?

13G7) For the Economics data in Data Exhibit **4** for Problems, let

$$y = \text{savings},$$
$$x_1 = \text{unemployment (all)},$$
$$x_2 = \text{Moody's Aaa interest rate},$$
$$x_3 = \text{short-term interest rate},$$
$$x_4 = \text{consumer price index},$$
$$x_5 = (\text{year} - 1965).$$

a) Regress y on x_1 through x_5, and try to interpret the resulting equation.
b) Which carriers are acting as proxies for which other carriers in the regression? for variables not in the regression? Remember that every carrier is increasing over time.
c) Can you guess the effect of a change in interest rates from this equation?

13G8) Do Problem 13G7 for the Education data (Data Exhibit **5** for Problems) where $y = $ expenditure per public-school child and all variables are included in the fit. Should some carriers be logarithms of variables? How are wealthier towns likely to differ on all carriers from poorer towns?

13G9) In Problem 13G7 or 13G8, why has the regression equation not gained in accuracy and validity due to control for so many background variables?

13H1) Describe a problem with carriers which cannot be resolved by even an arbitrarily large amount of data.

13H2) Explain how it can happen that coefficients b_1 and b_2 are very poorly determined, but that $b_1 + b_2$ is very precisely determined.

13H3) What is the general approach to finding difficulties as in Problem 13H2?

13H5) Regress y on x_1, x_2, and year in the data from Exhibit 3 of Section 12C. How do the techniques of this section help to illuminate the difficulty in getting stable coefficients?

13H6) Use the techniques of this section (a) to better understand the fit and (b) to simplify the regression equation in Problem 13G7.

13H7) Do Problem 13H6 using the regression from Problem 13G8.

13H8) How can the techniques of this section help to uncover when (a) several carriers are measuring essentially the same thing? (b) when 2 carriers are wholly different but strongly interrelated in the population under study? (c) when 3 or more carriers are measuring different mixes of two things?

Chapter 14

14A1) Show that the least-squares estimate of α in $y = \alpha + \beta x$ is $\hat{\alpha} = \bar{y} - \hat{\beta}\bar{x}$.

14A2) Show that

$$\text{Var}\,(\hat{\alpha}) = \frac{\sigma^2 \sum x^2}{n \sum (x - \bar{x})^2}.$$

Check this formula in the special case when each x is either 0 or 1 (not necessarily all alike).

14A3) For the fit $\mu + \beta(x - \bar{x})$ using least-squares, prove that the presence or absence of μ in the model has no influence on our estimate of β, and vice versa.

14A4) Suppose we are fitting $y = \alpha + \gamma x^2$ by equally weighted least squares. What is the variance of γ? How can we rewrite $\alpha + \gamma x^2$ so that the two coefficients can be fitted separately? What will the variances of the estimated coefficients be?

14A5) Suppose we are fitting $y = \alpha + \beta x$ and want to rewrite $\alpha + \beta x$ in terms of α and a new coefficient δ so that α and δ can be fitted separately. Express the condition for this in terms of the variances and covariances of α and δ. Find a rewriting that satisfies this condition.

14A6) Suppose we are fitting $y = \epsilon \cos \theta + \eta \sin \theta$ where the values of θ are irregularly scattered. Can we rewrite $\epsilon \cos \theta + \eta \sin \theta$ in terms of λ and η so that λ and η can be estimated separately? What is the natural condition? What rewriting satisfies it?

14A7) Suppose we are fitting $y = K_A e^{At} + K_B e^{Bt}$ where A and B are given. Can we rewrite $K_A e^{At} + K_B e^{Bt}$ in terms of K_A and, say, τ_B, so that K_A and τ_B can be separately estimated? How? Can we rewrite it so that K_B and, say, τ_A can be separately estimated? How? What are the variances of K_A after the first rewriting and of K_B after the second? What were they in the beginning?

14A8) (Uses 14A7) Suppose B is close enough to A so that

$$e^{(B-A)t} \doteq 1 + (B - A)t$$

is a satisfactory approximation; what is the approximate variance of K_A? What would happen if B were very close to A?

14B1) What is a matcher?

14B2) Why is the set of ones a matcher in fitting $y = \alpha + \beta x$ by least squares?

14B3) Show that in fitting $y = \beta x$ by least squares, multiples of x are matchers. Are they the only matchers?

14B4) Show that in fitting $y = \alpha + \beta x$ by least squares, $\{x(i) - \bar{x}\}$ is a matcher even if α and β are linearly dependent.

14B5) Why is it that in fitting $y = \alpha + \beta x$ by least squares, two matchers are needed if α and β are linearly independent?

14B6) Let $\hat{\beta}$ be the least-square estimator for β in the model $\beta(x - \bar{x})$. Let $\hat{\hat{\beta}}$ be another estimator for β not a least-square estimator, and let $h_i = x_i - \bar{x}$. Prove that if, for a given set of data, $\sum \hat{y}_i h_i > \sum \hat{\hat{y}}_i h_i$, then $\hat{\beta} > \hat{\hat{\beta}}$ for the same data.

14B7) Suppose that we are fitting $y = \beta x$, where the y's are integers and the x's come in eighths (fractions of an inch? stock prices?) What matcher wil' keep our arithmetic simplest? (Here, simplest means "no fractions" and "numbers" not unnecessarily large.)

14B9) Suppose that $x = 0, 1, 2, 3, 4, 5$ each once, and that we are fitting $y = \alpha + \beta x$. What two matchers does the discussion in the text suggest as particularly convenient? Why are they convenient?

14B9) Suppose that we are fitting $y = \alpha + \beta x + \gamma x^2$ and that someone has proposed

$$1 + x + 3x^2, \quad 1 + x + 5x^2, \quad 1 + x + 11x^2, \quad 1 + x + 17x^2, \quad 1 + x + 29x^2,$$

and

$$2 + x + 1023x^2$$

as matchers. How many do we need to choose? (Call this number k.) Which of the $\binom{6}{k}$ subsets of k can we choose? What is the simplest one? Do we like it? Suggest a better choice (not all from these 6 candidates)?

14B10) What would happen if we tried to fit

$$y = \alpha + \beta x + \gamma x^2 + \sigma(3x^2 - 17x + 12)?$$

How many matchers would we need? How could we find them?

14B11) Suppose we want to fit $y = \alpha e^{\beta x} + \gamma$, and that we have an approximation, β_0, to β. What is now the natural approximation to $\alpha e^{\beta x} + \gamma$? What are a corresponding set of matchers? Can we use the same set of matchers for more than one value of β_0? Why? Why not?

14C1) What is meant by a fit to the model $y = \sum_i \beta_i x_i$?

14C2) What is the value of c so that the matcher $\{1 + cx(i)\}$ tunes to α and tunes out β in fitting $y = \alpha + \beta x$ by least squares?

14C3) What is meant by a catcher for a fitted coefficient?

14C4) Show that

$$\frac{x(i) - \bar{x}}{\sum (x(i) - \bar{x})^2}$$

is a catcher for $\hat{\beta}$ in fitting $y = \mu + \beta(x - \bar{x})$.

14C5) Suppose we are fitting $y = \alpha + \beta x$. What is the catcher for β? For α? For μ (in the form $\mu + \beta(x - \bar{x})$)?

14C6) Suppose the values of x are 0, 1, 3, 6 and 10; what numerical forms do the catchers take that we have just asked for?

14C7) Suppose (i) the values of x are symmetric about zero, and (ii) we are fitting

$$\alpha + \beta x^2 + \gamma x^4 + \delta x^6 + \epsilon x^7 + \eta x^8 + \lambda x^{10}.$$

What is the catcher for ϵ? What is its arithmetic expression if $x = \pm 2, \pm 5, \pm 8, \pm 9$, and ± 10 (each once)?

14C8) How much would the last answer be changed if $x = \pm 2$ (each seven times), ± 5 (each five times), ± 8 (each twice), ± 9 (each twice), and ± 10 (each once)?

14C9) If $\{c_1(i)\}$ is the catcher for α_1 when we are fitting

$$\alpha_1 x_1 + \alpha_2 x_2 + \alpha_3 x_3,$$

and if $\{c_2(i)\}$ is the catcher for α_2, and $\{c_3(i)\}$ that for α_3, what will happen if we fit only

$$\alpha_1 x_1 + \alpha_3 x_3?$$

14C10) If $\{d_1(i)\}$ is the new catcher for α_1 and $\{d_3(i)\}$ that for α_3 in this latter fit, explain how to calculate the d's from the c's.

14C11) (For those with access to a computer only.) The 1962 *County and City Data Book* gives (among others) values of x_{203} (total population, 1960), x_{213} (% foreign-born), x_{223} (% completed less than 5 years of school), x_{233} (% male in labor force for unincorporated urban places of 25,000 or more population). In Pennsylvania, the names and values are given in Exhibit 1 for Problem 14C11.

Suppose that we are fitting

$$\alpha_0 x_{203} + \alpha_1 x_{213} + \alpha_2 x_{223} + \alpha_3 x_{233}.$$

What are the catchers for α_0, α_1, α_2, and α_3 (give answer in numerical form, one value for each of the 15 unincorporated places)?

Exhibit **1** for Problem 14C11

Data for large unincorporated urban places in Pennsylvania

	x_{203}	x_{213}	x_{223}	x_{233}
Abington	55,831	5.7	2.3	70.0
Bristol	59,298	3.4	2.3	74.6
Cheltenham	35,990	7.8	2.6	69.4
Falls	29,082	2.5	1.8	74.2
Haverford	54,019	6.7	2.8	71.3
Hempfield	29,704	3.8	7.8	74.6
Lower Merion	59,420	7.3	2.9	64.6
Middletown	26,894	3.9	1.1	77.5
Millcreek	28,441	2.9	3.8	72.5
Mount Lebanon	35,361	3.7	1.1	72.5
Penn Hills	51,512	4.3	2.9	75.3
Ridley	35,738	4.0	3.9	73.5
Ross	25,952	4.0	1.8	73.6
Springfield	26,733	4.7	1.8	73.7
Upper Darby	93,158	6.7	2.8	66.5

14C12) (Computer not required) The 1962 *County and City Data Book* also gives the same data for 13 places in the state of New York. If we were again fitting

$$\alpha_0 x_{203} + \alpha_1 x_{213} + \alpha_2 x_{223} + \alpha_3 x_{233}$$

to these 13 points, how would we expect the catchers for New York to be related to those for Pennsylvania? And for the 12 places in California?

14C13) (Computer needed.) Delete the one of the 15 data sets in Problem 14C11 assigned to you and repeat the calculation. How would you have expected the two sets of catchers to be related? How were they related?

14C14) (Class exercise; uses 14C13.) Collect the various results obtained for leaving out the various places and (i) examine as a whole, (ii) discuss in class.

14D1) Define the least-squares fit to $y = \sum_j \beta_j x_j$ in terms of residual sum of squares and matchers.

14D2) The set of matchers $\{x_1, \ldots, x_k\}$ yields the least-squares fit of $y = \sum_j \beta_j x_j$. Is this set unique? If not, give another set.

14D3) Why is $c_1 x_1 + \cdots + c_k x_k$ a matcher when fitting $y = \sum_j \beta_j x_j$ by least squares?

14D4) Find a catcher for $\hat{\beta}$, the least-squares estimate of β in $y = \alpha + \beta x$.

14D5) Use the idea of matchers to show that $\sum (\hat{y} - y)\hat{y} = 0$, where $\hat{y}$ is a fit of $y = \sum \beta_j x_j$ with matchers which are the linear combinations of the carriers $x_1, \ldots, x_k$.

14D6) Find the least-squares estimate of β_1 and β_2 in $y = \beta_1 x_1 + \beta_2 x_2$, assuming $(\sum x_1^2)(\sum x_2^2) - (\sum x_1 x_2)^2 \neq 0$.

14D7) (For those used to calculus only.) Starting from the representation of the sum of squared deviations as

$$\sum (y - \hat{\beta}_1 x_1 - \hat{\beta}_2 x_2 - \hat{\beta}_3 x_3 - \cdots - \hat{\beta}_k x_k)^2,$$

use calculus to find a set of conditions such that $\hat{\beta}_1, \hat{\beta}_2, \ldots, \hat{\beta}_k$ give the unrestricted minimum to this sum. Convert your conditions into statements that certain things are matchers. Relate your result to that in the text of Section 14D.

14D8) When will the result of ordinary least squares be unique? When not?

14D9) Why is $\sum (y - \hat{y})^2$ never negative?

14E1) Why is $x_{1 \cdot 2 \cdots k}$ a matcher for fitting $y = \sum_{j=1}^{k} \beta_j x_j$ by least squares?

14E2) Is $x_{1 \cdot 2 \cdots k}$ a catcher for β_1?

14E3) What are some catchers for β_1?

14E4) Why is $\sum x_{1 \cdot 2 \cdots k}^2 \leq \sum x_{1 \cdot \text{fewer}}^2$?

14E5) Suppose we are fitting $y = \beta_0 + \beta_1 x_1 + \beta_2 \sin x$ where

$x = \pm 1 \times 10^{-k}, \quad \pm 2 \times 10^{-k}, \quad \pm 3 \times 10^{-k}, \quad$ and $\quad \pm 4 \times 10^{-k}, \quad$ each once.

Using the approximation $\sin x \doteq x - \frac{1}{6}x^3$, calculate $x_{3.12}$ where $\sin x \equiv x_3$ to a corresponding approximation, $x \equiv x_2$, and $1 \equiv x_1$. How does $\sum (x_{3.12})^2$ compare with $\sum (x_3)^2$? For how large a value of k do you think you could do a respectable fit? How would you do the arithmetic for k relatively large?

14E6) Suppose that, as in the last example, we are fitting $y = \beta_0 x_0 + \beta_1 x_1 + \beta_2 x_2$, where $\sum x_0 x_1 = 0$ and $\sum x_0 x_2 = 0$. If $\sum (x_{3.12})^2$ is very small in comparison with $\sum (x_3)^2$, what can we say about $\sum (x_{2.13})^2$ in comparison with $\sum (x_2)^2$? Why?

14E7) (Uses 14E6) Suppose we are fitting $y = \beta_0 x_0 + \gamma_1 x_1 + \gamma_2 x_{3.12}$ under the same conditions as those of Problem 14E6. How will the variance of $\hat{\gamma}_1$ compare with the variance of $\hat{\beta}_1$? Why? What does this mean?

14E8) Suppose we are fitting $y = \beta_0 + \beta_1 x + \beta_2 \sin x + \beta_3 \tan x$, where

$x = \pm 1 \times 10^{-k}, \quad \pm 2 \times 10^{-k}, \quad \pm 3 \times 10^{-k}, \quad$ and $\quad \pm 4 \times 10^{-k} \quad$ each once.

Using the approximations

$$\sin x \doteq x - \tfrac{1}{6}x^3 + \tfrac{1}{120}x^5, \qquad \tan x \doteq x + \tfrac{1}{3}x^3 + \tfrac{2}{15}x^5,$$

what simple rewriting of $\beta_0 + \beta_1 x + \beta_2 \sin x + \beta_3 \tan x$ in the form $\beta_0 + \gamma_1 x + \gamma_3 x^3 + \gamma_5 x^5$ seems natural and revealing? How can we use this to get upper bounds for $\sum (x_{2.013})^2$ and $\sum (x_{3.012})^2$? What do these bounds tell us about var $\hat{\beta}_2$ and var $\hat{\beta}_3$?

14E9) (Continues 14E8) How do var $\hat{\beta}_2$ and var $\hat{\beta}_3$ in Problem 14E8 compare with (a) var $\hat{\delta}_2$ in fitting $y = \delta_0 + \delta_2 \sin x$ and (b) var $\hat{\eta}_3$ in fitting $y = \eta_0 + \eta_3 \tan x$? Comment and discuss.

14E10) Suppose we are fitting $y = \beta_0 + \beta_1 x + \beta_2 \sin x$, where $y \equiv \tan x$ and

$x = \pm 1 \times 10^{-k}, \quad \pm 2 \times 10^{-k}, \quad \pm 3 \times 10^{-k}, \quad$ and $\quad \pm 4 \times 10^{-k} \quad$ each once.

Using the approximations listed in 14E8, find the fit. To this approximation, what is $x_{2.01}$ (for $x_0 = 1$, $x_1 = x$, $x_2 = \sin x$)? And $y_{.01}$? Draw a picture of the points $(x_{2.01}, y_{.01})$. What do they tell you about $\hat{\beta}_2$?

14E11) (Continues 14E10) Suppose instead that

$$y = \tan x + \text{small random errors.}$$

About how large can the "small" random errors be and still have the sign of $\hat{\beta}_2$ reasonably well determined? For general k? For $k = 5$? For $k = 10$?

14E12) A chemist, fitting $y = \beta_0 + \beta_1 x_1 + \beta_2 x_2 + \cdots + \beta_6 x_6$, finds the following values for $(x_6, x_{6.543210}, y_{.543210})$: (1, .002, .0007), (4, −.001, .0001), (9, .001, −.0002), (16, .003, −.0004), (25, −.004, −.0002). Plot (a) $y_{.543210}$ against x_6, (b) $y_{.543210}$ against $x_{6.543210}$. What does the comparison of the two plots tell you? Do you think the sign of $\hat{\beta}_6$ is at all well determined? Suppose one were to fit $y_{\cdot012345} = \gamma_6 x_6$; would the sign of $\hat{\gamma}_6$ be reasonably well determined? How can all this be? What is your advice to the chemist?

14E13) (For computer users only) Go back to the data in Problem 14C11 and take $y = x_{203}$, $x_0 = 1$, $x_1 = x_{213}$, $x_2 = x_{223}$, $x_3 = x_{233}$, and find the values of $x_{0\cdot123}$, $x_{1\cdot023}$, $x_{2\cdot013}$, $x_{3\cdot012}$, and $y_{\cdot0123}$. Make the four plots of $y_{\cdot0123}$ against each of the other four, identifying any points that appear important. What does this tell you about which unincorporated places have a relatively big influence on which of $\hat{\beta}_0$, $\hat{\beta}_1$, $\hat{\beta}_2$, or $\hat{\beta}_3$? (Assume a fit of the form $\beta_0 + \beta_1 x_1 + \beta_2 x_2 + \beta_3 x_3$.)

14F1) What is a weighted least-squares fit to $y = \sum \beta_i x_i$?

14F2) Show that ordinary least-squares result when

$$w_{ij} = \begin{cases} 1 & i = j, \\ 0 & i \neq j. \end{cases}$$

14F3) If the precision of each measurement y_i is proportional to $1/x_i$ in $y = \beta x$ so that an appropriate weight for the ith observation is $1/x_i$, show that the weighted least-squares estimate of β is $\bar{y}/\bar{x}$.

14F4) In Problem 14F3, if the precision of each measurement y_i is proportional to $1/x_i^2$, show that the weighted least-squares estimate of β is

$$\frac{\sum_{i=1}^{n} \dfrac{y_i}{x_i}}{n}.$$

14F5) An astronomer wishes to fit $y = \beta_x$ with weight w to the following data, given as (w, x, y): $(4, 0, 0.13)$, $(9, 1, 0.27)$, $(4, 2, 0.43)$, $(1, 3, 0.69)$, $(16, 4, 0.91)$, $(25, 5, 1.32)$, $(9, 6, 1.50)$, $(1, 8, 2.03)$. Write down an equivalent set of equally weighted data. Then plot it.

14F6) (Continues 14F5) From the plot of Problem 14F5, pick an approximate value B for β and replot the residuals from $y = Bx$ in a similar way (still appropriate for an unweighted fit). What does this tell you? Does any one of the points seem out of line? How much would you have to change its weight to bring it back more or less in line?

14F7) (For computer users only) Go back again to the data given in 14C7 and, taking $w = x_{203}$, $y = x_{233}$, $x_1 = x_{213}$, $x_2 = x_{223}$, fit $y = \beta_0 + \beta_1 x_1 + \beta_2 x_2$.

14F8) (For anyone) Make the plots of $\sqrt{w}y$ against $\sqrt{w}x_1$ and $\sqrt{w}x_2$, for the data specified in 14F7. What conclusions do you reach?

14F9) (Continues 14F8) Make a visual estimate C_1 of γ_1 in $y = \gamma_0 + \gamma_1 x_1$ and plot $\sqrt{w}\,(y - C_1 x)$ against $\sqrt{w}x^2$. What do you conclude?

14F10) An analyst once fitted $y = \beta_0 + \beta_1 x$ to 47 data points using the following weights: thirty-two 1's, five 3's, five 10's, one 100, two 1000's, and two 1,000,000's. If you had the (x, y) pairs, how could you approximate his fit most easily?

14F11) In the example of the 100 points (3 sour) discussed in the text, suppose that x's went with the y's, that we wanted to fit $y = \beta x$, and that the three points that deserved the low weights had x's of just about the typical size. How much are we likely to reduce the variance of $\hat{\beta}$ by including the sour points (with correct weights)? Would the confidence interval for $\hat{\beta}$ shorten appreciably? Why/why not?

14F12) (Continues 14F11) What if the x's for the three "sour" points were about 100 times all the other x's? Answer the questions of Problem 14F11 then. How is it easy to see the result?

14G1) For the same 10 fixed values and one moving one, calculate the influence curve for the midmean, defined as the mean of the middle half of the values, taking this to mean (a) the middle five of eleven, (b) the middle five-and-a-half (the middle five with unit weight and the next on each side with weight $\frac{1}{4}$). How much do these two curves differ? Generally? At most?

14G2) Do the same as Problem 14G1 for two other trimmed means, (a) the mean of the middle three, and (b) the mean of the middle seven.

14G3) Do the same as Problem 14G1 for the trimean,

$$\frac{\text{Lower hinge} + 2(\text{median}) + \text{upper hinge}}{4}.$$

(Recall that in batches of 11, the hinges are at depth 3.5—are halfway between the 3rd and 4th from each end.)

14G4) Do the same as Problem 14G1 for the singly and doubly Winsorized means, where in singly Winsorized means (a) the highest value is replaced by a repetition of the 2nd highest (in doubly Winsorized means the two highest are replaced by repetitions of the 3rd highest) and (b) the lowest by a repetition of the 2nd lowest (in doubly Winsorized means, the two lowest are replaced by repetitions of the 3rd lowest).

14G5) (Uses 14G1, 14G2, 14G3, 14G4.) It is easy to classify influence curves by location into six categories, according as (a) they are or are not straight in the middle and as (b) they rise (fall) indefinitely, boundedly without returning, boundedly with returning, with the general properties shown in Exhibit 1 for Problem 14G5. Make a copy of the 2×3 table with marginal labels, and fill in, each in the appropriate cell, the mean, median, biweight (2 versions), the four trimmed means, the trimean, and Winsorized means just considered. Discuss what your result seems to suggest about which of these location indicators to use when.

14H1) Why is iterative least-squares fitting desirable?

14H2) Do the same analysis as that given in Example 1 (p. 356) but using

$$w(u) = \begin{cases} 1 - |u| & |u| < 1, \\ 0 & |u| \geq 1. \end{cases}$$

14H3) An analyst was trying to fit

$$y = \beta_0 + \beta_1 x_1 + \beta_2 x_2 + \beta_3 x_3$$

and tried to have a biweighted fit made by iteration, with the results in Exhibit **1** for Problem 14H3. What do you think happened during the calculation?

Exhibit **1** for Problem 14G5

Performance of estimates of location

	Indefinite rise	Bounded rise, no return	Bounded rise, with return
Not straight in middle	Unsafe; efficiency not very high for any reasonable distribution	Safe against medium-stretched tails; efficiency as ←	Very safe; but efficiency cannot be very high for any reasonable distribution
Straight in middle	Unsafe; efficiency very high for some reasonable distributions, but not for others	Safe against medium-stretched tails; efficiency very high for some, and may be high for a variety of distributions	Very safe; efficiency very high for some and may be high for a variety of distributions

Exhibit **1** for Problem 14H3

Biweighted fit by iteration

Iteration	$\hat{\beta}_0$	$\hat{\beta}_1$	$\hat{\beta}_2$	$\hat{\beta}_3$	$\hat{S}$
Start	0	0	0	0	7.5
One	3.1	2.4	1.9	7.3	3.0
Two	2.9	2.5	2.0	7.2	2.8
Three	4.9	2.5	2.1	7.1	4.0
Four	3.0	2.5	2.1	7.1	2.9
Five	2.9	2.5	2.1	7.1	2.7
Six	2.9	2.5	2.1	7.1	4.7
Seven	3.0	2.4	2.2	7.0	2.9
Eight	3.0	2.5	2.1	7.1	2.8
Nine	3.9	2.5	2.1	7.1	2.8

14H4) Make a plot of the weights against u given by the simple step-weights with $c = 5$ and on the same graph a plot of the bisquare weights with $c = 5$. (Multiply the latter by a constant to make it look more than the former!) How closely do the two versions compare?

14H5) Do the same for the (slightly more complicated) step-weighting with $c = 9$ and a bisquare weighting with $c = 9$.

14H6) What might happen because a particular (x, y) pair moved back and forth across a break in the step-weighting?

14H7) (For computer users.) Turn to the data in 14C11. This time take $x = x_{203}$ and $y = x_{233}$, and fit $y = \beta x$ (a) by least squares and (b) by biweight iteration with $c = 4$ starting with $\beta = 0$. Discuss the difference. (Plot the residuals if it is helpful.)

14H8) (For computer users; continues 14H7.) Do the same by biweight iteration, starting with $\beta = 0$ and $c = 6, 8$, and 10. Compare all five sets of results and discuss differences. (Plot the residuals if it is helpful!)

14H9) (For computer users; continues 14H7 and 14H8.) Do the same by biweight iteration, starting from $\beta = -.0002$ and using $c = 4$ and $c = 10$. Compare with earlier results and discuss. (Plot the residuals if it is helpful!)

14I1) Compare the pros and cons of least-squares fitting and least absolute deviation fitting.

14I2) What is an unfavorable feature of least absolute deviation fitting?

14I3) Let the matchers $w_\ell x_1, \ldots, w_\ell x_k$ for the ℓth iteratively weighted linear least-squares fitting of $y = \sum_{j=1}^{k} \beta_j x_j$ converge $(\ell \to \infty)$ to $w_{\text{last}} x_1, \ldots, w_{\text{last}} x_k$. Show that $w_{\text{last}} x_1, \ldots, w_{\text{last}} x_k$ are matchers for the least absolute deviation fitting of $y = \sum_{j=1}^{k} \beta_j x_j$.

14I4*) If we have only y's, no x's, what kind of a fit does least absolute deviations reduce to? In the light of 14G5, what might we fear for least absolute deviations?

14I5*) Go back to the example of 14G with 10 fixed points and one moving one, and find what value of $\hat{x}$ minimizes $\sum \Psi(x_i - \hat{x})$ where $\Psi(u)$ is u^2 or $k|u|$ according as $|u| \leq$ or $\geq k$ for a variety of values of the 11th point (what iterative weights are to be used?) What kind of influence curve does this kind of fitting seem to have?

14I6*) (For computer users only; a town to be assigned to each solver.) Go back to the data in 14C11, taking $x = x_{203}$, $y = x_{233}$, $c = 9$, and initial $\beta = 0$. Make flattened-least-absolute fits for (a) the data as is, (b) y for the town assigned to you altered by $\pm10, \pm8, \pm6, \pm4$, and ±2 (with unaltered y's for the other 11 towns). Plot the corresponding influence curve for $\hat{\beta}$. Compare the results for various towns in class.

14I7*) (For computer users only; a town to be assigned to each solver, continues 14I6.) Do the same with $y = x_{233}$, $x_1 = 1$, $x_2 = x_{223}$, $x_3 = x_{213}$,

$W = x_{203}$ (the weight of the first kind), and the same c, initial β's, and changes for one y.

14J1) An analyst fitted (or tried to fit) $y = \beta_0 + \beta_1 x_1 + \beta_2 x_2 + \beta_3 x_3$. All the β's came out fairly large, something anticipated, so instead of thinking of plotting, for example, $y_{\cdot 023} = y_{\cdot 0123} + \hat{\beta}_1 x_{1 \cdot 023}$ against $x_{1 \cdot 023}$, it was better to think of plotting $y_{\cdot 0123}$ (the final residual) against $x_{1 \cdot 023}$. For 237 data points (y, x_1, x_2, x_3), the analyst found the following pattern:

Number of data points	Value of $y_{\cdot 0123}$	Values of			
		$x_{0 \cdot 123}$	$x_{1 \cdot 023}$	$x_{2 \cdot 013}$	$x_{3 \cdot 012}$
211	small	small	small	small	small
13	small	small	large	small	small
10	small	small	small	large	small
1	small	small	small	small	large
1	small	large	small	small	small
1	large	large	large	large	small

What questions ought to be asked? Focusing on which data points? Which of $\hat{\beta}_0$, $\hat{\beta}_1$, $\hat{\beta}_2$, $\hat{\beta}_3$ would you suggest believing? Which data points would you suggest setting aside? Which $\hat{\beta}$'s would be most likely to be changed when this was done?

14J2) An analyst had carefully thought through reasons for fitting n data points with

$$y = \beta_0 + \beta_1 x_1 + \beta_2 x_2 + \beta_3 x_3 + \beta_4 x_4 + \beta_5 x_5$$

where $n/10 \le x_i^2 \le 10n$ for each of the x_i's. Before looking at y at all, however, she found that

$$\sum (x_{0 \cdot 12345})^2 = .27n, \quad \sum (x_{1 \cdot 02345})^2 = .38n, \quad \sum (x_{2 \cdot 01345})^2 = .0003n,$$

$$\sum (x_{3 \cdot 01245})^2 = .79n, \quad \sum (x_{4 \cdot 01235})^2 = .0001n, \quad \sum (x_{5 \cdot 01234})^2 = 1.23n.$$

She was unhappy about the two small sums of squares and wondered what kind of trouble she had. What ought you to tell her? (First, about the trouble. Second, about what can be done about it.) Do either of these depend on what the y's are?

14J3) A sports analyst had data on 1279 basketball players with y = lifetime average, x_1 = length of legs (inches), x_2 = length of arms (inches), x_3 = height (inches), x_4 = weight. Exploring the x's he found

$$\sum (x_{1 \cdot 234})^2 = 1279(.01 \text{ inch})^2, \quad \sum (x_{2 \cdot 134})^2 = 1279(.008 \text{ inch})^2,$$

$$\sum (x_{3 \cdot 124})^2 = 1279(.012 \text{ inch})^2, \quad \sum (x_{4 \cdot 123})^2 = 1279(13.2 \text{ lbs})^2.$$

He decided he would have to drop one or more x's. Was he right? Why/why not? If one or more x's are to be dropped, what is a good list to pick from? Can the numbers tell us which of these to try dropping first? Can common sense tell us? Which would you drop first?

14J4) Another sports analyst had data on 534 tennis players with y any of various performance measures, x_1 = mean velocity (miles per hour) of first serves when crossing the net, x_2 = % of first serves in, x_3 = mean velocity of second serves, x_4 = time to run 40 yards, x_5 = time to run the marathon (26 miles, a few hours). On examining the x's for the first time, she found

$$\sum (x_{1 \cdot 2345})^2 = 534(.07 \text{ mph})^2, \quad \sum (x_{2 \cdot 1345})^2 = 534(11\%)^2,$$

$$\sum (x_{3 \cdot 1245})^2 = 534(17 \text{ mph})^2, \quad \sum (x_{4 \cdot 1235})^2 = 534(0.47 \text{ seconds})^2,$$

$$\sum (x_{5 \cdot 1234})^2 = 534(32 \text{ minutes})^2.$$

Do you think these were reasonable answers? Why/why not?

14J5) Suppose we are considering only x_1 and x_2. If $x_0 = 1$ and r is the Pearson product-moment correlation between x_1 and x_2 [means allowed for, so that $r^2 = (\sum (x_{1 \cdot 0} x_{2 \cdot 0}))^2 / (\sum (x_{1 \cdot 0})^2)(\sum (x_{2 \cdot 0})^2)]$, show that

$$\sum (x_{1 \cdot 02})^2 = (1 - r^2) \sum (x_{1 \cdot 0})^2 \quad \text{and} \quad \sum (x_{2 \cdot 01})^2 = (1 - r^2) \sum (x_{2 \cdot 0})^2.$$

What does this mean about

$$\frac{\sum (x_{1 \cdot 02})^2}{\sum (x_{1 \cdot 0})^2} \quad \text{and} \quad \frac{\sum (x_{2 \cdot 01})^2}{\sum (x_{2 \cdot 0})^2} ?$$

14J6) (Continues 14J5) What about

$$\frac{\sum (x_{1 \cdot 2})^2}{\sum (x_1)^2} \quad \text{and} \quad \frac{\sum (x_{2 \cdot 1})^2}{\sum (x_2)^2} ?$$

14J7) (Continues 14J6) What about

$$\frac{\sum (x_{1 \cdot 2345})^2}{\sum (x_{1 \cdot 345})^2} \quad \text{and} \quad \frac{\sum (x_{2 \cdot 1345})^2}{\sum (x_{2 \cdot 345})^2} ?$$

And so on? Why/why not?

14J8) Are the following ratios possible together?

$$\frac{\sum (x_{1 \cdot 234})^2}{\sum (x_1)^2} = 10^{-11}, \quad \frac{\sum (x_{2 \cdot 134})^2}{\sum (x_2)^2} = 10^{-2},$$

$$\frac{\sum (x_{3 \cdot 124})^2}{\sum (x_3)^2} = 10^{-1}, \quad \frac{\sum (x_{4 \cdot 123})^2}{\sum (x_4)^2} = 1.$$

Why/why not?

Re14Q. These problems, appropriate for Section 10G, require the techniques of Sections 14G, 14H, and 14I.

14Q1) Take your own sample of 20 cities (drawn in Problem 3D6) and 3 variables from Data Exhibit 1 for Problems, of which two should be 320 (median family income) and 327 (% of college graduates) and go through the čob-b̂ar computations of Section 10G.

14Q2) Do the same with three variables, two of which are 331 (% in same house as in 1955) and 358 (moved in between 1958 and 1960).

14Q3) Do the same for variables x_1, x_5, and y from Data Exhibit 7 for Problems.

14Q4) Do the same for variables x_3, x_4, and x_5 of Data Exhibit 9 for Problems.

Chapter 15

15-1) What do we mean by guided regression?

15-2) Why does guided regression make sense only when we do not need to interpret individual coefficients? What modification seems at first to conquer this difficulty? Does it/doesn't it?

15A1) What are the five "ideal conditions" we assume for guided regression? If we're worried about whether they hold, what is usually the first step in checking them?

15A2) Explain what a stock is.

15A3) What is a minimand?

15A4) When there are many alternative stocks, several may turn out to have minimands nearly as small as the observed smallest minimand. How can this be exploited in practice?

15A5) What does PRESS stand for?

15A6) What is Anscombe's (Tukey's) $s^2/(n - k)$?

15A7) What is Mallows' C_p?

15A8) How are the quantities in Problems 15A5 through 15A7 used in practice?

15B1) What is stepwise regression? Why do we often need to do this rather than trying all subsets of carriers?

15B2) What are forward and backward steps? Why do we need a backward step?

15B3) How can step-by-step output of fits and residuals tell us about what's happening in the stepwise procedure? What if the coefficient for a carrier of interest changes a lot with each new step?

15B4) How might a resistant stepwise regression be done?

15B5) In "Woes of regression coefficients" (Chapter 13) we have warned against some interpretations of coefficients that Macdonald and Ward seem to be using successfully. Why may things work out for them and not for some other sorts of data?

15B6) Suppose we wish to predict expenditure per pupil in the Education data (Data Exhibit 5 for Problems). There are 6 carriers listed; for the 5 carriers assigned to you, do the following;

a) How many possible stocks of size $k = 2$ are there? Choose the stock you expect to be best.

b) Use a stepwise procedure to look for a "best" stock of this size; be sure each carrier has a chance to be deleted. How does your solution compare with (a)?

15B7) Compute all 10 regressions of stocks of size $k = 2$ in Problem 15B6. How many are nearly optimal? Did you find the optimal solution in parts (a) or (b)?

15B8) Repeat Problem 15B6 with stocks of size $k = 3$.

15B9) Compute all 10 regressions of stocks of size $k = 3$ in Problem 15B8. How many are nearly optimal? Did you find the optimal solution in parts (a) or (b)?

15B10) Use Mallows' C_p to guide choice of k in Problem 15B6, and use a stepwise procedure to yield a "best" stock. How does your answer compare with work done in the last 4 problems?

15B11–15B15) Repeat Problems 15B6–15B10 for a data set of your own.

15C1) What is the advantage of all-subset techniques over guided regression?

15C2) Can we gain by looking at stocks in any particular order? What order?

15C3) How does the work of Daniel and Wood or of Furnival and Wilson enlarge the applicability of this method?

15C4) How might a resistant version of all-subset regression techniques be done?

15C5) (Heavy computation.) For the Economics data (Data Exhibit **4** of Problems), let y = log of personal consumption, x_1 = log of income, x_2 = long-term interest rate (Moody's Aaa), and x_3 = short-term interest rate. Regress y for all possible stocks of x's, and discuss your results. What interpretation can be made of the regression coefficients?

15D1) How might we sort many carriers into 3 categories of importance for fitting?

15D2) What are the main steps of an analysis of this sort?

15D3) When removing key carriers, why is it especially unlikely that the deviations $(y_i - \hat{y}_i)$ are independent? uncorrelated?

15D4) The Boston weather data, Data Exhibit **6** for Problems, has at least 6 carriers to predict daily temperature. Others that may be important are, for example: date, yesterday's temperature, or change in station pressure, etc. List a set of carriers of possible interest for predicting temperature (y). Choose 2 carriers x_1 and x_2, and fit them resistantly to y using the method of Section 12C. What are your values of $y_{;12}$, and $x_{2;1}$? (Do not include carriers involving precipitation, pressure, relative humidity, or fog, as they are needed in later problems.)

15D5) Sort the ancient-warfare data, Data Exhibit **12** for Problems, into 3 categories of interest. Fit at most 2 of the key carriers, and compute residuals, using number of months at war as the y variable.

15D6) (Heavy computation) This is the first of a series of problems to predict inflation using the Economics data (Data Exhibit **4** of Problems). (See Problem 12F5 for a definition of inflation in terms of the consumer price index.) Choose 1 or 2 key carriers and divide the remaining ones into "interesting" and "long shots". You may wish to define some new carriers such as "last year's inflation" or "change in GNP".

15E1) What are the 2 costs of using many carriers in a regression?

15E2) What is the intuitive motive for combining the "football" carriers (modifying carriers in Section 15E) in the logarithmic scale?

15E3) If variables are combined after careful study of the data, the estimated residual variance is likely to be less than if judgment components are used *a priori*. Explain why this does not contradict the statement in Section 12E that the variance of $\hat{y}$ will be larger.

15E4) Why are linear combinations often sensible ways to combine carriers? What kinds of circumstances make nonlinear combinations preferable?

15E5) (Continues Problem 15D4) Create a judgment component for fitting the temperature residuals with a measure of "wetness" from precipitation, pressure, relative humidity, and fog. Add this component to the fit.

15E6) (Continues Problem 15D5) Create judgment components for the several categories of carriers in the ancient warfare data (Data Exhibit **12** for Problems). Discuss why we are forced to combine carriers in this problem. Fit these components to the residual months at war from Problem 15D5.

15E7) (Continues 15D6) Create judgment components for "interest" and for "unemployment" and add these to your regression.

15F1) What are principal components? Discuss the most important differences in analysis between using the principal component $.5x_1 - 2x_2 + x_5$ or the judgment component $.5x_1 - 2x_2 + x_5$.

15F2) Should principal components ever be used instead of *a priori* judgment components? What questions are raised about (a) the data or (b) the judgments behind the *a priori* components?

15F3) What is the distinction between $\hat{\text{var}}\,(x_j)$ and the judged measurement variance of x_j in Section 15F?

15F4) For principal components, what are the 4 categories of carriers for study? How do they compare with the 3 categories of Section 15D?

15F5) Why do we feel justified in reshaping principal-component carriers to be more interpretable? How much are we likely to lose?

15F6) (Continues Problem 15D4) What are the principal components for precipitation, pressure, relative humidity, and fog? How does the most important one compare with your judgment component? How do the fits compare?

15F7) (Continues Problem 15D5) Compare principal components with some of your judgment components. If we fit several principal components, how ought we respond to a markedly better fit than with judgment components?

15F8) (Continues Problem 15E7) How do principal components compare with judgment components for the "interest rate"? Which do you prefer?

15G1) Discuss the coefficients you have obtained in a regression analysis in the light of the remarks of this section.

15I1) What is the role of past knowledge and current analyses?

15I2) Since we are doing linear regression, fitting x_1, x_2, x_3, x_4, x_5 to y is algebraically equivalent to fitting x_1, x_2, x_3, x_4, x_5 to $(y - 17x_1 - 5x_3 - 3x_5)$. What do we gain by using the latter approach?

15I3) Suppose you were about to analyze the weather for Boston in April of 1976. How would you use the results from the corresponding 1975 data (Data Exhibit 6 for Problems) to strengthen your analysis? How do the ideas in Section 15H apply here as well?

Chapter 16

16A1) Why should regression analysis always include an examination of residuals?

16A2) Why is plotting $(y - \hat{y})$ against y a bad idea? What is a better idea?

16A3) If we find a pattern in a residual plot, there are at least two approaches to improving the fit. What are they?

16A4) What are convex-upward functions? Concave-upward functions?

16A5) What are some different ways of examining $(y - \hat{y})$ with respect to $\hat{y}$?

16A6) How would your choice among the possibilities in Problem 16A5 change if you had 10, 40, 100, or 1000 points?

16B1) What do we mean by a "re-expression" of x as opposed to the more general term a "function" of x?

16B2) What is folding?

16B3) What is the distinction between a carrier and a variable?

16B4) When two different sets of variables give rise to the same stock, identical fits will be obtained. Why then does the choice of variables matter?

16B5) Are we restricted to use variables in the currently most popular form? What do we lose by using new re-expressions? What might we gain?

16C1) When plotting $(y - \hat{y})$ against t_{old}, does it matter if t_{old} is re-expressed?

16C2) What is x_{dot}?

16C3) What can we do when there is not enough data to tell us about dependence of $(y - \hat{y})$ on t_{old}?

16C4) For what purposes would we want to plot residuals against t_{old} and not against x_{dot}, even if x_{dot} is on hand?

16C5) How do we smooth residuals? When can this help?

16C6) If a plot of $(y - \hat{y})$ against x_{dot} shows that t_{old} (or, equivalently, x_{dot}) should be added to the regression, how can we easily do this?

16C7) *Completeness of Death Registration.* In any population, there is a tautology:

$$\text{Death rate} = \text{Birth rate} - \text{growth rate.}$$

If we think of the series of populations defined as that part of a population who are older than x, we have the series of relations

$$\text{Death rate}(x) = \text{Birth rate}(x) - \text{growth rate}(x),$$

where

$$\text{Death rate}(x) = \frac{\#\text{ of deaths above age } x}{\#\text{ of people above age } x}$$

and

$$\text{Birth rate}(x) = \frac{\#\text{ of people ''passing'' age } x}{\#\text{ of people above age } x},$$

$$\text{Growth rate}(x) = \frac{\text{additional }\#\text{ of people above age } x}{\#\text{ of people above age } x}.$$

Demographers frequently use the concept of a "stable population," i.e., one in which each segment is growing at the same rate, so that growth rate(x) does not depend on x. Stable populations result from a long period of unchanging fertility and mortality.

In a survey of rural China, taken in 1930, it is suspected that fertility and mortality have been unchanged for a number of years, but that only a certain percentage, c, of deaths have been reported, this percentage not depending on age.

a) How would this affect the series of relations described above? How can linear regression be used to estimate c? Given in Exhibit 1 for Problem 16C7 are the birth and death rates, above age x, for males in the survey. What is your estimate of the "completeness of registration", c?

b) Does the population seem to be stable? Why/why not? Does completeness seem to be independent of age? Why/why not? Do you prefer to plot residuals against age or birth rate? Why/why not? Will both plots tell us more than one? Why/why not?

16C8) Let y = bond yield,

$\quad\quad x_3$ = population of issuer,

$\quad\quad x_4$ = total net debt,

for the Municipal Bond data (Data Exhibit **9** for Problems).

a) Regress y on x_3, y on x_4, and y on x_3 and x_4.

b) Use plots of residuals to diagnose what has happened, and the quality of the fit by $y = b_0 + b_3x_3 + b_4x_4$.

c) How would you continue your analysis?

16D1) When adding a new variable t_{new} to a regression fit, why will the carrier t_{new} often fail to improve the fit, while x_{dot} greatly improves the fit?

16D2) (Continues Problem 16C8) Use residual plots to add the best new variable to your best fit from Problem 16C8. Why will residual plots help in this situation when nearly any mechanical stepwise algorithm is likely to fail?

16D3) (Continues Problem 16D2) Finish developing a model for bond yield with the Municipal Bond data (Data Exhibit 9 for Problems). Discuss why you took the approach you chose, and discuss the meaning of your model.

16E1) Why do we consider multiplicative fits of the form $\hat{y}_{med} + (\hat{y} - \hat{y}_{med})(1 + u)$ instead of the form $\hat{y}(1 + u)$?

16E2) What is $Q_{\hat{y}}$? Why do we use it?

16E3) What is $q_{\hat{y}}$? How does it compare to $Q_{\hat{y}}$?

Exhibit **1** for Problem 16C7

Reported birth and death rates (in 1000's)

Age	Birth rate	Death rate
5	29.5	16.2
10	30.1	15.8
15	31.8	16.8
20	33.6	18.2
25	36.7	19.6
30	40.3	21.8
35	45.3	24.9
40	54.7	29.0
45	63.4	33.8
50	77.9	41.6
55	91.4	52.1
60	120.8	64.6
65	143.5	82.8
70	169.4	111.8
75	231.6	129.4

S) SOURCE

Data reprinted by permission of M. A. Stoto. For discussion of data, see G. W. Barclay, A. J. Coale, M. A. Stoto, and T. J. Trussel (1976). "A reassessment of the demography of traditional rural China," *Population Index*, **42**, 606–635.

16E4) Why do we plot residuals, rather than simply trying a new fit?

16E5) How do we assess whether to add products of carriers?

16G1) (Continues Problem 15F8) Complete a model describing how economic variables vary with the rate of inflation.

a) "Lagged" variables might allow a prediction of next year's inflation. How good a prediction can be made without knowing the values of other carriers in next year's fit? What risks are involved in such a prediction?
b) To what extent can the coefficients in your regression be interpreted? Of those variables that could be manipulated, what policy changes would you recommend to reduce inflation?
c) Discuss how the changes suggested in (b) might affect values of other carriers.

16P1) (Project) Redo the analysis of Problem 9A10, using regression to study the relationship of pressure and temperature throughout the year. Clearly, "month" is not a suitable carrier for regression without re-expression (why?). One way to start is to consider the problem as 12 smaller problems, pulling analyses together later (cf. Sections 12G, 12H, 15H, 15I).

16P2) (Project) Analyze the heart-transplant data (Data Exhibit **11** for Problems). Questions of most interest to a surgeon may include:
a) What factors seem important for longer survival?
b) How useful is the mismatch score for predicting survival time? Risk of rejection?
c) How would you advise a physician about trade-offs between accepting a poor mismatch score or increasing waiting time? Is age a factor?
d) What are the risks involved in changing policy on the basis of (c)? How would you explain this by example to a physician?
e) Do you need to control for a time trend in these data?

16P3) (Project) Analyze the Armed Forces Research data (Data Exhibit **10** for Problems). How does the amount appropriated compare with the amount requested? The amount requested next year? How is the process changing over time? How do the three branches of the military fare during this period?

16P4) (Project) (Continues Problems 15F7, 15D5, etc.) Complete your analysis of the ancient warfare data (Data Exhibit **12** for Problems). What do you conclude about the effectiveness of military strength as a deterrent in history?

16P5) (Project). *Demographic Transition*. Switzerland, in 1888, was entering a period known as the "demographic transition"; i.e., its fertility was beginning to fall from the high level generally found in undeveloped countries to the lower level it has today. If we agree to look at changes in "I_g", a common standardized fertility measure, they can be related to changes in various socioeconomic indicators. I_g and 5 of these indicators collected by Francine van de Walle for a book on the demographic transition in Switzerland, are given in Exhibit **1** for Problem 16P5 for 47 French-speaking provinces at about 1888. Analyze these data. Which variables seem most important? Comment on the "static" nature of this approach to a "dynamic" problem.

Exhibit 1 for Problem 16P5

Swiss fertility and socioeconomic indicators

ID no. of province	I_g	x_1	x_2	x_3	x_4	x_5
1	.802	.170	.15	.12	9.96	.222
2	.831	.451	.06	.09	84.84	.222
3	.925	.397	.05	.05	93.40	.202
4	.858	.365	.12	.07	33.77	.203
5	.769	.435	.17	.15	5.16	.206
6	.761	.353	.09	.07	90.57	.266
7	.838	.702	.16	.07	92.85	.236
8	.924	.678	.14	.08	97.16	.249
9	.824	.533	.12	.07	97.67	.210
10	.829	.452	.16	.13	91.38	.244
11	.871	.645	.14	.06	98.61	.245
12	.641	.620	.21	.12	8.52	.165
13	.669	.675	.14	.07	2.27	.191
14	.689	.607	.19	.12	4.43	.227
15	.617	.693	.22	.05	2.82	.187
16	.683	.726	.18	.02	24.20	.212
17	.717	.340	.17	.08	3.30	.200
18	.557	.194	.26	.28	12.11	.202
19	.543	.152	.31	.20	2.15	.108
20	.651	.730	.19	.09	2.84	.200
21	.655	.598	.22	.10	5.23	.180
22	.650	.551	.14	.03	4.52	.224
23	.566	.509	.22	.12	15.14	.167
24	.574	.541	.20	.06	4.20	.153
25	.725	.712	.12	.01	2.40	.210
26	.742	.581	.14	.08	5.23	.238
27	.720	.635	.06	.03	2.56	.180
28	.605	.608	.16	.10	7.72	.163
29	.583	.268	.25	.19	18.46	.209
30	.654	.495	.15	.08	6.10	.225

→

Exhibit 1 for Problem 16P5 (continued)

ID no. of province	I_g	x_1	x_2	x_3	x_4	x_5
31	.755	.859	.03	.02	99.71	.151
32	.693	.849	.07	.06	99.68	.198
33	.773	.897	.05	.02	100.00	.183
34	.705	.782	.12	.06	98.96	.194
35	.794	.649	.07	.03	98.22	.202
36	.650	.759	.09	.09	99.06	.178
37	.922	.846	.03	.03	99.46	.163
38	.793	.631	.13	.13	96.83	.181
39	.704	.384	.26	.12	5.62	.203
40	.657	.077	.29	.11	13.79	.205
41	.727	.167	.22	.13	11.22	.189
42	.644	.176	.35	.32	16.92	.230
43	.776	.376	.15	.07	4.97	.200
44	.676	.187	.25	.07	8.65	.195
45	.350	.012	.37	.53	42.34	.180
46	.447	.466	.16	.29	50.43	.182
47	.428	.277	.22	.29	58.33	.193

D) DEFINITIONS

x_1 Proportion of population involved in agriculture as an occupation;
x_2 Proportion of "draftees" receiving highest mark on army examination;
x_3 Proportion of population whose education is beyond primary school;
x_4 Proportion of population who are Catholic;
x_5 Infant mortality: proportion of live births who live less than 1 year.

S) SOURCE

Data used by permission of Francine van de Walle. Office of Population Research, Princeton University, 1976.
Unpublished data assembled under NICHD contract number No 1-HD-O-2077.

Problems where data exhibits are used

1. Middle-U.S. cities: 3D6 through 3F4, 3G1, 3G2, 14Q1, 14Q2
2. Smoking and health: 11A7 through 11A10, 11C3, 11E7, 11F6, through 11F9
3. U.S. population 1960–1970, by state: 3A2 through 3A7, 3B4, 3B6 through 3B8, 3C1
4. Economics: 12C8, 12D4, 12D5, 12D6, 12F5 through 12F9, 13B3, 13E5, 13G7, 13G9, 13H6, 15C5, 15D6, 15E7, 15F8, 16G1
5. Education: 13A8, 13G8, 13G9, 13H7, 15B6 through 15B10
6. Meteorology: 12D7, 12D8, 12E6, 13C4, 15D4, 15E5, 15F6, 15I3
7. Coleman: 12C9, 12D3, 13E6, 14Q3
8. Herniorrhaphy: 12A5, 12A6, 12B5, 12C7, 13B3, 13B4, 13E7
9. Municipal bonds: 13A7, 13E4, 14Q4, 16C8, 16D2, 16D3
10. Armed Forces Research: 12B6, 16P3
11. Heart transplants: 16P2
12. Ancient warfare: 12D9, 15D5, 15E6, 15F7, 16P4
13. Suicide (unused)

Data Exhibit **1** for Problems

Middle-U.S. cities

This exhibit gives various facts about the 152 cities of 325,000 population in 1960 in the three "middle" U.S. census regions: West North Central, West South Central, and Mountain. The variables listed are:

#	Serial number in this list (alphabetic by state and city),
301	Land area (sq. miles, rounded),
302	Rank among U.S. Cities (1 = largest),
303	1960 population (in hundreds),
306	% of population nonwhite in 1960,
$313\frac{1}{2}$	% of population either foreign-born or with at least one foreign-born parent (sum of columns 313 and 314),
320	Median family income, 1959,
325	% of persons (≥ 25 years of age) who completed less than 5 years of school,
327	% of persons (≥ 25 years of age) who are college graduates,
331	% of persons (≥ 5 years of age) living in same home 1955 and 1960,
352	% of persons (≥ 5 years of age) in one-unit housing structures (includes row houses),
357	% occupied housing units with ≥ 1.01 persons per room,
358	% of occupied housing units moved into by head of household during 1958 to 1960.

These are from a few of the 160-odd columns of information given by the 1962 *County and City Data Book*.

State and city	#	301	302	303	306	313½	320	325	327	331	352	357	358
Arizona													
Mesa	1	15	488	338	2.8	14.4	5598	5.5	8.7	33.7	86.4	16.9	47.3
Phoenix	2	187	219	4392	5.8	16.6	6117	6.0	8.9	34.7	87.2	12.9	46.5
Tucson	3	71	54	2129	4.4	22.3	5703	5.7	11.1	33.9	89.3	14.2	48.1
Colorado													
Aurora	4	9	317	485	1.2	12.6	6627	.6	13.5	26.4	78.7	13.4	58.7
Boulder	5	7	429	377	1.2	13.0	6726	1.0	29.9	26.4	70.5	7.0	48.6
Colorado Springs	6	17	204	702	5.0	13.7	5669	2.4	12.2	33.7	70.3	7.8	45.3
Denver	7	71	23	4939	7.1	18.7	6361	4.3	12.2	42.0	65.6	8.0	39.3
Englewood	8	6	498	331	0.8	12.4	6744	1.8	9.9	43.5	84.5	10.2	38.5
Ft. Collins	9	6	675	250	1.0	14.3	5409	4.2	18.4	28.6	78.5	7.1	48.8
Greeley	10	5	647	283	0.9	17.3	5351	5.2	14.2	35.4	61.7	7.5	45.2
Pueblo	11	17	147	912	2.6	17.9	5698	8.5	6.5	49.3	81.5	15.0	33.2
Idaho													
Boise	12	10	473	345	1.1	14.7	5851	2.3	11.2	43.4	71.6	6.9	42.0
Idaho Falls	13	8	506	332	.9	12.3	6844	2.1	14.7	36.5	79.1	16.4	44.9
Pocatello	14	8	595	285	2.6	13.7	6023	2.8	9.3	43.7	68.9	15.2	40.5
Iowa													
Ames	15	8	635	270	1.1	11.6	6191	0.7	32.9	29.5	79.6	8.0	45.6
Burlington	16	12	519	324	1.6	11.7	5848	2.5	6.5	50.6	77.0	7.3	30.7
Cedar Rapids	17	33	145	920	1.4	14.5	6687	2.9	10.4	46.9	76.0	8.6	33.3
Clinton	18	11	493	336	1.1	18.7	6146	3.3	6.5	54.6	76.8	8.4	26.7
Council Bluffs	19	16	276	556	1.1	12.6	5967	3.0	4.7	46.2	81.7	13.4	33.6
Davenport	20	47	155	890	2.1	16.4	6479	2.3	7.8	47.0	68.4	10.5	33.3
Des Moines	21	64	55	200	5.1	12.5	6436	3.1	9.7	47.3	74.9	6.8	34.8
Dubuque	22	14	271	566	0.3	14.0	6973	2.2	8.5	53.1	67.4	11.6	28.2
Ft. Dodge	23	7	598	284	1.1	16.8	6059	2.6	8.1	47.0	69.8	7.7	38.6
Iowa City	24	8	495	334	1.5	12.1	5769	1.6	29.4	30.6	60.7	8.5	44.8
Mason City	25	13	550	306	0.9	19.3	5979	3.3	7.8	46.2	86.2	8.7	30.9
Ottumwa	26	12	485	339	1.3	0.7	5647	2.9	5.8	52.7	85.2	9.7	28.8
Sioux City	27	49	153	891	2.1	17.5	5812	3.8	7.8	49.8	75.0	9.8	31.7
Waterloo	28	34	200	718	6.9	12.2	6526	2.9	6.5	46.6	81.4	10.3	32.0

Exhibit 1   553

Continued on next page

Data Exhibit **1** for Problems (continued)

State and city	#	301	302	303	306	313½	320	325	327	331	352	357	358
Kansas													
Hutchinson	11	432	376	3.3	8.2	5469	3.4	7.5	41.3	80.4	8.7	39.1
Kansas City	41	98	1219	23.2	11.6	5583	8.0	4.9	53.7	77.1	12.8	30.0
Lawrence	8	510	329	9.1	7.4	5427	2.6	23.2	30.5	77.7	6.9	45.4
Prairie Village	5	668	254	0.2	11.1	10225	0.5	30.9	43.0	100.0	2.4	31.4
Salina	8	367	432	3.7	10.8	5475	2.1	10.2	34.7	83.6	10.9	48.1
Topeka	36	100	1194	8.2	9.0	6039	3.1	11.5	42.6	77.6	9.5	37.8
Wichita	52	51	2547	8.3	6.4	6121	3.0	10.4	41.2	78.8	10.4	39.8
Louisiana													
Alexandria	10	394	403	43.4	4.4	3768	19.3	7.8	51.2	84.6	15.7	33.8
Baton Rouge	31	80	1524	29.9	4.3	5789	11.8	13.0	51.1	86.1	15.7	33.9
Bossier City	11	512	328	11.2	5.4	5043	7.1	7.1	23.6	80.7	16.5	62.7
Lafayette	7	392	404	28.3	2.9	4361	27.7	11.5	50.8	87.5	18.0	35.6
Lake Charles	16	236	634	21.7	4.3	5462	13.4	10.0	42.5	85.0	15.1	42.8
Monroe	18	295	522	43.8	3.0	3958	18.8	8.6	46.4	83.3	19.2	35.9
New Iberia	7	585	291	23.4	2.4	4663	28.1	6.4	50.4	90.9	22.2	33.7
New Orleans	199	15	6275	37.4	8.6	4807	13.7	7.7	50.3	49.5	18.2	32.4
Shreveport	36	76	1644	34.5	3.5	5205	13.2	10.8	45.5	83.2	14.7	37.6
Minnesota													
Austin	5	611	279	0.1	18.9	7618	2.1	7.6	51.1	81.1	9.8	28.6
Bloomington	35	308	505	0.3	15.7	7201	0.7	12.8	33.2	98.6	16.1	35.4
Duluth	63	122	1069	1.1	36.8	5877	4.7	8.5	53.3	62.3	7.2	29.6
Edina	16	596	285	0.2	20.6	12082	0.4	27.9	39.9	99.0	2.7	31.2
Minneapolis	56	25	4829	3.2	31.2	6401	3.5	9.6	49.5	48.7	6.3	33.6
Minnetonka	28	674	250	0.4	19.2	8180	0.9	18.3	39.9	96.7	8.5	36.1
Richfield	10	372	425	0.3	17.9	7721	0.7	13.6	51.5	94.5	12.6	27.8
Rochester	8	391	407	0.5	19.6	6638	1.9	15.0	41.4	66.7	9.3	40.5
St. Cloud	9	486	338	0.7	18.7	5592	4.0	8.6	50.0	79.9	15.2	30.9
St. Louis Park	10	365	433	0.5	26.9	7808	1.1	16.2	52.5	88.7	7.3	27.6
St. Paul	52	40	3134	3.0	27.9	6543	3.7	9.2	53.5	57.4	9.0	30.0

Exhibit 1   555

Missouri

City													
Columbia	56	11	445	366	7.9	5.2	5616	3.8	25.7	27.5	73.7	9.1	47.0
Florissant	57	4	419	382	0.1	8.2	7740	1.0	12.9	23.7	98.3	13.3	42.9
Independence	58	14	244	623	1.2	6.8	6535	2.6	6.1	46.0	84.2	8.9	36.2
Jefferson City	59	10	604	282	10.7	5.8	5876	6.0	9.1	41.0	64.7	7.8	36.5
Joplin	60	20	409	390	2.2	4.2	4915	5.4	6.9	46.1	85.7	7.9	33.8
Kansas City	61	130	27	4755	17.7	10.9	5906	5.4	7.9	43.2	58.2	8.9	37.6
Kirkwood	62	8	574	294	3.2	10.8	8753	1.9	22.8	52.8	94.8	6.1	25.8
St. Joseph	63	28	181	797	3.3	9.6	5522	6.1	5.0	50.7	72.9	9.8	33.0
St. Louis	64	61	10	7500	28.8	14.1	5355	9.1	4.5	45.0	35.2	16.4	35.6
Springfield	65	35	137	959	2.6	4.2	4955	5.0	7.8	40.4	82.4	8.9	38.4
University City	66	4	302	512	0.4	39.2	8105	6.4	16.3	54.8	63.9	3.6	24.6
Webster Groves	67	6	587	290	3.8	12.5	8750	1.9	22.4	62.7	95.4	4.8	19.0

Montana

City													
Billings	68	9	292	529	1.2	21.6	6638	3.4	13.1	38.3	71.4	9.3	42.7
Butte	69	5	615	279	0.9	33.9	5156	4.9	7.0	58.0	65.4	11.4	29.1
Great Falls	70	11	278	552	1.6	23.5	6257	3.3	8.9	38.5	69.2	13.8	45.4
Missoula	71	6	633	271	0.8	21.9	5847	3.8	13.9	39.1	72.3	11.2	41.4

Nebraska

City													
Grand Island	72	5	658	257	0.4	17.4	5109	3.2	6.4	46.6	81.1	9.6	33.1
Lincoln	73	25	95	1285	1.9	16.1	6032	1.7	14.6	39.9	68.9	7.3	39.7
Omaha	74	51	42	3016	8.7	21.0	6315	4.1	9.0	47.1	71.3	10.7	34.7

Nevada

City													
Las Vegas	75	25	231	644	15.8	17.8	7662	3.8	8.1	27.9	68.8	12.3	53.8
Reno	76	12	301	515	3.2	22.5	7433	2.9	11.6	34.2	70.0	8.5	48.2

New Mexico

City													
Albuquerque	77	56	60	2012	2.9	9.5	6621	4.7	14.7	34.5	86.9	14.0	48.9
Carlsbad	78	7	663	255	2.5	9.2	6293	9.0	8.1	42.0	95.7	12.1	40.3
Hobbs	79	11	650	263	8.0	2.9	6229	5.6	8.5	29.9	95.5	19.1	53.5
Las Cruces	80	9	577	294	2.9	16.0	5789	11.3	14.5	32.2	85.7	22.2	52.3
Roswell	81	13	403	396	4.6	7.1	5543	7.8	10.4	31.5	89.8	18.2	53.5
Santa Fe	82	26	499	334	1.7	6.3	5502	10.8	13.6	52.3	87.7	20.5	37.2

Continued on next page

Data Exhibit **1** for Problems (continued)

State and city	#	301	302	303	306	$313\frac{1}{2}$	320	325	327	331	352	357	358
North Dakota													
Bismarck	83	9	618	277	0.4	29.0	6094	5.8	11.8	36.5	55.3	16.7	45.5
Fargo	84	9	331	467	0.5	26.5	6522	2.6	13.0	41.4	61.6	12.1	39.8
Grand Forks	85	6	475	345	1.0	24.5	5849	3.5	11.9	34.3	61.3	16.7	45.4
Minot	86	7	551	306	0.8	25.7	5953	3.4	8.9	34.3	68.0	16.6	46.2
Oklahoma													
Bartlesville	87	6	612	279	4.5	4.3	6606	3.4	18.9	40.4	87.9	7.3	39.0
Enid	88	10	413	389	4.8	7.1	4956	5.0	8.3	43.2	84.9	8.5	37.3
Lawton	89	12	247	617	11.7	8.0	4633	4.7	8.2	29.1	85.3	18.1	58.9
Midwest City	90	24	454	360	2.5	4.8	5843	2.3	7.6	34.6	96.1	13.6	46.1
Muskogee	91	13	421	381	22.2	2.9	4449	10.0	9.1	46.5	82.9	8.9	33.5
Norman	92	10	497	334	1.3	4.5	5259	6.4	22.6	29.6	85.0	6.6	50.6
Oklahoma City	93	322	37	3243	13.0	4.5	5600	5.2	9.8	42.5	82.6	10.6	39.3
Tulsa	94	48	50	2617	10.0	4.6	6229	3.9	12.0	42.2	83.2	7.9	37.5
Oregon													
Eugene	95	14	305	510	0.5	14.7	6267	1.7	18.4	33.5	73.7	6.0	47.0
Portland	96	67	32	3727	3.5	25.8	6335	3.8	3.5	50.5	70.1	4.3	32.4
Salem	97	13	315	491	0.5	17.9	5859	6.0	10.0	42.1	81.2	4.1	40.3
South Dakota													
Rapid City	98	16	374	424	4.1	13.3	5694	1.2	9.9	29.5	74.6	16.5	51.9
Sioux Falls	99	17	229	655	0.9	18.7	6081	2.0	8.6	43.8	75.9	10.9	35.8
Texas													
Abilene	100	62	149	904	5.1	4.9	5460	7.2	10.9	28.8	86.5	13.8	51.4
Amarillo	101	55	188	1380	5.8	4.4	5877	3.6	9.8	33.3	84.9	12.7	48.4
Arlington	102	24	351	448	0.9	3.8	6574	2.4	12.1	31.1	94.9	10.1	45.3
Austin	103	49	67	1865	13.3	10.6	5119	11.7	15.1	39.2	84.3	12.7	42.7
Baytown	104	18	606	282	6.7	7.5	6937	7.9	9.9	50.3	92.4	9.0	32.2
Beaumont	105	71	102	1192	29.4	4.8	5577	12.4	8.8	47.6	84.6	13.5	35.5
Big Spring	106	10	542	312	5.2	6.8	5682	8.1	7.4	32.0	89.2	16.4	53.1

Exhibit 1   557

107	Brownsville	16	321	480	43.9	0.1	3021	36.9	5.9	51.8	90.3	37.2	35.6
108	Bryan	17	622	275	10.2	23.4	4258	15.7	13.1	43.2	92.9	15.7	41.1
109	Corpus Christi	38	74	1677	16.8	5.6	5221	17.6	9.4	42.1	90.1	20.7	41.0
110	Dallas	280	14	6797	6.9	19.3	5976	7.4	10.4	39.8	75.6	11.5	41.1
111	Denton	10	638	268	3.3	7.6	4994	6.2	19.4	30.8	83.0	10.5	44.0
112	El Paso	115	46	2767	39.7	2.7	5211	16.4	9.1	32.8	73.6	24.4	50.5
113	Ft. Worth	140	34	3563	5.8	16.0	5484	4.8	10.0	42.2	82.3	11.7	38.8
114	Galveston	20	217	672	18.1	27.5	4698	14.3	6.7	47.2	68.8	13.3	35.9
115	Garland	19	416	385	2.3	3.9	6792	3.1	9.5	32.0	97.9	10.6	46.3
116	Grand Prairie	15	555	304	3.0	7.0	5764	4.5	8.8	39.4	91.1	14.2	42.3
117	Harlingen	31	387	412	35.4	1.7	4167	26.8	9.2	36.1	83.5	28.6	47.9
118	Houston	328	7	9382	9.7	23.2	5902	8.9	10.7	41.9	80.1	12.7	39.3
119	Irving	21	339	460	3.7	0.1	6843	3.2	9.5	34.6	96.0	12.3	45.1
120	Kingsville	5	669	252	18.2	4.0	4366	22.8	11.4	42.7	85.1	23.2	45.5
121	Laredo	14	252	607	53.0	0.4	2935	38.4	5.0	56.0	90.6	40.1	34.3
122	Longview	19	400	400	1.7	23.0	5355	8.1	9.5	42.1	93.0	11.4	40.7
123	Lubbock	75	94	1287	5.1	8.1	5582	8.5	11.5	28.8	86.0	17.5	52.8
124	McAllen	10	514	327	44.2	0.4	3790	30.0	9.3	45.5	87.3	30.6	40.6
125	Mesquite	21	623	275	2.9	0.5	6241	3.5	6.6	13.5	99.1	13.0	60.8
126	Midland	23	243	626	5.2	10.0	7094	5.7	19.5	25.3	87.7	13.9	59.9
127	Odessa	16	179	803	4.2	5.8	6210	6.6	8.1	25.6	91.4	19.8	56.6
128	Orange	8	660	256	3.6	23.0	5510	10.8	9.3	38.7	86.5	16.4	42.0
129	Pasadena	22	262	587	5.3	0.1	7176	3.5	9.5	39.9	94.9	10.8	39.3
130	Port Arthur	46	224	667	7.9	30.8	5659	16.9	5.0	50.1	83.3	15.0	32.8
131	San Angelo	30	260	588	9.3	5.4	4650	11.6	7.8	43.5	89.9	13.9	38.2
132	San Antonio	160	17	5877	24.0	7.4	4691	19.5	6.8	47.3	83.7	21.5	36.0
133	Temple	19	554	304	8.5	18.9	4509	10.7	6.5	44.7	88.8	12.2	36.0
134	Texarkana	16	557	302	2.5	25.5	4353	10.6	6.9	48.8	90.1	10.8	31.7
135	Texas City	45	526	321	6.1	19.6	6101	8.2	8.3	41.6	90.7	14.2	37.1
136	Tyler	18	303	512	2.8	22.3	5478	5.6	12.4	41.6	90.6	9.4	38.7
137	Victoria	12	508	330	10.4	8.3	5279	16.9	9.3	42.7	89.2	16.2	39.9
138	Waco	37	134	978	7.1	18.5	4859	10.2	10.0	42.4	86.5	10.8	38.5
139	Wichita Falls	37	127	1017	6.0	8.4	5451	7.0	9.4	33.3	82.0	14.0	48.0

Continued on next page

Data Exhibit 1 for Problems (continued)

State and city	#	301	302	303	306	313½	320	325	327	331	352	357	358
Utah													
Ogden	140	19	203	702	3.5	17.4	6145	3.1	8.5	48.9	70.8	13.8	34.9
Provo	141	19	455	360	0.8	14.1	5310	2.8	14.9	37.7	68.1	14.8	44.8
Salt Lake City	142	56	65	1895	2.1	23.9	6135	3.2	12.4	48.9	62.4	10.3	35.4
Washington													
Bellingham	143	22	469	347	1.2	31.3	5735	3.3	8.5	51.1	82.6	5.2	30.3
Bremerton	144	5	589	289	4.3	21.9	6046	8.5	6.4	40.0	77.6	6.6	39.3
Everett	145	12	393	403	1.1	29.2	5843	3.1	7.1	46.3	72.8	5.5	34.8
Seattle	146	88	19	5571	8.4	31.4	6942	3.4	12.4	46.7	63.3	4.6	36.6
Spokane	147	43	68	1816	2.5	22.4	6044	3.4	9.0	49.2	71.9	6.7	34.0
Tacoma	148	48	83	1480	5.3	27.6	5993	3.8	6.8	50.4	77.0	5.8	32.6
Vancouver	149	10	518	325	1.5	18.5	6535	3.1	7.5	45.3	80.7	6.4	34.8
Yakima	150	9	366	433	2.8	20.0	5900	4.3	9.0	44.4	79.9	6.8	35.6
Wyoming													
Casper	151	7	410	389	1.6	12.6	1157	2.0	13.4	35.5	75.2	10.0	44.7
Cheyenne	152	10	364	435	2.5	15.3	6575	3.4	9.8	32.6	62.7	14.3	51.3

Exhibit 2   559

Data Exhibit 2 for Problems

Smoking and health

These data come from "A Canadian study of smoking and health—Second report," by E. W. R. Best and C. B. Walker, published in the *Canadian Journal of Public Health*, **55**: 1–1, 1964.

The study took the form of a questionnaire survey of Department of Veterans' Affairs pensioners with a followup of deaths of pensioners who responded to the questionnaire. The study lasted 6 years from July 1, 1956, to June 30, 1962. (Those who had smoked at least a total of 100 cigarettes or 10 cigars or 20 pipefuls of tobacco during their life-time were classified as smokers.) The study group consists of World War I and II pensioners and Korean War pensioners and their male dependents (sons and fathers). The age groups are based on age in 1956.

Age	Nonsmokers		Cigar and pipe only		Cigarette and other		Cigarette only	
	Pop.	Deaths	Pop.	Deaths	Pop.	Deaths	Pop.	Deaths
40–44	656	18	145	2	4531	149	3410	124
45–49	359	22	104	4	3030	169	2239	140
50–54	249	19	98	3	2267	193	1851	187
55–59	632	55	372	38	4682	576	3270	514
60–64	1067	117	846	113	6052	1001	3791	778
65–69	897	170	949	173	3880	901	2421	689
70–74	668	179	824	212	2033	613	1195	432
75–80	361	120	667	243	871	337	436	214
80+	274	120	537	253	345	189	113	63

S) SOURCE

Data reprinted by permission of the author and of the journal. Study conducted by the Department of National Health and Welfare and the Department of Veterans Affairs, Canada, and the Canadian Pension Commission.

Data Exhibit **3** for Problems

The U.S. population 1960–1970 by state and region

| Region and state | Population (in thousands) | | | Change in population 1960 to 1970 | | | | |
| | | | | Net increase | | | | Net total |
	1960	1965	1970	Number	Percent	Births	Deaths	migration
U.S.	179,979	193,526	203,806	23,912	13.3	39,033	18,192	3,070
N.E.*	10,532	11,329	11,883	1,338	12.7	2,169	1,147	316
Maine	975	997	997	24	2.5	203	109	−69
N.H.	609	676	742	131	21.5	133	71	69
Vt	389	404	446	55	14.1	85	45	15
Mass	5,160	5,502	5,706	541	10.5	1,040	574	74
R.I.	855	893	951	90	10.5	171	93	13
Conn	2,544	2,857	3,041	497	19.6	537	255	214
M.A.*	34,270	36,122	37,274	3,034	8.9	6,725	3,749	59
N.Y.	16,838	17,734	18,384	1,458	8.7	3,361	1,852	−51
N.J.	6,103	6,767	7,193	1,101	18.2	1,259	645	488
Pa	11,329	11,620	11,813	475	4.2	2,105	1,252	−378
E.N.C.*	36,291	38,406	40,313	4,028	11.1	7,832	3,652	−153
Ohio	9,734	10,201	10,664	946	9.7	2,047	975	−126
Ind	4,674	4,922	5,202	531	11.4	1,023	475	−16
Ill	10,086	10,693	11,128	1,033	10.2	2,153	1,077	−43
Mich	7,834	8,357	8,890	1,052	13.4	1,754	729	27
Wis	3,962	4,232	4,429	466	11.8	856	395	4
W.N.C.*	15,424	15,819	16,518	930	6.0	3,133	1,604	−599
Minn	3,425	3,592	3,815	391	11.5	744	327	−25
Iowa	2,756	2,742	2,832	68	2.4	541	291	−183
Mo	4,326	4,467	4,688	358	8.3	857	502	2
N. Dak	634	649	620	−15	−2.3	135	55	−94
S. Dak	683	692	668	−14	−2.1	146	65	−94
Nebra	1,417	1,471	1,488	72	5.1	291	146	−73
Kans	2,183	2,206	2,249	70	3.2	419	218	−130
S.A.*	26,091	28,743	30,805	4,700	18.1	5,965	2,598	1,332
Del	449	507	551	102	22.8	109	45	38
Md	3,113	3,600	3,938	822	26.5	740	303	385
D.C.	765	797	756	−7	−1.0	182	89	−100
Va	3,986	4,411	4,659	682	17.2	909	369	141
W. Va	1,853	1,786	1,751	−116	−6.2	339	190	−265
N.C.	4,573	4,863	5,098	526	11.5	1,032	412	−94
S.C.	2,392	2,494	2,597	208	8.7	573	216	−149
Ga	3,956	4,332	4,607	646	16.4	975	379	51
Fla	5,004	5,953	6,848	1,838	37.1	1,107	596	1,326
E.S.C.*	12,073	12,627	12,839	754	6.3	2,665	1,213	−698
Ky	3,041	3,140	3,231	181	6.0	647	313	−153
Tenn	3,575	3,798	3,937	357	10.0	755	353	−45
Ala	3,274	3,443	3,451	177	5.4	729	319	−233
Miss	2,182	2,246	2,220	39	1.8	534	228	−267

Exhibit 3   **561**

Data Exhibit **3** for Problems (continued)

Region and state	Population (in thousands)			Change in population 1960 to 1970				
				Net increase				Net total
	1960	1965	1970	Number	Percent	Births	Deaths	migration
U.S.	179,979	193,526	203,806	23,912	13.3	39,033	18,192	3,070
W.S.C.*	17,010	18,209	19,388	2,371	14.0	4,012	1,599	−42
Ark	1,789	1,894	1,932	137	7.7	401	193	−71
La	3,260	3,496	3,652	386	11.9	832	316	−130
Okla	2,336	2,440	2,567	231	9.9	461	244	13
Tex	9,624	10,378	11,236	1,617	16.9	2,318	847	146
Mt.*	6,916	7,740	8,348	1,429	20.8	1,724	602	307
Mont	679	706	698	20	2.9	144	66	−58
Idaho	671	686	718	46	6.9	146	58	−42
Wyo	331	332	334	2	0.7	70	28	−39
Colo	1,769	1,985	2,223	453	25.8	401	163	215
N. Mex	954	1,012	1,023	65	6.8	263	68	−130
Ariz	1,321	1,584	1,792	470	36.1	365	122	228
Utah	900	991	1,066	169	18.9	245	65	−11
Nev	291	444	493	203	71.3	91	31	144
Pac.*	21,368	24,464	26,600	5,328	25.1	4,808	2,028	2,547
Wash	2,855	2,967	3,413	556	19.5	591	284	249
Oreg	1,772	1,937	2,101	323	18.2	346	182	159
Calif	15,870	18,585	20,007	4,236	27.0	3,634	1,511	2,113
Alaska	229	271	304	76	33.6	73	13	16
Hawaii	642	704	774	137	21.7	164	37	11

* Region composed of states that follow.

S) SOURCE

Statistical Abstract of the United States, 1975, pp. 12–13.

Data Exhibit 4 for Problems

Economics data taken from the Economic Report of The President, February 1975. (We chose to present data covering the years 1955–1974.)

Year	Gross National Product	Consumer Price Index*	Unemployment			Interest rates			Personal (in billions)		
						Long-term: Moody's		Short term			
			All	Men > 20	Women > 20	Aaa	Bbb		Consumption	Savings	Income
55	438.0	69.0	4.4	3.8	4.4	3.06	3.53	1.89	39.6	34.1	310.9
56	446.1	70.0	4.1	3.4	4.2	3.36	3.88	2.77	38.9	36.0	333.0
57	452.5	72.5	4.3	3.6	4.1	3.89	4.71	3.12	40.8	34.1	351.1
58	447.3	74.5	6.8	6.2	6.1	3.79	4.73	2.15	37.9	33.0	361.2
59	475.9	75.1	5.5	4.7	5.2	4.38	5.05	3.36	44.3	34.7	383.5
60	487.7	76.3	5.5	4.7	5.1	4.41	5.19	3.53	45.3	32.3	401.0
61	497.2	77.0	6.7	5.7	6.3	4.35	5.08	3.00	44.2	31.7	416.8
62	529.8	77.9	5.5	4.6	5.4	4.33	5.02	3.00	49.5	37.1	442.6
63	551.0	78.8	5.7	4.5	5.4	4.26	4.86	3.23	53.9	39.1	465.5
64	581.1	79.9	5.2	3.9	5.2	4.40	4.83	3.55	59.2	49.5	497.5
65	617.8	81.3	4.5	3.2	4.5	4.49	4.87	4.04	66.3	55.4	538.9
66	658.1	83.6	3.8	2.5	3.8	5.13	5.67	4.50	70.8	65.1	587.2
67	675.2	86.0	3.8	2.3	4.2	5.51	6.23	4.19	73.1	65.0	629.3
68	706.6	89.6	3.6	2.2	3.8	6.18	6.94	5.17	84.0	68.3	688.9
69	725.6	94.4	3.5	2.1	3.7	7.03	7.81	5.87	90.8	60.6	750.9
70	722.5	100.0	4.9	3.5	4.8	8.04	9.11	5.95	91.3	76.2	808.3
71	746.3	104.3	5.9	4.4	5.7	7.39	8.56	4.88	103.9	87.4	864.0
72	792.5	107.7	5.6	4.0	5.4	7.21	8.16	4.50	118.4	97.9	944.9
73	839.2	114.4	4.9	3.2	4.8	7.44	8.24	6.44	130.3	120.2	1055.0
74	821.1	127.0	5.6	3.8	5.5	8.57	9.50	7.83	127.8	×	1050.4

* The consumer price index is relative to 1970.

Exhibit 5   563

Data Exhibit **5** for Problems

Education expenditure

The following data was contributed by the economist M. S. Feldstein. A more detailed reference is "Wealth, neutrality, and local choice in public education" by Martin S. Feldstein, *The American Economic Review,* **65,** No. 1, March 1975, pp. 75–89. Data reprinted by permission of the author and of the American Economic Association.

Variables:

MFI	Median family income,
SBG	State block grant,
FG	Federal grants,
RES	% of local tax base in residential property,
PSC	Public-school students per capita,
EEP	Educational expenditure per public-school pupil,
TVP	Taxable property value per public-school pupil,
P	The price to the town of obtaining 100 dollars for education.

Towns whose schools are partially supported by matching state funds need pay less than a dollar to obtain a dollar for educational expenses.

|MFI|	|SBG|	|FG|	|RES|	|PSC|	|EEP|	|TVP|	|P|
9890	181	308	68	0.180	873	19560	100
12247	65	84	82	0.180	864	34311	100
10904	1	102	71	0.185	874	24418	75
11292	1	84	76	0.208	758	26824	74
13030	1	99	80	0.256	841	28044	82
10377	146	111	64	0.212	719	17453	100
9815	47	146	45	0.103	1184	48012	100
9738	150	149	60	0.197	622	15950	100
14958	66	155	90	0.273	1008	32105	100
12516	1	101	76	0.211	842	25743	74
10086	68	117	35	0.175	949	50993	100
9881	1	220	70	0.294	859	39036	87
11982	1	158	71	0.248	903	19296	70
11550	1	128	70	0.232	782	16941	63
9750	96	144	62	0.200	894	26419	100
9756	1	196	29	0.181	861	22569	70
11031	113	101	85	0.256	871	25478	100
11645	162	106	60	0.220	574	18759	100
11278	1	164	79	0.539	854	22904	75
10871	1	70	55	0.228	850	23858	73
17558	112	136	88	0.299	1027	24570	100
9739	1	90	59	0.167	742	31847	80
11020	1	126	65	0.246	772	16347	66
10665	1	91	66	0.205	805	24561	75
12424	1	56	84	0.227	761	24857	77

➡

Data Exhibit 5 for Problems (continued)

MFI	SBG	FG	RES	PSC	EEP	TVP	P
9638	156	119	73	0.226	674	17915	100
12580	1	63	50	0.300	745	17616	68
12656	91	321	81	0.215	720	27058	100
13144	1	122	80	0.273	844	26744	82
9992	130	316	69	0.200	858	20570	100
8924	1	162	50	0.197	878	19348	67
12629	1	107	75	0.286	783	14790	63
12606	1	114	68	0.228	867	24950	74
10621	215	88	67	0.294	760	10848	100
11629	157	185	68	0.207	749	21630	100
9510	60	83	67	0.523	917	40645	100
11094	142	108	68	0.189	954	25794	100
13434	1	93	88	0.244	763	23991	74
10067	66	79	88	0.178	810	52291	100
11541	1	125	85	0.240	815	26512	82
14805	1	172	84	0.302	948	19009	71
9594	1	78	50	0.153	761	22433	67
9752	131	65	60	0.239	634	18598	100
12281	1	105	83	0.231	795	26689	77
9957	162	137	67	0.150	751	22710	100
12412	1	89	82	0.222	808	30002	83
9802	1	136	51	0.230	786	25084	74
9418	1	188	70	0.146	769	36782	81
12837	1	169	56	0.224	903	22132	74
11631	1	105	55	0.240	688	29349	82
9279	1	140	69	0.250	822	12658	61
11685	1	92	90	0.177	731	23856	71
10038	1	125	44	0.165	955	20115	66

Data Exhibit 6 for Problems

Data about the weather in Boston during April 1975

This data comes from the Preliminary Local Climatological Data and Surface Weather Observations of the National Weather Service, Forecast Office in Boston, Mass.

Any positive precipitation under .01 inches is rounded up to .01 inches. "Sunshine" is measured by a photoelectric cell and is the number of hours above a threshold brightness level. The pressure and relative humidity readings are not averages; they are all taken between 12:50 and 1:00 P.M. Fog is coded as 0 if there is no fog, 1 if fog is moderate, 2 if visibility drops below $\frac{1}{4}$ mile.

Exhibit 6   565

Data Exhibit **6** for Problems (continued)

Date	Ave. temp.	Precipitation (inches)	Ave. wind speed	Sunshine (hours)	Ave. sky cover (tenths)	Pressure (inches)*	Relative humidity*	Fog**
1	38	0	7.4	4.1	8	29.82	57	0
2	40	0	8.6	9.0	8	30.00	55	0
3	44	1.40	21.5	0.0	10	29.02	93	1
4	35	.23	17.0	0.0	10	29.10	85	2
5	36	.13	14.8	2.2	10	29.49	82	1
6	41	.01	14.6	2.2	10	29.52	58	0
7	41	.01	15.1	5.6	9	29.62	46	0
8	39	.04	12.4	10.1	6	29.76	47	1
9	40	0	13.3	5.1	9	29.85	43	0
10	43	0	8.6	13.0	1	29.81	35	0
11	42	0	9.8	13.2	0	29.82	42	0
12	40	0	8.6	11.2	8	29.85	68	0
13	42	0	13.0	11.5	4	29.98	30	0
14	48	0	11.6	13.3	0	30.13	25	0
15	44	0	7.5	10.0	9	30.00	52	0
16	48	0	13.6	9.9	10	29.85	45	0
17	52	0	11.0	10.8	5	29.77	40	0
18	54	.01	11.9	8.5	6	29.91	25	1
19	61	.12	16.2	1.2	10	29.45	63	1
20	53	0	21.4	8.1	5	29.62	30	0
21	45	0	15.8	11.8	3	29.98	17	0
22	47	0	10.1	13.7	1	30.32	25	0
23	51	0	11.9	12.6	2	30.27	28	0
24	54	.28	6.9	0.0	10	29.84	72	2
25	52	.04	7.2	0.0	10	29.89	74	2
26	45	.16	12.9	3.7	8	29.86	71	1
27	44	.01	14.5	2.8	10	29.92	31	1
28	44	0	9.4	8.8	7	29.81	34	0
29	43	0	8.3	14.0	2	30.11	46	0
30	54	0	7.7	14.0	1	30.17	34	0

* Between 12:50–1:00 P.M.

** 0 = none; 1 = moderate fog; 2 = visibility $\leq \frac{1}{4}$ mile.

Data Exhibit **7** for Problems

Random sample of 20 schools from *The Coleman Report,* **for Mid-Atlantic and New England states**

Variables:

y verbal mean test score (all sixth graders),
x_1 Staff salaries per pupil,
x_2 6th grade per cent white-collar fathers,
x_3 Socioeconomic status composite deviation: 6th grade means, for family size, family intactness, father's education, mother's education, per cent white-collar fathers, and home items,
x_4 Mean teacher's verbal test score,
x_5 6th grade mean mother's educational level (1 unit = 2 school years)

School Number	x_1	x_2	x_3	x_4	x_5	y
1	3.83	28.87	7.20	26.60	6.19	37.01
2	2.89	20.10	−11.71	24.40	5.17	26.51
3	2.86	69.05	12.32	25.70	7.04	36.51
4	2.92	65.40	14.28	25.70	7.10	40.70
5	3.06	29.59	6.31	25.40	6.15	37.10
6	2.07	44.82	6.16	21.60	6.41	33.90
7	2.52	77.37	12.70	24.90	6.86	41.80
8	2.45	24.67	−0.17	25.01	5.78	33.40
9	3.13	65.01	9.85	26.60	6.51	41.01
10	2.44	9.99	−0.05	28.01	5.57	37.20
11	2.09	12.20	−12.86	23.51	5.62	23.30
12	2.52	22.55	0.92	23.60	5.34	35.20
13	2.22	14.30	4.77	24.51	5.80	34.90
14	2.67	31.79	−0.96	25.80	6.19	33.10
15	2.71	11.60	−16.04	25.20	5.62	22.70
16	3.14	68.47	10.62	25.01	6.94	39.70
17	3.54	42.64	2.66	25.01	6.33	31.80
18	2.52	16.70	−10.99	24.80	6.01	31.70
19	2.68	86.27	15.03	25.51	7.51	43.10
20	2.37	76.73	12.77	24.51	6.96	41.01

Exhibit 8   567

Data Exhibit **8** for Problems

Herniorrhaphy data

The following table presents data on the experience of 32 patients undergoing an elective herniorrhaphy (there were no deaths). The outcome measures are:

LEAVE (condition leaving the operating room):

1. routine recovery,
2. went to intensive care unit for observation overnight,
3. went to intensive care unit; moderate care required,
4. went to intensive care unit; intensive care required.

NURSE (level of nursing care required 1 week after operation):

1. intense,
2. heavy,
3. moderate,
4. light.

LOS (length of stay in hospital after operation, in days)

Variables describing the patient's preoperative condition are, where not self-explanatory:

PSTAT (physical status, discounting that associated with the operation) on a scale of 1–5, 1 being perfect health and 5 being very poor health.

BUILD (body build):

1. emaciated,
2. thin,
3. average,
4. fat,
5. obese.

CARDIAC or RESP. (preoperative complications):

1. none,
2. mild,
3. moderate,
4. severe.

➡

Data Exhibit 8 for Problems (continued)

| Patient | Age (years) | Sex | PSTAT | BUILD | Preoperative complications | | LEAVE | LOS | NURSE |
					CARDIAC	RESP.			
1	78	M	2	3	1	1	2	9	3
2	60	M	2	3	2	2	2	4	—
3	68	M	2	3	1	1	1	7	4
4	62	M	3	5	3	1	1	35	3
5	76	M	3	4	3	2	2	9	4
6	76	M	1	3	1	1	1	7	—
7	64	M	1	2	1	2	1	5	—
8	74	F	2	3	2	2	1	16	3
9	68	M	3	4	2	1	1	7	—
10	79	F	2	2	1	1	2	11	3
11	80	F	3	4	4	1	1	4	—
12	48	M	1	3	1	1	1	9	3
13	35	F	1	4	1	2	1	2	—
14	58	M	1	3	1	2	1	4	—
15	40	M	1	4	1	1	1	3	—
16	19	M	1	3	1	1	1	4	—
17	79	M	3	2	3	3	3	3	—
18	51	M	1	3	1	1	1	5	—
19	57	M	2	3	2	1	1	8	3
20	51	M	3	3	3	2	1	8	4
21	48	M	1	3	1	1	1	3	—
22	48	M	1	3	1	1	1	5	—
23	66	M	1	3	1	1	1	8	4
24	71	M	2	3	2	2	2	2	—
25	75	F	3	1	3	1	2	7	—
26	02	F	1	3	1	1	1	0	—
27	65	F	2	3	1	1	2	16	3
28	42	F	2	3	1	1	2	3	—
29	54	M	2	2	2	2	2	2	—
30	43	M	1	2	1	1	1	3	—
31	04	M	2	2	2	1	1	3	—
32	52	M	1	3	1	1	1	8	3

S) SOURCE

B. McPeek and J. P. Gilbert of the Harvard Anesthesia Center. Data reprinted by permission of the contributors

Exhibit 9   569

Data Exhibit **9** for Problems

Municipal bond data for 20 cities

Variables:

y	Bond yield,
x_1	Block offer size (no. of $1000 bonds),
x_2	Term to maturity (100's of months),
x_3	Population of issuer (100,000's of people),
x_4	Total net debt,
x_5	College students/population.

No.	City	x_1	x_2	x_3	x_4	x_5	y
1	Birmingham	30	1.81	3.61	.280	1.03	335
2	Oxnard	10	1.93	.29	.012	.00	365
3	Salinas	30	2.79	.24	.023	4.29	315
4	Danbury	15	1.81	.40	.036	2.38	325
5	New Haven	15	1.87	1.65	.186	6.54	283
6	Norwalk	40	2.17	.59	.145	.15	300
7	New Orleans	15	2.34	6.40	.710	1.91	327
8	Baltimore	10	1.85	9.74	1.827	2.24	290
9	Detroit	10	2.09	19.25	1.703	1.81	317
10	St. Louis	55	2.03	8.73	.533	3.09	273
11	Clifton	5	2.37	.81	.088	.00	356
12	New York City	5	2.33	82.00	20.720	2.25	314
13	North Hempstead	35	1.93	2.05	.075	2.10	345
14	Tulsa	25	2.53	2.54	.193	2.57	315
15	Philadelphia	80	2.14	22.00	3.861	2.36	305
16	Memphis	90	1.93	4.53	.465	1.55	285
17	Hopewell	15	2.16	.22	.023	.00	350
18	Norfolk	10	1.90	2.99	.356	.47	320
19	Madison	100	1.93	1.17	.205	21.09	270
20	So. Milwaukee	25	1.81	.17	.027	.00	305

Data Exhibit 10 for Problems

Armed Forces Research and Development (in millions of dollars)

The attached table gives the amounts (in millions of dollars) requested by the Army, the Navy, and the Air Force for research and development for the years 1953 through 1973, and the corresponding amounts appropriated by the U.S. Congress.

Year	Army		Navy		Air Force	
	Request	Appropriation	Request	Appropriation	Request	Appropriation
1953	450.0	440.0	75.7	70.0	525.0	525.0
54	475.0	345.0	74.9	58.6	537.0	440.0
55	355.0	345.0	61.0	419.9	431.0	418.1
56	333.0	333.0	439.2	439.2	570.0	570.0
57	410.0	410.0	477.0	492.0	610.0	710.0
58	400.0	400.0	505.0	505.0	661.0	661.0
59	471.0	498.7	641.0	821.2	719.0	743.0
60	1046.5	1035.7	970.9	1015.9	750.0	1159.9
61	1041.7	1041.2	1169.0	1218.6	1334.0	1552.9
62	1130.4	1203.2	1267.0	1301.5	1637.0	2403.2
63	1329.0	1319.5	1474.0	1475.9	3439.0	3632.1
64	1474.6	1390.2	1578.4	1530.5	3627.9	3458.7
65	1401.5	1344.1	1456.3	1377.5	3210.9	3117.3
66	1442.7	1410.6	1478.1	1444.2	3153.9	3109.4
67	1522.2	1531.9	1752.5	1762.4	3058.1	3116.8
68	1544.0	1514.2	1863.9	1826.5	3293.6	3251.2
69	1661.9	1522.6	2146.4	2141.3	3364.7	3570.3
70	1849.5	1596.8	2211.5	2186.4	3561.2	3060.6
71	1717.9	1618.2	2197.3	2165.1	2909.7	2762.1
72	1951.5	1839.5	2431.4	2372.3	3017.0	2912.9
73	2122.7	1829.0	2813.8	2545.3	3262.2	3122.5

S) SOURCE

James R. Capra, "Analysis of data describing Congressional responses to DOD budget requests," Ph. D. thesis, Naval Postgraduate School, Monterey, California, June, 1974.

Exhibit 11   571

Data Exhibit 11 for Problems

Heart transplants

After a patient is admitted to the Stanford program, a donor heart, matched on blood type, is then sought.

We chose to present here only patients who have been followed until their death, because that avoids some problems arising from dealing with censored data. (For patients still alive, we know only that their survival time will be longer than it is to date, so the observation is incomplete or "censored".) The data reported here cover the deaths following heart transplantation during the period January, 1968 through April, 1974.

The variables in the table are:

Survival time	Number of days the patient survived after the operation,
Age	Age at time of operation,
Waiting time	Number of days from entry into the program until the operation was performed,
Calendar time	Number of days after January 1, 1968 that the operation was performed,
Mismatch T5 score	A measure of the degree to which donor and recipient are mismatched for tissue type.

Survival time (days)	Reject? Yes = +	Mismatch T5 Score	Age (years)	Waiting time (days)	Calendar time
15		1.11	54.3	0	6
3		1.66	40.4	35	123
624	+	1.32	51	50	244
46	+	0.61	42.5	11	235
127		0.36	48	25	253
64	+	1.89	54.6	16	279
1350	+	0.87	54.1	36	300
280	+	1.12	49.5	27	327
23		2.05	56.9	19	325
10	+	2.76	55.3	17	412
1024	+	1.13	43.4	7	405
39	+	1.38	42.8	11	454
730	+	0.96	58.4	2	469
136	+	1.62	52	82	565
1		0.47	54.2	70	594
836	+	1.58	45	15	612

(Continued)

Data Exhibit **11** for Problems (continued)

Survival time (days)	Reject? Yes = +	Mismatch T5 Score	Age (years)	Waiting time (days)	Calendar time
60	+	0.69	64.5	16	623
54	+	2.09	49	45	870
47	+	0.87	61.5	18	864
44		0.0	36.2	0	1101
994	+	0.81	48.6	1	1107
51	+	1.38	47.2	20	1149
253	+	1.08	48.8	31	1210
51	+	1.51	52.5	9	1326
322	+	1.82	48.1	20	1382
65	+	0.66	49.1	2	1420
551		0.12	48.9	32	1525
66	+	1.12	51.3	11	1538
65	+	1.68	45.2	2	1561
25	+	1.68	53	4	1634
63	+	2.16	56.4	26	1742
12		0.61	29.2	4	1727
29	+	1.08	54	66	1862
48		3.05	53.4	31	1882
297	+	0.60	42.6	36	1893
50	+	2.25	46.4	59	1966
68	+	1.33	51.4	138	2060
26		0.82	52.5	159	2082
161	+	1.2	43.8	3	2087

S) SOURCE

Stanford Heart Transplantation Program.

Described in more detail in Rupert G. Miller, Jr., "Least-squares regression with censored data," *Biometrika,* **63**, 449–464, 1976. Data reprinted by permission of the author and of the journal.

Exhibit 12   573

Data Exhibit **12** for Problems

Ancient warfare: Does military deterrence work?

These data consist of a sample of influential higher civilizations* throughout history, for which adequate records remain to determine nearly all the carriers below. For each civilization chosen, a randomly chosen decade within several centuries is chosen. The Swiss confederation was added so that the data set would include some republics.

Variable descriptions (deleting those defined as functions of other variables):

	Column	
War	1	Months at war in second sampled decade,
Territory	2	Territory gained by conspicuous state (% of original area),
Stance	3	Defensive military stance,
Strength	4	Superior numbers in army,
Mobility	5	Superior horses or ships,
Quality	6	Army known superior by historians,
Fortifications	7	Fortified border with rival,
Prestige	8	*Everybody* agrees army better, even rivals,
Propinquity	9	Has common border with rival,
Barriers	10	At least $\frac{2}{3}$ of common border has natural barrier,
Capital city	11	Capital city of conspicuous state within 300 miles of border,
Benefits	12	One-sided benefits conferred by one side on the other,
Culture	13	Cultural exchange common,
Trade	14	Trade exchange common.

➡

Data Exhibit 12 for Problems (continued)

Historical identification	Decade beginning	Conspicuous state	Conspicuous rival	(Raw scores)		Military						Geographical			Cultural		
				Months of war	Territorial gain	Defensive stance	Strength	Mobility	Quality	Fortifications	Prestige	Propinquity	Natural barriers	Capital city	Benefits	Cultural exchange	Trade exchange
Chinese	125 B.C.	Former Han Dynasty }Huns		105	3.6%	A	P	A	P	P	A	P	P	A	P	A	0
	25 B.C.			0	0%	A	P	A	P	A	P	P	P	A	P	A	0
	776 A.D.	Tang Dynasty	Tibetans	51	0%	P	P	A	A	P	A	P	P	P	P	P	0
	1076	Sung Dynasty	Tanguts	34	0%	A	P	0	0	P	P	P	P	A	P	A	A
	1376	Ming Dynasty	Yunnanese	1	6%	A	P	A	A	A	P	P	P	A	A	P	A
Islamic	776	Abbasids	Byzantines	72	.01%	A	P	0	P	P	P	P	P	A	P	A	0
Russia	1476	Muscovy	Novgorod	6	25.2%	A	P	P	P	A	A	P	A	P	P	P	P
Graeco-Roman	225 B.C.	Rome	Carthage	34	−19.9%	A	P	A	A	A	A	P	P	P	P	A	0
	25 B.C.		Parthia	0	1.5%	P	P	A	P	A	P	P	P	A	A	P	P
	176 A.D.		Marcomanni Quadi	34	0%	P	A	P	P	P	P	P	P	A	P	P	P
	376		Visigoths	66	0%	P	A	A	P	0	P	P	A	P	P	P	P
	576	Byzantines	Persia	120	0%	0	0	A	P	P	0	P	P	A	A	A	A

Data Exhibit 12 for Problems (continued)

Variables (carriers)**

Historical identification			(Raw scores)		Military						Geographical				Cultural	
Decade beginning	Conspicuous state	Conspicuous rival	Months of war	Territorial gain	Defensive stance	Strength	Mobility	Quality	Fortifications	Prestige	Propinquity	Natural barriers	Capital city	Benefits	Cultural exchange	Trade exchange
Western																
1276	France	England	0	−1.3%	P	P	A	A	P	P	P	A	P	P	P	P
1376	England	France	90	−25.6%	0	A	A	P	P	A	P	A	P	A	P	P
1576	Spain	Netherlands	120	−2.27%	P	P	P	P	A	P	A	P	A	A	A	A
1676	France	France	46	4%	A	P	A	P	P	P	A	A	P	P	P	P
1776	England	France	67	−39%	P	P	A	A	A	P	A	P	P	A	P	P
Swiss																
1376	Swiss Confederation	Kiburg	17	13%	P	P	A	P	P	P	P	A	P	A	P	P
1476	Swiss Confederation	Burgundy	12	11.5%	A	A	A	P	P	A	A	P	0	A	P	P
1576	Swiss Protestant	Swiss Catholic	0	0%	P	A	A	A	0	A	P	P	0	A	P	0

* A society with at least one city of fixed dwellings housing at least 25,000 people within a 6-km radius, and whose literature has been translated extensively into another script.

** For the carriers, the code used is: P = trait present; A = trait absent; 0 = no data available.

S) SOURCE

R. Naroll, V. L. Bullough, and F. Naroll (1974) *Military Deterrence in History*. **State University of New York Press: Albany, New York; p. 377, Appendix C.**

Data reprinted from *Military Deterrence in History* by permission of the author and of the State University of New York Press.

Data Exhibit **13** for Problems

Suicide and death rates

A) Male suicide rates

	Age					
Country	15–24	25–34	35–44	45–54	55–64	65–74
Bulgaria	8.3	7.9	10.9	19.1	25.2	40.2
Finland	15.5	35.2	53.2	62.2	68.6	68.9
Hungary	30.5	44.1	56.0	65.2	70.7	76.4
Israel	4.8	10.2	11.4	16.4	18.2	23.0
Netherland	4.1	7.5	8.4	13.9	18.8	24.5
Sweden	13.8	28.3	40.8	51.0	50.5	47.4
U.S.A.	9.9	17.3	22.4	28.4	36.0	35.7

S) SOURCE

D. S. Hamermesh and N. M. Goss (1974). "An Economic Theory of Suicide," *Journal of Political Economy*, 82, p. 83. Data reprinted by permission of author and of the publisher, the University of Chicago Press. © 1974 by The University of Chicago. All rights reserved.

See opposite page for panel B.

Data Exhibit **13** for Problems (continued)

B) Death rates in seven countries for 1964 or 1965

Country	Sex	Population (in thousands) Age						Deaths Age					
		15–24	25–34	35–44	45–54	55–64	65–74	15–24	25–34	35–44	45–54	55–64	65–74
Bulgaria	Male	661.0	609	650.6	413.4	419.3	212.7	758	306	1620	2602	6283	8528
	Female	650.0	605.2	647.7	441.8	421.6	249.7	334	521	1050	1783	3184	7743
Finland	Male	428.3	304.5	287.4	233.3	200.1	995.1	514	638	1366	2698	5338	5958
	Female	413.5	293.1	305.2	279.2	253.3	155.6	185	234	543	1207	2767	5562
Hungary	Male	781	677	713	534	546	310	874	1182	2215	3830	10415	15209
	Female	766	712	774	606	621	417	413	607	1499	2749	6998	14085
Israel	Male	174.7	132.3	128.0	119.1	91.6	41.4	191	146	251	644	1448	1716
	Female	163.9	136.3	134.1	120.6	83.1	41.5	90	100	223	561	1074	1449
Netherlands	Male	1074.8	812.0	753.4	641.2	538.6	348.6	376	819	1607	3891	2023	13961
	Female	1021.0	769.7	766.7	671.6	595.1	413.1	414	322	1020	2272	4831	10352
Sweden	Male	624.6	466.9	511.9	526.1	457.0	291.2	629	585	1154	2968	6401	1136
	Female	595.0	449.3	505.6	520.4	477.0	338.9	260	285	763	1794	3948	8505
U.S.A.	Male	14467	10866	11923	11601	8011	5146	22406	21553	45831	102426	183919	256867
	Female	14591	11246	12438	11132	8664	5259	9081	12097	29079	58279	99227	174175

S) SOURCE

Keyfitz, N., and W. Flieger (1968). *World Population*. University of Chicago Press; pp. 162, 222, 276, 310, 372, 424, 508.

Exhibit **13** (B)   577

Index

1) References to chapter summaries have been distinguished by terminal "s".
2) References to exhibits are not distinguished (only page number)
3) Words and terms not part of the language of statistical methodology usually refer to examples (pages 1-466) or to problems (pages 467-577).